EDMOND HALLEY

Portrait of Halley in about 1687 by Thomas Murray (in the Royal Society). See notes on portraits of Halley, p. xvi. (Reproduced by permission of the President and Council of the Royal Society).

The endpapers are a reproduction of Halley's planisphere of the southern stars. Reproduced with permission from the Bodleian Library, University of Oxford; taken from Halley's *Catalogus Stellarum Australium*, shelfmark Savile G.11.

EDMOND HALLEY

*Charting the Heavens
and the Seas*

ALAN COOK
Selwyn College, Cambridge

CLARENDON PRESS · OXFORD
1998

Oxford University Press, Great Clarendon Street, Oxford OX2 6DP
Oxford New York
Athens Auckland Bangkok Bogota Bombay Buenos Aires
Calcutta Cape Town Dar es Salaam Delhi Florence Hong Kong
Istanbul Karachi Kuala Lumpur Madras Madrid Melbourne
Mexico City Nairobi Paris Singapore Taipei Tokyo Toronto Warsaw
and associated companies in
Berlin Ibadan

Oxford is a trade mark of Oxford University Press

Published in the United States
by Oxford University Press Inc., New York

© Sir Alan Cook, 1998

All rights reserved. No part of this publication may be
reproduced, stored in a retrieval system, or transmitted, in any
form or by any means, without the prior permission in writing of Oxford
University Press. Within the UK, exceptions are allowed in respect of any
fair dealing for the purpose of research or private study, or criticism or
review, as permitted under the Copyright, Designs and Patents Act, 1988, or
in the case of reprographic reproduction in accordance with the terms of
licences issued by the Copyright Licensing Agency. Enquiries concerning
reproduction outside those terms and in other countries should be sent to
the Rights Department, Oxford University Press, at the address above.

This book is sold subject to the condition that it shall not,
by way of trade or otherwise, by lent, re-sold, hired out, or otherwise
circulated without the publisher's prior consent in any form of binding
or cover other than that in which it is published and without a similar
condition including this condition being imposed
on the subsequent purchaser.

A catalogue record for this book is available from the British Library

Library of Congress Cataloging in Publication Data
Cook, Alan H.
Edmond Halley: Charting the heavens and the seas/Alan Cook.
Includes bibliographical references and index.
1. Halley, Edmond, 1656–1742. 2. Astronomy–England–History.
3. Astronomers–Great Britain–Biography. I. Title.
QB36.H25C66 1997   520′.92–dc21   97–12596[B]
ISBN 0 19 850031 9

Typeset by EXPO Holdings, Malaysia

Printed in Great Britain by
Bookcraft (Bath) Ltd
Midsomer Norton, Avon

Preface

A book on the life and works of its subject, some may consider old-fashioned. I have written it for reasons that I think forceful. The social and communal aspects of the developments of science are important, but individuals make up the community. Their separate actions, thoughts, and achievements must be known and understood if arguments of a general nature are to be soundly based. There were, and to some extent still are, sufficient problems about Halley's life to warrant a study of it as a matter in itself. Secondly, I maintain that Halley's life is of such inherent interest as to justify a book about him. He was a man of remarkably varied attainments with a European reputation in his own times. Thirdly, his personal circumstances and interests provide excellent instances of factors that seem to have been important in the social context of science in a period in which the physical sciences underwent major intellectual development.

Independent gifted individuals, however notable, are not isolated. It is unrealistic to write history as if they were. Philosophical, technical, economic, and other movements of the time made possible and contributed to the explosion of observation, experiment, and theory in the sciences in the late seventeenth and early eighteenth centuries. Halley would not have been Halley, nor would he have done what he did, had he lived in some other day. Nor would Newton nor Hooke, Huygens nor Leibniz, nor the Bernoullis nor Wren, have followed their actual courses. The conditions were ripe for a great florescence of science and some very remarkable individuals were there to take advantage of them and ensure that the florescence occurred. The progress of science depends on asking the right questions and having the technical ability to make a reasonable attempt at an answer. Who set the questions in Halley's day; how was it that the techniques were available?

In so far as science is a social activity, Halley was particularly significant on account of the correspondence that he carried on as an officer of the Royal Society, keeping members of the international community of scientists aware of what each was doing. Halley's life indeed exemplifies in a number of ways how science is a communal activity, how the achievements of one depend on those of another. The outstanding instance is Halley's part in the conception, development, printing, and publication of Newton's *Principia*—no one but Newton could have written it, but Newton would not have done so as he did but for Halley. Halley, on the other hand, clearly derived his ideas for his own projects from what were commonly perceived as significant problems, many of which, in his case, were related to navigation and especially to the determination of longitude.

If ideas, however they arise, are to result in a real development of science, they must be more than socially acceptable or commercially relevant. The 'construction' of science is not a subjective undertaking; it must agree with the empirical structure of the world around us. Scientists try to make sense of the world as it is, not as some might like it to be. The other condition is that the means to pursue the ideas must be available: ideas without the means to pursue them remain pipe dreams. Halley had the independence and the connections to put his ideas into practice.

Most of the great natural philosophers of Halley's day were distinctive characters. Newton, Flamsteed, Hooke, and Wren all figure in Halley's life and add to its interest, and Halley himself was a man of notable qualities. Ideas and preconceptions that Halley's contemporaries held were very different from ours today. We find it difficult to think ourselves back into the minds and feelings of people of that time and to imagine what drove them. We can read what they read, but unless some gossip puts it down in a diary we miss the talk that went on in the coffee houses, the informal converse that formed habits and assumptions. Society, especially London society, was very compact. People in different avocations, as we would see them, were in daily touch, and the same person might be a natural philosopher and a musician and a Court official. We are apt to see the social circles in which those different interests were pursued as separate, but they embraced the same people.

Halley's life and family present some intriguing mysteries to tantalise. We do not even know for sure exactly when he was born, nor where. When did his father remarry after his mother's death, and whom? Why did Halley cut short, if he did, his tour in Italy and marry rather soon after his return to London? How did his father meet his end? What were Halley's connections with the East India and Levant Companies? Why was it Halley who presented Newton's *Principia* to James II; and what were his religious views and politics? Some of those questions no doubt relate to his family's position in the City of London, and thus to the means he had and the patronage that he enjoyed that enabled him to undertake the immense project of his Atlantic cruises. Would the answers to those questions, could we find them, help us to understand the course of Halley's career, and through it, some of the factors that affected the development of the physical sciences in his day? Unfortunately Halley has left little personal correspondence from which we can reconstruct his social and family life. His scientific activities are for the most part well documented but he left no diary, and almost no letters to friends or relatives. There are less than half a dozen references to his wife Mary in some sixty-four years together. It is sometimes possible to make inferences or to guess at people he knew outside his professional acquaintances but there is little hard fact, so that some of what has been published about him is no better than romantic fancy, and some of the speculations indulged in here may turn out to be baseless.

Important though the large impersonal questions are, when all is said and done, Halley is not a lay figure on which to hang sociological notions: he was a large man in a grand time, well able to stand not only with fellow scientists, but with such as Pepys and Evelyn and Wren and Kneller and Dryden and Locke, and well worthy of our attention.

Just as Halley's merchant adventurer contemporaries went out to seek trade in unknown lands and waters, so did he in pursuit of natural knowledge. From his very first expedition to St Helena, for the next fifty years he was venturing into the unknown and bringing back great profits.

Acknowledgements

My debts are great to many persons over many years for information, guidance, and criticism. My first forays into Halley's Adriatic journeys were guided by Dr A. Hollaender of the Public Records Office in London, by Dr Christiane Thomas of the Haus-, Hof-, und Staatsarchiv of Vienna, by Dr Tiepolo and Dr Elena Zolla of the Archivio di Stato in Venice, and by Dr Purkarthofer and Dr Speitzhofer of the Steiermärkisches Landesarchiv in Graz. Professor Legovic and Professor Munic of Zavod za Povijesne i Drustvene Znanosti graciously showed me the harbour of Buccari. Professor Hattendorf of the US Naval War College enlightened me on the Blenheim campaigns. Dr Valerie Pearl and Dr Mark Goldie have led me through seventeenth-century London and its labyrinthine politics. I am grateful to the staff of the British Library; of the Guildhall Library, London, and Mr P. W. A. Cain of the Records Office of the City; of the Bodleian Library, Oxford. The Librarians of King's College, Cambridge, Tim Munby, Peter Croft, and Peter Jones, have been most helpful, as have the Librarians of the Royal Society and their staff especially Mary Sampson. I thank also the archivists of Northamptonshire County Archives, of the Canterbury Cathedral Archives and of the Essex Record Office. I am grateful also to Mme. Suzanne Débarbat and the Librarian of the Observatoire de Paris. Above all I am deeply indebted to staff in the Rare Books and Manuscript Rooms of the University

Library, Cambridge for their invariable courteous and effective assistance, and especially to Adam Perkins for the Royal Greenwich Observatory archives.

Three editors of correspondence have been very helpful, Dr Fritz Nagel for the Bernoulli papers in Basel, Dr James O'Hara for the Leibniz papers in Hanover, and Dr Frances Willmoth for the Flamsteed papers in Cambridge.

Dr Mark Goldie, Miss Fiona Greenwood, and Dr N. E. Emerton have allowed me to quote from unpublished material, and Professor Owen Gingerich has discussed with me his copy of the *Historia coelestis* with Flamsteed's annotations. Professor G. A. Gresham advised me on the wounds of suicides.

I am most grateful to the Revd Dr N. Cranfield, Dr Mark Goldie, Mr and Mrs S. Lennane, Dr Valerie Pearl, Professor N. J. W. Thrower, Dr David Smith, and Professor D. T. Whiteside, who have read and criticised drafts of chapters. It has been a pleasure to be published by Oxford University Press, the Press with which Halley was so much involved, and I am particularly indebted to Susan Harrison, Keith Mansfield and Julia Tompson.

I record with deep gratitude that I was originally stimulated to undertake my studies of Halley by Sir Edward Bullard, although many years after he had been my graduate supervisor, and I also recall the friendship of Antonio Marussi of Trieste who encouraged my investigations there and in Venice and Graz. I owe much to my wife Isabell who has had to compete with Halley for many years, and even follow him around on occasion.

Cambridge A.C.
October 1996

Contents

List of plates	xiv
Portraits of Halley	xv
A note on dates	xvi

PART I
The young astronomer
(1656–1687)

1	*Halley's world*		3
	1.1	Times of change	3
	1.2	Halley's London	4
	1.3	England	11
	1.4	The Arts in England	15
	1.5	The wider world of Europe	17
	1.6	The Americas and the East	19
	1.7	Navigation and astronomy	20
	1.8	Mathematics, technology, and natural philosophy	23
	1.9	Natural philosophy and natural theology: views of the world	26
	1.10	Halley: an agent of change	30
2	*Formation of an astronomer*		32
	2.1	Forbears	32
	2.2	Parents	34
	2.3	Family property and possessions	36
	2.4	The Tower and Sir Jonas Moore	38
	2.5	Neighbours	39
	2.6	The Levant Company	42
	2.7	Schooldays: St Paul's	43
	2.8	Oxford: the Queen's College	50
	2.9	Undergraduate publications	54
	2.10	Off to sea	59
3	*Skies of the south*		61
	3.1	Into the unknown	61
	3.2	Apprenticeship of a navigator	64

	3.3	The state of astronomy	65
	3.4	A year in St Helena	72
	3.5	A new star and a new constellation	77
	3.6	How far is the Sun?	79
	3.7	Matters of the Moon	83
	3.8	Consequences	84
	3.9	Halley's achievement	86
4	*Into Europe*		89
	4.1	The sage of Danzig	89
	4.2	Halley at Danzig	93
	4.3	After Danzig	101
	4.4	Six months in France	105
	4.5	On to Rome	119
	4.6	Astronomical observer	124
	4.7	A traveller's harvest	126
5	*A wedding and two funerals*		128
	5.1	A City wedding	128
	5.2	The sky at Islington	130
	5.3	*Those bloody and barbarous men*	132
	5.4	*Such an impertinently litigious Lady*	142
	5.5	Science with distractions	144
6	*Achilles produced*		147
	6.1	The origin of *Principia*	147
	6.2	Halley in 1686	151
	6.3	Printing *Principia*	153
	6.4	Previews and reviews	165
	6.5	Other pursuits	170
	6.6	Achievement and preparation	176

PART II
Often at sea
(1688–1703)

7	*Improving natural knowledge*		181
	7.1	The Royal Society, twenty-five years old	181
	7.2	Clerk to the Royal Society	184
	7.3	Mathematician	187

	7.4	Geomagnetism and meteorology	190
	7.5	Physics, metrology, and optics	197
	7.6	Demography	198
	7.7	The student of antiquity	200
8	**Celestial architecture**		203
	8.1	The paths of heavenly bodies	203
	8.2	Comets	205
	8.3	The problem of the longitude	217
	8.4	How far is the Sun?	219
	8.5	The Moon is speeding up	225
	8.6	Halley's ways	228
9	**Use and practice of the contemplation of nature**		230
	9.1	A glorious revolution	230
	9.2	*Hally a sayling*	232
	9.3	Diving and salvage	236
	9.4	Hydrodynamics	243
	9.5	Astronomy at Oxford: an unsuccessful candidate	245
	9.6	The Mint at Chester	249
	9.7	Looking forward	254
10	**Far seas and new prospects**		256
	10.1	Setting out	256
	10.2	Preparations	260
	10.3	The Atlantic	269
	10.4	The magnetic chart	281
	10.5	Tides	284
	10.6	*At the publick charge*	290
11	**Upon the Dalmatian shore**		292
	11.1	War over Europe	292
	11.2	The record of Halley's survey	295
	11.3	Maps of Dalmatia	296
	11.4	Halley surveys Trieste and Buccari	297
	11.5	Fortifying Buccari	308
	11.6	Sites and construction of Halley's works	310
	11.7	Consequences	313
	11.8	Surveyor and cartographer	315
	11.9	Sea captain and engineer	317

PART III
Scholar and sage
(1704–1742)

12	**In the Savilian Chair**	**321**
12.1	The Electors meet	321
12.2	A life in Oxford	325
12.3	The Royal Society again	332
12.4	The Greek geometers	333
12.5	Lights in the heavens	345
12.6	The eclipse of 1715	351
12.7	A mathematician among mathematicians	353
13	**The matter of the Moon**	**354**
13.1	The preoccupation of a lifetime	354
13.2	Observations of the Moon	355
13.3	Principles of lunar theory	356
13.4	Halley's lunar tables	366
13.5	Halley's comparisons with Newton's theory	368
13.6	Jupiter and Saturn	373
13.7	Conspectus	375
14	**Astronomer Royal**	**377**
14.1	The Royal Observatory	377
14.2	Halley and Flamsteed	380
14.3	Two *Histories*	388
14.4	Halley returns to the Observatory he saw set out	393
14.5	Knowledge of the longitude	395
14.6	Tidying up	398
14.7	Last days	401
15	**The improvement of natural knowledge**	**405**
15.1	Halley in his own day	405
15.2	Religion and politics	407
15.3	Two revolutions	413
15.4	The natural philosopher	414
15.5	The navigator	419
15.6	The practical man	421
15.7	Halley in his time	422
15.8	Servant of knowledge, servant of the realm	423

Appendices 426
    1   The ellipse and the parabola 426
    2   Genealogies 431
    3   The personal estate of Halley's father 435
    4   Chronology of Edmond Halley 437
    5   The southern stars 439
    6   Halley's Ode to Newton 442
    7   Correspondence of Halley not listed by MacPike (1932) 444
    8   The sale of Halley's books 447
    9   The manuscript ULC RGO 1/74 of Flamsteed's *Catalogue* and its implications 452
   10 Halley's *Memoriall* to the Emperor 456

Notes to chapters 458

Bibliography 500

Index 525

The plates section can be found between pages 272 and 273.

List of plates

I	Sir Joseph Williamson (Sir Godfrey Kneller).
II	Henry Oldenburg (Jan van Cleef).
III	John Wallis (Derard Soest).
IV	John Flamsteed.
V	J.-D. Cassini.
VI	Sir Isaac Newton (Charles Jervas).
VII	Sir Hans Sloane (Sir Godfrey Kneller).
VIII	Halley in the uniform of a captain RN (Sir Godfrey Kneller).
IX	Halley's chart of the magnetic variation over the Atlantic.
X	Halley's tidal chart of the English Channel.
XI	John Arbuthnott (Robinson).
XII	George Stepney (Sir Godfrey Kneller).
XIII	G. W. Leibniz (artist unknown).
XIV	Halley in a fresco commemorating the transit of Venus of 1882 at the Observatoire de Paris (El.-L. Dupain).
XV	Halley at Oxford before 1713 (Thomas Murray).
XVI	The solar eclipse of 1715.
XVII	Halley in 1722 (Richard Phillips).
XVIII	Halley aged 80 (Michael Dahl).

Portraits of Halley

There are five known portraits of Halley. The first (the frontispiece) is that by Thomas Murray in the Royal Society. Halley holds a diagram related to his paper (1687*b*) on cubic and quartic equations and the picture therefore probably dates from about that year. The next portrait (Plate VIII) is by Sir Godfrey Kneller and is in the National Maritime Museum. Halley is in naval uniform and the portrait may date from 1702 when he was in London between his Channel cruise and his Adriatic mission. The third portrait is again by Murray. There are two versions, one in the Bodleian Library (Plate XV) and one in the Queen's College, Oxford. Halley points to the Atlantic Ocean on a large globe. He wears a furred gown over what may be naval uniform. The Bodleian portrait was presented to the Library by Murray in 1713 but the iconography suggests that the occasion of it was Halley's LLD degree. The National Portrait Gallery has a portrait by Phillips (Plate XVII) dated to 1722 so that it may have been painted just after Halley had become Astronomer Royal. Finally there is a portrait of Halley at eighty by Dahl (Plate XVIII), also in the Royal Society to whom it was presented by his daughter, Mrs Katherine Price. The diagram that he holds may refer to his shell model of the Earth.

Halley appears in a fresco in the Observatoire de Paris that commemorates the transit of Venus of 1882 (Plate XIV). An angelic putto holds a silver medallion with a portrait of Delisle and a gold one with a portrait of Halley (see Débarbat 1990).

A note on dates

Two calendars were in use throughout Halley's life, the older, Julian, followed in England and the revised, Gregorian, that had been adopted in western Europe. The difference between them was 10 days in the seventeenth century and 11 days in the eighteenth century; for example, 10 June in the Julian calendar would be 20 June in the Gregorian before 1700 and 21 June after 1700. Both forms of date, such as 10/20 June, might sometimes be given, in particular on letters sent from or going to places overseas. In general the difference of calendars is immaterial and does not cause confusion, so that it is not worthwhile to give all dates in the Gregorian form. To do so would be misleading when certain events are associated with a particular date. Important associations would be lost if Christmas were 4 January instead of 25 December, or the execution of Charles I were 9 February instead of 30 January, or if the anniversary meeting of the Royal Society (St Andrew's day) were 10 December instead of 30 November. Dates have therefore been kept in their original form. Where it seems important, the calendar has been designated by (os) meaning old or Julian style or (ns) meaning new or Gregorian style.

The civil year in England began on 25 March but astronomers began the year on 1 January. All dates are given here in the latter form as now generally followed. An English civil date such as 10 February 1685 will be given as 10 February 1686 and designated (os) if that matters. It would correspond to 20 February 1686 in western Europe.

A further complication that sometimes arises is that the civil day starts at midnight but the astronomical day starts at noon.

PART I

The young astronomer

(1656–1687)

I
Halley's world

> The world's great age begins anew,
> The golden years return,
> The Earth doth like a snake renew
> Her winter weeds outworn...
>
> SHELLEY: HELLAS, 1, 1060–3

1.1 Times of change

EDMOND HALLEY lived in pregnant times. In his lifetime Europe emerged from the Renaissance world of humanism into the Enlightenment and Ancien Régime. Renaissance people saw nature quite differently from us; they thought of it quite differently. To them knowledge of the world was mystical and secret; to us it is rational and open. In the late seventeenth and early eighteenth centuries knowledge and thought, manners and technology, became the precursors of ours today.[1] Halley was one of the virtuosi who changed our understanding of the world around us.

Halley was born in 1656 when the Commonwealth had yet four years to run. He was a schoolboy during the Dutch wars, the Great Plague, and the Great Fire of London. He was twenty-seven at the time of the Rye House Plot in 1683, thirty-two when William of Orange invaded England, and fifty-one at the Union of England and Scotland. For more than twenty-five years of his most active career England was at war with France, first in the Nine Years War of 1689–1698, then in the War of Spanish Succession of 1702–1713. Halley saw the Jacobite rebellion of 1715 and died four years before the greater one of 1745. He was Astronomer Royal from 1720 to 1742, almost exactly the tenure of Sir Robert Walpole as Prime Minister.[2] When Halley was a young man, Pepys was reforming the administration of the Royal Navy. The first steam engine worked in his middle years.[3] Purcell, then Handel, bestrode English music, Wren and Hooke built, Dryden wrote, Pepys and Evelyn kept their diaries, Nell Gwynn played, and Locke cast a sceptical eye upon all.

Cities changed: Rome, Paris, and London were rebuilt more elegantly and commodiously. North America expanded, trade by sea spread world-wide, music and painting took forms that lasted until early this century. Mathematics was relatively primitive when Halley was born; at his death many problems that still occupy mathematicians today had been set and much had been done to explore them.

Halley was no passive observer of those profound developments of the human mind and spirit. He was an agent, a powerful agent, not alone nor in isolation, but as a man of his time, in the first place as a Londoner. He was born and lived most of his life in London. He was also an Englishman and conscious of being so, familiar with Europe and Europeans and affected by the military and political and economic turmoils of his day. When Halley was a Londoner, London was aware of, and open to, and affected by the political and philosophical and artistic upsurges in Europe.

1.2 Halley's London

Halley was nine when the Great Plague devastated London in 1665 and ten at the time of the Great Fire next year. His father had property in Winchester Street which escaped the Fire but his school, St Paul's, was destroyed along with the medieval cathedral (Fig. 1.1). As a boy and young man he would have seen much of the City rebuilt in brick, the first stages of the new St Paul's, and a multitude of churches rising up from Wren's office. Rebuilt though London was, its geography and institutions were scarcely altered by the Fire; in them and under them Halley lived.

The City proper was enclosed by the Roman and medieval walls with gates shut at night. The Tower loomed to the East, dominating the City, the greatest fortress in England. The Norman fortress and palace was now more used as offices, Mint and stores, and as a State prison. Halley's father was a yeoman warder, a sinecure for the most part. His early patron, Sir Jonas Moore, was the Surveyor General of the Ordnance, with his office in the Tower. Ordnance artificers probably made some of Halley's first instruments.

Quays stretched westward from the Tower along the river to the other fortress, Baynard's Castle. Beyond that lay the Inns of Court, where the father and uncles of Halley's wife were lawyers. The lines of the main medieval streets followed the Roman plan and the Fire did not change them; they are little changed today.

Building extended far beyond the City walls. Southwark south of the river and part of Cripplegate north of the walls had been settled early in medieval times and were City wards. Moorfields, just north of the

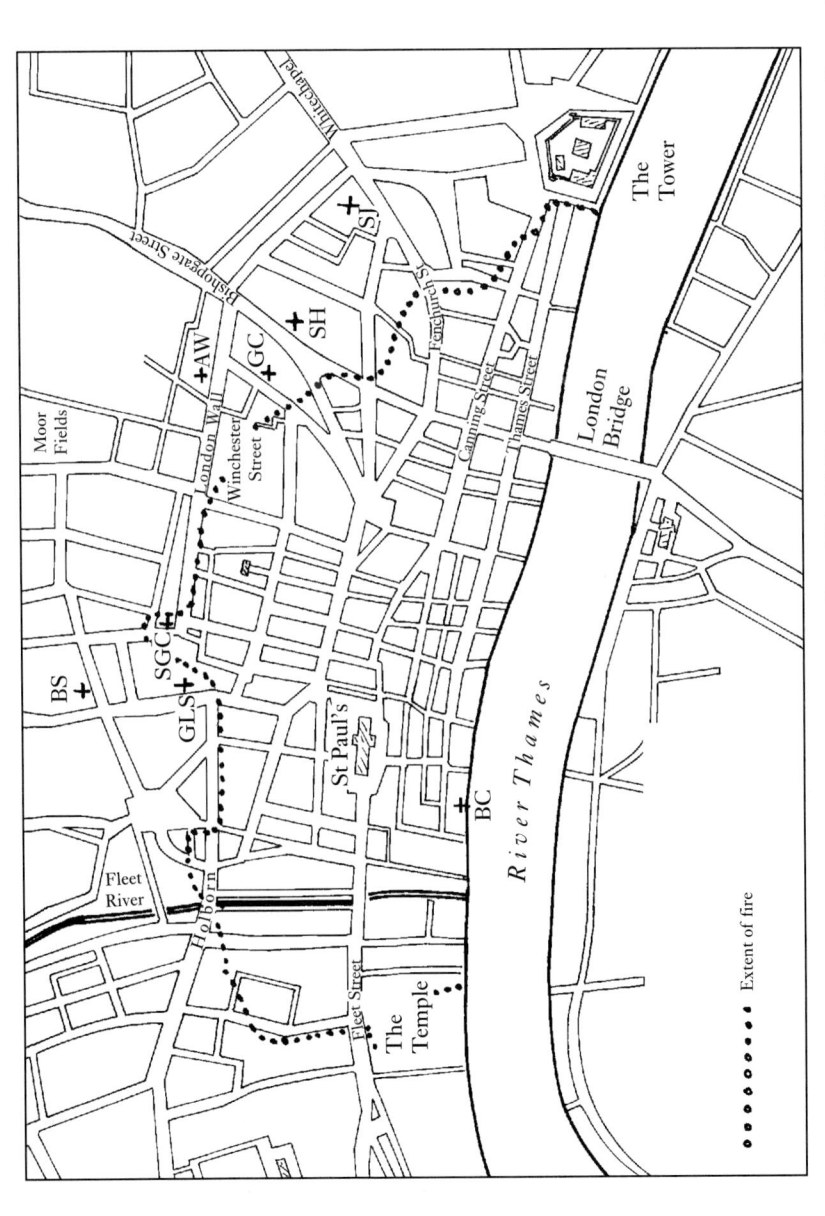

Figure 1.1 Plan of the City of London and the extent of the Great Fire. *Key:* AW: All Hallows on the Wall; BS: Bridgewater Square; BC: Barnard's Castle; GC: Gresham College; GLS: Golden Lion Square; SGC: St Giles's, Cripplegate; SH: St Helens, Bishopsgate; SJ: St James', Duke Place.

Wall, was developed in Halley's time.[4] Shoreditch, where Halley was probably born, Islington where he lived in the first years of his marriage, and Hackney with the houses of Levant Company merchants and other well-to-do citizens, were further north. Grand houses of magnates stood upon the river between the City and Westminster; theatres such as Drury Lane had reopened after the Commonwealth. Fire finally destroyed the Palace of Whitehall in 1690, leaving only the Banqueting House of Inigo Jones, where Charles I had gone to his execution on the thirtieth of January 1649.

London was by far the greatest port in the realm. The Royal Navy was administered from offices in the City. Ships were built, maintained, and laid up just down river at the Navy Yard at Deptford, the principal yard for stores. The greater dockyard at Chatham was not far off. Wheat from East Anglia, coal from Newcastle, silks, spices, and currants from the East, ivory and gold from Africa, sugar from Jamaica, tobacco from Virginia, were landed on London quays; and London exported woollens and tin and other valuable products of England. London merchants traded with many countries: with the Baltic for the pitch and masts and planks without which the Royal Navy could not put to sea; with the Low Countries for books and pictures; with Venice and with the North American colonies and the West Indies.

The livery companies, still to a large extent guilds of merchants, traders, or artisans, ruled the City. It was not necessary to pass through the stages of apprentice and journeyman to become a freeman and liveryman. Halley's father, in business as a soap boiler, was a Salter, although the Tallow Chandlers was possibly a more natural company for soap boilers, and his father had been a Vintner. The companies regulated the trades of their members and promoted charitable works, hospitals, for example, or schools such as St Paul's which was supported by the Mercers. An elected Court of Assistants managed the affairs of a company. The Master, who served for a year, and officers such as wardens, were chosen from the Assistants. Twelve companies, the oldest and wealthiest, ranked as the great companies, among them the Drapers, the Mercers, to whom the grandfather of Halley's wife belonged, the Fishmongers, and the Salters. Vintners were a lesser Company.[5]

The City was divided into twenty-six wards. The wealthiest families congregated in the inner wards, with the poorer sort mostly in the outer ones.[6] Winchester Street, the Halley family home for twelve years or more, was in the inner ward of Broad Street. Relatives and associates of Halley lived in the adjacent parishes of St-Peter-le Poer and St Helen's Bishopsgate, all wealthy districts.

Each ward returned one alderman to the Court of Aldermen, and a number of councillors to the Common Council. Most wards also had a deputy to the alderman. Aldermen and councillors had to be free of the City. The principal City officers, the Sheriffs and the Lord Mayor, were chosen from the aldermen and served each for one year. They policed the City and had criminal jurisdiction within it. In normal times the Watch, bands of constables, maintained order in the streets. The City also maintained four regiments of Trained Bands in which many prominent citizens were officers (*John Gilpin was a citizen of credit and renown,/ a Trained Band captain eke was he of famous London Town*). A few select citizens served instead as Yeoman Warders of the Tower: Halley's father was one. Their duty was to the Sovereign in person when he was in the City.

Religion dominated the life of the City, as it did England. Almost all political issues were intertwined with religion. The questions of popery and tyranny that had divided people before and during the Civil Wars still disturbed civic life. Nonconformists who would not accept the Church of England were placed under severe disabilities by acts of the Restoration Parliament, but blind eyes were often turned to their practices, especially of presbyterians and independents. The Church of England itself had both strictly puritanical and strongly High Church members, the one sympathetic to Puritan Nonconformists, the other to Roman Catholics. Architecture demonstrated the political dominance of religion. The numerous city churches were the centres of the townscape, visual, religious, political, and social. The cure of souls in an important City parish was often the stepping stone to a bishopric or other major preferment.[7] Halley was still a young man when most of the medieval churches lost in the Fire were rebuilt by Wren, and he was just sixty when the new St Paul's was completed.

The Church of England was dominant but it was not all. Whatever Acts of Parliament may have ordained, meeting places were tolerated for Dissenters and Puritans. The Chapels Royal of the queens Catherine of Braganza and Mary of Modena, and of the Queen Dowager, Henrietta Maria, served Roman Catholics long before James II had public Roman Catholic services. There were churches of foreign protestant communities that did not conform to the Church of England. The Dutch Calvinist church in Austin Friars was not far from the Halleys in Winchester Street. The largest Calvinist community, the one that caused the established church the most concern from time to time, was the Huguenots.[8]

Most Huguenots were refugees from religious persecution in France. They first came after the massacre of St Bartholomew in 1572. Descendants such as the princely Houblon merchants, friends of Samuel Pepys and neighbours of the Halleys in Winchester Street, were well established by the 1650s. Almost five hundred thousand Huguenots left France after the Revocation of the Edict of Nantes in 1685. About equal numbers went to the Netherlands and to England. Many were very able and distinguished. Henri Justel was one, an amateur of science, a steady and copious correspondent of the Royal Society, whom Halley first met in Paris in 1681 and who came to England about 1688. The mathematician Abraham de Moivre was another whom Halley may have met in 1681, when he visited the Huguenot academy at Saumur. Huguenots could be a source of discord in the City. Most citizens welcomed them as victims of Romish tyranny, although less enthusiastically when they competed effectively in City trades. The royal government was cooler. It disliked their Calvinism and for most of the reigns of Charles II and James II the monarchs were covertly or openly allied to Louis XIV. The welcome that Londoners extended to the Huguenots accordingly fed discord between Crown and City: it was not a prime cause, but it reflected and exacerbated the opposing views and policies of those two major elements in the polity of the land.

The City was indeed cool towards the Crown throughout the reigns of Charles II and James II, and at times downright hostile. On the whole it was opposed to France and favourable to the Netherlands, inclined to Puritanism and hostile to Rome. It was far, far richer than the monarch, who could not offend it too deeply when he often looked to it for large loans. The City on its side sought diplomatic and military support for overseas trade, and had to retain the forbearance if not the goodwill of the monarch.

The Restoration of Charles II in 1660 by no means resolved the issues of the Civil War. They were the root of the crisis of around 1680 (the Exclusion Crisis) when Parliament attempted to free itself from control by the monarch and itself tried to set fetters on the king. Algernon Sidney and his associates obtained control of the City government, which from 1679 to 1681 was almost an independent 'republic by the King's side' that exerted financial blackmail upon the King and exacted vengeance on the Earl of Stafford.[9] Then in 1682, with greater financial freedom, Charles turned upon the City, imposed his own sherrifs, and in 1683, after the Rye House Plot, had Sidney and the Earl Russell tried for treason, condemned, and beheaded. Sidney's other associate, the Earl of Essex, met a violent end while in the Tower awaiting trial. Halley's father was on duty as a yeoman warder that day.

Some Londoners were viciously partisan, others, especially the Quaker William Penn, were more tolerant. Most were fearful of renewing the Civil Wars, almost all joined in the annual commemoration of the execution ('martyrdom') of Charles I on the thirtieth of January, kept as a major festival of the City. At the very time of the Exclusion Crisis, a church dedicated to Charles the Martyr was founded in

Tunbridge Wells with donations from subscribers who seem to have been of all persuasions and, in particular, to have included everyone who was anyone in the City from all sides. Some eighty Fellows of the Royal Society subscribed, among them all the presidents of the day, save only Sir Christopher Wren, whose office, however, may well have been involved in the design.[10] For all the political passion and tension of which we are now so conscious today, the concerns of Halley, his friends, his family, and their colleagues, so far as we know them, were trade, science, navigation, litigation: the affairs of a community devoted to commerce. As Athenians did in classical times, so did Londoners retreat from savage political conflicts to the elegance and ease of domestic and social life.

Trade was the great business of the City. Provincial cities had their tradesmen such as goldsmiths and silversmiths, and printers and stationers selling books, but they were few. Almost all foreign trade went through London, for the companies that had monopolies of foreign trade were London companies, the East India Company for the Indian trade, the Levant Company for Venice and Asia Minor, the Royal African Company for West Africa and the Americas (based on slaves), and the Baltic Company for the north-east.[11] Immense fortunes were made in them and some of the most influential City magnates were members of two or more. They all controlled their memberships tightly. Applicants had ordinarily to be free of the City, and either belong to the family of an existing member or have been apprenticed to one. The Levant Company also admitted persons who were 'mere merchants', that is, they did not engage in any of the other trades of the City companies. Sometimes, but rarely, the Company would admit a man of great influence, such as a Secretary of State, who might be useful to the Company, although not otherwise qualified. The Companies were very powerful in politics, but their trade depended on English control of the seas over which they sailed. Halley, probably a member of no Company, was involved with all four in some way.

No Londoner could overlook the importance of overseas trade and of maritime affairs. Many owned ships. Halley took passage to St Helena on *Unity* owned by the Leithullier family, Levant merchants and neighbours of the Halleys in Winchester Street (Chapter 2). He later tried to salvage a ship of the Royal African Company, most probably at the behest of Sir Gabriel Roberts and Abraham Hill, governors of the Company well known to him (Chapter 9). Pepys's office from which he administered the Royal Navy was in Seething Lane close to the Tower, and the Pay Office was in Broad Street by Winchester Street. The Secretaries of State and senior admirals determined naval strategy and policy, the Navy Office, effectively Pepys until 1689, saw to the construction and manning of men-of-war and the administration of the dockyards. The Ordnance Office in the Tower of London supplied guns and ammunition.

London was the financial hub of the country. Goldsmiths in the City offered deposit and credit facilities and the government of the day called on them for money to prosecute war. The heavy demands of the wars of William III with France led in 1694 to the establishment of the Bank of England which was able to mobilise credit for the War of Spanish Succession after 1700.

London was the legal centre of the country. The Lord Chancellor's Court and the three that sat in Westminster Hall, King's Bench, Common Pleas, and Exchequer were the principal courts and dealt mostly with civil matters. Although King's Bench might take the most serious and politically driven criminal matters, local jurisdictions handled most criminal cases—in London the courts involved were those of the Lord Mayor and the Middlesex Sessions at Old Bailey. There was a host of other courts besides. Wills, probate, and inheritance were the main business of the surviving ecclesiastical courts, especially of the Prerogative Court of Canterbury. Halley became involved in legal matters on a number of occasions over the years in various of those courts.

Lawyers congregated in the legal Inns, the Inner and Middle Temple, Grays and Lincoln's Inns, Sergeants' Inn for the higher advocates, and Doctors' Commons under St Paul's for the ecclesiastical lawyers. The Inns of Court had once been the principal academies for the noble and fashionable youth, but in the time of Charles I they had begun to yield that ground to the universities of Cambridge and Oxford, and to some extent to Gresham's college in the City and the short-lived Chelsea College. In Halley's day new thought and learning, especially in the sciences and philosophy and theology, were springing up in the universities. Even those who would pursue a legal career now usually went first to Oxford or Cambridge.

London was no arid society constrained by narrow accountants' mentality or by political passion or religious bigotry, important though all those were in its life. Londoners were prosperous and spent money on elegant and comfortable furniture and furnishings. Three diarists of the day, to all of whom Halley was well known and in whose pages he appears, show us how Londoners spent their time: with Pepys in devoted administration, making music, at the theatre, womanising, dining with friends; with Evelyn disapprovingly at Court, organising relief for victims of war, at Oxford and the Royal Society, in his garden and among his trees, advising on and patronising artists; with Hooke in his contorted daily life, among coffee-house habitués, surveying and rebuilding London, providing experiments for the Royal Society, buying books. London was where music was made, where plays were performed, where poets and essayists were published. We know from Hooke's diary that Halley dined out and took tea with his friends and entertained them to a bottle of wine together. We do not know if he took the interest in music or theatre that some of his friends did, but he was painted five times, he knew amateur musicians and political satirists, he deplored the decline of the arts in Rome when he went there, and he lived in the city that was one of the liveliest for the arts in his day.

1.3 England

Halley's world was not only London. He was an undergraduate at Oxford and later Savilian Professor of Geometry. The family had a connection with Barking in Essex; Halley had uncles and other relations in and near Peterborough and Huntingdon. Maritime affairs took him to Harwich and Portsmouth and he had to visit family and university estates. He had for a short while an office at the Mint in Chester and observed the barometer on Snowdon. He himself left no picture of his England but Pepys, Evelyn, Defoe, and Celia Fiennes give us some notion of it at various stages of his life.

England was still primarily agricultural. Great quantities of corn were grown in East Anglia and were shipped through Cambridge and King's Lynn to be carried to London by river and sea. Wool was the staple of the Cotswolds. Mining and industry were none the less becoming more important. London imported coal from the north-east—Halley's father had invested in the Coal Office that funded the rebuilding of the City after the Fire from taxes on imports of coal. The Weald of Kent with iron ore and charcoal was a major source of iron, as place names such as those related to hammer ponds still tell us. Forestry was important, providing timbers for building and even more, for merchant ships and the great men-of-war. Ship-building and ship-fitting and repair were the greatest industrial activities of the age, and as the Arsenal of Venice surpassed all other enterprises in Europe in the Middle Ages, so in Halley's day did the shipyards of the Thames, of Brest, and of Toulon. Power came from natural sources, wind or water or horses or men, but the first steam engine worked before Halley was fifty and there were well over a hundred in use when he died.

The ships that plied the trade that made the wealth of London sailed from the River of Thames across the North Sea or through the Downs

Figure 1.2 East Anglia and the Thames approaches.

and into the Channel (Fig. 1.2). That was the highway to riches; that was the theatre of great battles. It was exceptionally hazardous. The North Sea and the grey waters around the east of England are some of the most dangerous anywhere. Shifting banks of sand and mud guard the approaches. The Goodwin Sands that shelter the Downs between the North and South Forelands bear the gaunt skeletons of many wrecks. North of the Thames the Gunfleet shelters Clacton, and is another trap for the unwary. Strong currents swirl around the banks in tortuous streams as the tides ebb and flow through a great range. The weather is often forbidding. Clouds driven by strong winds across grey skies scud over brown waters whipped into spume. Wrecks upon wrecks cover the sea bed.

The very great storm always menaces those turbulent seas, when upon a high spring tide surging waters are driven southwards by tempests in the north. Three or four storms destroyed Dunwich in the fourteenth century; another did immense damage to the coast of East Anglia and the lands of Holland in 1953. The storm of November 1703 destroyed four times as many ships in one night in the North Sea as a French fleet did of a Levant company fleet off Lagos in 1693. Somewhat later two hundred colliers bringing coal to London were lost in a storm off the north Norfolk coast.[12]

Sailing out of the River was never simple: if the wind blew from the east ships could hardly get out, and if westerly, then it was difficult to make the Downs and stand down the Channel. The passage to France from Dover was very unsure; Halley once took forty hours in the packet from Dover to Calais. When the government of James II realised that William of Orange was planning to invade, Pepys worked long hours in 1688 to bring out the battle fleet. Its commander anchored behind the Gunfleet ready to intercept William's fleet as he came out of Texel. A storm delayed William in October but then the wind blew from the east and held the English fleet inside the Gunfleet while William passed down the Channel to Torbay. On such acts of God did the fortunes of freedom and tyranny turn in the days of sail. On such contingencies did science also hang, for when Halley took *Paramore* to sea to observe the magnetic field in the Atlantic, many days passed before he could get out of the River into the Channel (Chapter 10). Halley knew those dangerous waters well. He crossed to the Baltic for Poland and to Calais for France, he surveyed the approaches to the Thames estuary, he took a ship in and out of the Thames a number of times.

Travel was difficult in England. Roads were poor and often impassable in winter. Cambridge was within a day's journey from London by horse, but a coach or a wagon would take two days. Difficulties notwithstanding, people travelled a great deal. Evelyn, Pepys, Defoe, and Celia Fiennes all tell of major excursions. The Court, while far more sedentary than under earlier monarchs, was often on the road over the short distance to Hampton Court or the longer journey to Newmarket or Euston. Londoners travelled to see estates from which they drew rents, many had relatives in the country, and others travelled just to see splendid country mansions or the mysterious wonders of the past such as Stonehenge.

Antiquarian interest in the past of the country and its people had become widespread in the seventeenth century. The age was very conscious of the debt it owed to the Republic and Empire of Rome and was keenly interested in the relics of Roman times still to be seen, in London, at Silchester, Bath, and many other sites. An age which looked to Rome for instruction in architecture naturally sought out the memorials of the times when Britain was a part of the Empire. England also holds monuments of far earlier times which are still a mystery to us and excited the most imaginative speculations from people who were consciously seeing themselves as antiquarians, Halley among them. Stonehenge cannot fail to impress today and excite wonder, although busy roads embrace it and the debris of tourism surrounds it; how much more mysterious must it have been when it loomed alone on the rolling uplands of Salisbury Plain.

Other more naturally mysterious and isolated places lay not far from London. The coastlands of Essex, the haunts of waterfowl and smugglers, would become known to Halley, for he surveyed around them and one of the estates that supported his Oxford professorship was on them. Strangest of all English regions were the Fens. The Dutch engineer Vermuyden, employed by the Duke of Bedford and with Jonas Moore as his lieutenant, had tamed them to some extent in the Dutch manner, but left large tracts as shallow pools and reed marshes, remote and hidden, where an isolated people lived by fishing and wildfowling and did not welcome outsiders. Over those desolate expanses, swept by winds or shrouded in fog, there rose the ruins of once great monasteries, Ramsey, Crowland, Thorney. Peterborough, its abbey refounded as a cathedral, home of Halley uncles, lay on the fringe, not far from Alconbury, another Halley family site and close to Pepys's and Cromwell's home town of Huntingdon.

Above all, in good weather, Ely cathedral could be seen from miles around, shimmering upon its island in the haze. Ely was too remote to administer its diocese so that much of its legal and other business was done at Cambridge. Cambridge held the medieval Stourbridge Fair, the greatest in England, at the most northerly bridge between East Anglia and the Midlands. Defoe described it in 1723:

...which is not only the greatest in the whole Nation, but in the World: nor if I may believe those who have seen them all, is the fair at Leipsick in Saxony, the Mart at Frankfort on the Main, or the fairs at Nuremberg or Augsburg, any way to compare to this fair at Sturbridge. It is kept in a large Corn-field at Casterton, extending from the River Cam towards the [Newmarket Road] for about half a Mile square.

Booksellers and stationers came there and Newton bought a prism in 1665 to try the phenomena of colours.

The officials of the bishop of Ely in that small, rather remote town were the nucleus of the University which knew some modest fame and achievement in the middle of the seventeenth century. The nursery of Horrocks and Flamsteed, it was home to the Cambridge Platonists such as Cudworth, More, and Whichcote, and above all to Newton. It had a strong Puritan spirit; and in the reign of James II Newton was prominent in resisting the imposition of Roman Catholics into the University and colleges.[13]

Oxford was very different, more a metropolitan city, much closer in spirit and influence to the Court. Parliaments met in Oxford and it was a royalist stronghold in the Civil Wars. James II intruded Roman Catholics into its colleges also, and Oxford resisted so far as it was able. Oxford had in Bodley's Library munificent gifts of classical and other manuscripts, the bases for splendid scholarship. Mathematics and natural philosophy flourished in both Oxford and Cambridge, and both had presses that nourished and disseminated their learning.[14] Oxford, along with Gresham College in the City, was the nursery of the Royal Society at the Restoration.

1.4 *The Arts in England*

The Civil War and the Commonwealth had ended the exuberant artistic life of the early Stuart reigns. Developments in Europe such as the music of Lully or the theatre of Molière then passed England by. After the Restoration the arts in England flourished as they rarely have before or since.

Halley lived in the age of Baroque music. He could have heard Lully when in Paris. Couperin, Scarlatti, and Vivaldi were more his own age; Telemann and Bach were younger men who outlived him by a few years. In Halley's lifetime two very great composers worked in London. Purcell was born in 1659 three years after Halley, became a chorister in Westminster Abbey in the 1660s, organist in 1679, and died in 1695. He wrote some of his finest music for the annual birthday celebrations of Queen Mary and then for her funeral in 1694, a ceremony that Halley could have attended.[15] Handel settled in London in 1712 and died there in 1759. He brought an Italian spirit into English music with his operas, of which many were staged while Halley was Astronomer Royal at Greenwich. Handel and Halley had friends in common and may have known each other.

Theatres reopened at the Restoration; writers such as Dryden, Congreve, Wycherley, and Vanbrugh gave full rein to their wit and their theatrical invention and the licentiousness of the Court. Women appeared on the stage (Nell Gwynn, Anne Bracegirdle, Elizabeth Barry) and as playwrights (Aphra Ben).[16] For some, the stage was the threshold to greater things—Moll Davis and later Nell Gwynn became mistresses of Charles II. Many disapproved of the morals of the Court and theatre and of the actors and actresses, but many, like Pepys, disapproved and remained ardent theatre-goers.

Sir Peter Lely, among other painters, captured the beautiful women who appeared on the stage and at Court. His proud likeness of Charles II presides over the meetings of the Royal Society in London. Portraits of Halley and many of his contemporaries, Evelyn, Locke, Pepys, Sloane, Southwell, and Williamson, all of whom figure in Halley's career, some painted by Sir Godfrey Kneller, also hang in the Royal Society (Plates I, VII).[17] Kneller's Kit-Kat Club portraits of Whig supporters of William III have a more tenuous connection with Halley. They include Daniel Finch, Earl of Nottingham, the Secretary of State who sent Halley to the Dalmatian coast in 1703, George Stepney who was ambassador to Vienna at that time (Plate XII), and Sir Samuel Garth the physician and poet who was a friend of Halley.

Ricci and other immediate precursors of Tiepolo visited and painted in England in the 1690s. Italian paintings were sought by collectors in England, by William III who acquired the Raphael cartoons originally at Hampton Court and now in the Victoria and Albert Museum, as well as by more modest cognoscenti such as Sir Ralph Bankes of Kingston Lacy, a patron of Lely.[18] Architecture, even more than painting, was in debt to Italy; it had been so ever since the time of Inigo Jones. Wren was inspired by Italian models after the astronomer Auzout introduced him to Bernini.[19] The Fire of London, happening as the country was growing in affluence, gave Wren and other English architects opportunities that in France depended on the concepts of grandeur of Louis XIV.

Great English authors wrote in Halley's lifetime. A boy when *Paradise lost* and *Pilgrim's progress* appeared, he was in his prime when Dryden was writing. Defoe's *Robinson Crusoe*, based on Alexander Selkirk,[20] *Moll Flanders*, and *The journal of the plague year*, came somewhat later. Addison, Swift, and Pope were all contemporaries of Halley and quite probably known to him. Robert Hooke, a fervent collector of books, shows us in his diary the London booksellers selling those works, along with scientific, philosophical, and technical books.

1.5 The wider world of Europe

Halley went three times to Europe. As a young astronomer with a precocious reputation, he visited the venerable Hevelius in Danzig. He spent six months in Paris with a schoolfriend, Robert Nelson, and travelled with him through France into Italy and Rome. Some twenty years later he went to the Dalmatian coast by way of the Netherlands, German lands, and Austria. John Ray among his contemporaries made a number of visits to France; Evelyn studied in Padua during the Civil Wars. Evelyn and Pepys visited France; and Locke studied in Saumur and in Montpellier where he probably qualified in medicine. They and others have left accounts of their impressions of those countries.[21] Europe, vibrant like England with new thought and art and science and technology, was racked by almost continuous warfare, although that never seems to have prevented English people travelling abroad, especially to France and Italy.

Adverse winds and bad weather often gave English travellers a hard time crossing the Channel. When they reached France, three things impressed them all: how backward agriculture was compared to England, the poor and miserable state of the peasants against the splendour and riches of the higher nobility and wealthier professional people, and the empty roads, even close to Paris. Poor roads restrained economic progress, as did the small size of subdivided agricultural holdings and the complex system of internal customs duties and other heavy taxes. Language was diverse. Some form of the langue d'oc was spoken south of the Loire, Breton in Brittany, a form of German in Alsace, Flemish in Flanders, and English in the neighbourhood of Calais. The foreigner who came to France well prepared in metropolitan French was usually in difficulties outside Paris.

France was torn by religious wars at the beginning of the seventeenth century and their effects lasted to its end. The suppression of all Huguenot activities by the Revocation of the Edict of Nantes concerned English visitors, who had a fellow feeling for a Protestant community and an apprehension of increasing danger to Protestantism generally. The Huguenot exodus greatly diminished French intellectual, artistic, and technical life and correspondingly increased that in England and the Netherlands, and was very significant in Halley's world.[22]

Great cities lay outside Paris—Lyon, Bordeaux, Marseille, Strasbourg—but Paris drew every traveller. What matter that the Court was about to move to Versailles and that the abandoned Louvre was unfinished and in part a roofless shell: much administration continued in the Châtelet on the Île de la Cité. What matter that the North Bank was rather decrepit, its grand buildings set in very mean streets inhabited by a turbulent urban proletariat of minor trades, thieves, and prostitutes, repulsive to visitors: the amateur of the arts and the sciences was drawn to the Left Bank. There lay the homes of distinguished lawyers and administrators, physicians and scholars, in hôtels now marked by plaques. There was the great abbey of St Germain-des-Prés with its renowned library. Many streets are little changed from Halley's day. Visitors today will see between St Germain-des-Prés and the river more of the Paris that Halley saw than they will of his London, where little remains after wartime bombs and drastic reconstruction but the lines and names of streets.

If France, despite the absolutism and centralisation of Louis XIV, was a divided country, Italy was yet more fractured, divided between four major powers, the Papal States, Venice, Milan, and the Kingdom of Naples controlled by Spain, with a few lesser principalities. Rome, Venice, and Florence had all declined from their intellectual and artistic peaks of the early seventeenth century. The papal reconstruction of Rome was over; the major building of the villas and palaces of Venice was ended. Rome was not as tolerant to original thought in philosophy as it had been until the condemnation of Galileo. It still had active correspondents of the Royal Society, but no longer held the prime in natural philosophy as it did when the Accademia dei Lincei printed the works of Galileo. Padua no longer led in anatomy as it had when Harvey studied there. Venice, that once held empire over the East, retained her lands on the *Terrafirma*, but she had declined in the face of the Ottoman empire and trade to America. When Halley was in Dalmatia in 1703, the Serenissima was powerless to prevent the naval forays of the French who burnt English ships even inside the lagoon at Malamocco. In one thing did all Italy excel in those days—in music and especially in opera, in Naples and Rome and Venice. Handel came to Italy three years after Halley's tour in Dalmatia and prospered there, notably in Venice with *Agrippina* at the Teatro San Giovanni Grisostomo (now Malibran), perfecting his craft of oratorio and opera that he would practise so magnificently in London in Halley's later years.

Halley visited other countries besides France and Italy, countries where important correspondents of the Royal Society were also living. English connections with the Netherlands were always close and

Huygens and others visited London and were well known to fellows of the Royal Society. In 1703 Halley twice passed through the Netherlands and German lands; on his way to Vienna and Dalmatia he met Leibniz at the Court of the future George I at Hanover.

Continual warfare impoverished the participants, but it also opened up new lands. After Austrian victories over the Turks at the end of the seventeenth century, English people, and especially fellows of the Royal Society, could learn of the geography of Hungary and Slovenia from diplomats such as St George Ashe with whom Halley corresponded as Clerk of the Royal Society.[23] Warfare did not stultify the things of the mind, and there were vigorous universities in Europe at the beginning of the eighteenth century. Although the Huguenot academies had been closed down in France, the Sorbonne and the College d'Harcourt were lively. Some Huguenots migrated to Geneva where French learning was pursued, some went to Leiden in the Netherlands, where oriental studies flourished. Basel, the home of the Bernoulli family, was another centre of learning. The Bernoullis were correspondents of English and Scottish mathematicians, especially de Moivre, and entirely familiar with Halley's achievements.

1.6 The Americas and the East

Many Londoners were closely associated with the Americas and the East. Halley spent a year on St Helena, an East India Company possession, he sailed to America and had acquaintances there, and friends and probably family were members of the Levant Company.

North America was expanding rapidly, both intellectually and geographically. The New England colonies, Massachusetts in the lead, were showing the independence that would culminate in the Continental Congress a century later. Boston was a substantial port. Harvard College was founded ten years before Halley was born. William Penn received the grant of Pennsylvania from Charles II in 1681 and settled the city of Philadelphia in the same year. The name of the city is an indication that London Levant Company merchants took part in Penn's projects, for it may have reminded them of Philadelphia in Asia Minor.[24] The English colonies were intellectually alert and in close touch with London: the Royal Society had fellows there, and Newton included in Book III of the *Principia* observations of comets made in New England. From beyond India in the east to the West Indies and the eastern lands of America in the west, in the Mediterranean, across the Indian Ocean, and on both sides of the Atlantic, the tentacles of the

London merchant companies stretched far and wide, and involved directly or less directly many London families. Younger members of trading families, sometimes as young as seventeen, went out for a few years to manage the family interests or oversee the business of one of the companies.[25] They would be out of touch with their principals in London for months at a time, and the principals in turn had but a general knowledge of distant events and conditions. The military and even more, the naval affairs of the realm, were carried on under the same constraints. Those sent to distant parts, merchants, admirals, or generals, could be given only rather general orders and were entrusted with great responsibility and had to display imagination and resource while still quite young. Halley was very young when he went to St Helena and when the Royal Society asked him to visit Hevelius, but that was the habit of the times.

1.7 *Navigation and astronomy*

All connected with the sea, as were many inhabitants of London and of all England, knew well that navigation was very important and very deficient. At sea or in remote lands for weeks or months at a time, often out of touch with London, it was essential to know where one was. Positions on the surface of the Earth can be specified by the angles between the local vertical (the radius vector from the centre, to a first approximation) and axes fixed in the Earth. One axis is the North pole, the axis about which the Earth rotates, and the angle between it and the vertical is the colatitude; the other angle, longitude, is the angle between the meridian plane through the site and some other reference meridian plane (Fig. 1.3). There is no natural choice of reference

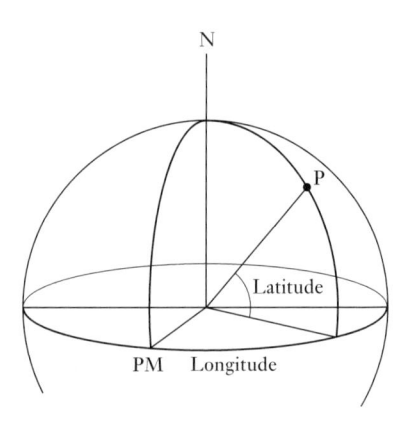

Figure 1.3 The definition of latitude and longitude. P: local site; PM: prime meridian.

meridian as there is of the polar axis and it has to be settled by convention; nowadays we adopt the meridian of Greenwich by international convention. Greenwich was already the English choice in Halley's day; the meridian of the Observatoire in Paris was the French choice and there was yet another in Rome. The meridian of Herreiro, the most westerly of the Canary Islands, had been an earlier origin when venturers first sailed to America.

One minute of arc in latitude is about 1.8 km over the ground, and navigators would like to know their position at least as well as that when near to land and in danger of wreck. Sailors who needed to estimate how far they were from land, or how long it might take to sail a given course, had to know the distance corresponding to an angle over the Earth's surface and their speed over the ground. Picard and Cassini in France, Snel in the Netherlands, and Norwood in England were making the size of the Earth, or the distance corresponding to a degree of latitude, quite well known. Speed through the water was found from a towed log, but estimates of speed over the sea bed were very uncertain because currents and tides were poorly known in the seventeenth century.

Atlantic seaman sailed to a straightforward plan. They knew the latitude in which they had to arrive to make their landfall and since most voyages across the Atlantic were from east to west or west to east, they simply got into the correct latitude and stayed in it for the crossing. Simply: but they could make only a rough estimate of when they would make land, and if clouds had allowed them only a few rare sights of the Sun they might be far away from the correct latitude and in danger of shipwreck, which did occur frequently. A ship would often find herself on the wrong side of Cornwall and at risk of wreck on the Scilly Isles, as happened to Sir Cloudesley Shovell and his squadron returning from the Mediterranean in 1707. Problems of navigation were still greater when the course was not one of constant latitude.

Finding latitude was not a great intellectual problem, for it is $90°$ minus the angular distance from the North pole, and so is equal to the elevation of the Pole star above the horizon (Fig. 1.4). That is difficult to observe from a rolling and pitching ship. Latitude may also be found from the altitude of the Sun at noon if its distance from the pole is known for the day of the observation. The polar distance of the Sun varies with the season because the path of the Sun relative to the Earth, the ecliptic, is inclined to the equator. Observations of the Sun, as of the stars, were far more difficult at sea than on land. None the less, the principles were clear and the rather simple observations could be made anywhere by men of modest ability and knowledge.

Longitude was a different matter. It was a major practical and intellectual challenge of the day, the goad and inspiration for a very great

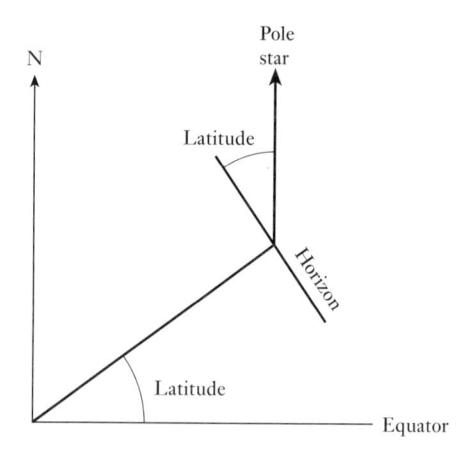

Figure 1.4 Latitude and the stars.

deal of what Halley did. It is proportional to the difference between the observer's local time and local time at some fixed reference place such as Greenwich or Paris. It is found astronomically by comparing the time when a star or the Sun passes across the local meridian with the time when the same star or the Sun passes across the reference meridian. That presupposes that there are clocks keeping the same time at the two places. Until Harrison made his timekeepers in the middle of the eighteenth century no mechanical clock could do so, especially on shipboard. Instead navigators used natural astronomical clocks, the Moon itself, and the moons of Jupiter.

The Moon goes round the Earth in a little over twenty-seven days as against the period of the Earth around the Sun of $365\frac{1}{4}$ days. The angular distance of the Moon from the Sun (lunar distance) is then a measure of time which can be determined at any meridian. The application is far from straightforward and was never wholly successful. The angle between the Sun and the Moon cannot be measured directly when the Sun and Moon are too close (new Moon) nor when they are too distant (full Moon). The Sun and Moon must therefore be located among the stars, the positions of which in turn have to be known to deduce the distance between the Moon and the Sun. The positions of the stars were not well enough known in the middle of the seventeenth century. The Royal Observatory at Greenwich and the Observatoire in Paris were set up to improve them and Halley went to St Helena for the same purpose.

The Moon, attracted by the Sun as well as by the Earth, does not move in a simple ellipse about the Earth but in a much more complex path. The Greeks had found some of the anomalies; Tycho Brahe had found another a century before Halley's day. They amount to a degree or more, and with other vagaries of the Moon's motion, mean that a very detailed theory of the motion is needed if longitudes are to be found from the distance of the Moon from the Sun. No adequate theory had been developed in Halley's lifetime. The theory of the Moon, Newton had said, had made his head ache. A theory good enough for practical navigation had to wait for many years after his death.[26]

The moons of Jupiter were more satisfactory timekeepers, as Galileo had himself suggested. Cassini had produced tables of their motions but they were not easy to observe. The method worked well on land and French astronomers had made a number of observations that, among other results, reduced the length of the Mediterranean from 60° to 40°, but it was wholly impractical from an unsteady ship at sea. Halley used the satellites of Jupiter for longitude on land in the course of his observations of the magnetic field over the Atlantic, but never at sea.

The problem of the longitude drove improvements in astronomy, in methods and in theory. When Halley was young, French and English astronomers such as Auzout and Picard, Towneley and Hooke, were improving astronomical instruments, supplying them with telescopic sights, with eyepiece micrometers, and with good pendulum clocks. With such instruments on mural quadrants, the absolute celestial position of an object could be found from its altitude and time as it crossed the meridian, but mural quadrants with telescopic sights were not erected in Paris until 1681 nor in Greenwich until 1689.[27] Most astrometric observations were not absolute; angular distances between objects were measured with sextants. Hevelius at Danzig, Flamsteed in his first years at Greenwich, and Halley on St Helena, all used sextants.[28] The main theoretical problem that occupied natural philosophers was that of the elliptical orbits of the planets. Kepler's laws were well known and Horrocks had shown how the motion of the Moon could be accommodated by a moving elliptical orbit, but the grounds for all that in the universal attraction of gravitation had to wait upon Newton—and that is part of the story of this book.

1.8 *Mathematics, technology, and natural philosophy*

In Halley's day there were no more than half a dozen men in all Europe paid to practise science, just the two astronomers at Greenwich and Paris, the curators of the botanic gardens at Padua and Leiden, and perhaps one or two anatomists. In the seventeenth century much mathematics was developed by practical men for practitioners. They studied geometry for its uses in perspective, astronomy, architecture, fortification, land survey, ballistics, and optics. The first chairs of mathematics in England were the Gresham professorships of astronomy and geometry, and the astronomical professor had to lecture on navigation and nautical instruments. Henry Briggs, the first professor of geometry (to 1620), Henry Gellibrand, an early professor of astronomy, and Christopher Wren, were all concerned with the practical uses of mathematics, while making considerable advances in pure mathematics as well. Wren studied logarithms and the geometry of the cycloid, he drew the Moon's surface on a globe, and showed that the changing appearances of Saturn, first observed by Galileo, came from rings about it.[29]

Natural philosophy uses mathematics to formalise the logical order perceived in the empirical observations of Nature. The observations depend upon the technology that students of Nature have to hand. Technology developed rapidly in the seventeenth century. Microscopes were made to observe the very small; telescopes for the very distant. Divided scales for the measurement of angles were introduced and improved; the eyepiece micrometer was devised. Barometers and thermometers, at first combined in a rather indefinite way, were distinguished and developed separately. Air pumps were constructed. Instruments and apparatus of those sorts were for laboratories and observatories but technology advanced no less in the world outside, in naval architecture, in gun-founding, in wind and water power and in steam power.

The most rigorous parts of natural philosophy at the end of the seventeenth century were the dynamics of collisions, ballistics, and planetary theory, all owing much to Galileo. William Gilbert initiated the study of magnetism and the magnetic field of the Earth in his great work, *De magnete* of 1600. He was followed by Gellibrand, Wren, and others.[30]

Biology was essentially descriptive. Ray made considerable progress with taxonomic classifications and Linnaeus introduced the binomial nomenclature of genera and species that is still used. Leeuwenhoek and Malpighi exploited the new microscope to reveal many remarkable structures and phenomena. Some mechanical ideas were being successfully applied in physiology. Jean Bernoulli advanced a mechanical theory of the action of muscle and Stephen Hales measured blood pressure in mammals and the pressure of sap in plants.[31]

Chemistry as such scarcely existed. What we now consider chemistry was divided between pharmacology and industrial activities including mining, tanning, and brewing, on the one hand, and on the other, the speculative pursuit of alchemy, much though not all of which was a dead end. Newton, Boyle and Locke were all adepts. Newton held very strongly the common alchemist's principle that matter evolved under the influence of an inherent spirit or essence. He was idiosyncratic when he identified the alchemist's essence with the Holy Spirit of Christianity. Newton was not an orthodox Christian but an Arian who rejected the Trinity and saw Christ and the Holy Spirit as creatures subordinate to God. He thought of the Spirit as the agent of God acting on matter just as in Christian belief the Spirit acts in the believer and in the Church: Columbus, for example, had believed that the Holy Spirit was leading him to America. Newton's devotion to alchemy came from his desire to gain some understanding of the material action of the Spirit, a desire that also drove his dynamical studies.[32] Ironically, the success of the formal mechanical approach that Newton developed in the *Principia* would lead in later decades to the conviction that the world ran by itself and that at the most God's contribution was to have set it going.

The Royal Society stated as one of its principal aims the improvement of natural knowledge through experiment, and its members pursued that aim more or less assiduously in its early years, in which Halley took a great part as Clerk to the Society and in presenting experiments at meetings.[33] Many of the demonstrations were ephemeral, but others were forerunners of important studies. Newton's experiments on optics and on what we now call the weak principle of equivalence in gravitation are outstanding instances, for Newton was a most imaginative and exact experimenter, by far the most effective of his day.

The Royal Society, founded in 1662, was one of the institutions and communities set up to pursue natural knowledge. Two of the first such institutions were Gresham College and the short-lived Accademia dei Lincei in Rome, founded by Prince Federico Cesi, to which Galileo was elected in 1611.[34] The Académie Royale des Sciences in Paris followed the Royal Society in 1666 and Leibniz had plans for others.[35] Natural philosophers met in those academies to discuss their findings, to plan projects, and show experiments. Astronomy, which depended on observations over many years, needed the devotion of observers who would pursue it for a lifetime and would have the instruments for it. Kepler's investigation of the orbits of the planets depended on Tycho Brahe's observatory; the first up-to-date catalogues of the positions of the stars were made by Hevelius at his observatory in Danzig. Tycho and Hevelius were rich. Tycho was of the highest Danish nobility and maintained at Uraniborg a forerunner of a modern university institute with many students and assistants dominated by a haughty and aristocratic professor.[36] Hevelius was a wealthy brewer and town councillor; he worked with a few colleagues, his wife among them.[37] Their observations were highly important, yet if an observatory were to continue beyond the lifetime of a wealthy and gifted individual, as seemed necesary to support navigation, it had to be a public institution; and so the Observatoire in Paris and the Royal Observatory at Greenwich came into being in 1667 and 1675 respectively.

1.9 *Natural philosophy and natural theology: views of the world*

Halley and his contemporaries were heirs of the humanists of the Renaissance. They learnt Latin, Greek, and Hebrew at school. They read the classical works that humanists had restored from incomplete and corrupt copies, and had spread abroad by the printing press. They knew that the Bible also was a work of art and that its versions might be imperfect. They had Aristotle free of accretions, and not only Aristotle, but also Plato and Epicurus and the Stoics and the Sceptics were now open to them. Aristotle was found wanting. Galileo led the attack on his natural philosophy and so incurred the condemnation of the Roman Catholic Church. Gassendi adopted Epicurus and the atomic picture of matter, and his metaphysics had a great influence on English natural philosophers from Francis Bacon to Newton. They, like Gassendi, were empiricists and held an atomic view of physics. Halley would take atoms sufficiently seriously to try to set experimental limits to their size.[38] Natural philosophy, in the sense of trying to place our perceptions of the world around us in some logical structure, was greatly influenced by Descartes, once known to 'every schoolboy' for analytic geometry. He maintained that one could be so certain of some ideas that they must be true and that all knowledge could be founded on them. Mind and matter were distinct for him and did not interact; there was no such thing as a vacuum but all space was filled by some substance through which bodies would interact. Descartes and his followers insisted on a strictly mechanical scheme of interactions between bodies in their account of motion. They denied the possibility of action at a distance and devised a system of vortices to explain the motions of the planets.[39] Newton at first depended heavily on Descartes's ideas, but he and Locke came to oppose them. Locke developed a thoroughgoing empiricist philosophy, arguing that all our knowledge must derive from impressions on our senses. Newton abandoned vortices for the inverse square law form of action at a distance across empty space, a theory that Leibniz condemned as occult and almost atheistical. Newton, following Gassendi, espoused an atomic theory of matter. In his reaction against Descartes, Newton went so far as to reject Cartesian analytic geometry.

The debate between the *a priori* metaphysics of Descartes and Leibniz on the one hand, and the empirical critical approach to natural knowledge and scriptural interpretation on the other, was not the whole of the intellectual turmoil from which the Enlightenment emerged. The Christian fideists such as Montaigne criticised the rational basis of all rational enquiry. Pierre Bayle in his *Dictionnaire historique et critique* of 1697 developed the fideists scepticism. English versions soon came out. The form of the *Biographia britannica*, with its rather brief text surrounded by extensive supplementary and critical notes, is that of the *Dictionnaire*. The *Dictionnnaire*, critical of Leibniz and Locke both, was one of the influential works that, with critical methods in science, religion, and philosophy, generated Enlightenment thought and principles.[40]

Almost everyone considered that the harmonisation of natural and divine knowledge was a serious issue. St Augustine had said that scripture should be interpreted in a manner consistent with secure natural knowledge. That implicitly was Dante's view: in the *Commedia*, revelation is consistent with the cosmology of his day. The heliocentric system of Copernicus and, even more, atomic theories of matter, placed the concordance of Nature and Revelation under great strain, such that eventually the Roman Catholic Church firmly asserted that natural science must be subordinate to theology. The Reformed Churches, especially in Holland, adopted the same position under the constraint of the inspired literal sense of the Bible. Few in England or in France took those extreme positions, yet Newton none the less continued to demonstrate how the Bible, especially the Biblical account of the Creation, agreed with natural knowledge.

Most people saw natural philosophy as grounded in natural theology. God had created the world and He sustained it; the true end of the natural philosopher was to expound the wonders of God in His creation and continuing activity. In his books, *The wisdom of God manifested in the works of the creation* and *Miscellaneous discourses concerning the dissolution and changes of the world*,[41] John Ray gave many examples of detailed apparently purposive adaptations of living creatures and considered the primitive chaos and creation of the world, the cause and effects of the general deluge, and the dissolution of the world and future conflagrations.

Physical scientists held the same views. Newton is the prime exponent. In Query 31 of the *Opticks* and again in the *Scholium generale* that he added to the second and third editions of the *Principia*, Newton set out his argument that the orderly nature of the physical world, both on the celestial and the atomical scales, had its origin in an 'Imperator', a 'Pancreator' outside that world. He described his conception of God in some detail in the *Scholium*. The *Queries* were not in the first edition of the *Opticks* of 1704; Newton added them to the second edition of 1717. They were almost contemporaneous with the *Scholium generale*, which was not in the first edition of the *Principia* but was added to the second edition of 1713. *Queries* and *Scholium* alike remained in the last, third, editions of *Opticks* and *Principia* that Newton himself supervised. The Enlightenment later in the eighteenth century might sideline God; Newton in the middle of the century did not.

Addison expressed the view of the age in his well known poem 'The spacious firmament on high', a paraphrase of: Ps. 19, vv. 1–9, with its final verse:

> What though in solemn silence all
> Move round the dark terrestrial ball;
> What though no real voice nor sound
> Amid their radiant orbs be found;
> In reason's ear they all rejoice,
> And utter forth a glorious voice;
> For ever singing as they shine,
> 'The hand that made us is divine'.

The geocentric solar system was long out of date when Addison wrote, and yet Halley himself, in the *Ode* he prefixed to the *Principia* in 1687, had expressed the same thought in the same context of an out-of-date geocentric world (Chapter 6). Dual sets of inconsistent beliefs were not new. Galen, for example, dismissed figures of Greek mythology as 'Gorgons and Pegasuses and an absurd crowd of other impossible and fabulous creatures', yet traced the history of medicine back to 'the Centaur Cheiron and heroes of whom he was the leader'—and Newton based his chronolgy on Cheiron's astronomical activities (see Chapter 14). Metaphor is deeply ingrained in our use of language and most scientists still follow Addison and Halley when, for instance, they use such anthropomorphic terms as 'molecular recognition'.

Addison expressed a view of the natural world that goes back to Isaiah and the Psalmist. Hevelius placed in the frontispiece of his *Selenographia* the words of the prophet:

Atollire in sublime oculos vestros, et videre qui creaverit ista.

[Lift up your eyes on high and behold who hath created these things.]
Isaiah 40:26

—an injunction echoed by the Psalmist.

Hevelius placed other words of the Psalmist in a scroll above his engraving of the full Moon in *Selenographia*:[42]

Magna opera Jehovae exposita omnibus qui delectantur iis;

[Great are the works of the Lord,
Pondered over by all who delight in them.]

Psalm 111:2 [43]

Some English theologians and philosophers held millenarian ideas, that the world would come to an end, and that right soon. The Cambridge Platonists, More, Cudworth, and others, thought so. Newton may have, and Whiston, his Arian disciple, most certainly did. They interpreted the prophecies of the Apocalypses of the Old and New Testaments, in particular the books of Daniel and Revelation, in very literal ways. Timing was crucial in millenial interpretations of the Bible, so Newton and others like him gave great attention to chronology. Halley towards the end of his life would defend Newton's chronology (based on Cheiron's decrees and including the fall of Troy and the voyage of the Argonauts) against clerical criticism from France (Chapter 14).

Thirty years later the world would look very different. The very success that Newton and his successors in France had in demonstrating the dynamical basis and predictability of celestial mechanics, on the sole assumption of the inverse square law of universal gravitation, convinced people that Newton's God was not needed to sustain the world: as Laplace said in answer to Napoleon's question, he had no need of that hypothesis. When God was removed from the day to day performance of celestial machinery it was not long before He was set aside as constructor of the machine. Natural philosophy cut itself adrift from natural theology and became autonomous. At the same time the literal interpretation of the Bible, which had been for a century or more the foundation of natural theology, especially Reformed theology, was put at risk when the critical methods of philology that had been developed for the study of classical literature were turned to the sacred scriptures. Even by 1550, they had shown that the pointing of the Hebrew text for vowels had not been devised in Old Testament times but far later when Hebrew had fallen into disuse among Jews themselves: it could not be part of the literal inspiration of the Bible. Newton likewise applied the methods of literary criticism to texts claimed to support the doctrine of the Trinity. He asserted they were very late interpolations—it turned out in that instance that while his method was valid, his conclusion was wrong.

God did not figure in the Enlightenment, or not much, yet religious metaphysics remained in the background. It lies there still.

Natural philosophers must assume that there is order in the natural world, for otherwise our enterprise, the rational understanding of our observations of the natural world, would be futile. Belief in a world that is rational and comprehensible is especially Jewish and Christian. Our contemporary world, no less than the Enlightenment, continues in that respect to be the heir of Augustine and Dante, of Newton and Ray. We rightly see great changes in men's view of the world brought about by the achievements of Newton and those who followed him, Halley among them, but should not discount common underlying metaphysics.

It is right to dwell on the changes that came about in Halley's lifetime, brought about by Newton and his contemporaries, and developed in the particular circumstances of England in the late seventeenth century. That alone would be an incomplete story. In earlier years English physics was not remarkable and Galileo and Descartes, Gassendi, Mersenne and Huygens, to mention just a few, prepared the way for Newton and his followers. Halley was not a biologist and so biology hardly figures in an account of his life, but the profound changes in the ways people thought of the natural world also owe much to the systematic studies of John Ray and Linnaeus; and to the practice of anatomical dissection and such consequent discoveries as the circulation of the blood by William Harvey and the fallopian tubes by Gabriele Fallopio; and to the microscopic studies of Leeuwenhoek and Malpighi.

1.10 Halley: an agent of change

In Halley's lifetime rapid changes came about in many fields of human thought and endeavour. Much, it is true, did not change, especially transport and travel which remained lengthy, difficult, and usually disagreeable. Yet technology did develop, and great shifts in the way of life were on the way. Halley lived at the time when the foundations of our present practice of natural philosophy were formed. Robert Boyle enunciated the empirical approach to nature in his statement of what an effective hypothesis should be:

> That it enable a skilfull Naturalist to foretell future Phenomena by their Congruity or Incongruity to it; and especially the Events of such Experiments as are aptly devised to Examine it; as Things that ought or ought not to be Consequent to it.[44]

Newton married to that the idea that hypotheses should be cast in rigorous mathematical forms.

Halley did not just experience those changes of thought and belief and attitude, he contributed to them. He was among those who brought about the evolution of Renaissance casts of minds and views of the world into those of the Enlightenment and so into those of our times.

Halley was the first to pursue what we now call geophysics, the study of the solid Earth, the oceans, and the atmosphere by the methods of physics and according to physical principles. He publicised isolines in cartography and he was the first to persuade the Royal Navy to provide a ship for a purely scientific enterprise. By his prediction of the transits of Venus that would occur when he was dead, and by showing how they could give improved values for the size of the solar system, he prompted the Royal Society to propose James Cook's first voyage to the South Pacific, which led on to Cook's other great expeditions and the first modern knowledge of the southern seas. He applied scientific methods to the study of history and historical methods to scientific problems.[45] Above all else, he prompted Newton to compose the *Philosophiae naturalis principia mathematica*; he encouraged him to complete it; he was critic and editor; he oversaw the press; and he published the *Principia* and distributed the presentation copies. Later he calculated cometary orbits for Newton and predicted the return of 'his' comet, all in the spirit of Boyle's definition. Less well known but as significant, he studied the Moon extensively on the basis of Newton's theory (Chapter 13). Throughout Halley's work there runs the firm belief that Nature can be understood rationally and that from what we know we may predict what is yet to be observed.

What manner of man was Halley, how came he to have such a part in the construction of our present world? Who were his family, his friends and acquaintances, into what society was he born, what his education? What so gripped him to leave Oxford early, to step from a London quay upon the *Unity* East Indiaman bound for the southern seas, to embark upon a lifetime of mental voyaging, an adventurer for natural knowledge?

2

Formation of an astronomer

> At length they all to merry London came,
> To merry London, my most kindly nurse,
> That to me gave this life's first native source.
>
> SPENSER: *PROTHALAMION*, 1, 127–9

2.1 *Forbears*

THE *Memoir* of Edmond Halley gives his birth as 29 October 1656 (os) at Haggerston in the parish of St Leonard's, Shoreditch, a country suburb of the City. Parish records of the time of the Commonwealth are often incomplete; those for St Leonard's for 1656 are missing. No record of Halley's birth there or anywhere else survives. Halley recalled that he was sixteen when he matriculated at Oxford in June 1673. Flamsteed in a letter to Towneley of 8 June 1675 says that Halley is scarce nineteen and Halley, writing to Hevelius on 11 November 1678 (os), says he has only just passed his twenty-second birthday. All would agree with his birth in October 1656, save that his parents may have been married less than two months before. October 1656 is the date usually accepted but the authority for it is now unknown. The same date and place are given by Aubrey and in the *Biographia britannica*, but they may not be independent.[1]

The First Protector, Oliver Cromwell, was still alive in 1656 and Halley would have been about four when the Commonwealth collapsed in 1660 and Charles II came back to England and his throne.

Aubrey said that the astronomer was descended from a Derbyshire family of Halleys and a Humphrey Halley of Bakewell has been suggested as his great-great-grandfather. That is unproven, but Halley did have relatives in Derbyshire, one of whom, Luke Leigh, did calculations for Flamsteed.[2] Halley's paternal grandfather, Humphrey I, (Appendix 2) was a citizen of London. He had Yorkshire connections and possibly relatives. There were other Halleys in London contemporary with the astronomer and his father. A Dr Halley, probably a civil lawyer, lived in the parish of St Benet, Paul's Wharf, where one of Halley's daughters was baptised. Halley's name was often spelt *Hawley* and there were many Hawleys in London who may or may not have been his relatives. One was Major Thomas Hawley, the gentleman porter of the Tower (Chapter 5). Halley's genealogy, with common family Christian names, such as Edmond, Humphrey, and Katherine, is not always certain.

Humphrey I married Katherine Mewce, daughter of Nicholas Mewce, in St Margaret's Church, Barking in Essex, in 1617. Nicholas held the estate of Hedgemans in Essex which survived in the eighteenth and nineteenth centuries as a farm to the north-east of Barking; there are today Hedgemans Road and Hedgemans Way in Dagenham. Katherine's brother, Francis, married Elizabeth of the Washingtons of Sulgrave, the family of the first President of the United States of America.[3] The Halley connection with Barking continued up to 1684 at least (see below and Chapter 5).

Barking Abbey, a house of nuns, was one of the great monasteries of England in the middle ages, wealthy and privileged. Its ruins remain and the church of St Margaret's is next to them. At the dissolution of the monasteries, many of the estates of the Abbey were purchased by City merchants. The Leithullier family, neighbours of the Halleys in London in Winchester Street, were some of the greatest landowners. Barking was a country resort of rich citizens of London and the Halleys may have been in the company of influential persons from early in the seventeenth century and perhaps related to some through Mewce forbears; they may even have held land there.

Humphrey I was trading as an haberdasher in 1631, and as a vintner in the Unicorn in Lombard Street, which belonged to the Fishmongers' Company. He was churchwarden of St Clement's, Eastcheap, from 1639 to 1642 and in 1642 he contributed £25 to the forced loan that the City made to Parliament. John Pearson, the author of *Exposition of the creed* and later Bishop of Chester, was a lecturer at St Clement's and Humphrey was one of the parishioners to whom he dedicated his book.[4] Humphrey was Deputy for Candlewick Ward in three periods, 1646–1647, 1648–1649, and 1651–1665, and he was Master of the Vintners Company in 1661. The Chamberlain's Loan Accounts for the City show that Humphrey contributed on a number of occasions to loans raised in the City for the Crown, about half of the total on his own acount (£500) and about half on behalf of other residents of Candlewick Ward (£534, 10s. 4d.).[5]

Humphrey seems to have been at one time in partnership with a haberdasher in York. In 1636 he acted for the mayor of Huntingdon in transmitting a portion of the ship-money charged on Huntingdon to the Exchequer in London. Humphrey and Katherine moved to Alconbury in Huntingdonshire in 1665, and there they died, but their connections with Huntingdonshire go back for at least 30 years.[6]

Humphrey had five children, all baptised in All Hallows, Lombard Street. Three sons survived, Halley's father, Edmond (senior), of London, Humphrey II, also of London, and William, of Peterborough. William married and had children; Humphrey II did not marry and died at Alconbury. Humphrey I also had a daughter, Elizabeth, who married John Cawthorne and had four children, Katherine, Anne, William, and Humphrey.[7] Anne is probably the Anne Cawthorne who married Joe Chomat at St James's, Duke Place, in 1681 (Chapter 5). Alconbury will figure in the history of modern science.

Humphrey I died in 1672 at Alconbury and was buried in St Margaret's, Barking, together with Katherine who had died earlier, her remains having presumably been moved from Alconbury. A week later, Halley's mother, who had just died, was also buried there, and so again, in 1684, was Halley's father (Appendix 2).

2.2 Parents

Halley's father was Edmond (senior). He had three children by his first wife, Anne Robinson: Edmond himself, another son Humphrey III, younger than Edmond, and a daughter Katharine who was born in 1658 and died young. Edmond (senior) was a soapboiler and a member of the Salter's Company. He seems to have held no public office. Reputedly 'very rich', according to a pamphlet published just after his disappearance and death (Chapter 5), he was 'well known throughout the whole City':

> ...no less esteemed for his own Merits, as a Person always forward to the Service of his King & Country, than he was respected as a Gentleman of considerable Estate, and plentifully happy in all the Goods of Fortune.[8]

His personal estate at his death was about £4000 (about £1.6m in today's value).[*] That was almost certainly not the full extent of his wealth. There would have been the assets of his business and there are indications that he owned real property in London and outside, some of which he may have transferred to Halley in his lifetime, as was a common practice. Just before the Great Fire the father paid hearth tax on a substantial house in the parish of St Giles's, Cripplegate, and he may have owned the house in Islington where Halley set up married life in 1682.[9] Some family properties were destroyed in the Great Fire,

---

[*]The Bank of England paid its clerks £50 p.a. in 1694, the same as Pepys's salary in 1662 and Halley's as Clerk to the Royal Society in 1686. The Bank has said that a salary of £50 in 1694 was equivalent to £20 000 in 1994 (*The Independent*, 10 May 1994). When the Venetian painter Amigoni left England in 1730 with £4000 to £5000 he was accounted a wealthy man.

although others in Winchester Street were unharmed.[10] Properties in Cannon Street may have been the valuable houses that were later part of the father's estate (Chapter 5). Edmond (senior) supported his son's voyage to St Helena and his tour in France and Italy by £300 each, and after that, and for two years before his father's death, Halley had an annual allowance of £60 from him.[11]

Edmond (senior) was married twice. Halley's mother may have been the Anne Robinson who was married to an Edmond Halley in 1656 at St Margaret's, Westminster, apparently just less than two months before Halley was born. Anne's father may have been a William Robinson, and she may have been the daughter who was baptised in St Margaret's, Westminster, in 1628, but nothing else is known of her and the only reason for the identification is that it gives the most likely age for her. Anne died in 1672 in the parish of St Giles's, Cripplegate, when Halley was 16 or 17. The registers of St Giles's are unusually informative and Anne's entry is designated L, meaning that her house was in the Lordship of the parish, that is to say outside the City Wall; the cause of her death is given as 'Stopt Stomach', which is taken to mean a gastrointestinal infection, a common cause in those days of imperfect sanitation.[12]

At some time after 1672 Halley's father remarried; the record has not been found. The second wife, Joane, is even more shadowy than the first. She had a particular friend, Joseph Chomat, a merchant, who owed the father £148 in 1683. A Joe Chomat married Anne Cawthorne in 1681, at St James's, Duke Place.[13] The following passage occurs in a document of a Chancery action following the father's death:

> And this defendant Joane further sayeth that she was acquainted with the said Joseph Chomatt in this bill named as often removing to her house in the lifetime of the Intestate occasioned by his wife being a relation of the Intestate and there residing sometime and denyes that the said Joseph Chomatt was the defendant Joane's—friend or acquaintance as in the bill is charged other than as aforesaid.[14]

Apart from the suggestion that aspersions had been cast on Joane's relation with Chomat, the passage, taken with the marriage record, identifies Chomat almost certainly as the husband of Halley's cousin, Anne Cawthorne, who, like Halley, had a legacy of £100 from uncle Humphrey II.[15]

When in 1682 members of the House of Commons were trying to exclude the Duke of York from the throne as James II, when Shaftesbury and Algernon Sidney were raising London in near revolt, an Edmond Halley submitted a petition claiming that Yeoman Warders of the Tower should not be called on to serve in the City Train Bands because their duty was to the body of the Sovereign.[16] Halley himself had recently returned from Paris and was about to be married. Could the warder have been Edmond (senior), a well-to-do citizen? He it was, for among the assets of his personal estate listed upon his death was an item of two quarters' salary from the Tower. The Dividend Book of the Tower records his service from the 1660s.[17] To be a Yeoman Warder was for some rather like belonging to a fashionable City club, as it is now to be a member of the Honourable Artillery Company in time of peace. The Captain of the Yeoman was often a nobleman. A pamphlet of 1690 set out the father's position: 'This Mr Hawley [Halley's father] was very rich, and a warder only to exempt him from parish services but he never waited, unless it were on very solemn occasions'.[18] He could well have seen or guarded notable State prisoners such as Pepys or Russell or Algernon Sidney: his duties may have led indirectly to his bizarre death in April 1684, in circumstances that were never explained (Chapter 5). He left no will, and disputes arose between Halley and Joane and Joane's second husband, Robert Cleeter, which led to a number of actions in Chancery that afford particulars of the father's personal estate and other matters.

Edmond (senior) owned thirteen houses in Winchester Street and had lived in one of them from 1679 at least. Halley lived there until his marriage in 1682, when he moved to Islington. Winchester Street (see Fig. 1.1) is an L-shaped street between London Wall and Broad Street, close to the parish church of All Hallows, London Wall. Bethlehem Hospital, built by Robert Hooke after the Fire on a long site on the north of London Wall, is near by. The medieval church of All Hallows escaped the Fire, but was torn down and rebuilt in its present form after Halley's day, as was Winchester Street. Halley's uncle, Humphey II, lived in the parish when he made his will. After the settlement of the father's estate, Joane and Cleeter lived on in the Winchester Street house; there a daughter, Mary, seems to have been born to them, and all three were recorded in the parish of All Hallows in 1695.[19]

2.3 *Family property and possessions*

Sir John Buckworth, a neighbour of Pepys in Crutched Friars, was the ground landlord of the Winchester Street houses which yielded the Halleys a net income after ground rent of almost £200. Edmond (senior) also owned two houses in Canning (Cannon) Street, where the rents were much higher, as well as the Dog Tavern in Billingsgate. He had invested in the General Lottery and in the Coal Office of the City and had made loans to Joseph Chomat and the Rector of Sawtry near Alconbury, among others. He himself owed money to Halley's mother-in-law, Margaret Tooke. The Dog Tavern seems to have made a loss, but the father's net income at his death from his personal estate, houses, and investments, was about £400.[20] Pepys's salary in the 1660s was much the same, but Pepys also enjoyed various commissions, while the father's investments would have supplemented his profits from trade. When Newton was an undergraduate his mother had an income of £700. The Halley family had other property interests in and around the City, although they are not so well documented. In 1666 the father had paid tax on nine hearths for a substantial house in St Giles's, Cripplegate.[21] Halley lived first at Islington after his marriage, but later from about 1684 at Golden Lion Court in St Giles's, Cripplegate, and then after he was elected to the Savilian Chair of Geometry at Oxford, he had a house in Bridgewater Street by Bridgewater Square in the same parish. The father had purchased property at Bushey, Hertfordshire, for £150 in 1665,[22] and in his own will Halley bequeathed to his daughters jointly the remainder of the lease on two houses in Cannon Street—perhaps at some time he acquired them from Joane.[23] In 1694 Halley engaged in what seems to have been a rearrangement of family interests in a property in Mincing Lane on ground originally purchased from Susanna Sandwith, formerly of London, and later of Alconbury.[24] She had been involved with Humphrey I in a Chancery suit in 1670 and had received a legacy from Humphrey II.[25] Humphrey I, Humphrey II, and uncle William held property in Alconbury and Sawtry and probably also in Peterborough. Edmond (senior) had inherited property from both Humphreys and was the executor and residuary legatee of Humphrey II.[26] Since Halley was the only surviving child and heir at law, all that no doubt passed to him on the father's death, if not earlier.

After the death of his father Halley received the houses in Winchester Street, save for the one occupied by his stepmother Joane and her new husband. Joane took the rest of the father's personal estate, including the Cannon Street houses. That gave Halley and Joane each an income from personal property of between £150 and £200 a year. Halley's wife Mary had a legacy of £100 from an uncle and inherited a share of lands in Norfolk and houses in London from her father. The Halleys seem to have been comfortable, but not extremely rich like their ground landlord, Sir John Buckworth, or other members of the great trading companies, or as prominent government servants such as Pepys.

2.4 The Tower and Sir Jonas Moore

Halley came while quite young to the notice of prominent scholars and scientists. As an undergraduate at Oxford, he joined Hooke and Flamsteed in viewing the site and settling the design for the new Royal Observatory,[27] and when he sought support for his expedition to St Helena, Sir Jonas Moore and Sir Joseph Williamson (Plate I) gave it him. Moore was Surveyor-General of the Ordnance, his office and stores of guns and ammunition were in the Tower where Edmond (sen.) was a Yeoman Warder. He was very influential in the early years of the reign of Charles II. Before and during the Civil War he had been Vermuyden's surveyor for the draining of the Fens promoted by the Duke of Bedford. He was tutor to the Duke of York, later James II, when as a boy of thirteen James was imprisoned by the Parliament at Syon House near London in 1646–1647. That seems to have been the start of a close relation with James and later Charles II. After the Restoration, in 1663, Moore became Surveyor-General of the Ordnance. He wrote *an Arithmetick* (1660), but he himself says he owes almost all of it to Oughtred, although it may be that he did make original contributions. Moore was essentially a practical mathematician, with a great interest in applications. In 1663 he took part in an expedition to Tangier which had come into English possession as part of the dowry of Catherine of Braganza, and he drew up a map of the city and its surroundings. In the same year he made a map of the Thames, and after the Great Fire of London he carried out a survey of the City. He was elected to the Royal Society in 1674 and was a councillor from 1675 to 1678.[28]

The Master-General of the Ordnance was the head of the Ordnance Office. He and the Lieutenant-General were courtiers. The three chief professional officers, the Clerk, the Storekeeper, and the Surveyor-General, were often able mathematicians. The many artificers on the staff of the Office were among the most skilled in the country. They certainly made a quadrant for Flamsteed and very probably a sextant as well when the Royal Observatory was set up; they probably also made the quadrant and sextant that Halley took to St Helena. The Ordnance Office had connections with the Royal Society on account of its involvement in surveying, especially of fortifications, and in gunnery. Moore and others may have introduced Halley to those topics in his formative years, as also to the diving bell of Ralph Greatorex which Moore promoted for use on the Tangier expedition. Halley would later work on all those matters. Halley surely knew Moore already by 1675 when Greenwich was chosen as the site for the Royal Observatory. Thereafter his career was closely connected with it.

Moore, more than anyone else, was responsible for founding the Observatory. Moore supported Flamsteed in his early observations, obtained his appointment as Astronomer Royal, and then when the Royal Observatory came into being, provided its instruments. Flamsteed greatly lamented Moore's death in 1679, for it delayed the provision of a mural circle and other instruments that Moore had intended to obtain for him.

2.5 *Neighbours*

Winchester Street was close to the City Wall in the well-to-do Ward of Broad Street where, together with the adjacent wards of Cornhill and Bishopsgate Within, the most prosperous and powerful City merchants lived. Robert Hooke, the Curator of the Royal Society and Surveyor to the City of London, lived in Gresham College (Fig. 2.1) where the Royal Society met. Hooke was well known to Charles II and was a friend of Sir Jonas Moore, and it is little surprise that Halley should come to know him well and owe much to him. They were close colleagues until Hooke's death in 1703. Hooke came early in life to the notice of Robert Boyle while at Oxford and assisted Boyle in many of his projects. He then moved to London as Curator and operator at the Royal Society, responsible for procuring and demonstrating experiments at the meetings. He was Gresham Professor of Geometry. He was one of the surveyors of building sites after the Great Fire; his plans for rebuilding the City were accepted by the City Council, and he and Wren together did most of the work. It is a little difficult to distinguish Hooke's buildings from Wren's, for they were in effect a partnership, even if informally, and they did not limit their joint activities to London. They were both involved in the Royal Observatory, although that was a rather minor undertaking and most of the construction seems to have been done by Tower workmen. Halley would have known one of Hooke's most spectacular buildings, the Bedlam Hospital on London Wall close to Winchester Street, with its attractive gardens to the north, visited by Evelyn on 18 April 1678.

Hooke was gregarious and always in company. When Boyle moved to London, Hooke saw him often and was a friend of his sister, Lady Ranelagh, in whose house in Pall Mall Boyle lived in his last years. Hooke was always meeting friends and colleagues in coffee houses, especially at Jonathan's in Exchange Alley. Wren was one of his constant companions. He knew Londoners from the King to mechanics and booksellers, and was on good terms with most of them; but two quarrels

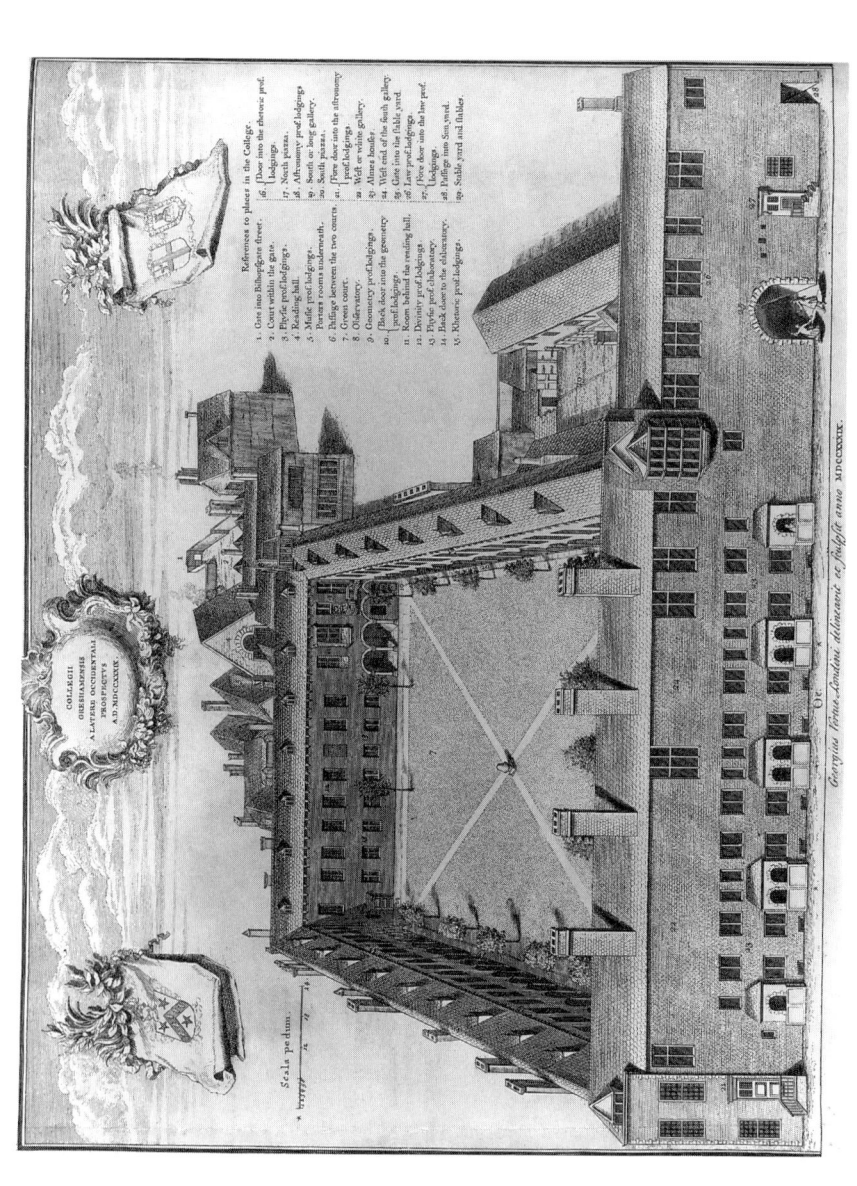

Figure 2.1 Gresham College in Broad Street. (Reproduced by permission of the President and Council of the Royal Society.)

ran for much of his life, with Newton and with Flamsteed. He resented both of them, they disliked him, and they themselves were eventually on very bad terms with each other. The origin of both of Hooke's great antagonisms lies probably in one of his characteristics as a scientist. He had a very lively mind and fertile imagination, so that he was forever putting out ideas and plans, but he was not good at applying himself to working them out in any detail, apart from a few notable exceptions. Since others also were considering the things he thought about, he found from time to time that something he had suggested years before was now being worked out by another. The outstanding instance is the hypothesis of the inverse square law of gravitation and the explanation of the orbits of the planets (Chapter 6). Hooke and Flamsteed were at odds over the first instruments for the Greenwich Observatory and Hooke wrote sharply of Flamsteed, as *Vulponi*, presumably referring to Ben Jonson's play, and as conceited and ignorant.

Halley and Hooke seem to have got on well, although Hooke is sometimes critical of Halley's ideas. They corresponded when Halley was in France in 1681–1682, and in 1686 they joined in observing eclipses of Jupiter by the Moon, as reported in *Philosophical Transactions*. Hooke saw Halley's will witnessed in 1693 at Jonathan's.[29] Already as an undergraduate at Oxford, Halley was acquainted with Wren and Hooke and would often join them at Garraway's or some other coffee house. Hooke in his *Diary* recounts some of Halley's activities.

Sir John Buckworth (1623–1687), who was later to mediate in Halley family disputes (Chapter 5), was a City grandee. He was a Levant Company merchant and for many years Deputy Governor. He held important public offices such as Commissioner of Excise and Commissioner of the Mint (which, like the Ordnance Office, was in the Tower) and was Alderman of Coleman Street ward. He was a member of the Royal African Company and held office in it. According to his funeral eulogy preached by Dr John Scott in St Peter-le-Poer, he was born a gentleman and admitted a merchant. He was a Prince among merchants and an Oracle of Trade, and a Deputy Governor of the Turkey Merchants. He was known as a peacemaker and an arbitrator of quarrels and a 'hearty protestant of the Church of England'.[30] He it was whom Halley chose to arbitrate in his disputes with his stepmother. The stepmother chose Richard Young, probably the Levant Company merchant who was an Assistant of the Company from time to time. Buckworth and Young acted together on another occasion when they jointly stood surety for someone going out to the east on Company business and both subscribed to the construction of the Church of King Charles the Martyr at Tunbridge Wells (see Chapter 1). A Richard Young, the only one listed in the City, lived in the parish of

St Andrew, Undershaft, in 1695.[31] Halley and his stepmother were evidently well enough acquainted with prominent members of the Levant Company that they could call on them to help resolve their disputes.

Sir James Houblon was a neighbour of the Halleys in Winchester Street. A merchant and friend of Pepys, he lived *en Prince*, so Pepys said. Two cousins of Pepys also lived in Winchester Street. As a schoolboy and undergraduate Halley did not seem to have met Pepys, at least he does not appear in the *Diary*, but in later life they came to know each other well. Halley knew Lord Brouncker, the colleague of Pepys, courtier, and first President of the Royal Society, for they observed a solar eclipse together at Greenwich in 1676. Halley similarly knew Henry Oldenburg (Plate II), the first Secretary of the Royal Society, before he went up to Oxford. He owed his introduction to Wallis (Plate III) to him. Halley's early acquaintances in London evidently had much to do in setting him off on his career.

The Pay Office of the Navy in Broad Street, where crews were paid off when a ship returned from sea to the Thames, stood close to Halley property. The sea and seagoing, seamen and seamanship, the Navy and trade, were all vital to the life of London. At the quays on the Thames, through family acquaintances in the trading companies or as shipowners, from seeing seamen attending the Pay Office, or by the concerns of Moore at the Ordnance office, maritime matters may rarely have been far from Halley's mind.

2.6 The Levant Company

Halley and his family must have been well acquainted with the Levant Company through neighbours and friends who were members. The Company had a monopoly of the trade to Venice and Asia Minor. English goods, especially cloths and tin, were traded for silks and spices and above all for the currants of Xante, for which the citizens of London had an apparently insatiable appetite. Levant Company merchants were among the wealthiest in the City, powerful in the government of the City as well as in national affairs. The father of Robert Nelson, Halley's friend from schooldays, had been a Levant merchant and Robert himself became a member of the Company; Robert's uncle Sir Gabriel Roberts, of the parish of All Hallows on the Wall, was another. Halley surely knew them well. Tookes also, possibly relatives of Mary Halley, were Levant merchants.

In his diary for Monday 24 July 1693 Hooke records '2 East India ships said to be taken by French in India. Hot, clear. Hallys trade taken by French'.[32] That might seem to mean that Halley was trading with the East India company and suffered loss by French action. However, Halley does not appear in any of the very full records of that Company, neither as a stockholder nor as a trader with the Company, nor in any accounts, although he could have been an undersharer. The date of 24 July shows that a greater disaster may have caused his loss: on that day it became generally known in London that a fleet of some 400 English and Dutch ships sailing to the Levant had suffered heavy loss from a French fleet off Lagos in the south of Portugal. The English battle fleet had convoyed the merchantmen as far as Ushant. Then because the admirals thought that the main French fleet was still in Brest, the merchantmen had continued into the Mediterranean protected only by a small squadron commanded by Sir George Rooke. The Brest fleet, in fact at sea, joined the Toulouse fleet and attacked the merchant ships. While Rooke avoided action, more than 80 merchantmen were sunk, burnt, or captured. It was the greatest naval disaster of the Nine Years War and the loss to London and to Dutch merchants was heavy. Both Houses of Parliament held enquiries into the actions of the Secretaries of State and the admirals.[33]

The records of the Levant company are by no means so complete as those of the East India Company. Halley was not a member and it is not possible to say if he was or was not trading with it.[34] Hooke's entry could mean that Halley was associated with relatives of his wife in the Company in trading to the Levant.

2.7 Schooldays: St Paul's

Halley grew up in London in the days of the Great Plague of 1665 and the Great Fire of 1666; Pepys in his *Diary* gives a lively impression, full of immediate detail, of what it was like to live then.

Households were relatively large; there were maids to do the housework, there might be a companion for the mistress and a footboy for the master. After Edmond's mother died in 1672, his father would probably have had a housekeeper until he remarried. Some at least of the father's apprentices would have lived as family, and undertaken personal services. Servants were part of the family, if Pepys's household was at all typical. Often, while Pepys was working, or arranging his books, Elizabeth and her 'people' would be at games, or dancing, and the whole company went on outings and made music.

The very rich and the less well off all lived next door to their butcher or hairdresser or tailor or joiner or bookseller. Through the ears of Samuel or Elizabeth we hear the cries of the street sellers as in Purcell's settings. Food was plentiful. Meat, such as beef or venison pasty, was preponderant. There were fish that are now delicacies, oysters and salmon, but there were few vegetables and little fruit. Beer and wine were the staple drinks: when Pepys thought that the Fire might spread to his quarters in Axe Yard, he buried his gold, his wine, and his Parmesan cheese. After a token breakfast most families dined about midday, with a later less formal supper.

Clothes for people of Pepys's standing were rich, often of silk, and gowns for ladies took many yards of material as shown by a commission of Halley for Mme Hevelius (Chapter 4). Mens' fashions changed, but ladies outdid them—it was said that women should not buy new dresses before Easter because the fashion would change then—*plus ça change*. Men wore swords and footboys carried short swords, but mostly for show, although footpads were a danger and duelling still occurred, but not usually among City people. A duel did threaten at the country Mint in Chester when Halley was there in 1697. Houses had no piped running water supply nor drains—Lady Ranelagh set up an outside privy for her brother, Robert Boyle, when he moved from Oxford to join her in Pall Mall; Halley threw the water in which he washed his hands on to a rosemary bush. Pepys was particular and made a special note of washing and tidying himself after his exertions in the Fire.

Booksellers abounded. There must have been many voracious buyers of books like Hooke and Pepys to support the considerable trades of printing and bookselling, engraving and map selling and instrument making, that clustered around St Paul's. Halley himself may have had a very substantial library by the end of his life (Appendix 8). The loss of the booksellers' stores that they had placed for safety in St Faith's, the crypt of St Paul's, but were all destroyed, was one of the great disasters of the Fire.

Pepys worked hard at the Navy office and Elizabeth spent much time supervising her household, but they had frequent diversions. On summer evenings they might take a coach into the country north of the City, or go by boat on the river. Their people would go with them and there would be singing or other music. They were both keen theatre-goers. Samuel often disapproved of the morals of a piece, or of the actresses, or of the audience, but was quite unable to resist the fascination of stage, singers, actors, or actresses. In all that he seems to have been at one with many of his associates, professional and administrative people and merchants. Unless the Halley family was quite exceptionally puritanical, we may suppose that they diverted themselves in much the same ways. Certainly the character that everyone seems to have given Edmond, genial, good natured, sociable, and fond of company, suggests that he was at home in the society of such as Pepys.

Because he told all, or almost all, in his *Diary*, Pepys seems exceptional to us now, but may not be entirely so. In his avid pursuit of women he matched the Court. Lord Brouncker, his colleague in the Navy Office and first President of the Royal Society, was rapacious. Anthony Hamilton, in his tales of the Court of Charles II, writes of two maids of honour who disguised themselves as orange girls but were detected by Brouncker who accosted them in the street; they recognised and repulsed him. Brouncker, it was said, kept quiet about the incident, because he could never bear to confess that Miss Jennings had left him in the same state that she met him.[35] Pepys's passion for music was deeper and longer lasting than even his pursuit of women. Again he was not alone. Sir Gabriel Roberts possessed a manuscript of instrumental music for two trebles and bass by Lawes, Matthew Locke, and other English musicians, as well as twenty Italian sonatas.[36] London afforded many opportunities to hear music, then and in Halley's later years. Many families made music at home—Pepys noticed when people were leaving the City by the river during the Fire that above one in three had a pair of virginals—but few can have outdone him in his attachment.

London and Westminster were small enough to walk from place to place, but on business or when he was with Elizabeth, Pepys would take a coach or a wherry on the river. It was a great day for him when he bought his own coach and two horses, to the disapproving envy of his colleagues. The Halley family probably had their own coach, if only because the father's soap boiling factory would have been outside the City walls with other noxious trades like tanning and dyeing. Horses' hooves and wheels of coaches were very noisy and the City was not quiet.

People often had to go into the country, as Pepys would to his relatives at Brampton, or Halleys to theirs at Peterborough or Alconbury. Courtiers and officials would follow the King to Newmarket, others like Celia Fiennes travelled just to see the country and its houses. Pepys might go by coach when Elizabeth was with him, but Elizabeth was a good horsewoman and sometimes they rode together. The roads were bad, coaches were scarcely sprung, and in fine weather it was probably pleasanter to ride than to go in an uncomfortable coach. By whichever means and by whichever route Pepys or Halley went from London to Brampton or Alconbury or Cambridge or Huntingdon, it usually took them two days and they had to stay at some inn on route. We do not know if Halley made such excursions with his family as a schoolboy, but the average Londoner was not confined to the City and at first or second hand, knew something of the rest of England.

Such then was the society, such the family life, that Edmond knew when he first went to school. He might possibly have entered St Paul's before the Plague of 1665, for boys entered the first form as young as seven or eight,[37] but he cannot have been long there before the Great Fire destroyed St Paul's and the school with it. Evelyn, Defoe, and especially Pepys have left vivid accounts of those two devastating visitations which drove most of the citizens out for more than a year. The streets were almost deserted throughout the plague summer and autumn. Some stayed to bury the dead in great communal graves on Moorfields and elsewhere, some devoted doctors and courtiers cared for the sick. The mortality bills rose close to 10 000 a week, the red crosses upon house doors grew ever more numerous. Business was near a standstill. Death stilled the City, its people, the horses' hooves and coach and cart wheels, the only sounds the bells' insistent tolling. Pepys sent Elizabeth away early and later the Navy Office and the Board moved to Greenwich. Evelyn also sent his family out of London, but stayed on at great risk to look after the prisoners of war for whom he was responsible. If Edmond (senior) could not carry on his business, he probably took his family to Islington or Barking or elsewhere in the country.

The plague had abated by December and life resumed. Halley perhaps went back to St Paul's. There was war with Holland in the Second Dutch War. In the first days of June 1666, Halley could have heard the great guns as the English and Dutch fleets fought the inconclusive Four Days Battle off the mouth of the Thames, and again possibly from the St James's Day Battle in July in much the same area when the English fleet was victorious.

> September. 2 *Lord's day*. Some of our maids sitting up last night to get things ready against our feast today, Jane called us up about 3 in the morning, to tell us of a great fire they saw in the City. So I rose, and slipped on my nightgown and went to her window, and thought it to be on the back side of Marckelane at the furthest, but being unused to such fires as followed, I thought it far off, and so went back to bed and to sleep.

So Pepys, and so no doubt many other citizens who thought it far off and so went back to bed and to sleep.

But later that morning he

> walked to the Tower and there got up on one of the high places, Sir J Robinson's[*] little son going with me; and there I did see the houses at that end of the bridge all on fire, and an infinite great fire on this and the other side the end of the bridge.

The Fire burnt for four fierce days. It spared only the extreme north-east with Winchester Street and Gresham College. It reached

---

[*]the Lieutenant of the Tower

St Giles's, Cripplegate, and may have destroyed the house which the Halley family occupied there earlier in the year.[38] They would again have gone into the country, taking valuable possessions with them or placing them somewhere they thought safe, and joining the crowds with carts or on foot who fled from the City. Edmond would have seen the hundreds of thousands of refugees camped out in Moorfields in the hot nights, the immense column of smoke by day that spread as far as Oxford, and the inferno that lit the sky by night. He would have heard the roar of the flames and the sharp explosions of hay, oils, wine, tar, that fed the fires. When they could venture back, past stark ruins, by lanes choked and obscured with rubbish, over streets that burnt the soles of their shoes, they would find their properties in Billingsgate, and Fenchurch Street, and Cannon Street, were cinders. Many churches burnt. St Paul's and the school beside it, the booksellers' stocks in St Faith's and the library of the school, all were ashes. Most houses were built leasehold on rented land with the condition that the owner of the house had to rebuild it. The City fathers enacted ordinances to reduce the burden, so that life might get back to normal as quickly as possible, and much commerce resumed within the year. It is said that Halley's father suffered considerable losses in the Fire, but probably no more than most other citizens. The records of the Fire surveys show him and Humphrey I paying for staking out properties after the Fire and his two houses in Cannon Street were yielding substantial rents in 1683. His rather smelly trade was probably carried on outside the walls and may not have been interrupted; the Fire no doubt fed an exceptional demand for soap if Pepys's experience was typical. Certainly, the father was able to give Halley £300 for his voyage to St Helena ten years later and another £300 for his later tours in Europe. But whatever the financial consequences, whether in the short or long term, the Plague and the Fire were long remembered by all Londoners, and no doubt especially by such as Halley whose school had been destroyed.

Yet another disaster alarmed the City. The English fleet had been laid up in the Medway after the campaigns of 1666 and it was not brought out in 1667 because peace negotiations were under way. The Dutch did put a fleet to sea and in the middle of June they broke the chain across the Medway and burnt or sunk or captured many ships. Londoners feared a Dutch incursion and made ready to move their money and goods yet again. Rumours swirled around in the streets and coffee houses. Pepys noted on 13 June:

> In the evening comes Mr Pelling and several others to the office, and tell me that never were people so dejected as they are in the City all over at this day, and do talk most loudly, even treason; as that we are bought and sold, that we are betrayed by the papists and others about the king ...

Those years of disaster and rumours of disaster must have burnt into the minds of children just becoming aware of the world outside the home, and, bright intelligent schoolboy that he was, Halley can hardly fail to have been impressed by the connection between the troubles of London and the well-being of the fleets. Perhaps he recalled those three terrible years when, many years later, he remarked that it was those who survived who were terrified by plagues and fires.[39]

Dean Colet had founded St Paul's School in the humanistic flowering of the early sixteenth century. Like its brother school at Westminster, it had attracted distinguished headmasters and nurtured eminent pupils. Samuel Pepys was a Pauline and retained a deep interest in the well-being of the school throughout his life. Milton and John Churchill, the Duke of Marlborough, were former pupils. The school stayed closed after the Great Fire until it had a new building under the walls of the rising cathedral. The masters were paid a retainer but might find other posts. The headmaster, Samuel Cromleholme, opened a school at Wandsworth, which some of his St Paul's pupils may have attended. Nothing else is known of masters or pupils until the school reopened at Easter 1671.

Cromleholme died in 1672. Thomas Gale, the Regius Professor of Greek at Cambridge, immediately succeeded him. Gale had just married Barbara, the daughter of a cousin of Samuel Pepys. He was a scholar of European standing and a fellow of the Royal Society. He left St Paul's in 1697 for the Deanery of York. He was one of the Secretaries of the Royal Society in 1686 when Halley was elected Clerk. Evelyn was a friend, and thought highly of him.

Edmond must have found 1672 another disturbed year, for his mother died and he had a new headmaster. He can have spent no more than two years at St Paul's after the Fire for he went up to Oxford in the summer of 1673.

The Gale papers in Trinity College, Cambridge, include *The constant method of teaching in St Paul's School, London*. The whole school began with 'A Chapter of the Bible and set prayers in Latine every morning at 7 of the clock'. The four lowest forms had Latin grammar every morning, with Friday morning given to 'a Repetition of what hath been said ye whole weeke'. Pupils in the fourth form were reading Ovid's *Metamorphoses* and *Epistles*.* The fifth form began Greek, with Virgil, Martial, and Sallust in Latin, and made Latin verse versions of psalms. Every morning the sixth form had Greek grammar, they also read Martial, Virgil, and the Greek Testament, and composed moral,

---

*When this author was at a comparable stage at school, he also read *Metamorphoses*, along with *De Bello Gallico*—the forms of the English classical tradition are long-lived.

rhetorical, and divine 'themes'. The seventh form began Horace, the minor Greek poets, Apollodorus, and Cicero's *Orations*, and composed themes, while finally the eighth form had 'A part in the Hebrew Psalter or Grammar' each morning, with the afternoons given to Homer, Persius, Demosthenes, and Juvenal and the composition of themes.

That was an impressive curriculum of classical literature and thought. That Halley was grounded in it is plain. He often used Latin for correspondence and private notes and for many of his important papers and books. He showed that he was familiar with Caesar and Dio Cassius when he located the site of Caesar's landing in Kent (Chapter 7). He composed a Latin ode in celebration of Newton for the first edition of *Principia*,[40] and was not pleased when Bentley altered ('improved') it for the second edition. He certainly knew Greek well as a school subject and would have had some Hebrew. He corresponded in French and Italian and it was said that he spoke German well, but probably did not learn them at school.

The *Biographia britannica* says of Halley's school days,

> He not only excelled in every branch of Classical learning, but was particularly taken notice of for the extraordinary advance he made at the same time in 'the Mathematicks', insomuch that he seems not only to have acquired almost a masterly skill in both 'plain' and spherical Trigonometry, but to be well acquainted with the science of Navigation.

That was written some ninety years later, but the contemporary teaching of mathematics at other schools, notably Eton, Westminster, and Christ's Hospital, and the fact that Edward Cocker, the writing-master at St Paul's, was a distinguished mathematician, support the *Biographia*. Halley could of course have spent only two years on mathematics at St Paul's and the story that an apprentice of his father's introduced him to mathematics at least fits in with the fact that St Paul's was closed until rebuilt in 1671. The combination of basic and applicable mathematics is characteristic of textbooks of the day, such as that of Moore. Halley made astronomical and magnetic measurements while a schoolboy and later recorded the result of his measurement of the variation of the compass, the magnetic declination as it is now called, in 1672.[41] He wrote a substantial mathematical paper when an undergraduate at Oxford, and he recalled his early interests in astronomy in his *Catalogus stellarum Australium*. Plague and Fire seriously interrupted Halley's school-days but pupils and teachers can surmount such troubles, as many did whose schools were displaced during the years of the Second World War without lasting damage to their university and later careers. Halley became head of the school in his last year and as such would have delivered a Latin oration at the annual ceremony of Apposition.

Robert Nelson was the only schoolfellow of Halley's of whom anything is known. Halley wrote to him some fifty years later, recalling their continued friendship since their childhood.[42] Nelson's father died while Nelson was still at school. Robert was brought up by his mother, Delicia, and her brother, Sir Gabriel Roberts. He went up to Trinity College, Cambridge, and he and Halley went to France and Italy together in 1681–1682 (Chapter 4).

2.8   Oxford: the Queen's College

Halley went up to the Queen's College as a commoner in the Vacation Term of 1673, being admitted on 25 June.[43] Queen's was one of sixteen colleges and about seven halls at that time. Its members were prominent in university and national affairs. The Provost was Thomas Barlow, the Lady Margaret Professor of Divinity and Bodley's Librarian, afterwards Bishop of Lincoln. He left a substantial library to be shared between Bodley's and Queen's. Sir Joseph Williamson (Plate I), a fellow of the Royal Society and President in 1677, was matriculated in 1650, had been elected a fellow in 1657 and was a Principal Secretary of State during Halley's time at Queen's. He had been responsible after the Restoration for continuing and developing the intelligence networks at home and abroad that had been set up by Thurlow during the Commonwealth, and he was for many years at the centre of the Government surveillance of spies, the disaffected, and radicals, notable whigs among them. He was dubbed knight in 1672. He was a considerable benefactor to Queen's, which owes one of its buildings to him, and where he founded a lecturership in Anglo-Saxon. Anthony à Wood said he was 'a great friend to Queen's Coll. men'.[44] He of all Queen's men had most influence on Halley's career, for through him Halley received the Royal support that enabled him to go to St Helena (Chapter 3).

Sir Robert Southwell, a near contemporary of Williamson, who matriculated from Queen's in 1652 and was five times President of the Royal Society, was closely associated with Halley in the Society. Henry Compton, who matriculated from Queen's in 1654, was tutor to Mary and Anne, the daughters of James II. He became Bishop of London. He was the only bishop to sign the invitation to William III to come to England in 1688. He was an elector to the Savilian Professorship when Halley was unsuccessful in 1691. John Hudson came up to Queen's in 1677 just after Halley had gone down. He became Bodley's Librarian in 1701 and was the *bête noire* of the antiquary and annalist, Thomas

Hearne. He was a vigorous librarian, and had much to do with the University Press.

Noblemen and gentlemen-commoners came first in the social hierarchy in Oxford. Noblemen, the sons of peers and bishops, were always few and were mostly at Christ Church. Gentlemen-commoners and commoners, the most numerous undergraduates, were distinguished by Provost Barlow:

> ...we have two ranks of Gentlemen in the Colledge. 1. Those we call *Communars*, which are Gentlemen of inferior quality usually (though many times men of higher birth and fortune, will have their sonnes and heires in that ranke). 2. *Upper Communars*, which are usually Baronnets or knights sonnes, or Gentlemen of greater fortunes...

Noblemen and gentlemen-commoners had elaborate gowns, dined at the high table with the fellows and rarely took degrees. Commoners paid their own way through college and about half would take a degree with the aim of being ordained into the Church, but those with some other career in view, or who expected to remain independent gentlemen, might not proceed to a degree. Foundationers, later known as scholars or exhibitioners, were supported to a greater or less extent from the endowments of the college. More menial undergraduates were paid servants of the college, or of senior members, or of wealthier undergraduates.[45]

The lives of all members of the University and its colleges were formal, highly organised, and strictly disciplined, if only in principle. Morning and evening prayers of the Church of England were said daily, sometimes in Latin, and undergraduates had to attend most days—at Queen's they were fined if they missed two days a week. Holy Communion was celebrated in most colleges on the first Sunday of Term and on the great festivals. All undergraduates, noblemen and sometimes gentlemen-commoners apart, had to dine each day in the college hall about midday. Gowns and square caps had always to be worn in public places, in college, in university buildings, and when calling on college and university officers. Such was the formal framework of undergraduate life; within it most managed some serious study, and found opportunities for parties, making friends, going on excursions, and indulging in mild riots. A few carried study or sybaritism or rioting or vicious behaviour to extremes: universities may change but students do not.

The accounts of Henry Fleming, who was up at Queen's in 1680, give some idea of what Halley's expenses would have been a few years earlier:[46]

Caution*	£5
Tutorage	10s. per quarter
Rooms	6s.8d.
Bedmaker	2s.
Laundress	2s.
Barber	2s.6d.
Batells*	£2

Fleming bought books on logic, ethics, arithmetic, Euclid, Greek and Latin classics, and Hebrew. The curriculum for the BA degree followed closely that laid down in the Laudian statutes of thirty years before. There were supposedly compulsory lectures by regent masters, the newly graduated MAs, on grammar, antiquities, rhetoric, logic, and moral philosophy. In their later years undergraduates were expected to attend the lectures of the Savilian Professor of Geometry and the Regius Professor of Greek. However, the system of public lectures was not at all effective and most teaching was done in the college. At the end of four years, the undergraduate would be examined for the BA degree orally by regent masters and by public disputations with bachelors. The curriculum and the disputations were close to the practices of the medieval university.

Halley almost certainly went to the lectures of the Savilian Professors of Astronomy (Edward Bernard) and Geometry. John Wallis (1616–1703), a founder member of the Royal Society, was the professor of Geometry (Plate III). He was an accomplished cryptographer who had been regularly employed by the Government to decode letters of spies, ambassadors, and others. Halley was to succeed him as Savilian Professor of Geometry in 1704. Wallis published in 1673 the posthumous works of Jeremiah Horrocks, which Halley thus had a good opportunity to get to know. Horrocks was a graduate of Emmanuel College, Cambridge, and ordained to a curacy in the north of England. He had made outstanding contributions to astronomy in the few years he had still to live after leaving Cambridge. Newton and Flamsteed greatly admired his works, which were in a number of ways the precursors of Halley's.

Only a week after Charles II had appointed John Flamsteed 'our astronomical observator' at the beginning of March 1675, and before the King had ordered an observatory to be built, Halley wrote his first letter to Flamsteed. He said that he was

> reasonably well provided in instruments in which I can confide to one minute without error by means of telescopicall sights and a skrew for the subdivision ...

*Caution money was a refundable deposit. Batells were food and similar supplies bought from the college butler.

Halley had been observing with his friend Charles Boucher who had already been in correspondence with Flamsteed in 1674.[47] Boucher matriculated at Magdalen in 1670, three years before Halley went up to Queen's, and as he left for Jamaica early in 1675, he would have been with Halley in Oxford for about a year and a half.[48] He seems to have been a close friend of Halley's at Oxford: he writes 'Dear Edmond' and concludes 'your most assured friend'. He apparently went to Jamaica on account of family lands and interests there and would remain for many years. He was Clerk of the Assembly at various times between 1678 and 1691 and was Clerk of the Court of Common Pleas from 1687. Jamaica had a great trade in sugar with England. It had been captured only ten years earlier by Sir William Penn, the admiral father of the Quaker founder of Pennsylvania, and Boucher accompanied Sir Henry Morgan who was going out as Lieutenant-Governor. His voyage was eventful, as he described in a letter to Halley (see below and Note 49). His ship was captured by privateers and he lost some of his books and equipment.

When Boucher left for Jamaica there was no one else in Oxford with whom Halley might observe. So he applied to Flamsteed, of whom he may also have heard in London. He had probably seen Flamsteed's account of Horrocks's theory of the Moon in the *Opera posthuma* of two years before. He explained that Boucher had gone to Jamaica, and so had not replied to Flamsteed's last letter. He told Flamsteed of a number of his own results, especially that the times of Jupiter and Saturn differed considerably from those given by Thomas Streete in the *Astronomia Carolina*. He had observed the recent lunar eclipse of 1 January with Streete in London, but imperfectly because errors in the tables of the Moon had made them too late to catch the start. He also discussed atmospheric refraction and the tables of G. D. Cassini. Flamsteed replied quite promptly with a substantial discussion of atmospheric refraction.[49]

The warrant for building the observatory at Greenwich is dated 22 June 1675[50] and the same day Hooke noted in his *Diary* that he and Wren were to direct, that is, design and build, the new Observatory for Flamsteed. Next day Flamsteed and Halley attempted to observe an eclipse of the Sun. But the sky was cloudy.[51]

Two weeks later Halley, then apparently in London, again joined Flamsteed when they observed a lunar eclipse from Moore's house in the Tower and used his instruments, as Oldenburg wrote to Hevelius, introducing Halley's name to the outside world for the first time.

Flamsteed wrote in the *Philosophical Transactions*,

> Edmond Halley, a talented young man of Oxford was present at these observations and assisted carefully with many of them.[52]

Halley also appears in Hooke's *Diary* for the first time that day when he and Flamsteed asked Hooke for his assistance with an iron quadrant, probably, as it turned out, the one that Flamsteed had at Greenwich the next year and which he found quite unsatisfactory. Next week Halley accompanied Hooke and Flamsteed to Greenwich to view the site of the new Observatory, and then on 2 July Hooke described the Observatory, his design that is, to Flamsteed, Halley, and others. Hooke went to Greenwich to set out the Observatory on 28 July.[53] Even so early there were signs of friction, for Hooke noted on 7 January 1676, 'Flamsteed & Halley fallen out'.[54]

Flamsteed moved to Greenwich in July 1675, and began to observe on 19 September (Chapter 13). Halley meanwhile sent him observations from Oxford.[55] Again on 24 March 1676 he sent observations of the Moon and on 29 April observations of Venus, all made at Oxford.[56]

The first of June 1676 was a notable day at the new Observatory. There was to be a partial eclipse of the Sun and the King intended to be there to see it. In the event he did not come. Nor did Sir Jonas Moore see the eclipse because he left Greenwich early when he thought it would be cloudy. Lord Brouncker did come, and saw it, and assisted Flamsteed and Halley with their observations.[57]

Halley must have returned to Oxford shortly after the eclipse for he observed occultations at Oxford on 19 and 20 June, 22–23 July, and 21 August 1676.[58] He was back to visit the Observatory in company with Hooke, Sir John and Lady Nott, and the King's plumber, Lingren, on 15 September.[59] He observed Mars at Oxford in September and October. Then his observations in Oxford came to an end for he was going to St Helena. He was associated with Flamsteed from the very origin of the Royal Observatory and they continued to collaborate for some years afterwards (Chapter 13).

2.9 Undergraduate publications

Halley published three papers in his last year in Oxford. The English translation of the title of the first is *A direct geometrical method to determine aphelia, eccentricities and proportions of the orbits of the primary planets, without assuming equality of the angular motion at the other focus of the ellipse.*[60]

Proportions here means the ratio of the major semi-axes of the orbits of the Earth and the planet; the primary planets are those beyond the Earth.

Halley sent his first version to Brouncker, possibly before they were at Greenwich for the solar eclipse. Brouncker suggested some improvements. Halley made them and returned his paper to Oldenburg in a letter dated July 0/8 1676, saying how he came by his method. He wrote for the first time of going to St Helena; he may already have mentioned it to Brouncker at Greenwich (Chapter 3). He also referred to observations of the Moon and estimates of longitude.

He had delivered to Wallis a letter that Oldenburg had given him, and so met Wallis for the first time (Chapter 2):

> ...who entertained me very kindly, and I had a great deal of discourse of astronomical nature with him; and he at my departure told me he would gladly see me some other time.

Halley sent a further draft of his paper in his next letter, dated 11 July 1676.[61]

It was difficult to determine planetary orbits from observations. The observer was on the Earth moving in its own path about the Sun and the first step was therefore to determine the places of the Earth at the times when a planet was observed. Only then could the astronomer turn to the determination of the orbit of the planet. Lacking a dynamical theory of the orbits, it is not surprising that the most able mathematicians were puzzled. Halley assumed only that the orbit of the planet was an ellipse, as Kepler had shown.

If an orbit is nearly circular, then (Appendix 1) the angular motion about the second, empty, focus of the ellipse is almost uniform. Seth Ward, the bishop of Salisbury, had taken it so in 1656.[62] It was a convenient alternative to Kepler's second law (that the areal velocity is constant) for small eccentricity. People at first paid little attention to the second law and only understood its crucial position after Newton had shown that it holds for any central force (*Principia*, Book I, Prop. 1, Th. 1). Halley assumed nothing about angular velocity.

Halley reduced the problem in three dimensions to one in plane geometry by considering the projection of the actual orbit of a planet on to the plane of the Earth's orbit. The first step is to obtain the orbit of the Earth—it is revealing that it was apparently not well enough known. Halley took the position of the Earth at four revolutions of the planet, the first being when the Earth, A, the Sun, S, and the planet, P, are in line (Fig 2.2). He considered Mars, with a period of 687 d. After one revolution of Mars, the Earth (B) will have travelled not quite two complete revolutions of $730\frac{1}{2}$ d. The angles in the triangle SPB are all known from observation and hence the ratio (distance of Earth : distance of planet) can be calculated. After one and two further complete revolutions of the planet, two more distances of the Earth from the Sun in terms of the distance of the planet are found.

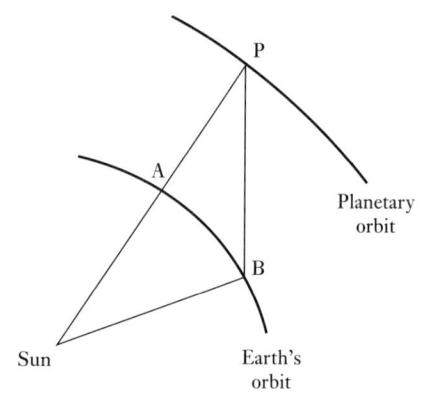

Figure 2.2 Determination of places on the orbit of the Earth.

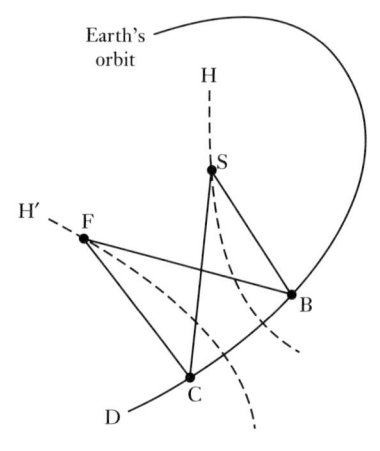

Figure 2.3 Determination of the orbit of the Earth.

Halley next found the second focus of the orbit of the Earth from the distances of the Earth from the Sun at three positions of the Earth, B, C, D (Fig. 2.3). The sum of the distances of any point of an ellipse from the two foci is a constant, while on an hyperbola the difference of the two distances is a constant. The Sun, S, is one focus of the elliptical orbit of the Earth; if F is the other focus,

$$SB + FB = SC + FC = SD + FD,$$

whence

$$SB - SC = FB - FC$$

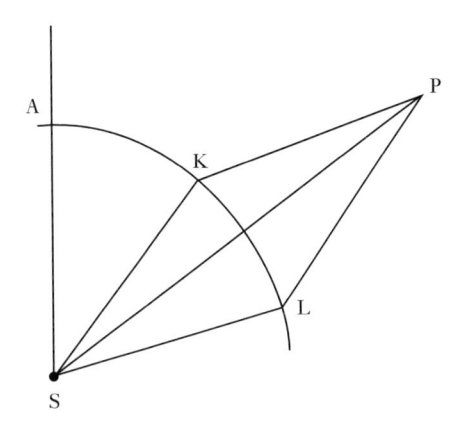

Figure 2.4 Determination of the orbit of a planet.

and

$$SC - SD = FC - FD.$$

Thus an hyperbola drawn with the foci at B and C and with the difference of distances equal to (SB–SC) will pass through the second focus, as will the other hyperbola based on C and D. The eccentricity of the Earth's orbit and the direction of its major axis follow from the position of F relative to the Sun, and the length of the major axis is equal to the sums (SB + FB), (SC + FC), and (SD + FD). The major axis of the Earth's orbit can now be taken as the unit of distance.

In his third step Halley established three positions on the orbit of the planet. Let K and L be two positions of the Earth at times separated by one complete revolution of the planet, P (Fig. 2.4). Again S is the Sun. The Earth's orbit having been determined, the distances SK and SL, and the angle KSL are known. The distance SP is then known in terms of the major axis of the Earth's orbit. The angle ASP is also known. If three positions of the planet are determined in that way, its orbit can be found by the method already used for the orbit of the Earth. Halley also gave an algebraical treatment that is equivalent to the geometrical solution. Both depend upon the orbits being exact ellipses.

De la Hire, the professor of mathematics at the Collège de France, was critical of Halley's treatment because he had used two conic sections, the hyperbolae, to determine a third, the elliptical orbit, and he also remarked that the algebraic formula involved an equation with many terms.[63] Halley's paper and de la Hire's comments show that in the years before Newton and the *Principia*, astronomers were still unsure how to find the elements of orbits from observation. They could calculate positions from elements, but the inverse problem was not so simple, and it was not difficult to devise a method that gave an indeterminate result. Mairan, in his *Éloge* of Halley, emphasised that Halley was the first to devise a way of finding an elliptic orbit from observations alone:

> He was scarcely nineteen years old when he gave his direct geometric method for finding the aphelia and eccentricities of the planets, a work that the most acomplished Astronomers of that time might envy and which ended a celebrated dispute among them on that subject.[64]

In Halley's second paper, *An extract of an account given by Mr Flamsteed of his own and Mr Edmund Halley's observations concerning the spots in the sun appearing in July and August 1676*,[65] he joined Flamsteed in describing the results of observations they had made at Greenwich and Oxford with eyepiece micrometers from 27 July to 4 August. They inferred that the Sun's axis of rotation was close to the pole of the ecliptic, they estimated the positions of the nodes on the ecliptic, and they derived a period of rotation of 25d 9h 5'.

Cassini observed the spots at Paris and although he made no measurements he published a theory of the rotation of the Sun similar to that of Flamsteed and Halley. Flamsteed recorded that it was Halley who first explained the tracks of the spots and estimated the parameters of the Sun's rotation. He did not want Halley to lose the credit of that discovery on account of his absence on St Helena.[66]

Halley's third undergraduate paper was *Mr Edmund Halley's observations concerning the same occultation of Mars by the Moon, made at Oxford, Anno 1676, Aug. 21 P.M.*[67] Hevelius (pp. 721-2) and Flamsteed (p. 723) described their observations at Danzig and Greenwich in the same issue of *Philosophical Transactions*, but while these two just reported their results, Halley put them all together and deduced the differences of longitude in time:

> Between Greenwich and Oxford: 0h 04' 58".
> Between Greenwich and Danzig: 1h 14' 46".

The two latter papers, along with correspondence with Flamsteed, show that Halley made many astronomical observations at Oxford. His account in the preface to *Catalogus stellarum Australium* of how he became gripped by astronomy reveals that he had made observations of Jupiter and Saturn, and found the current tables of their places defective, before he went to St Helena, while he was still at school or at Oxford.[68]

Halley spent three complete years at Oxford, one less than normally required for the BA degree, and then on 4 October 1676, the Court of Committees of the East India Company received a letter from Charles

II asking them to carry him to St Helena to make astronomical observations. He left Oxford without the BA degree. Halley's undergraduate career and its conclusion were not in that exceptional, they were exceptional in the original work he did as an undergraduate and in the reason for his going down.

2.10 *Off to sea*

In the latter half of 1676, Halley heard from Charles Boucher in Jamaica.[69] Boucher gave a detailed account of his eventful voyage early in 1675 when he was shipwrecked in the West Indies, and rescued by a former privateer whose ship was later captured by another privateer. He saved his glasses (lenses for telescopes), his eyepiece micrometer, and evidently a quadrant with which he later made observations. He also saved his copy of Tycho's *Historia coelestis*, but lost Kepler's *Epitome*, books on geometry and algebra, and above all, Streete's *Astronomia Carolina*, so that he had no means of calculating the positions of planets. He had heard of Flamsteed's appointment as King's Astronomer and he sent Halley some astronomical observations to pass on to him. He asked for information about Hevelius's lunar studies and also about the predicted eclipse (sic) of Mercury due in 1677 according to Streete's *Astronomia Carolina*. He desired to maintain a correspondence with a number of acquaintances in England, Flamsteed among them, both on astronomy but also about plants. Halley may not have replied to this letter, which could have reached Oxford only after he had sailed for St Helena, but Flamsteed did write to Boucher much later on 24 January 1678.[70] He had sent off the lenses and maps of the Moon for which Boucher had asked. He gave him news of the Observatory at Greenwich. The large quadrant that Hooke had prepared for him was unserviceable, despite all that Flamsteed could do to rectify it. Hooke had boasted about how splendid it would be and Flamsteed is not dissatisfied: 'this has shown the difference between boasts and performance'. He gave Boucher the latitude of Greenwich he had found with his sextant. He had observed Venus in daylight by mounting his sextant on a polar axis.[71] Finally, he sent Boucher news of Halley in St Helena and when he expected him back.

In the correspondence with Boucher, in the paper on sunspots, and in Halley's account of what he took to St Helena (Chapter 3), we read of the newest instruments then just coming into use, Towneley's eyepiece micrometers, the mural quadrant with improvements in the divided scales, better telescopes of long focal length, and accurate pendulum clocks. Flamsteed, Cassini, Halley, all saw that they gave opportunities to tackle long-standing fundamental problems of astrometry that had to be solved before it was possible to construct better star catalogues or improve the orbits of the planets and the Moon. Flamsteed had told Boucher how with a good pendulum clock he had been able to get better values of atmospheric refraction from observed altitudes of the Sun at known times, and he expected to do better with a larger quadrant. Halley saw how to obtain a reliable distance of the Sun from the duration of the forthcoming transit of Mercury. Flamsteed and Halley had used micrometers to study sunspots. None of those observations could have been made only a few years earlier. The staid letters of the age convey some sense of the excitement of the times and of the new instrumental astronomy on the threshold of which they stood. Halley, as it turned out, would be the first to make a star catalogue with the new instruments. Flamsteed's first mural quadrant was useless. He did not get another until 1689 when he could afford his own by Sharp. Only then did he make a serious start upon his Greenwich catalogue. Cassini, likewise, did not have a mural quadrant until 1681. By then Halley had completed his southern catalogue, though not as he had wished, on account of poor weather on St Helena. He had almost certainly begun to think about his St Helena expedition in 1675 and probably had discussed it with others, for only just over a year after his visits to Greenwich, the Court of the East India Company received the letter from the King recommending that the Company should carry Halley to St Helena.

Astronomy and navigation were Halley's dominant concerns at Oxford, but he also made some observations of the magnetic field of the Earth. Astronomy and geomagnetism, his lifelong interests, already possessed him as an undergraduate.[72]

The circumstances of Halley's family, their informal position in the City, their influential acquaintances in the trading companies, Halley's own early associations with foremost astronomers and natural philosophers such as Brouncker, Moore, Hooke, Wren, Streete, and Flamsteed, his schooldays, his time at Queen's, and his access to the Court through Moore and Williamson, would together afford him strong support for any plan he might have to pursue and improve natural knowledge. They were great advantages, but he was unlikely to have been entirely alone in them among men of his age in the City or at Oxford. What plants in some mind that overpowering desire that leads her to say, *I want to know, to understand*? Halley has told us of how astronomy laid hold of him early and would not let him go,[73] but we cannot guess, we cannot know, what had brought him late in the autumn of 1676 to a London quay about to step on board the *Unity* East Indiaman outward bound for St Helena.

3

Skies of the south

Macte nova virtute, puer, sic itur ad astra.
[The way to the stars is by your youthful courage.]

VIRGIL: *AENEID*, IX 641

3.1 *Into the unknown*

EDMOND HALLEY, just twenty years old, embarked on *Unity* one October day in 1676 and upon a lifetime improving natural knowledge, in the spirit of an age when 'The World was all before them'. He was bound for the island of St Helena to observe the positions of southern stars invisible from Europe, and to observe eclipses and a transit of Mercury across the Sun. He was going hopefully. He knew little of the conditions of St Helena, only recently acquired by the East India Company. His plans were too optimistic and fulfilled only in part. For all that, by being in the right place at the right time, he reaped an unanticipated harvest that would, in years to come, open up new ways of seeing the natural world.

Halley's leap into the unknown was not then perhaps so precocious, when Londoners of his age went out to the factories of the East India and Levant Companies to bear heavy responsibilities. It was more notable that influential men gave him their support by arranging to have him taken to St Helena and accommodated there, and that his father paid for his instruments. People must already have had confidence that he would get things done.

Halley recognised that his patrons were Sir Joseph Williamson and Sir Jonas Moore. Moore spoke to the king, Charles II, who wrote to the East India Company to ask them to carry Halley to St Helena and to maintain him there. Lord Brouncker also supported him according to Wallis.[1] Halley was already thinking about an expedition to the south when he was preparing his first paper for the Royal Society (Chapter 2). At the same time he asked Oldenburg about a book that he has heard is in press in Paris, the *Peregrinationes astronomicae* of Richer. Does it contain a complete catalogue of the southern stars?

> ... for if that work be yet undone, I have some thoughts to undertake it myself, and go to St Helena or some other place, where the south pole is

considerably elevate, by the next East Indian Fleet, and to carry with me large and accurate instruments, so as to be able to make a most accurate sphere of the fixed stars, and complete our globe throughout: nor will that be all; but by comparing observations made there and here, the proportions of the Moon and the Earth, with their distance, will be more exactly than any other found.

I would willingly do something to serve my generation, and here I can do nothing but what will be rendered wholly inconsiderable by the greater accurateness of the three great promoters of the astronomical science in our age. I will willingly venture myself upon this enterprise, if I find the proposition acceptable, and that the East Indian Company will cause me to be kindly used there, which is all I desire as to myself, and if I can have any consideration for one to assist me. This, Sir, I propose to you, desiring your advise as to what conveniences there may be, and if you think what I propose may meet with any encouragement.

Halley was to sail on a ship owned by neighbours, the Leithulliers, in Winchester Street; no doubt he knew the East India Company's plans for the sailing of their ships.

Halley wrote again to Oldenburg on 8 August. He had been delayed in returning his own paper

... had I not been hindered by some exercise I was obliged to do in the house, for my degree.

He would like to have been told, like some modern authors, who had refereed his paper, and guesses it was Collins.

I pray Sir signify the receipt of this by a line or two, and let me understand what you hear from France about the southern stars, which in your last you promised to enquire into, and if there be any appearance of encouragment for my friend that will go along with me. I design to be in London within one month to fit instruments and necessaries for my intended voyage, in which I promise myself I shall be able to do something accceptable to the learned world.[2]

Halley was enthusiastic and ambitious. His spirit was patriotic like that of Purcell's *King Arthur*. It was the spirit of the age. Halley may already have had his major instruments in his possession, for it seems unlikely that a large sextant, not to speak of other instruments, could have been made in the short time from August until he set sail at the end of October or early in November.

Sir Joseph Williamson, the Secretary of State and former provost of Queen's College, received a memorandum early in September:

Edmund Halley, student of Queen's College, Oxford, having been for some years a diligent observer of the planets and stars, has found it absolutely necessary, besides the continuation of observations here, that in some place between the Tropics, where the sun, moon and planets will pass near the zenith without refraction, their motions will be much better ascertained and navigation perfected, and that St Helena will be a fit place, where the celestial globe may be finished, the stars in the southern hemisphere being very much out of their places. He humbly desires His Majesty's letter of recommendation to the East India Company, that they will cause the ship ready to go to St Helena to transport him and his friend thither, and that they may be received and entertained and have fitting assistance.[3]

The idea of reducing the effect of atmospheric refraction by observing the Sun, Moon, and planets as they passed through the zenith is interesting but is not mentioned again.

So it was that on 4 October 1676, the Court of Committees of the East India Company received a letter from the King, delivered by Sir Jonas Moore, and read by the Governor:

> ... recommending unto the Court that Mr Edmund Hally a Student of Queen's College in Oxford, with a friend of his might have their passage in the first ship bound for St Helen's whether they are desirous to go & remayn for some time to make observations of the planets & starrs, for rectifying and finishing the celestial globe, being a place (he conceives) very fit and proper for that design; and that they may be received and entertained there, and have such assistance and countenance from the Compas officers as they may stand in need of. On consideration whereof had, It is ordered that Mr Hally with his friend doe take their passage for St Helena on the Unity with their necesary provisions free of charge; and that a lre be written to the Governor & Council of the said Island to accomodate them with convenient lodging during their stay there, and afford them such assistance and countenance as may be for their encouragement, to proceed in so useful an undertaking.[4]

Halley had powerful support, and the East India Company, dependent on shipping and navigation for its trade, would have readily recognised that the undertaking was 'so useful'.

Two days later, the Court confirmed the order of the Committee

> ... that it be referred to the Comtees. for Shipping to give directions for the accomodation of Mr. Edmo. Hally and his friend on board the *Unity* in order to the transportation of themselves and their necessities for St Helena, and that they be treated in their passage with all civilitie.

The Court also wrote to the Governor and Council of St Helena on 27 October 1676:

> His Matie, haveing been pleasd to recommend unto us, That mr Edwd. Halley, a Student of Queen's College in Oxford might with a friend of his have their passage for our Island St Helena whither they are desirous to goe & remaine for about two years to make observation of the Plannets & Stars

for rectifying & finishing the Caelestiall Globe, being a place he conceives proper for that design; Wee have thereupon ordered their passage on the Unity free of Charge, and doe order & direct that you accomodate them with convenient lodging in the howse of the Governor or his Deputy or in some other fitt & convenient habitation dureing their stay, & to afford them such assistance and countenace as may be for their encouragement to proceed in so useful an undertaking, They being to provide diet at their own charge, Wee also recommend it to our Comandrs. to assist Mr Halley upon his request in what they can for his furtherance in making his observations, And wee doe order That when Mr Halley & his freind shall desire to return back for England; That you recommend him to some of our Comanders that are bound from our Island homewards, and in the mean time, while he shall inhabit with you to use him with all respect & kindness.[5]

Halley acknowledged that the Company was very helpful in submitting to the will of their Sovereign and refused nothing that he asked for.

The friend who assisted Halley in St Helena was called Clerk. Nine Clerks or Clarks were contemporaries of his at Oxford, two of whom matriculated at Queen's, but it is unlikely that any of them could have been Halley's friend, who need not have been an Oxford contemporary.[6]

3.2 Apprenticeship of a navigator

As *Unity* dropped down the River, Halley had three months at sea ahead of him, months of enforced leisure. He does not say how he spent them. He tells us in the introduction to his *Catalogus* that the voyage to St Helena was pleasant enough ('satis foelicem'). East India captains kept good tables and there might have been interesting company of other passengers going to India, some perhaps known to him. Halley no doubt put his sea time to good use. He could have seen how a great ship was worked, the responsibilities of the captain, the place of the mate and how he navigated the ship, the duties of the other officers, the bosun, the carpenter, the cook and surgeon, how watches were kept, how the apprentices, the midshipmen, were taught their business and how to keep discipline among the crew. He could have studied the rigging of the ship and the setting of sails, and the close attention to wind and weather. After he returned from St Helena he had little opportunity to study seamanship until he went to sea on his own account some ten years later, already apparently competent. His passages to and from St Helena must have been his schooling in seamanship and navigation and command.

Unity probably sailed in company, for single ships might always be the prey of pirates, especially the Sallymen from the island of Sal in the Cape Verde group. They would have called at Madeira, the Canaries, and the Cape Verde Islands, to take on water, wine, fuel, and food, as twenty years later Halley would do, sailing south in his own command (see Fig. 10.2).

Halley was a passenger. In the long voyage out he had ample time to reflect upon the state of navigation and astronomy and to plan his observations. He had clear ideas of what he would do, of the state of positional astronomy, why it mattered, and how to improve it. He would observe the southern skies for a year from a remote and not very hospitable island, he would construct the first star catalogue of either hemisphere since that of Tycho, and he would use telescopic sights and other modern instruments in an extensive programme for the first time. Each was novel, each had ample scope for failure.

3.3 *The state of astronomy*

Halley says in his *Catalogus stellarum australium*[7] that he was naturally inclined to mathematics and that from his most tender youth he gave himself entirely to astronomical speculation with pleasure such as only those who have experienced it themselves can appreciate. Astronomy, he saw, was not to be practised 'in the cabinet' but with instruments to measure 'celestial arcs' and by passing many nights making observations. He had found at school and Oxford that the current tables of the Moon and the planets were defective; he thought of correcting the tables, but realised that he would be wasting his time unless he first corrected the places of the fixed stars. Distinguished astronomers were doing that, Hevelius in Danzig, Flamsteed in Greenwich, and Cassini in Paris, and he would not compare himself with them, but their works were as yet incomplete and he would have the satisfaction of discovering unknown stars in the southern sky.

The southern skies were by no means unknown to Europeans. Ptolemy had some stars in his catalogue, Dante in the *Commedia* knew something of them, and Kepler took the southern stars from Ptolemy for his catalogue. Frederic Hourman of the Netherlands had observed from Sumatra, and Blaeu published a celestial globe with his stars placed on it: Halley says he will confine himself to the comment that the observer was very little practised in that study.[8] Meurisse and Richer went to Cayenne but Halley did not know with what success, for the work had not been published. Cayenne lies north of the equator, stars of Chameleon cannot be seen from there, and Meurisse and Richer observed only a few bright stars. There was work for Halley to do. He consulted friends experienced in astronomy and sought their advice about instruments. He did not say whom he consulted, but it is easy to guess at a few. He already knew Flamsteed, had observed with him while at Oxford, and with him had come to see the possibilities that new instruments and methods were affording. Their relation was then unclouded by Flamsteed's later bitterness and disapproval. Halley writes elsewhere of Thomas Streete, the author of *Astronomia Carolina*, as his good friend, and they had observed a lunar eclipse together. Those two and Jonas Moore could have given him good advice. He had observed with Lord Brouncker at Greenwich, and he had corresponded with Henry Oldenburg. In the end it came down to his drive and conviction for, as he later wrote to Molyneux in Dublin in relation to a misrepresentation by Hevelius, 'At my own motion and charges I undertook that voyage to St Helena'.[9] Richer and Meurisse, who had somewhat different aims, were sent from Paris by the Académie des Sciences at royal expense.

Halley's own correspondence with Oldenburg and that of Flamsteed and Oldenburg with Hevelius, and of Oldenburg with Cassini, show that from the start he had other aims beside observing stars, and that Cassini and Hevelius were well aware of them. Cassini not only told Oldenburg what Richer and Meurisse had done in Cayenne, but he also considered that the parallax of the Moon might be found from simultaneous observations of her meridian altitude at St Helena and Greenwich or Paris, for all three were close to the same meridian. Cassini sent details of eclipses of Jupiter's satellites with which a better longitude for St Helena might be found. Halley seems not to have used the Jovian satellites but he did plan to observe a lunar and a solar eclipse by which he could determine the longitude of St Helena. He also intended to observe the upcoming transit of Mercury with the aim of determining the solar parallax as James Gregory had suggested.[10] He left Oxford when he did, without his BA degree, to catch the transit and the eclipses.

Halley intended to determine the positions of fixed stars in the sky, that is to say, the directions in which they are seen from the Earth. Directions from a centre are usually specified by polar coordinates, colatitude, and the azimuth (Fig. 3.1). Celestial directions are expressed in two systems of angular coordinates, equatorial and ecliptic. In the equatorial system directions are referred to the north pole of the rotation of the Earth, colatitudes are north polar distances and their complements are *declinations*, while azimuths are *right ascensions*. Colatitudes are unambiguous, but the origin for azimuths is arbitrary. The right

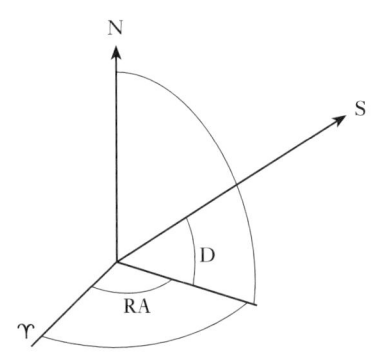

Figure 3.1 Colatitude and azimuth. S: star; ϒ: First Point of Aries; RA: right ascension; D: declination.

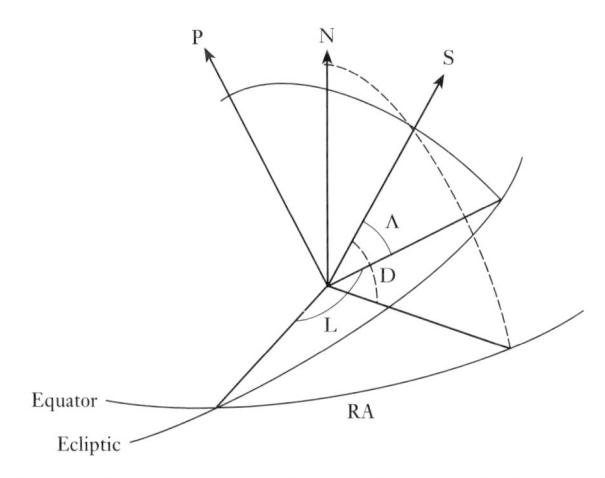

Figure 3.2 Equatorial and ecliptic coordinates. N: north pole; P: pole of the ecliptic; S: star; D: declination; L: longitude; Λ: latitude.

ascension is measured from the vernal equinox, or First Point of Aries, that is the intersection, or *node*, of the equator upon the ecliptic, the plane of the orbit of the Earth about the Sun (Fig. 3.2).

The reference direction for the ecliptic system is the normal to the plane of the Earth's orbit through the Sun. Colatitudes are *ecliptic polar distances* and their complements are *latitudes*. Azimuths measured around the ecliptic are *longitudes* and are usually referred to the First Point of Aries. The ecliptic is divided into twelve equal sectors, signs of the Zodiac, each of 30°, with the origin at the First Point of Aries (Fig. 3.3). In Halley's day it was usual to give longitudes from 0° to 30° within a sign of the Zodiac, rather than as now from 0° to 360°.

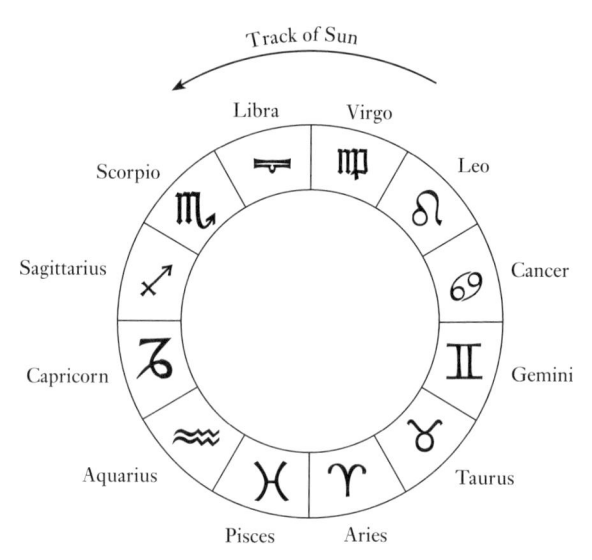

Figure 3.3 Signs of the Zodiac.

The relation between the equatorial and ecliptic systems is more than a matter of elementary spherical trigonometry. A reference system should so far as possible be invariant, based on a self-contained dynamical system subject to no external forces. Neither the equatorial nor the ecliptic system is invariant. It had been known since antiquity that the polar axis of the Earth, which is at an angle of $23\frac{1}{2}$ deg to the pole of the ecliptic, is not fixed in space but rotates (precesses) about the pole of the ecliptic once in 25 800 years, so that the relation between equatorial and ecliptic coordinates varies with time. The cause of the precession was not known until Newton in the *Principia* (Book III, Prop. 39, Prob. 20), showed that it arose from the gravitational attraction of the Sun and the Moon on the equatorial bulge of the Earth.[11] The ecliptic system is much more nearly absolute, but the motion of the Earth in its orbit is perturbed by the attractions of the other planets and the orbit is not confined strictly to a fixed plane; the common plane (the invariable plane) of all the planets is a better reference. The history of reference systems has been that of using more and more distant objects, now very distant radio sources, the directions to which are found by radio interferometry. That was all far in the future in Halley's day.

The centre of the equatorial system is the centre of the Earth, that of the ecliptic system is the centre of the Sun. The fixed stars are so distant that Halley and his contemporaries could not detect the consequent differences; annual parallaxes of the closer stars, which are no more than a second of arc, were only seen much later. Either system could therefore be used for the fixed stars. The equatorial system is more convenient for navigation because it relates celestial objects to terrestrial positions, while the ecliptic system is better for planetary and cometary studies because positions in it reveal more directly the orbits of the planets and comets in space; Halley and Hevelius both used it in their catalogues.

There are advantages and disadvantages to both reference systems, but the primary measurements of position have to be made in the equatorial system because the polar axis of the Earth is the only reference direction readily accessible to terrestrial observers. Fundamental measurements involve the determination of the angle between the north pole and an object at the instant when it crosses the meridian of the observer, and the observations are made with a telescope that rotates in the plane of the meridian. The typical instrument of the late seventeenth and eighteenth centuries was the mural arc, a framework subtending at least a quarter of a circle (or quadrant) permanently mounted vertically on a wall in the meridian and carrying a telescope (Hevelius and earlier observers used open sights). Mural arcs were difficult to set up and the observations were lengthy. The direction of the pole should be determined each night from observations of the upper and lower transits of circumpolar stars. The north polar distance of such a star is half the difference of the two directions, while the direction of the pole is the mean of the two. Picard in Paris had set out the requirements for precise astrometry: a mural arc with telescopic sights, properly adjusted for collimation and with micrometers in the focal plane of the telescope (which therefore could not be a Galilean telescope but had to have a convex ocular eyepiece). When Halley went to St Helena no observatory had such an arc. The quadrant in the Observatoire in Paris was not set up until 1681 and Flamsteed had to wait for his arc until 1689. Richer in Cayenne had used a portable octant to observe meridian altitudes and times of meridian passage of stars.

Halley's principal instrument on St Helena was a sextant. A sextant is just a frame subtending 60 degrees, carrying a graduated scale of angle and with telescopic or other sights attached to it. If it is provided with two sets of sights and is conveniently mounted, it can be adjusted so that the two sights are set simultaneously on two objects to measure the angle between them. Flamsteed, without a mural arc, also used a sextant when Halley was on St Helena. Halley also had a pendulum clock with which he could establish the time of meridian crossing and hence the right ascension. He could have made fundamental observations with the instruments he took to St Helena, but as he later wrote to Hevelius, he found that the constant interruptions by clouds made them impractical.[12] He evidently intended to determine the fundamental positions of some stars from meridian altitudes, for he says in the *Catalogus* that it was because of the brief periods in which he could observe that he restricted his observations to distances from Tycho's bright stars, without determining meridian altitudes. Hevelius had done so before him and so would Flamsteed for some years to come. They took the positions of two prominent stars as given, measured the angular distances of a third object from them with a sextant, and calculated the position of the third object from those two distances (Appendix 5).

Hevelius, Flamsteed, and Halley all had sextants of about 6 ft radius (1.8 m). The engraving of Hevelius with Mme Hevelius at his large sextant (see Fig. 4.1) shows how unwieldy it was. It had a fixed sight and a movable one: Elizabeth Hevelius is at the fixed sight and Hevelius is at the movable one. The whole sextant had to be pointed at one object and moved as the heavens rotate. Flamsteed made a considerable improvement by mounting the sextant on a polar axis and having two movable sights. His seems to have been the first description of an equatorial mounting.[13] He just had to rotate the whole instrument about the axis in time with the rotation of the heavens while setting the sights to fixed declinations to observe two objects. Halley mounted his large sextant on two toothed semicircles at right angles, turned by Archimedean screws so that the telescopes could be pointed at two stars to observe them simultaneously. He does not say if it was on a polar axis, although it is difficult to see how he could have observed with it by himself unless it was so mounted. Hooke's *Diary* shows that Flamsteed and Halley consulted him about instruments in June 1675 and their instruments probably shared similar features.

The radius of Halley's sextant was $5\frac{1}{2}$ ft (1.7 m), similar to that of Hevelius (see Fig. 4.1). The frame was iron and the scales were brass.[14] Many years later de L'Isle visited Halley in London, learnt of his procedures and instruments, compared some of his observations with others made in Paris, and made a sketch of the sextant (Fig. 3.4).[15] It was designed and made in a crucial period in the history of astronomical instruments. In 1673 Hevelius had published the first part of his *Machina coelestis* with detailed descriptions of the many instruments at his observatory at Danzig. Hooke in his *Animadversions* of 1674 was very critical of them, especially of the method of dividing a graduated arc and of Hevelius's use of open instead of telescopic sights.[16] Hooke's pamphlet offended Hevelius and would lead to Halley's visit to Danzig in 1679 to see Hevelius's instruments and methods (Chapter 4). Halley must have been very well aware of the issues before he went to St Helena and was, as a later letter of his to Hevelius shows, already con-

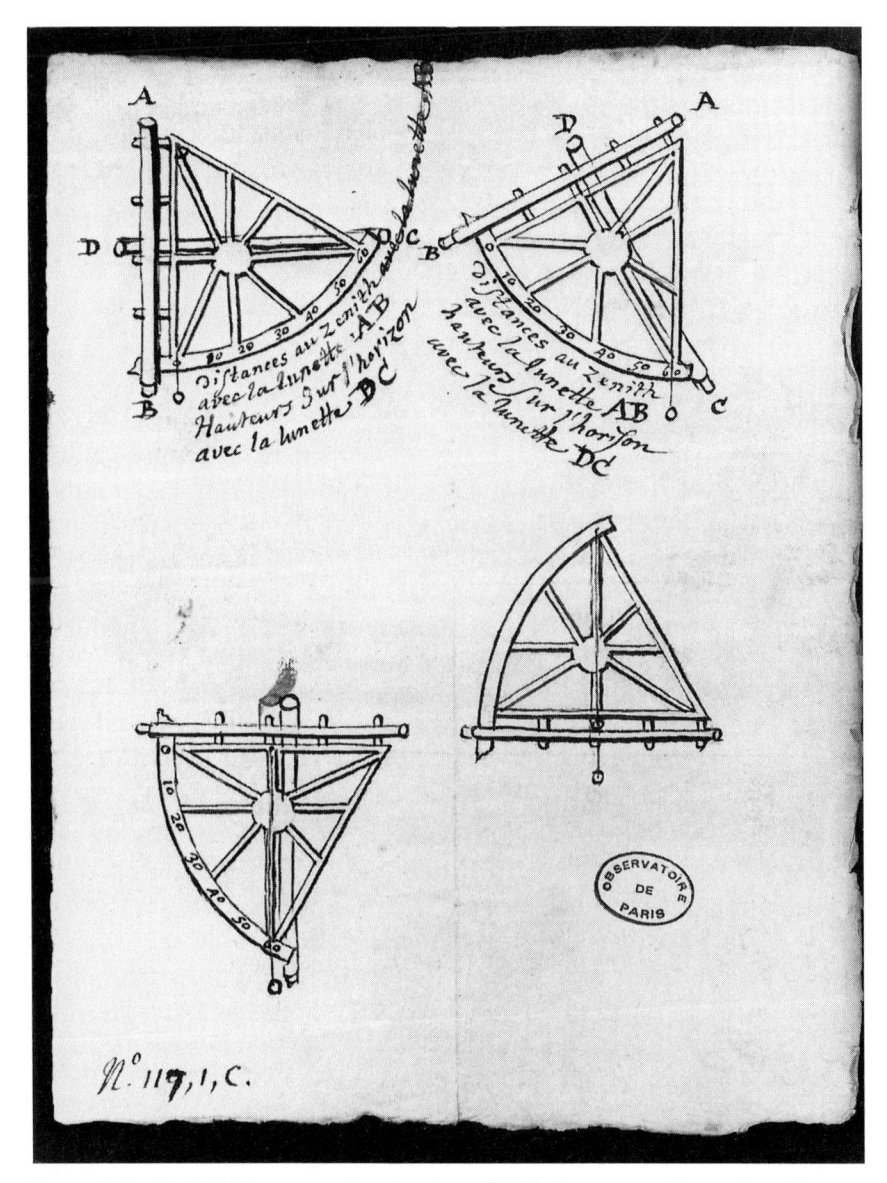

Figure 3.4 De L'Isle's manuscript sketches of Halley's sextant. (Reproduced by permission of the Director of the Observatoire de Paris.)

vinced of the advantages of telescopic sights.[17] He would be the first to use telescopic sights in any extensive programme.

Halley used the same instrument at his house in Islington in the years 1682–1684 when he made systematic observations of the Moon and other objects, and he possibly used it on later occasions as well. He also had a quadrant of 2 ft (0.6 m) radius that he had used before and with which he observed the meridian passsage of the Sun to determine the correction of his pendulum clock. Subsequently he took this quadrant to Danzig when he observed with Hevelius (Chapter 4) and also to France and Italy. He had a number of long telescopes, the longest of 24 ft (7.3 m), micrometers to measure angles in the fields of telescopes, and a pendulum clock. All were the most advanced instruments then available.

Halley derived his star places from those of reference stars (Appendix 5) in the catalogue of Tycho Brahe, the only one then available to him. Hevelius's catalogue did not appear posthumously until 1690, and Flamsteed and Cassini were just beginning their programmes. Tycho's catalogue was the first to be drawn up in early modern times and he (and his assistants) had made the observations for it at Uraniborg (Denmark) in the 1590s. His instruments were the most advanced of his day but he had no telescopes. His clocks were unreliable and he could not depend on them to determine the vernal equinox, the origin of longitude; he used instead the motion of Venus as a clock. He discovered atmospheric refraction. Most of his positions he found from distances to known stars and they have errors of the order of 40″.[18] Halley referred his own observations to bright stars that were close to the equator and so visible from both Uraniborg (lat 51° N) and St Helena (lat 15° 55′ S). He published his catalogue as a supplement to Tycho's.

It is easy to see that coordinates derived from those of other stars may be very uncertain. The variances of the calculated latitude and longitude of the unknown star depend in a complicated way on the variances of the positions of the reference stars and of the observed distances (Appendix 5). The most favourable condition is that all the angles should be about 45° and that the variances of the positions of the reference stars and of the observed distances should also be about the same; then the variances of the coordinates of the unknown star are comparable to those of the reference stars. In less favourable geometry those variances could be much greater. The further south a star lies, the greater the uncertainties will be.

3.4 *A year in St Helena*

Where should Halley set up his observatory? At first he thought of Rio de Janeiro and then of the Cape of Good Hope, but he would have had to learn the language, Portuguese at Rio and Dutch at the Cape, and learn to live among people with different ways. That would take too much time. He chose St Helena, the most southerly land under British rule. A Portuguese ship had found it on the feast day of St Helena, 1502. The Portuguese kept it secret until about 1600 when Dutch and English ships began to call for water. The Dutch and the English disputed possession until 1672 when the English East India Company finally settled it.

St Helena is a volcanic island like others off the west coast of Africa. Many valleys run down to the sea from the central peak. The hills are steep and the sea cliffs high. There is just one adequate anchorage at Jamestown at the end of the James (or Chapel) valley. The original thick forests had been destroyed by goats brought in by Portuguese visitors, and soil erosion had gone far by Halley's day. The population at Halley's visit was between three and four hundred, of whom some eighty were black slaves. The settlers or the garrison were in almost continual mutiny and the governors were either too weak or too tyrannical. Halley hints that he had difficulties, not just from the weather, and it would be surprising if his year there had been entirely uneventful.[19]

Isobel Gill and her husband, the astronomer David Gill (later Sir David and Astronomer Royal at the Cape), spent a week on St Helena in the nineteenth century on their way to Ascension Island where Gill would measure the parallax of Mars. Mrs Gill described how her husband went to 'Halley's Mount' and found a low wall, overrun with wild pepper, that he took to be a relic of Halley's observatory. Mrs Gill found the island very agreeable and describes the interesting plants and the sharp stars in the clear sky at night. St Helena was then notorious as the place where Napoleon had lived out his exile after Waterloo and many visitors and pilgrims came to see his house. It had also been the site of the observatory from which Johnson, in the mid 1820s, observed positions of southern stars.[20] Halley was assured by people from the West Indies that the skies would be clear, and so Isobel Gill found during her visit. Her experience was exceptional and Halley's choice was not good.

Halley cannot have sailed much before the end of October when the letter to the Governor and Council of St Helena was written, and so he would have reached St Helena about the beginning of February as the southern summer was ending. He presumably at once set in hand the construction of some sort of observatory, at the least supports for his larger instruments, probably with the help of Clerk and perhaps with the labour of Company servants on the island. He does not say where his observatory was, but he later told Hooke that he did not select the highest peak of the island because he found it too cloudy. Instruments there were continuously coated with dew and he saw the stars better at a lower site. He left no diary of his activities on St Helena but gave some idea of how he spent his time in a letter to Sir Jonas Moore [21] and in the preface to his *Catalogus*. In May 1677 he observed an eclipse of the Sun and one of the Moon, from which he hoped to determine the longitude of the island, but clouds prevented complete observations of the eclipses, and strong winds shook his telescopes. Rainy weather during the winter gave him great trouble and he could not observe at all during August and the first half of September. He did however observe the transit of Mercury across the Sun on 28 October. He explained his delay in not writing to Moore until 22 November by his hope for clearer weather as summer approached:

> I hoped still that we might have some clear weather when the Sun came near our Zenith, that so I might give you an account that I had near finished the Catalogue of the Southern Stars, which is my principal concern; but such hath been my ill fortune, that the Horizon of this Island is almost covered with a Cloud, which sometimes for some weeks together hath hid the Stars from us, and when it is clear, is of so small continuance, that we cannot take any number of observations at once*; so that now when I expected to be returning, I have not finished above half my expected work.

Despairingly, Halley says that he will continue for another two or three months and then return. He also wrote to Flamsteed who passed on his news to Boucher in Jamaica,

> I expect not Mr Halley from St Helens till the middle of the next Summer, the last news we had of him being that he was much troubled with Clouds and had not finished above more than half his work but I much fear whether I have joined any observations of the Moon made at the same time with him, our weather has been so Cloudye this last 12 Months.[22]

Halley used his experience of the weather on St Helena when he drew up his chart of the trade winds (Chapter 7). Twenty years later, on his second Atlantic cruise, he encountered similar cloudy conditions (Chapter 10).

Moore had other letters from Halley that Hooke presented to the Royal Society. On 13 December 1677 Hooke said that Moore had told him that the barometer was being observed on St Helena, which Moore confirmed, without however mentioning Halley by name, although the letter must have been his. Then on 14 February 1678 Moore reported that Halley had observed the transit of Mercury. Next week Hooke spoke of Gallet's observation of the transit at Avignon, where he was Provost of the church of St Symphorien, as reported in the *Journal des Sçavans* of 20 December 1677. Hooke mentioned Halley's letter again and there was a discussion on deriving the solar parallax from the two

---

*In the *Catalogus* he says that clear skies lasted for less than one hour.

observations. The Society discussed the transit three times more, on 28 February and 7 and 14 March. It was appreciated that the transit of 1677 together with the earlier ones of 1631 and 1671 should enable the orbit of Mercury to be greatly improved, and the apparently oval form of the image of Mercury on the Sun led to considerable discussion.[23]

Halley observed the magnetic variation on St Helena and used the result when he discussed the global field in 1692.[24]

Halley was back in London some time in May 1678, as Sir Jonas Moore reported to the Council of the Royal Society on 30 May 1678, when he 'gave an account that Mr Edmund Halley, who went to the Island of Saint-Helena ... was newly returned to England'. The same day, Hooke 'met Hally from St Helena with S. J. Moore & Colwall at Toothes'.[25] Moore said that Halley's observatory was 100 yards above the level of the sea and that he had observed above 400 stars (an exaggeration). The island was usually covered with mist and cloud that blew up from the sea while the sea itself around the island was clear. The climate was temperate and Halley cited as evidence of its benificent effects the case of the wife aged 52 of an husband of 55 so far without children, who were going to settle on St Helena and who were expecting a child when Halley left. There were wild partridges and turkey hens on the island.

Halley's report of the advanced age of the husband and wife about to have a child, which was the sort of curiosity often brought before the Royal Society, was subsequently garbled. Aubrey has it thus:

> There went over with him (amongst others) a woman, and her husband who had no child in several years, before he came from the Island the woman was brought to bed of a child.

Others later thought that meant that Halley was the father of the child. How rumour and slander grow.[26]

Flamsteed had written to Hevelius on 23 May to say that Halley had sent him the results of his observations of the lunar eclipse of 6 May 1677, the solar eclipse of 21 May, and the transit of Mercury of 28 October.[27] The letter may mean that Flamsteed had received Halley's results by letter before he, Halley, was back in London, or it may mean that Halley had passed them to Flamsteed as soon as he was back. Halley must have been back in London about the middle of May, and possibly as early as 7 May, but if so it is surprising that Hooke did not notice his return until the end of May and that it was not mentioned to the Royal Society earlier. However that may be, Halley would have left St Helena at the beginning of March or the end of February 1678, some three months after he wrote to Moore, and just about the time that Moore received his letter in London. According to his notes in the papers of the Royal Observatory, he observed an occultation on 4 February, presumably from St Helena, while on 14 September he observed another occultation from Greenwich.[28]

Halley told Moore that there are three stars of the first magnitude that are never seen in England, but that there are no bright stars near the south pole. He wrote of the Magellanic Clouds:

> The two Nubeculae called by the Saylors the *Magellanick* Clouds, are both of them exactly like the whiteness of the milky way lying within the Antarctick Circle; they are small, and in the Moon shine scarce perceptible; yet in the dark the bigger is very notable.

As to the weather, he says that they have had none hotter than the summer of England is ordinarily.

Halley greatly appreciated the company of Clerk:

> ... a person wonderfully assistant to me, in whose company all the good fortune I have had this voyage consisteth, all other things having been cross.

Not only the weather was cross. The Court of the East India Company received several complaints about the Governor of St Helena, C. Gregory Field; others beside Halley found him disobliging and difficult. On 20 February 1678, the Court had a report upon the letters of complaint that they had received and ordered that the Governor 'be removed from the said employment, & that he take his passage on the first shipping that shall touch there after ye Johannah's arrival'. They appointed Major John Blackmore as Governor from the same date.[29] Halley would have left the island before the change of governors.

Halley and Clerk took passage home on *Golden Fleece*. The whole expedition concluded to the satisfaction of the Court of the East India Company:

> On reading a Report from the Comttes for Shipping, touching Mr Edmo Halley & Mr Clerke; the Court being satisfied by Certificates of their civil demeanor during their being on the Island of St Helena, & in the passage home, doe think fit to direct that the Ownrs and Commander of the *Golden Fleece* be desired to deliver up the bond of 20£ that was taken for their transportation; and this Court will accompt with them for their passage, according to Chra-party.[30]

As on the outward voyage they would probably have sailed in company and would have called at the Cape Verde group, the Canaries, and Madeira, and Halley again no doubt profited from watching how the ship was sailed. He must also have put the three months at sea to good use in working up his results and preparing them for publication. Doubtless on St Helena the interruptions by clouds enabled him to keep up to date with the reduction of his observations, but he showed his results so quickly when he got back to London that he must have written some of his account and prepared the planisphere at sea on *Golden Fleece*.

3.5 *A new star and a new constellation*

Halley's achievements were rapidly recognised. Hooke 'shewed the planisphere and description of the stars of the southern hemisphere made by Mr Halley' to the Royal Society on 7 November 1678 and the Society was so impressed that it ordered copies of the planisphere and book to be sent at its charge to foreign astronomers, Gallet, Butterfield, and the Abbé de la Roche.[31] Well might the Society be impressed, for Halley's results were outstanding among those to come before it in its early years. The planisphere with the delineation of a new constellation, *Robur Carolina* (Charles's Oak) in honour of the monarch, must already have been presented to Charles II, for on 18 November Charles recommended to Oxford that they confer the degree of Master of Arts on Halley. Moore proposed Halley as a fellow of the Royal Society, and he was elected at the Anniversary meeting on 30 November 1678.[32]

The planisphere must have been engraved by 16 July 1678 when Flamsteed wrote to Sir Jonas Moore,

> I have now one of the prints of Mr Halleyes plate of the stars in the southern hemisphere in my hands; I have also seen part of his catalogue whereby I am satisfied he has done all that lay in his power towards their rectification. I would not therefore have you understand anything I shall here write concerning his works to his disadvantage;

and continues:

> He has made a long voyage to observe the southern constellations, and been unexpectedly crossed with ill weather. At his return, his friends, however, expect an account of his pains. He has not had an opportunity, nor time, to examine the Sun's motions, nor his distances from the fixed stars, nor their latitudes, by due observations: 'tis a work rather of years than months. He is afraid it would be said he had done nothing if he make not something immediately appear. To shun this imputation, therefore, he assumes the places of some of Tycho's fixed stars, and the altitudes of many of them to be true, *which may be erroneous*; and by the distances of the rest observed from these, he determines their places and latitudes. So that whatever errors were committed by Tycho in the fundamental stars, are transmitted by Mr Halley into this his new catalogue, which yet will be more accurate than the Tychonic, if considered all together; since he could determine his observed distances with more certainty than the noble Dane could possibly;

and therefore for our saylors, his catalogue will be exceedingly useful. Further we need not enquire.[33]

Flamsteed was to repeat those criticisms on subsequent occasions, and generally he did so without bias even when he held a very low opinion of Halley's character and behaviour. He himself used the same method of measuring distances with a sextant in his first years at Greenwich up to 1689, having until then no mural arc. He gave the fullest account of his critique in a letter of 15 November 1704 to James Pound (1669–1724) who was about to go abroad as a chaplain to the East India Company. Flamsteed urged Pound to measure meridional distances of southern stars:

> ... which (without derogateing from Mr Hally's previous labors) I must tell you are still wanteing his paines, will be a good guide to you, but 1st he left a many fixed stars unobserved and even some intire constellations. 2. he had no good assistant. 3d with such an one as had no relish of his worke it was almost impossible for him to avoyde faults in writing downe what he observed & I fear a many are committed not so much through his hast as his assistants negligence & Indifferences for his business, but 4th he calculated their places from his observations I fear but once himselfe & had no skilfull calculator to repeate them, the Calculations in his Cases were long perplext & troublesome so yt it was almost impossible for him to avoyd faults & what he has committed are on yt account excusable. I have heard him say he had a mind to repeate his Calculations. he took no Zenith Meridionall distances. when you have these taken it will be easy examineing his observed places.[34]

At about the same time as he wrote that letter, Flamsteed was proposing to include Halley's catalogue in his *Historia coelestis Britannica* (Chapter 14).

Halley called his *Catalogus stellarum Australium* a supplement to the catalogue of Tycho. He dedicated it to Charles II.[35] It consists of an introduction, an historical account of stellar catalogues, with reference to the work of Cassini, Flamsteed, and Hevelius in progress, and the need for southern observations; an account of his voyage and instruments; the main table of results giving the observed distances from known stars in Tycho's catalogue and the derived latitudes and longitudes; a table of right ascensions and declinations together with magnitudes of stars principally useful for navigation; an account of the transit of Mercury; and a discussion of lunar parallax and lunar theory. A planisphere of the southern heavens accompanied it.

The French version, translated by Jean Baptiste van Luer, was published in the same year as a supplement to the *Carte du ciel* of August Royer. Later Gottfried Kirch printed the catalogue in his almanack for 1681.[36]

A typical entry gives the measured distances of a star from two reference stars in Tycho's catalogue, followed by the longitude and latitude calculated from them. The longitude is given to the nearest 30° within the appropriate sign of the Zodiac. Halley also lists values from older catalogues where those are known, and he gives the magnitude.

The positions in Tycho's catalogue were reduced to the year 1678 and were derived with a value of 23° 21′ 30″ for the obliquity of the ecliptic, a value that Hevelius had found to be too great. Both in his preface to the *Catalogus* and in a letter to Hevelius of 11 November 1678 from Oxford,[37] Halley explains that he kept Tycho's value for consistency; it would be straightforward, he said, to convert to a different value, in particular that for the catalogue that Hevelius was himself preparing.[38] Flamsteed's criticism that Halley's latitudes would be afflicted with Tycho's errors relates principally to this limitation in Halley's procedure.

The table of right ascensions and declinations for use in navigation contained 37 stars and Halley gave the changes in the coordinates over the next 100 years.

The planisphere of the southern hemisphere was a splendid production. Its most novel feature was the delineation of the new constellation, *Robur Carolina*, which Halley set up as a compliment to the king who had recommended him to the East India Company: it referred to Charles's escape from the battle of Worcester by hiding in an oak tree. Oak leaves were widely sported on the anniversary of that day, Oak Apple day, to show loyalty to Charles after his restoration, a custom that continued until some sixty years ago, at least in rural Essex. Few other astronomers accepted the new constellation, but Hevelius has it on his chart of 1690.

Halley's was not the only chart of the southern hemisphere. Andreas Cellarius had published a handsome celestial atlas in 1660 in which, as in many others, depictions of the mythical figures of the constellations almost obscured the stars. In the second edition of 1661 he included a Christianised heaven in which the classical constellations were replaced by Christian figures, the signs of the Zodiac were marked by the Apostles, and St Helena was placed in the southern hemisphere with the Cross. The southern hemisphere was evidently familiar to Reiner Ottens in 1729. When he compiled his atlas Halley's was still the most comprehensive catalogue of the southern stars. Ottens possibly made use of it.[39]

3.6 *How far is the Sun?*

Most astronomers would no doubt see the first reliable catalogue of the southern stars, for all its limitations, as a great achievement and most valuable. It was also the first catalogue of any part of the sky derived from observation with the most up-to-date telescopic instruments. There would be nothing comparable for many years to come. Yet it would be superseded, its usefulness would be ephemeral. It was one of a number of catalogues then under construction: it did not start anything radically new. Halley did start something really new at St Helena when he observed the transit of Mercury across the Sun, as he wrote to Moore:

> I have notwithstanding had the opportunity of observing the ingress and egress of [Mercury] on the [Sun], which compared with the like Observation made in *England*, will give a demonstration of the Sun's Parallax, which hitherto was never proved, but by probable arguments.[40]

In that sentence, Halley set a new course in the astronomy of the solar system, and hence step by step, in the determination of the size of the visible cosmos.

The solar parallax, the angle subtended by the Earth at the Sun, or the distance of the Sun from the Earth, was both of philosophical interest and of practical importance in astronomy. The angle (obliquity) between the ecliptic and the equator, and the longitude of the node of the ecliptic upon the equator, are fundamental data for the celestial reference systems. They change with time (precession of the equinoxes) and it is an important task for astronomers to determine them. The obliquity can be found from the differences in the zenith distances of the Sun at the summer and winter solstices, but those angles are subject to two systematic effects. One is that of the refraction of the atmosphere, which is less when the Sun is near the zenith than when it is far from it. On the other hand there is the effect of the solar parallax. At the equinoxes the line of sight from an observatory on the equator to the Sun lies in the ecliptic, but at the solstices, the line of sight lies either above or below the ecliptic. It turns out that the observations of zenith distance at any one place may be satisfied by a wide range of postulated values of parallax and refraction, but to obtain consistent values of the obliquity and node from observations at different latitudes, the true values of parallax and refraction have to be employed because the corrections due to refraction vary with latitude (Fig. 3.5). Richer and Meurisse undertook their expedition to Cayenne to settle that issue by comparing their results with those of Cassini in Paris, and the upshot was clear—in contrast to previous estimates and speculations, the solar parallax could not exceed 10″. Important as that conclusion was, it was not independent of the refraction which was still not adequately understood. Newton would develop a theory of refraction about

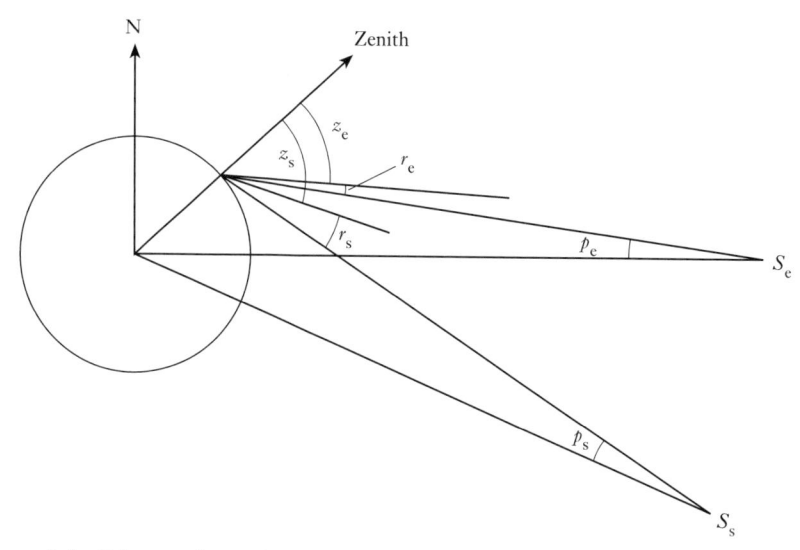

Figure 3.5 Solar parallax and refraction. *Key*: S_e, S_s: directions of Sun at equinox and solstice; r_e, r_s: atmospheric refraction at equinox and solstice; p_e, p_s: parallax of Sun at equinox and solstice; z_e, z_s: observed zenith distance of Sun at equinox and solstice.

which he and Flamsteed corresponded, and Halley's friend Boucher made observations of refraction in the West Indies.[41]

Cassini and Flamsteed independently observed the diurnal parallax of Mars with the intention of deriving that of the Sun by the use of Kepler's third law. The periods of the planets are well known, their relative distances from the Sun are therefore also well known by Kepler's third law, and a single measured distance such as that of Mars from the Earth at closest approach would give the scale of the solar system and the distance of the Sun from the Earth. The observations of parallax were however unsatisfactory, for although they did seem to give consistent values of about 10″, the errors were comparable with the parallax and Halley, who took part in Flamsteed's observations, saw that a different approach was needed. Instruments good enough to determine the parallax of Mars with the necessary precision did not become available until the middle of the nineteenth century and Gill went to Ascension Island to make use of them.[42]

Transits of Mercury and of Venus across the face of the Sun in 1631 had been predicted by Kepler in 1627 and although that of Venus could only be seen in America and was not observed, Gassendi saw the transit of Mercury in Paris, as did Cysat in Ingolstadt, and Remus Quietanus in Ruffach. Gassendi, like others, was very surprised by the small size of the image of Mercury, only about 20″ according to his estimate: indeed, the main use to which those first observations were put was to estimate the sizes of Venus and Mercury, so much smaller than people had supposed. Horrocks made the first observation of a transit of Venus in 1639, and later Hevelius observed the transit of Mercury in 1661.[43]

James Gregory had proposed that transits could be used to determine the distance of the Sun in his *Optica promota* of 1663, and presumably Halley knew of that. Halley however worked out the idea in practical form for the first time, prompted no doubt by the happy opportunity of observing the transit of Mercury from St Helena. The other new matter in his letter, although it may not be very obvious, is that for the first time he observed the interior contacts of the planet on the Sun at ingress and egress, instead of just the exterior contacts.

The time of transit of an inner planet across the Sun as seen from a place on the Earth depends upon the radius of the small circle swept out by the observatory rotating about the axis of the Earth during the transit. The differences from place to place can amount to about fifteen minutes in a total transit time of some six hours, rather greater for Venus and less for Mercury. The distance between two terrestrial observatories is found from the difference of latitude and the radius of the Earth, while the ratio of the distances of the planet to the Sun and of the planet to the Earth follows by Kepler's third law from the periods of the orbits of the Earth and the planet about the Sun. Consequently the distance of the Earth from the Sun is found in terms of the radius of the Earth from a combination of the periods of the Earth and planet about the Sun, and the times of transit of the planet across the face of the Sun as seen from the two sites of known latitude. The scheme will be discussed more fully in Chapter 8 with Halley's later presentation of his analysis to the Royal Society.

Halley was fortunate in his observations on St Helena. The tables that he had for predicting the transit were not sufficiently accurate, as may be seen from an almanack produced by Flamsteed for 1674. It was adapted from one of Hecker that predicted a transit of Mercury that year which, by Kepler's Rudolphine tables, should have been observed in daylight at Brandenburg. According to Streete, however, it would have taken place at night time in England. Halley no doubt used Streete's tables for the prediction of his transit, but it must have been unexpected good fortune that it occurred in daylight on St Helena and not, as might well have been, at some less convenient time.[44]

It was windy and cloudy the night before the transit; it became calm as day dawned, but still cloudy. Fortunately Halley was able to observe both the ingress and egress of Mercury. He had his 24 ft telescope directed to the Sun and waited for gaps in the clouds. The total time of the transit, from the entry to the exit of the centre of the planet was 5h 14min 20s. The entry of Mercury on the Sun was cloudless:

h	min	s	
9	20	35	Sun brilliant, no cloud
9	26	17	Mercury at edge of Sun, a little dot about 10° from nadir of Sun on the right side
9	27	30	Whole of Mercury in the Sun, moving to east. Clouds at 10.30 obscured Mercury until they cleared about 2p.m., when Halley saw that Mercury was soon about to disappear.
2	38	39	Edge of Mercury about 1 diameter of Mercury from edge of Sun
2	40	8	Edges touch
2	41	0	Exit of centre of Mercury, 30° from nadir of Sun
2	41	54	End of transit.

Gallet, at Avignon, saw only the egress, for clouds obscured the ingress. Clouds covered all English sites.[45] Halley was thus disappointed of his hope of a definite value for the solar parallax, although he did make an estimate. He also pointed out that the geometry of the transits of Venus is better.

3.7 Matters of the Moon

The orbit of the Moon, a great puzzle in the last years of the seventeenth century, would preoccupy Halley throughout his life. Newton in the *Principia* would show that its vagaries were the result of the gravitational attraction of the Sun, although he could not demonstrate them in all their richness and complexity. Universal gravitation was as yet unimagined when Halley went to St Helena. The challenge in 1678 was to describe the motion of the Moon—physical explanation was for the future. The greatest advance had been made by Jeremiah Horrocks, whose *Opera posthuma* had been published by Wallis just as Halley went up to Oxford. Horrocks took the orbit to be an ellipse but with a variable eccentricity and a rotating oscillating apogee. His predictions from that model agreed with observation to better than 8′, a considerable divergence but very much better than anything that had gone before. Wallis included in his collection an account of Horrocks's theory by Flamsteed together with some lunar tables calculated by Flamsteed.[46] Halley was surely familiar with Horrocks's model when he went to St Helena.

Halley's essays in his *Catalogus* advanced the study of the Moon's orbit in three ways. In the first he discussed how to find the parallax of the Moon—'Modi quiddam pene geometrici pro parallaxi Lunae investigandum' ('a certain rather geometrical method for finding the parallex of the Moon')—and concluded that the best way was to compare the meridian altitude of the Moon at St Helena with that in Europe at the same time, as Cassini had suggested. In his second section, 'Quaedam Lunaris Theoriae emendationi spectanti' ('an improvement to a certain lunar theory that requires examinations'), he showed that as compared with an ellipse, the orbit of the Moon was more curved at quadrature than at syzygies[*], a result that had not previously been published although Picard had known of it in 1668. Later Newton would show how it followed from the net attraction of the Earth and the Sun towards the centre of the Earth being reduced at syzygies as compared with at quadrature.

Halley's third discovery was to identify an inequality in the motion of the Moon that depended on the varying distance of the Earth from the Sun in its elliptic orbit. It is known as the parallactic inequality, from the varying parallax of the Sun, and Newton would show how it arose from the varying gravitational attraction of the Sun at the Earth and the Moon. Halley modified Horrocks's description of the lunar orbit to allow for the parallactic inequality by placing the focus of Horrocks's rotating ellipse upon an epicycle, as Newton noted in the *Principia*.

Halley analysed the lunar orbit well before Newton wrote the *Principia* so that he was guided by no gravitational nor dynamic theory, but when those became available, especially in the second edition of *Principia*, it was seen how Halley's discoveries might be explained. That is the topic of Chapter 13.

3.8 Consequences

Halley's achievements, restricted though they had been, were immediately recognised in England. The Royal Society elected him a fellow.[47] Oxford conferred the MA degree upon him at the King's desire. The Provost of Queen's College, Dr Timothy Halton, set out Halley's status at Oxford in a letter to Sir Joseph Williamson. Williamson had evidently proposed that Halley should receive a degree:

> I have spoken with Mr Vice-Chancellor about procuring a degree for Edmund Halley, of our college. He is very ready to comply with your desires, but suspects to find opposition among the Masters, who are difficult both to remit exercise and give time, so that if you will procure the King's

---

[*]Syzygies are the positions when the Earth, Moon, and Sun are all in line, quadrature when the Moon, seen from the Earth is in a direction at right angles to the direction of the Sun.

letter, it will be both effectual and not unpleasing to the University. He is now of almost 6 years standing and less than a master's degree cannot be conferred on him. He is now in the college and depends solely on your assistance in this matter.[48]

Members of Oxford and Cambridge will easily find present parallels for 'opposition among the Masters'.

The King obliged in a letter from Whitehall of 18 November which was read to Convocation at a meeting to receive royal letters:

Trusty and Wellbeloved We Greet you Well. having received a good account of the proficiency in Learning of our Trusty Wellbeloved Edmund Hally of Queens College in that our University especially as to the Mathematick and Astronomy, Whereof he has (as We are informed) gotten a good Testimony by the Observation he has made during his abode in the Island of St. Helena; We have thought fit for his Encouragement hereby to recommend him effectually to you for his Degree of master of Arts; Willing and Requiring you forthwith upon the receipt hereof (all Dispensations necessary being first granted) to admit him the sd. Edmund Hally to the said Degree of Master of Arts without any Condicion of performing any previous or subsequent Exercises for the same any Statute or Statutes of that Our University to the contrary in any Wise notwithstanding And so We bid you farewell Given at our Court at Whitehall the 18th day of November 1678 in the Thirtieth year of our Raigne.

Convocation agreed that Edmund Halley should be admitted to the degree of Master of Arts at the next Congregation according to the King's letter. The degree was conferred on 3 December.[49]

The modern reader may be struck, even taken aback, by the speed with which the administrative machinery worked. This may be the first instance of a degree conferred for research, and not only that, another modern convention seems to have been anticipated. Halley may have been of six years standing, but he certainly had not resided in Oxford for that time nor for long enough to have qualified for any degree. Did the University deem his passages to and from St Helena, and the time he spent on the island, to be residence in Oxford, just as nowadays an astronomer can spend a number of terms at some overseas observatory, or a high-energy physicist conduct his experiments at CERN, and have the observatory or CERN deemed to be Oxford or Cambridge for satisfying the conditions for a degree?

The King often recommended someone for a degree who had not strictly fulfilled the requirements: Flamsteed had similarly benefited at Cambridge.[50] Halley's claim stands out for the very influential support he had, for the precise grounds of the royal command, and for the spectacular nature of his absence from Oxford. It was not the last time he would have the effective support of a Secretary of State in academic matters (Chapter 11).

For all his substantial criticisms of Halley's work, Flamsteed wrote in 1680 in the Preface to his *Doctrine of the sphere*:

> I find myself obliged to give the following account of its [*The doctrine of the sphere*] Original that I may not hereafter be accused of injustice to my two singular kind friends, the admirably ingenious Sir Christopher Wren (Master Surveyor of His Majesties Buildings) and our southern Tycho, Mr Edmond Halley.[51]

Hevelius had known of Halley's southern plans already in 1676 when Flamsteed wrote to him in August with news of the proposal for the stellar catalogue, for the observation of the transit of Mercury and of the eclipses for the longitude of St Helena. Flamsteed sent news of Halley's results in May 1678 and Hevelius replied on 13/23 August in a letter that Flamsteed showed to Halley. Later in a letter to Kirch, Hevelius said how valuable Halley's results had been to him.[52] Halley wrote to Hevelius on 11 November to say that he hopes soon to send him a copy of his catalogue; he no doubt did so for there were two copies in Hevelius's library. Hevelius praised Halley's work in the *Annus climactericus* of 1679, although Halley became annoyed that he garbled the story in some respects.[53] Hevelius had been corresponding with the Royal Society about the precision of his methods. He hoped someone could visit Danzig to see what he was doing and Halley, with his recent experience of modern instruments, was the obvious person to go. So he did, with the commendation of the Society, in the summer of 1679 (Chapter 4).

Oldenburg had kept Cassini in Paris aware from the first of Halley's intentions and Cassini had made helpful comments. Flamsteed had previously sent Cassini details of the eclipse that he and Halley had observed in Greenwich in 1675 which gave the difference of longitude between Greenwich and Paris, and after Oldenburg's death he kept Cassini *au fait* with Halley's progress. Thus Cassini had early news of Halley's catalogue and was no doubt sent a copy. The publication of the French translation in the same year as Halley's Latin original would make it known to French astronomers.

3.9 Halley's achievement

Halley's main achievements were the catalogue of the southern stars and his observation of the transit of Mercury, but he also recorded other natural phenomena to which he was to refer in later years. His notes of the winds went into his chart of the trade winds; he observed the depression of his barometer at the height of his observatory; Newton used his record of the change required in the length of the pendulum of his clock to keep time as evidence that gravity was less in low latitudes and therefore that the Earth was flattened at the poles. He made magnetic observations at sea, he found that the dip was zero at about 15° N and he found that when an iron rod was held vertically the lower end would attract a south pole when they were north of 15° N, but a north pole when they were more southerly. That must be an early observation of magnetic induction.[54]

Halley's observations of some 300 stars in the course of a single year, by methods which depended on the results of others, especially Tycho, cannot justly be compared with the labours of Hevelius, or Flamsteed over many years. Hevelius, like Halley, observed the distances between stars with a sextant, as did Flamsteed up to 1689 when he at last got his mural quadrant, but Flamsteed, again like Halley, had a sextant with telescopic sights. Halley's methods and instruments were quite similar to those of Flamsteed up to 1689 but his programme differed in that he spent only a year and did almost everything himself, observations and calculations. It was heroic.

Halley's was the first catalogue of any part of the sky, north or south, to use up-to-date instruments with telescopic sights and the eyepiece micrometer. Richer in Cayenne had used telescopic sights but could not see all the southern stars and seems never to have compiled a catalogue. Halley printed his catalogue rapidly and almost 45 years before Flamsteed's own publication. Halley was the pioneer, both of the southern hemisphere and of the systematic use of state-of-the-art astrometry. He was also, now and in the future, conscious of the value of speedy publication, in sharp contrast to Flamsteed. He would publish his magnetic and tidal charts as rapidly (Chapter 10).

Flamsteed's attitude to Halley's work was ambiguous. He praised him as the 'Southern Tycho' and always recognised the hard work and application that had gone into his observations, but he was also critical of his methods. His criticisms, repeated from time to time, were always the same, that Halley had not had a good assistant, and that he had not had the calculations done independently to check for errors. He also regretted that Halley had based his catalogue on the southernmost stars of Tycho, rather than making absolute measurements with a mural quadrant. Flamsteed wanted a much longer course of observations made in a more satisfactory site. At one time he said he preferred the observations of French Jesuits published in 1688, but later, when he came to prepare his own work for publication, he included in his proposals the reprinting of Halley's catalogue along with the earlier ones from

Ptolemy to Hevelius (Chapter 14). He wanted thereby to show how his own work was related to earlier catalogues, but of course Halley's catalogue did not overlap with his. He proposed to print no other southern catalogue, so presumably he thought in 1709 that Halley's had not been superseded.

Flamsteed did not see his revision of Halley's catalogue through the press. He died before he could complete the catalogue volume of his *Historia coelestis britannica* and it was brought out in 1725 by his former assistants, Joseph Crosthwait and Andrew Sharp. Sharp recalculated Halley's catalogue positions from Flamsteed's positions for the northern reference stars instead of Tycho's; Halley's catalogue thus amended appears at the end of Volume 3 of the *Historia coelestis Britannica* of 1725 (see Chapter 14 for an account of that publication). Halley was said to be very angry when he saw it, but he had really invited such a revision when he wrote that he was publishing his observed distances between stars so that others might recalculate his positions when better values for Tycho's stars were available. Some comparisons between Halley's original positions and those recalculated by Sharp are given in Appendix 5. They show that Halley was observing differences of position with his sextant to better than 1', and that greater differences between his values and Sharp's arise from changes in the adopted positions of Tycho's stars and, in particular, a difference of about 41' in the longitude of the vernal equinox. No other southern catalogue was to be published until that observed from the Cape of Good Hope by La Caille.[55] Was not Sharp's revised catalogue of 1725 the greatest compliment to Halley and the most solid recognition of his achievement in St Helena? The work was worth updating.

Halley's observations of the transit of Mercury and his remarks on the motion of the Moon did not attract so much attention, but they were at least as fruitful and influential in the long run, for the former led on to his study of the transits of Venus and all that followed from it, while the latter foreshadowed the great attention that he would pay to the motion of the Moon in later years after Newton's attempts to place the theory on a basis of gravitational attraction by the Sun.

In his first and early major campaign Halley showed an outstanding ability to plan and persuade, to organise and to carry through his project. He showed theoretical insight into fundamental matters of astrometry, and acquired more practical experience as an observing astronomer with up-to-date instruments than anyone else of his day. He had accomplished the three things he set out to do, to observe in the south, to produce the first catalogue after Tycho, and to use modern instruments in a major systematic campaign. He was well prepared for his visit to Hevelius in Danzig.

4

Into Europe

> Per correr migliori acque alza la vele
> omai la navicella del mio ingegno
> [Now the little boat of my genius raises her sails to run over better seas]
>
> DANTE: *PURGATORIO*, I, 1,2

4.1 *The sage of Danzig*

JOHANN HEVELIUS of Danzig (1611–1687) was the most distinguished astronomer in Europe when the Royal Observatory was founded at Greenwich. His family were rich brewers and he was a city councillor and a Consul, active in public affairs. Danzig (Gedanus in Latin, now Gdansk) was prosperous, one of the four principal Hanseatic ports. Many Londoners would know it well for the Baltic trade was a great part of London's business. The palaces of the rich merchants were grand and the civic buildings notable. Danzig must have looked opulent to travellers coming from a London just rebuilding after the Fire. It was no small matter to be a councillor and Consul of Danzig.

Hevelius had studied law at Leiden and had visited England and France as a young man. After his first wife died he married in 1662 Elizabeth, the sixteen-year-old daughter of a Danzig merchant. He had built an observatory on his own property and Elizabeth assisted with his observations (Fig. 4.1). He mapped the surface of the Moon and discovered its libration in longitude. He discovered a number of comets and suggested that they moved in parabolic orbits. He was the first to detect a variable star, which he named Mira. He observed transits of Mercury (after Gassendi in 1631 and Shakerley in 1651) and of Venus (after Horrocks in 1639). He was a Fellow of the Royal Society[1].

Hevelius corresponded with many astronomers, especially Cassini, and with Flamsteed, Oldenburg, and others in the Royal Society. Johann Erik Olhoff (Olaf), the secretary to the City Council of Danzig, published some of Hevelius's correspondence as *Excerpta ex litteris J. Hevelius.*[2] Hevelius published his work in six main books, *Selenographia*,

Figure 4.1 Hevelius and Mme Hevelius observing (from Hevelius, *Machina coelestis*). (Reproduced by permission of the University Librarian, Cambridge.) This is an historic picture, almost certainly the first to show a woman engaged in science.

Cometographia, the *Machina coelestis* in two parts (1673, 1679), the *Annus climactericus* (1685), and the posthumous *Prodromus astronomiae* (1690). He distributed them widely, in particular to the Royal Society, to the Universities of Oxford and Cambridge*, and to Flamsteed.[3]

Hevelius used open sights, or pinnules, to observe directions with the naked eye, instead of telescopes. His main instruments at the time of Halley's visit in 1679 were a quadrant for the determination of meridian altitudes (Quadrans parvo orichalcico, or Q.p.o.: small brass quadrant) and a sextant for the measurement of angular distances between objects (Sextans Magno Orichalcico, Sex M.O.: large brass sextant). Hevelius described them and his other instruments in the first part of the *Machina coelestis*. Halley wrote to Flamsteed that the quadrant was 'not above 10 Inches [25 cm] Radius with such a perpendicular, as we used at the time of the Lunar Eclipse we observed at Sr Jonas's'. The sextant was of six foot (1.8 m) radius, permanently mounted in the observatory. Hevelius also had a twelve foot (3.7 m) telescope which he used for timing occultations and observations of lunar features, but not for angular measurements. He had two pendulum clocks: 'his Clocks are not extraordinarye, he hath but 2, and those with very short pendulums, and such as are made to stand on tables'.[4]

When the first part of *Machina coelestis* appeared in 1673, Hooke had severely criticised Hevelius's instruments.[5] He argued that Hevelius's method of dividing the circles of his instruments was defective, that telescopic sights were much better than open sights, and that Hevelius's objections to them were invalid. Hooke considered that Hevelius's precision would be limited to 1', or possibly 0.5' at the very best, and that his procedures were no better than Tycho's. Hevelius had claimed that a set of measurements of the angular distances between eight stars on the ecliptic which added up to exactly 360° demonstrated his precision. Hooke argued that the agreement was spurious and indeed evidence against Hevelius, for the different refraction at the various altitudes of the stars in the sky, would lead to a total measured difference other than 360°. Hooke described his own method of dividing a scale of angles, and set out the designs of instruments that he proposed, with a full description of telescopic sights having graticules of hairs at the common focus of the object lens and eye glass. He also argued that a graduated scale of angles should be read with a lens and vernier. All that was as Flamsteed and Halley were to put into practice at Greenwich and on St Helena. Hooke based his estimate of the precision

---

*The copy of *Annus climactericus* in the University Library at Cambridge is the presentation copy inscribed by Hevelius and the copy of the first part of *Machina coelestis* may also be the presentation copy, although Hevelius's correspondence with Flamsteed suggests that it was never acknowledged by the University (*nostra culpa*).

of open sights on the size of features on the Moon that could be distinguished by the naked eye, but the precision of setting a mark on an optical image is often better, sometimes much better, than the resolution, and it is quite likely that Hevelius could make settings on an image of a star, which might be as much as a minute in diameter, to far better than a minute. Hooke gave an hostage to fortune by describing his own designs, for when he came to make a mural quadrant for the new Observatory at Greenwich, Flamsteed could not use it (Chapter 3).

Hevelius complained to the Royal Society about those criticisms from a fellow member, he challenged Hooke to measure eight distances between specified stars, and urged the Royal Society to send one of its members to Danzig.[6] The Society agreed to send someone who could observe the same objects with Hevelius's instruments and with telescopic instruments. Since the former were very large and permanently installed at Danzig, the Royal Society's observer had to go there and take with him some portable telescopic instruments. Halley had just returned from making observations with telescopic instruments at St Helena and had compiled a catalogue; he was the obvious person to go.[7]

It is not clear that Halley went at the express request of the Society, for he had apparently planned a visit himself. He wrote to Hevelius in November 1678 to say that he would shortly send him his catalogue of the southern stars. He thanked Hevelius for his opinion of his work and said how gratified he was that it should be linked with that of Hevelius. He went on to explain how clouds had restricted his observations so that he had to base his catalogue on that of Tycho and in particular, on Tycho's fundamental elements. His own catalogue could be recalculated when better values become available from Hevelius's work. Hevelius was familiar with Halley's work on St Helena as he showed in his letter of 24 April 1679 sent with copies of the second part of his *Machina coelestis* to a number of correspondents in England, Halley among them.[8] He also included an account of Halley's expedition to St Helena in *Annus climactericus*.

Halley was not just the tactful junior sent by his seniors, Hooke and Flamsteed, to calm the irritation of the elderly Hevelius. Halley was indeed a tactful ambassador; Hevelius was the doyen of observers. Open sights might be out of date, but Hevelius had notable publications to his credit and his reputation was solid and deserved. The achievements of Flamsteed and Cassini by 1679 were not comparable. Flamsteed had made a very careful study of the Sun's motion, but little else, while Cassini, with a considerable reputation from his Bologna days, still awaited good instruments at Paris. Both had yet to show results from the new instruments that they and their colleagues advocated and were designing. Halley had results. He had taken instruments with telescopes and micrometers and pendulum clocks to the south and had produced a catalogue of 300 stars, the first made anywhere with the new methods. Halley came to Danzig with experience of telescopic instruments under difficult conditions and wearing the laurels of achievement. Hevelius might have felt flattered that he had such a visitor.

4.2 Halley at Danzig

When Detlerus Cluver wrote to Hevelius from Hamburg on 4 March 1679 to tell him that Halley was preparing his voyage to Danzig, he might have given the impression that the Royal Society was sending Halley. A month later in a letter of 3 April 1679, Croone, a Secretary of the Royal Society, introduced Halley to Hevelius, as did Wallis. Croone said that after St Helena, Halley proposed to undertake a second astronomical peregrination and visit the Prince of Astronomers of the Age—Hevelius. The Society commended him as of exceptional astronomical merit, but it does not seem that he was going as a formal representative of the Society.[9]

Halley began to observe with Hevelius as soon as he arrived in Danzig on 26 May.[10] They used three of Hevelius's instruments, the quadrant (Q.p.o.), with which Hevelius made regular observations of the Sun, the large sextant (M.O.—see Fig. 4.1) for angular distances between stars, planets, and the limbs of the Moon, and the 12 ft telescope for occultations. Halley brought with him what Hevelius called a telescopic sextant, but which was in fact the 2 ft quadrant that Halley had used on St Helena; as William Molyneux remarked in a later letter to Halley, it was not really a sextant, but a quadrant. That misunderstanding, together with others by Hevelius in his account of St Helena in *Annus climactericus*, irritated Halley.[11] Six observers participated, Hevelius and Halley, together with J. E. Olhoff, Nathaniel Buthner, Mme Hevelius, and the printer (Typographus) who regularly assisted Hevelius in observing.

Halley first watched Hevelius at his instruments. On 5 June they observed together and timed the occultation of Jupiter by the Moon with the 12 ft telescope. Halley observed angular distances from 9 June onwards, with the sextant M.O. and with his own telescopic sextant, and on 6 July Hevelius and Buthner observed with Halley's instrument. Halley and Hevelius observed further occultations on 24 and 25 June, and on 12 July they both observed meridian altitudes of stars, Hevelius with the Q.p.o. and Halley with his telescopic quadrant. All six observers measured two sets of distances on 15 July and finally, on

17 July, Hevelius and Buthner used Halley's sextant. Before he left on 18 July Halley wrote to Hevelius with his assessment of the accuracy of Hevelius's instruments and methods.[12]

Halley plainly did not confine his activities to astronomy for he excused some delay in writing to Flamsteed on 7 June 1679:

> ... had I not been wholly taken up with the Curiosityes of this Place, which has caused me to send no more than 2 Letters till this day.

There would indeed have been much to see in Danzig in daytime when they were not observing: private and public palaces, and ships and wharves and warehouses, and the famous great crane.[13]

Some of the observations permit estimates of the performances of various instruments. One or more observers repeated observations of angular distances with the same instrument. They used open and telescopic sights to measure angular distances between two stars and between a star and a planet, and also meridian altitudes of stars. They observed the angular diameter of the Moon with open and telescopic sights. Some of their stars are easily identified, for example, the bright stars Vega in Lyra and Altair in Aquila. Other stars cannot now be identified. For that reason, the angular distances that were observed cannot easily be compared with modern values and are identified here simply by letters A, B, and so on.

The relevant observations of angular distances taken from the lists in *Annus climactericus* are summarised in Table 4.1. One distance, D in

Table 4.1 Angular distances between celestial objects as observed at Danzig, June–July 1679

Date	Distance	Observers	Instruments	Result ° ′ ″
1 June	A	He, T	M	35 21 55
	A			35 22 30
7 June	B	He	M	38 11 10
	B			38 11 00
9 June	C	He, T	M	39 41 50
	C	B	M	39 42 00
	D	He, Ha	M	33 31 30
	D	He, T	M	33 31 40
12 June	E	He, T	M	23 52 45
	E	He, T	M	23 52 40
	E	Ha, B	ST	23 47 30
	F	He, T	M	55 19 00
	F	Ha, T	M	55 19 00
	F	He, Ha	M	55 19 00
	F	Ha	ST	55 11 00

Table 4.1 *Continued*

Date	Distance	Observers	Instruments	Result ° ′ ″
22 June	G	He, T	M	45 52 50
	G	He, T	M	45 52 55
	G	Ha, T	M	45 53 10
	G	(Hevelius, 1661		45 52 00)
	G	Ha, B	ST	45 50 00
	H	Ha, B	ST	39 23 30
	H	(Hevelius, earlier observation		39 27 40)
June	J	Ha, T	M	53 48 20
	J	Ha, T	M	53 48 30
	K	Ha, T	M	43 47 30
	K	Ha, T	M	43 47 20
	K	He, T	M	43 47 30
	K	He, T	M	43 47 20
30 June	L	Ha, T	M	47 47 00
	L	B, T	M	47 48 00
	L	He, T	M	47 48 10
	L	He, Ha	M	47 48 10
6 July	M	He, B	ST	33 25 00
	M	He, B	ST	33 28 00
	M	He, B	ST	33 27 00
	M	He, B	ST	33 27 30
15 July	N	Ha	M	33 31 55
	N	Ha, MH	M	33 31 35
	N	Ha, MH	M	33 31 45
	N	MH, T	M	33 31 35
	N	He, MH	M	33 31 45
	N	Ha, O	M	33 31 50
	O	He, T	M	26 52 50
	O	Ha, B	ST	26 54 00
	O	–, T	M	26 52 55
	O	–, T	M	26 52 55
17 July	P	He, B	M	26 50 00
	P	Ha	M	26 48 00

Identification of observers: Hevelius: He; Halley: Ha; Typographus: T; Nathaniel Buthner: B; Mme Hevelius: MH; Olhoff: O.
Identification of instruments: Quadrant parvo orichalcico: Q; Sextans Magno Orichalcico: M; Halley's telescopic sextant: ST.

that table, between Lucida Aquilae and Caput Serpens, had been observed by Hevelius with the large sextant, M.O., on a number of previous occasions with different colleagues:[14]

    1659 2 June with Kretzner   33° 31′ 30″
                                                  33° 31′ 40″

		33° 31′ 40″
		33° 31′ 40″
1660	20 June with Marquado	33° 31′ 40″
		33° 31′ 35″
1665	2 July with Mme Hevelius	33° 31′ 40″
		33° 31′ 50″
		33° 31′ 40″
1674	1 August with Fullenius	33° 31′ 35″
		33° 31′ 40″.

The mean value, 33° 31′ 40″, seems very secure for observations with the large sextant at midsummer, the same period as Halley's visit, when the effect of refraction would be much the same. The distance was given particular attention at Halley's visit.

Table 4.2 gives the meridian altitudes of eight stars observed on 12 July, by Halley with his telescopic quadrant and by Hevelius with his quadrant (Q.p.o.). Three observed by Halley were not visible from the fixed instrument in the observatory. Again the identification of the stars is not always certain and they are given roman numerals in Table 4.2.

The figures in the tables indicate that measurements were recorded, and presumably made, to the nearest 5″. The measurements of angular distances are not easy to analyse because observations with Halley's quadrant were repeated only when it was used by Hevelius and Buthner, who had no previous experience of it. Observations with the Magno orichalcico were often repeated, but in no systematic way, so that it is not possible to say more than that observations with it appear to have been reproducible to about 10″. The key distance, D, between Lucida Aquilae and Caput Serpens, was observed with the Magno orichalcico on three nights, with the following results:

9 June	Hevelius, Halley	33° 31′ 30″
	Hevelius, Typographus	33° 31′ 40″
23 June	Halley, Typographus	33° 31′ 40″
15 July	Halley, Mme Hevelius	33° 31′ 55″
	Halley, Mme Hevelius	33° 31′ 35″
	Halley, Mme Hevelius	33° 31′ 45″
	Mme Hevelius, Typographus	33° 31′ 35″
	Hevelius, Mme Hevelius	33° 31′ 55″
	Hevelius, Olhoff	33° 31′ 50″.

The mean values are, for 9 June, 33° 31′ 35″, and for 15 July, 33° 31′ 45″ (to the nearest 5″), both in reasonable agreement with the earlier results. Taking the results of all three days together, the standard deviation of a single measured distance is 10″.

Table 4.2 Declinations derived from meridian altitudes observed by Halley and Hevelius

Star	Cat dec. (1660)			Halley						Hevelius						D	
			O		R		Δ		O		R		Δ				
	deg	min s	deg	min s	deg	min s	min s	deg	min s	deg	min s	min s	min	s
I	−2	54 53	−2	53 08	−2	54 03		+50	−2	55 08	−2	56 03	−1	10	+1	60
II	38	30 46	38	31 52	38	30 57		+11	38	31 22	38	30 27		−19		+30
III	26	35 27	26	35 13	26	36 09		+42 (refraction of 2′ 35″)								
IV	13	26 00	13	27 52	13	26 39		+39								
V	2	31 05	2	34 52	2	33 01	+1	56	2	32 52	2	31 01	−04	+2	00
VI	9	52 16	9	55 52	9	53 24	+1	08	9	53 52	9	51 24	−52	+2	00
VII	8	02 16	8	07 32	8	04 52	+2	36	8	04 52	8	02 22	+06	+2	30
VIII	5	38 15	5	42 52	5	40 30	+2	15								
Means							+1 17						−38	+1	48
Standard deviation of single difference (s)							45					33		45

Identification of stars: I: Cauda Serpentis; II: Lyrae; III: Humeri sinistri; IV: Cauda Aquilae; V: Brachii antinoi dextra (of Aquila); VI: Humeri Aquilae; VII: Lucidae Aquilae; VIII: Pectore Aquilae.
The declinations correspond to the elevation of the equator as seen from Danzig being 36° 27′ 38″. They are reduced to 1660 by the precession in declination.
Cat. dec. 1660: Value in catalogue of 1660;
O: observed declination; R: declination reduced to 1660;
Δ: (R-cat. 1660); D: Halley minus Hevelius.

Halley's telescopic instrument gave a very different result:

6 July	Hevelius, Buthner	33° 25′ 00″
		33° 28′ 00″
		33° 27′ 00″
		33° 27′ 30″
15 July	Halley	33° 27′ 00″.

Discounting the first observation, which was the first time that the instrument had been used by Hevelius, the mean value is 33° 27′ 20″ (to the nearest 5″) and the standard deviation is about 30″.

The overall conclusion is that observations with the large sextant are consistent to within about 10″. There is however a considerable difference of about 4′ between results with that instrument and Halley's, and even greater differences appear in other observations as the following summaries show:

12 June	E	Hevelius, M.O.	23° 52′ 45″
		Hevelius, M.O.	23° 52′ 40″
		Halley, his quadrant	23° 47′ 30″
	F	Hevelius, Typogr., M.O.	55° 19′ 00″
		Halley, Typogr., M.O.	55° 19′ 00″
		Hevelius, Halley, M.O.	55° 19′ 00″
		Halley, his quadrant	55° 11′ 00″

On 22 June three observations of distance G with the M.O. gave a mean value of 45° 52′ 55″, while Halley's result with his quadrant was 45° 50′ 00″, and the mean of three measurements of distance O on 15 July with the M.O. was 26° 52′ 53″, whereas the mean of three measurements of the same distance with Halley's instrument on 15 and 17 July was 26° 50′ 50″. Hevelius and Buthner made the observations on 17 July. The differences between the instruments are substantial and evidently there were large systematic errors in one or both.

The observations of meridian altitude (Table 4.2) are more consistent, as the comparison of the means and standard deviations shows. The uncertainties of single observations are much the same for the two instruments, about 45″, and the discrepancies are close to 2′, but now Halley's quadrant gave the larger angles instead of the smaller. The systematic errors of Hevelius's quadrant (Q.p.o) evidently differed from those of the Magno orichalcico. The systematic errors of Halley's small telescopic quadrant seem not to have been much greater than 1′ 30″.

Halley gave his own opinion in his letter to Flamsteed of 7 June before the comparisons between his quadrant and the Magno orichalcico.

Table 4.3 The diameter of the Moon, Danzig, 23 June 1679

	Time h	min	s	Distance from limb deg	obs min	s	residual sec
Limb A	11	12	30	35	52	00	−15
	11	15	00	35	51	00	+15
	11	16	30	35	50	30	+10
	11	18	30	35	49	00	−25
	11	21	30	35	48	20	+20
Limb B	11	25	30	35	13	40	0
	11	30	00	35	11	40	+10
	11	31	30	35	10	35	−15

The residuals are calculated from the following parameters: distance of centre of Moon at mean time: 35° 48′ 00″
diameter of Moon: 32′ 30″
decrease of distance of centre of Moon: 0.472′/min

But as to the distances measured by the Sextans, I assure you I was surpriz'd to see so near an agreement in them, and had I not seen, I could scarce have credited the Relation of any; Verily I have seen the same distance repeated severall times without any fallacy agree to 10″, and on Wednesday last I myself tryed what I could doe, and first I at the moveable sight, and the Printer at the fixt, did observe the distance of Yed Ophiuchi from Lucida Aquila 55°-19′-00″; then we removed the Index, and my Lord at the moveable sight, and I at the fixt, did observe the same 55°-19′-05″, and you will find the same distance six times observed in Page 272 of ye fourth book of his Machina Coelestis, so that I dare no more doubt of his Veracitye.[15]

On 23 July they determined the diameter of the Moon by measuring the angular distance of each limb from a star in Aquila. The list of values for the diameter is not presented in such a way that uncertainties of the observations can be estimated. The times of observation and the measured distances are summarised in Table 4.3, which also lists the residuals from a least squares fit of three parameters, a mean distance between the star and the centre of the Moon, the diameter of the Moon, and the rate of change of the distance from the star (0.47 arcmin/min). The results show once again the consistency of the large sextant, with a standard deviation of 15″ comparable to those of other observations, although the limbs of the Moon are not as well defined as stars. The apparent diameter of the Moon depends upon its distance from the Earth, but the estimated rate of change of distance is a little low, the true value corresponding to the lunar period of 27.32 d being about 0.55 arcmin/min.

The observations of Halley and Hevelius and their Danzig colleagues give a rare insight into the methods and precision of observations before mural quadrants with telescopic sights came into general use. Flamsteed was the first to study systematic errors carefully, to urge the fundamental importance of meridian altitudes, and to insist on having assistants to repeat observations and calculations, as well as arguing strongly for telescopic instruments. He would have been confirmed in his views by what Halley would have told him of Danzig. Even such perceptive observers as Hevelius had the most rudimentary ideas of how to arrange observations to eliminate systematic errors and of how to estimate best values of parameters from observations. A quite elementary point is that as the Moon changes its position in the sky at a regular rate, the observations for its diameter should have been completed with repeated observations on limb A following those on limb B, which would have greatly improved the determination of the motion of the Moon and of its diameter.

When Halley was about to leave Danzig on 18 July, Olhoff told him that Hevelius would appreciate a written statement of his opinion of the methods and instruments he had seen. Halley then wrote the assessment mentioned above, in effect a certificate that Hevelius's results were highly reliable; Hevelius printed it in *Annus climactericus*.[16] Halley confessed that when he first began to pursue astronomy he thought that collimation with bare sights would be uncertain by some minutes, and did not know why Hevelius opposed the use of telescopes to measure angles. He did not, as a young man, dare to express his doubts but while he was still worried by them, reports came that Hevelius was going to publish his observations and that would allow a much improved catalogue of the fixed stars to be made. He expressed his pleasure at seeing the instruments that Hevelius had described in the first part of *Machina coelestis*. He can testify to the certainty of the instruments against all who would cast doubt on Hevelius's observations. He had seen with his own eyes that not just occasional observations but many by different observers with the large sextant agreed incredibly well together and any discrepancies were very small, which he greatly admired.[17]

Although he said he was impressed by the performance of the large sextant, Halley did not quote any numerical values, nor did he make any comparisons with other observers with different instruments. Hooke must also have been confirmed in his criticisms of Hevelius by Halley's account of his visit, for when Halley was back in London on 14 August, he noted in his '*Diary*, Halley returned this day from Dantzick (Hevelius Rods in pisse)'.[18]

4.3 After Danzig

Two months after Halley's departure a fire that Hevelius thought was started maliciously by a servant destroyed his observatory. They thought in London that Hevelius might have perished in it. In that belief Halley wrote to Olhoff 'Words fail me to express the sense of sorrow I have at the news of the death of Lord Hevelius', and he reported the death of Sir Jonas Moore about the same time. Before he left Danzig Elizabeth Hevelius had asked him to have a gown made for her in London, and he continued 'I know that his very sad wife must be walking in mourning clothes'. Nevertheless, for a number of reasons he would send her the dress: he is not sure that her husband is dead; and he would lose a lot of money if he tried to sell it in London. 'It is of silk and of the very latest fashion and I am confident it will please the Lady Hevelius if she will consent to put it on'. Finally, it would keep without damage until the time of mourning is over. The dress was made of ten yards of outstanding silk and the petticoat of eight yards of silk. The total cost, with the tailor's time and minor charges, was £6 8s. 4d., or $27\frac{1}{2}$ Imperiales*. Halley had eighteen Imperiales in hand to buy books to send to Hevelius, and so he was owed $9\frac{1}{2}$ Imperiales. If the Lady Hevelius would be pleased to send him a copy of the *Selenographia* and one of the *Cometographia* he would consider his account settled. He hopes to hear whether Hevelius is in fact dead for he has to send him some other books and telescope lenses that he has bought for him. He sends his condolences to Mme Hevelius and his duty to Mme Olhoff and Olhoff's parents-in-law and warm expressions of affection to Olhoff.[19]

If ten yards of silk for a gown and eight for a petticoat be thought excessive, the picture of Mme Hevelius observing (Fig. 4.1) may make it more reasonable. The waist was then low and the skirt was open at the front to display the petticoat, which had to be of as fine material as the gown.

Hooke, according to his *Diary*, heard by 20 November that Hevelius was alive, but Halley was then in Oxford[20] and evidently thought, when he left for France in December 1680, that Hevelius was dead. Hevelius had written to Halley in 1681, acknowledging Halley's letter to Olhoff and saying that happily he was alive. He asked Halley to obtain some lenses for long telescopes and other items that he was wanting for rebuilding his observatory. He wrote of the settlement of Halley's

---

*The Imperialis was presumably the currency of Danzig and one pound sterling would be about $4\frac{1}{3}$ Imperials.

expenses for Mme Hevelius's gown and reported his observations of the comet of 1680 on 2 December, the comet that Halley had by then observed in France. The letter reached Halley's father in London after Halley had left for France and it was passed on to Flamsteed who wrote to Hevelius on 7/17 October 1681 to say that he would get the telescope lenses and other items that Hevelius wanted.[21]

When Halley reached Rome in October 1681 he learnt from correspondents in England that Hevelius had not died. He wrote to him on 5/15 November to say that he had heard with joy that the report of his death was incorrect, and concluded with greetings to Mme Hevelius and others: 'Saluez de ma part (je vous en prie) Madame votre femme, la famille de M Olhoff et les deux Butners père et fils' ('I beg you to greet Madame your wife on my behalf, Mr Olhoff's family and the two Butners, father and son'). After he was back in London, and immediately before his marriage, he sent Hevelius lenses, a micrometer, books, and charts as he advised him in his letter of 7/17 April.[22]

Halley, Hevelius, Mme Hevelius, and the Olhoff family must all have found his visit to Danzig very agreeable.

Many years later, when Halley was Savilian Professor in Oxford, Thomas Hearne, Bodley's Assistant Librarian who knew him well, noted the arrival of his portrait by Thomas Murray (Plate XV) in the Gallery of the Bodleian Library:

> It hangs by Hevelius whom Dr Halley, when he was young, had visited at Dantzick, and for that reason, as well as his skill in Astronomy, Hevelius hath mentioned him in one of his books. And some Persons say that he is very justly placed by Hevelius, because he made him (as they give out) a Cuckold, by lying with his wife when he was at Dantzick, the said Hevelius having a very pretty Woman to his Wife, who had a very great Kindness for Mr Halley and was (it seemed) observed often to be familiar with him. But this story I am apt to think is false.[23]

The circumstances in which Hearne came to record that story are discussed in Chapter 12. Elizabeth Hevelius was very much younger than Hevelius but some ten years older than Halley; she took part in observations with Halley along with Hevelius and Olhoff, as she had done with other visitors in the past. Back in London Halley bought her a dress for which he expected to be paid—he did her a service but did not make her a gift. When he wrote to Hevelius from Rome he sent his greetings to Elizabeth in the same phrase as those to the Olhoffs. No doubt the worst construction could be put on those meagre records (and if they concerned Samuel Pepys or Lord Brouncker it might be reasonable to do so), but on the face of it they amount to a friendly but not an adulterous acquaintance. A pleasant time may have been had by all, but not obviously improperly.

Halley corresponded amicably with Hevelius for some years after his visit, but his views of the instruments and procedures that he had seen at Danzig were much more reserved than he had given Hevelius to believe.[24] Even so, the favourable account that he gave of them when he returned displeased some in England. Wallis discussed Hevelius's observations in 1686 and wrote to Flamsteed:

> As to Mr Halley, if you think (as you seem to intimate) that he hath been too lavish in his commendations, you must needs think that Mr Hook hath been so in his reprehensions.

William Molyneux wrote similarly to Flamsteed from Dublin.[25]

Halley evidently agreed with Molyneux that Hevelius was a peevish old man who did not understand the principles of telescopic sights. He was vexed that Hevelius had described his own instruments incorrectly in the *Annus climactericus*, calling his small quadrant a sextant, and in other ways giving a misleading account of his expedition to St Helena. In a letter to Molyneux of 26 March 1686, in which he set out the difficulties over the secretaries of the Royal Society that led to his being appointed Clerk, he wrote:

> ... the Controversy between Mr Hevelius and Mr Hook, as you very well observe does, as Hevelius manages the matter, affect all those observers that use Telescopic sights, and myself in particular, and it is our common concern to vindicate the truth from the aspersions of an old peevish gentleman, who would not have it believed that it possible to do better than he has done, and for my own part I find myself obliged to vindicate my observations made at St Helena and to rectifie some mistakes, whether wilfull or not I cannot say; first he sais I was sent by R.S. to St Helena at his request to observe the Southern Starrs (pag.14) whereas it is very well known to all our Astronomers that at my own motion and charge I undertook that voiage above two years, before I had the honour of being a member of the R.S. all of which I have declared in the preface to my Catalogue. Again he says pag. 18 of the Preface that I was sent with a sextant with Telescopic sights for no other purpose but rigidly to examine his instruments ...

He again wrote to Molyneux on 27 May 1686:

> As to Mr Hevelius we heare as yet no farther from him, and I am very unwilling to let my indignation loose upon him, but will unless I see some publick notice taken elsewhere, let it sleep till after his death if I chance to outlive him, for I would not hasten his departure by exposing him and his observations as I could do and as I truly think he deserves I should.[26]

Nine years had passed since Halley was in Danzig and he was still indignant at the garbled account that Hevelius had given of his southern expedition in subsequent publications. As the first person to have compiled a stellar catalogue from telescopic observations, any misrepresentation of his work cast doubt on his scientific competence and on the new methods.

Hevelius thought that the collimation length of a telescope with cross hairs in the eyepiece was the distance from the observer's eye to the cross hairs whereas, as William Molyneux remarked, it is the distance from the cross hairs to the objective. Thus, the effective length of the telescope on Halley's quadrant was about 2 ft (0.6 m), much less than those of Hevelius's instruments, but compensated by the magnification of the telescope.

Errors of pointing were not the only uncertainty; there were also considerable errors of reading the scales of the quadrant or sextant, and the scale of Halley's small quadrant would almost certainly be less precise than that of Hevelius's large sextant. Errors of mounting the scales would lead to further systematic errors of distance or meridian altitude.

Halley's 2 ft quadrant might have been no better than Hevelius's 6 ft sextant.

Hevelius did obtain remarkably reproducible results, probably the best possible with open sights, but his instruments and procedures were out of date and suffered from substantial systematic errors.

Halley wrote to Hevelius in the autumn of 1686 in reply to letters of 9 July and 17 September that seem no longer to exist; those are the last known letters between them. Halley sent details of an occultation of Jupiter by the Moon that took place on 10 April, as observed by Cassini and others at Paris, and of a recent one of Saturn and its satellites. He says that books of Hevelius that are on the way are eagerly expected and, with compliments from the Royal Society, ends;

> Vale, Vir dignissime, ac eodem, quo sideralem Scientiam affectu, meque Societatemque nostram tui amantissimam prosequi digneris.
>
> Devinctissimus tuus
> E.H.[27]

Farewell, most honoured Sir, and, just as you have endowed the science of the Heavens, so may you also deign to honour both me and our society which holds you in such great affection.

> Your most devoted servant
> E.H.

A polite letter at the least; next year Hevelius would be dead.
Halley had written to Flamsteed from Danzig on 7 June 1679,

> I hope about 3 weeks hence to goe for Denmark, ther to observe somethinge for the difference of Meridians of Uraniburge from London: wherfore I entreat you to be more than ordinarye intent upon the occultations and appulses dureinge my absence ...

There is no record of his going to Tycho's observatory.[28]

It seems that Halley and Flamsteed had some arrangement in those years to make joint observations of lunar positions through occultations and close approaches (appulses) of the Moon to stars.

4.4 Six months in France

The King's astronomer in Paris was J.-D. (or Gian Domenico) Cassini (1625–1712), who had been in charge of the Observatoire since 1669 (Plate V). He was born near Naples in 1625, became professor of astronomy at Bologna in 1650, and held responsible engineering posts in the city and in the Papal states. He had measured the periods of rotation of Venus, Mars, and Jupiter, and had seen that Jupiter was flattened at the poles. He discovered four satellites of Saturn, the elements of which Newton was to use in the *Principia*. He had observed the Medicean satellites of Jupiter and had produced tables of their orbits with the idea that they might be used as clocks in the determination of longitude—as Galileo had already proposed when he discovered them. Cassini and Römer found anomalies in the observations of the periods that Römer suggested were the consequence of a finite speed of light; Cassini was not convinced.[29] Cassini, like Hevelius, had made careful observations of the Moon and had produced a map in considerable detail, sometimes spurious. Although Halley met the most committed students of the Moon of his day, he himself never seems to have taken any great interest in its appearance or structure or libration; it was with the Moon's motion in its orbit that he was so deeply concerned throughout its life.

The new Observatoire, to the design of Perrault, was part of the Académie Royale des Sciences, recently set up by Colbert, to which he had brought four notable astronomers, Picard, Richer, and Huygens and Cassini (Fig. 4.2). Picard had determined the diameter of the Earth from measurements of the length of a degree in France in 1669–1670.[30] Richer, besides his observations of stars from Cayenne, had noted the change in length of the pendulum beating seconds at Paris, Cayenne, and on Gorée, near the Cape Verde Islands (Chapter 3). Newton knew of the geodetic results of Picard and Richer and used them in *Principia*; Halley may have told him of them (see below). Picard had set out how a mural quadrant should be used for fundamental measurements of celestial positions, and Halley was visiting the Observatoire just at the time when the mural quadrant there was being installed and brought into use in 1681. His own experience on

Figure 4.2 Louis XIV and Colbert at the Académie des Sciences, with the Observatoire (unrealistically) in the background (from Académie des Sciences, *Recueil d'observations faites en plusieur voyages par ordre de Sa Majesté*). (Reproduced by permission of the University Librarian, Cambridge.)

St Helena, followed by his visit to Hevelius and now his presence in Paris, must have given him unrivalled knowledge of the most advanced techniques of astrometry.

Cassini was very well known to English astronomers and maintained a regular correspondence with them. He would have been familiar with Halley's observations in St Helena: the French version of Halley's account, *Catalogues des estoilles Australes*, had appeared already in 1679 and there is a copy in the library of the Observatoire. Hevelius had kept him informed of Halley's stay in Danzig. Römer had visited London from Paris in 1679 and had met Newton and Flamsteed.[31] Cassini would surely have welcomed Halley's visit.

Halley set out for France at the end of 1680.[32] His father allowed him £300 for his expenses. He carried with him a quadrant for the observation of latitude, probably the one he had on St Helena and at Danzig. He travelled with his schoolfellow, Robert Nelson (1656–1715), with whom he would maintain a lifelong friendship, as he wrote to him fifty years later of 'the uninterrupted course of friendship which has always subsisted between us since our childhood'.[33]

Nelson was the same age as Halley. His father, who died when he was young, was a Levant Company merchant, his mother, Delicia Roberts, was the daughter of another Levant merchant and the sister of Sir Gabriel Roberts, also a Levant merchant, who was Robert's guardian.[34] Robert was admitted to the Levant Company in 1676[35], was elected to the Royal Society in 1680, and was later a member of the Council at the same time as Halley and Sloane. He became very well known as the author of *A companion for the festivals and fasts of the Church of England*,[36] and as a philanthropist. He was a founder of the Society for the Promotion of Christian Knowledge and of the Society for the Propagation of the Gospel in Foreign Parts, he contributed to libraries for country clergy and helped to relieve foreign Protestants in Canterbury. He and his mother each contributed 1 guinea to the rebuilding of St Paul's. In 1703 he was soliciting subscriptions for the construction of an English church in Rotterdam for the benefit of military and civilians in the Netherlands with the Duke of Marlborough.[37]

In Rome with Halley, Nelson met his future wife, Theophile, the daughter of Sir Kingsmill Lacy and the widow of the first Earl of Berkeley. He married her in England on 23 November 1682; she later became a Roman Catholic. From 1688 to 1691 the Nelsons travelled in Europe for Theophile's health. When they returned to England, Nelson became a nonjuror. Pepys, also a nonjuror, asked Nelson in March 1703 to help him find a non-juring clergyman in his neighbourhood (Clapham Common) to minister to him. Nelson recommended Nathaniel Spinkes who lodged with a glazier in Winchester Street and managed a fund for the relief of non-juring clergy who had been ejected from their livings. Halley surely knew Spinkes.[38]

Apart from Nelson and Pepys, Halley knew a number of other nonjurors, such as Sir Anthony Deane, the shipwright and friend and colleague of Pepys and Evelyn, and Dodwell who had been ejected from the Camden Chair of History at Oxford. Although he could not swear allegiance to William and Mary, Nelson was active in the interests of the established Church of England, which suggests that some of the religious divisions that figure so largely in later accounts of the period were not so strong at the time.[39]

Nelson and his mother and uncle were close to John Tillotson, then Dean of Canterbury, subsequently Dean of St Paul's after the accession of William III, and Archbishop of Canterbury when the non-juring Sancroft was ejected.[40] He married a granddaughter of Oliver Cromwell; his mother-in-law, Cromwell's daughter, married as her second husband Bishop Wilkins, a founder fellow of the Royal Society. Tillotson was one of the clergy who worked for the accession of William. Nelson looked after him in his last illness and he died in Nelson's arms.[41] His correspondence with Nelson in France contains some news of Halley.

Halley must have been approved by Nelson's family and friends such as Tillotson for them to accept that the two of them should travel together to France and Italy. While they were in France, Nelson had an offer of a place at Court, which Tillotson and others urged him to turn down on account of the moral danger of the Court. Would they have agreed to his travelling with Halley if Halley was then the flippant atheist he was later supposed to be? Tillotson in fact shows by his letters that he had a high regard for Halley, who must also have been well known to the Roberts and their circle. Hooke's *Diary* shows us Halley as the coffee house habitué for more than twenty years in the company of Hooke, Wren, Gale, Aubrey, and others of the Royal Society. Halley's tour with Nelson and the letters of Tillotson show us Halley in the circle of devout, serious, cultured, and cheerful members of the Church of England who looked to Tillotson as their spiritual guide.

Halley and Nelson set out for France on 1 December 1680. They stayed with Tillotson in the Deanery at Canterbury before going to Dover for the packet to Calais, but did not reach Paris until Christmas Eve:

> I got hither the 24th of the last month after the most unpleasant journey that you can imagine, having been 40 hours between Dover & Calais with wind enough.[42]

Tillotson gives more details in a letter to Nelson:

London, 5 January 1681

Your letter was most welcome to me because I was in great paine till I heard of yo^{re} safe arrival at Paris, wch together with the sight of the Great

King must needs make amends for all the difficulties and distresses of your journey. When I saw the change of weather wch happen'd immediately after yo^{re} departure from us I was not a little pleas'd with myself that my reason had prevail'd against my inclination so far as not to importune you for a longer stay with us, for had you tarried till Tuesday you had been upon the Sea in that storme in which three Merchant Ships in the Downs were cast away. I thank God heartily that you escaped that danger. The Comet hath appeared here very plaine for several nights with a stream much the length you describe it. What it portends God knows; the Marquess of Dorchester and my Ld Coventry dyed soon after ...

My humble service to your learned Friend and Companion. I have not yet received his favo^{re} but shall be glad to see any thing of his & and much more to be able to understand it.[43]

That last remark is often echoed by non-scientists who are given learned works by natural philosophers.

According to the '*Memoir*' of Halley which was apparently written by Martin Folkes,

> ... I remember to have heard him say, he was on the road between Calais and Paris when he first saw the Famous Comet of that year in its return from the Sun; he had already seen it in its going down to the Sun in the month of November before his setting out.[44]

Halley first saw the comet in France just before Christmas on the road to Paris

> for Mr Hally writes me word in his first letter that Dec8/18 in ye Morneing being on his journey to Paris near Bologne before Sun rise, *hee saw the tayle of the Comet riseing as it were perpendicularly from ye horizon.*[45]

Newton has a fuller report of Halley's observation in his Waste Book, a common-place book of his stepfather with unused pages on which he entered various notes and other material. After a summary of the observations of the comet of 1682 up to 9 March there are detailed accounts of some of the observations of the comet of 1680, Halley's among them:

> On December 8 old style in the morning, our Halley, travelling towards Paris near Boulogne ...

and goes on to describe how Halley saw the vertical tail. A fuller account concludes the reports:

> Halley has told me that on his journey to Paris on Dec 8 old style he saw the tail of the comet rising perpendicularly from the horizon ...

The way in which Newton records Halley's account—'Halleius mihi narravit' 'Halley has told me',—may indicate that Halley met Newton to

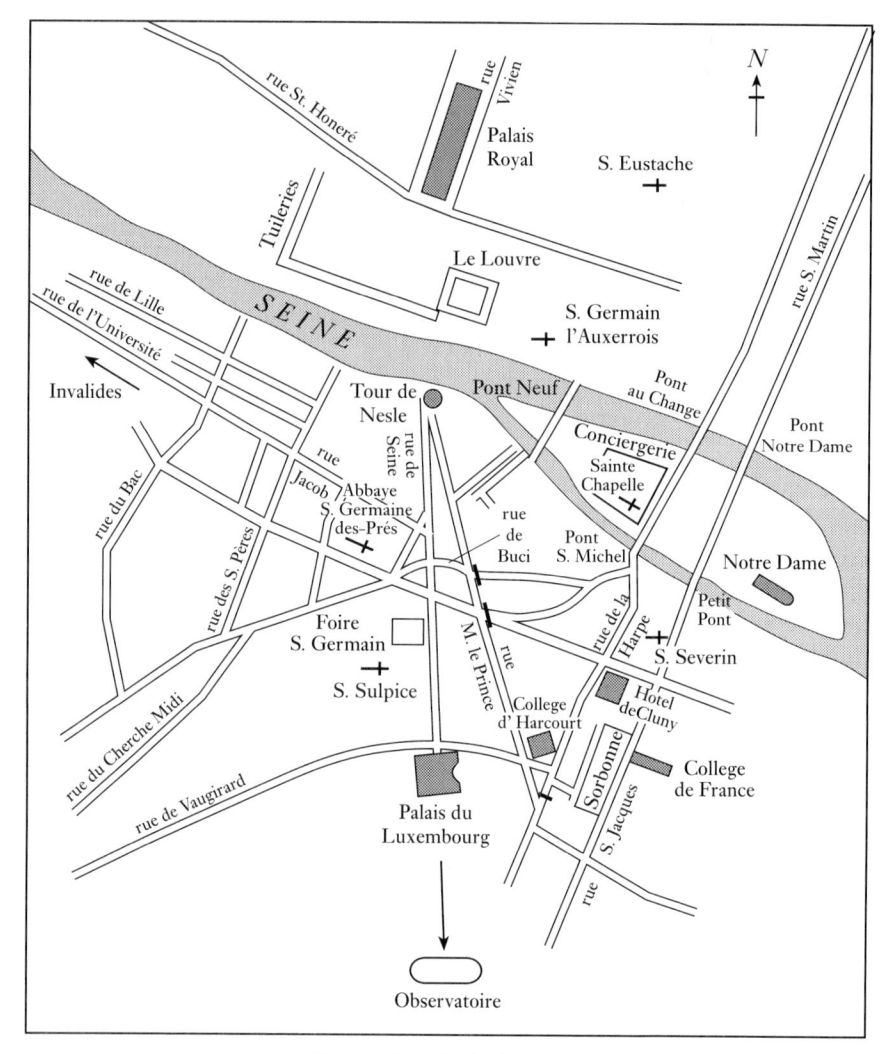

Figure 4.3 Halley's Paris.

report what he had seen and did so sometime early in 1682 after he returned from Paris. If so it would have been Halley's first meeting with Newton.[46]

Halley arrived in Paris on 24 December 1680 (ns) and spent the winter and spring of 1681 in the quartier St Germain (Fig. 4.3). Tillotson's letters to Nelson in Paris were addressed 'rue de Seine, au Faubourg St Germain, à l'hostel de Baviere, à Paris' and perhaps Halley stayed there also. When he left Paris for Saumur and Italy, Halley asked that letters should be sent to him addressed, 'A Monsieur Monsieur Mareshall à la Ville de Venise, rue de Bussy au Faubourg St Germain pour fair tenir a Mons. Halley à Paris';[47] it may be that he stayed there separately from Nelson or that he intended to stay there when he returned to Paris on his way back to England. When Nelson was in Paris in July 1682 he stayed in the hostel Imperial in the nearby rue du Four.[48]

The quartier St Germain was outside the walls of Paris south of the Seine. The rue de Seine (6e) runs south from the Seine near the site of the notorious Tour de Nesle. The rue de Buci is close to St-Germain-des-Prés and runs from the Carrefour de Buci, opposite the old Porte de Buci, where there was a gallows in the seventeenth century, to the rue du Four which it meets on the present Boulevard St Germain. A pillory stood there in Halley's time. The Ville de Venise, at the present No. 25, close to the junction with the rue de Seine, was an inn. John Evelyn had stayed there for a month when he first went to Paris during the English Civil War 28 years earlier,[49] so English visitors may have patronised it. Today the ground floor is occupied by the Restaurant d'Arbuci. In Halley's day the rue de Buci, now a busy street full of splendid food shops in tempting variety, was inhabited by a great variety of people and avocations. There were two other inns, a tennis court at No. 12, and the first coffee house in Paris was opened at the present No. 28 in 1675.

Although the rue de l'Université dates from the middle ages and the present Hotel de l'Université at No. 22 incorporates remains of a house of the Knights Templars, the district began to be developed in the sixteenth century. Molière's first playhouse was in the rue de Seine, where the grandest building was the Hotel de Liancourt. The hostel de Baviere cannot now be located. Joseph de Montesquiou, comte d'Artagnan, a cousin of the famous d'Artagnan and also a Musketeer, lived in the rue de Seine when Halley was in Paris. The English ambassador lived not far away, probably at the site of the present Hôtel d'Angleterre on the rue Jacob.[50]

The quartier St Germain stretched as far south as the Palais du Luxembourg and west to les Invalides. A guide book published just after Halley's time said of it:

> LE QUARTIER SAINT GERMAIN est sans contredit plus beau et plus grand que les autres à cause de son étendue, du nombres de ses belles Maisons & de la quantité de peuple que s'y trouve ...
>
> The St Germain quarter is undoubtedly larger and more beautiful than the others because of its extent, the number of fine houses and the number of people one finds there ...

and observed that strangers preferred that part of the town to all the rest of Paris. Some still do for its streets with distinguished book shops and expensive art and antique shops.

On a quelque fois conté dans un Hyver douze Princes Étrangères, & plus de trois cens Comtes ou Barons, sans un bien plus grande nombre des simples Gentilshommes, que la reputation de France attiroit pour apprendre notre Langue avec un soin extrême.[51]

In a single winter there have often been counted twelve foreign princes, and more than three hundred counts or Barons, besides a considerably larger number of ordinary Gentlemen, whom the reputation of France has attracted to learn our language with extreme care.

The remains of Roman Lutetia, the baths in the mansion of the abbots of Cluny, the arena, and an Imperial palace lay to the east of St Germain. The grandiose Hôtel des Invalides rose at the west end of the rue de l'Université. The new Observatoire was south of the Luxembourg palace.

The Abbey of St Germain-des-Prés was the chief house of the reformed Benedictine congregation of St Maur. It held one of the greatest libraries of France until the Revolution destroyed it, and the collection of manuscripts exceeded anything else in France. Stationers and booksellers flourished around it. The Maurists were distinguished for their critical scholarship, especially in palaeography, diplomatic, and antiquities. Fathers such as Mabillon and Montfaucon had a high reputation; Newton, for example, had books of theirs in his library. Their scholarship, distinguished and impressive as it was, was not alone. The University Press in Oxford had a somewhat similar, though far, far smaller, programme for the publication of editions of the Greek and Latin classics including Greek mathematical works (Chapter 13). John Aubrey's classification of medieval architecture by age and style was akin to Mabillon's treatment of palaeography. Halley, in Rome, investigated Greek and Roman weights and measures in the manner of Montfaucon and Muratori and returned more than once to antiquarian studies, whether for their own sake or for what they could tell of the natural world. (In 1720 he became a fellow of the Society of Antiquaries.) The many editions of the Maurists were in the forefront of the intellectual movements of the age and, together with Newtonian natural philosophy, were a main source of the Enlightenment. Halley may or may not have met any of the fathers but he could have seen their books in the shops and almost certainly heard something of their activities.[52]

No doubt it was from the printers and booksellers who clustered around the Abbey that Halley purchased the books and charts that Hooke, the Royal Society, and other correspondents asked him to obtain. He was evidently active in doing so. He got to know booksellers, for he writes to Hooke from Saumur of '*an acquaintance of mine, a Bookseller there*', and called in the aid of Cassini and Henry Savile, the English ambassador, whom he came to know.

The Luxembourg Palace (Palais d'Orleans) with its attractive gardens lies almost due south of the rue de Buci, and south of it again is the Observatoire, founded in 1669 and a grand building (Fig. 4.2). It stands at the edge of a plateau which falls away to the south. Now it is at the end of a wide tree-lined avenue and surrounded by houses; then it was in open country. It was approached by the rue St Jacques, where the printer to the King, who had published the French version of the catalogue of southern stars, had his shop. At the Observatoire Halley observed the great comet with Cassini; he probably would have met Picard, Richer, and Huygens who were all in Paris in 1681, and he might have met Römer again.

Such then was the agreeable and distinguished quarter in which Halley spent almost six months at the beginning of 1681. Comets apart it was an interesting time. *Tout Paris* had been intrigued and alarmed by the Affair of the Poisons a little earlier. Some disreputable women had provided aphrodisiacs and poisons to a wide range of acquaintances. After an extensive enquiry and the execution of the principal purveyors, the matter was dropped because it came too close to royal circles, especially to Mme de Montespan and her associates. The move to Versailles was the other exciting matter. For some time the Sun King had been preoccupied with Versailles and now it was almost ready. In May 1682, not long after Halley had returned to England, Louis moved the Court permanently there. Halley surely saw the sights as he did in Danzig, and Versailles must have attracted his visits.

Halley was in Paris just as Louis XIV had obtained control of the ten cities of Lower Alsace by legal manipulation and of Strasbourg by force. His dragoons were intimidating French protestants and bringing about forced conversions that prepared the way for the Revocation of the Edict of Nantes four years later. Those events would between them bring to England notable Huguenots such as Henri Justel and Abraham de Moivre, both to become distinguished fellows of the Royal Society and well known to Halley.

Hooke introduced Halley to Justel.[53] Justel had for many years exchanged scientific news between England and France through his correspondence with Oldenburg at the Royal Society. It was one of his many activities. He held weekly meetings at his house for residents in Paris and visitors from foreign countries as well as from the French provinces. Later, when Halley was Clerk to the Royal Society, Justel maintained a considerable correspondence with him. Justel settled in England a few years after Halley's stay in France, and William III appointed him his librarian.[54]

Very many Englishmen had been in Justel's circle, often with a letter of introduction from Oldenburg. Christopher Wren, William Croune,

Samuel Pepys, the Earl of Clarendon, Algernon Sidney, John Locke, and John Covell, the chaplain of the Levant Company, all had called on him. In Justel's house Halley would have met many notable French savants. He mentions one only in his letter to Hooke, Nicolas Thoynard of Orleans, but that was only shortly after he had arrived in Paris. He might well have met, for instance, the astronomer Auzout, who was one of the first to use telescopes in astrometry.

The friends moved in exalted circles. They came to know the English ambassador, Henry Savile, possibly an acquaintance from London. Tillotson's letter to Nelson of 7 March 1681 reveals their acceptance in the highest Parisian society:

> I thank you for the particular account you are pleased to give me of your occurrences, among which nothing pleases me more than the condescension of your great Cardinal in honouring yo^{re} St Bartholomew Faire with so secular a kind of preference and deportm't.

The great Cardinal was the Cardinal d'Estrées, about to return to Rome as the French ambassador to the papal court. Aubrey reported that he had introduced Halley to his brother, the Admiral[*], and that he had 'caressed' him, that is, in the sense of the time, treated him with great courtesy and distinction.[55]

In the same letter, Tillotson has news of the comet:

> I put you some charge by the enclosure, wch yet I hope will be acceptable to you. It is the observations of our Canterbury Astronomer, Mr Hill of the late Comet, wch he told me within this last fortnight appear'd still, but was very little. He is not a learned man but very industrious. I submit it to Mr Haley's better judgem't …
>
> My mother and my wife pres'nt their humble service to you. Mine I pray to Mr Haley.[56]

The comet of 1680–1681 was spectacular and according to Mairan 'si remarquable par sa grandeur & si terrible aux yeux d'un vulgaire' ('So remarkable for its size and so dreadful in the eyes of ordinary people'), while Mme de Sévigné wrote of its beautiful tail and of the alarm of great personages.[57] It was not Halley's comet, which was that of 1682, but it did lead Cassini among others to speculate that comets might return, and many astronomers were active making observations and trying to fit orbits. As soon as he arrived in Paris, Halley joined Cassini in observing the comet and they continued to do so together whenever it was visible.[58] He wrote to Hooke on 15 January 1681

> The generall talk of the virtuosi here is about the Comet, which now appears, but the cloudy weather has permitted him to be but very seldom

---

[*]The admiral was a distinguished sailor who had commanded French squadrons in alliance with the English in the Dutch wars, notably at Sole Bay.

observed, Whatever shall be made publick about him here, I shall take care to send you, and I hope when you shall please to write to me you will do me the favour to let me know what has been observed in England.[59]

Cassini collected observations of the comet in a pamphlet that he presented to Louis XIV and he included Flamsteed's, which apparently came to him through Halley. Halley may have referred to Cassini's tract when he wrote to Hevelius from Rome about a collection of all the observations, his own as well as those made in England, Germany, and Italy. He wrote to Flamsteed about 22 January with details of some of the observations of the comet that had been made in Rome by Cellio and Gallet (who in fact observed in Avignon) and in France by Cassini and himself.[60] The letter is lost but on 17 February Flamsteed replied with his idea that 'the Sun attracts all the planets and all like bodies that come within our Vortex'.[61] Notions of comets at this time were very varied. Was the comet seen at the end of 1680 the same as that seen early in 1681? Newton and Flamsteed debated whether the comet passed behind or in front of the Sun between the two apparitions. Of what form were the orbits, straight lines, great circles, parabolas, or ellipses, and did comets reappear after some time, as Cassini suggested?

After leaving Paris, when he was at Saumur, Halley wrote to Hooke on 29 May:

> Monsieur Cassini did me the favour to give me his booke of ye Comett Just as I was goeing out of towne; he besides the Observations thereof, wch. he made till the 18 of March new stile, has given a theory of its Motion wch. is, that this Comet was the same with that that appeared to Tycho Anno 1577, that it performes its revolution in a great Circle including the earth wch. he will have to be fixt in about 2 yeares and a halfe ... as likewise to that of April 1665. I know you will with difficulty Embrace this Notion of his, but at the same Tyme tis very remarkable that that 3 Cometts should so exactly trace the same path in the heavens with the same degrees of velocity.[62]

All those and many other ideas were in the wind as Halley and Cassini observed the comet in January and February 1681. Cassini's book was in the sale of Halley's books in 1742 (Appendix 8).

In 1695 Halley calculated the orbits of more than twenty comets by the methods Newton had set out in Book III of the *Principia*. The ideas about comets that were current before then will be described in Chapter 8 as an introduction to his contributions. The theory of the orbits of comets and the discussion of the observations of the comet of 1680 take up a substantial part of Book III of the *Principia* and greatly extend the concept of orbits under the inverse square law of attraction. Newton certainly obtained some of the observations from the collection that Halley made in France, and Halley's identification of the elliptical orbit of 'his' comet of 1682 must surely owe something to his stay in France in 1681.

Comets were not the only topic to occupy Halley in Paris. Entries in the Halley papers in the University Library, Cambridge (ULC RGO 2/9, ff. 25–6), show that he observed the Moon early in 1681—it must have been in Paris. He almost certainly heard at first hand of Richer's expedition to Cayenne, the results of which had been published the year before. Immediately after the comet entries quoted above, Newton wrote in his Waste Book:

> Monsr. Richer sent by ye French King to make observations in the Isle of Cayenne (North Lat 5^{gr}) having before he went thither set his clocke exactly at Paris found there in Cayenne that it went too slow as every day to loose two minutes and a half for many days together and after his clock had stood & went again it lost $2\frac{1}{2}$ minutes as before. Whence Mr Halley concludes that y^e pendulum was to be shortened in proportion of—to—to make y^e clock true at Cayenne. In Gorea y^e observation was less exact.

(The gaps are in the MS.)

Richer's own published account is slightly different, that the length of the pendulum adjusted to beat seconds was $1\frac{1}{4}$ ligne less in Cayenne than at Paris, where it was 3 ft $8\frac{1}{4}$ ligne.[63] Halley had Richer's book but it is not in the catalogue of Newton's library.[64] Five years later Newton would explain Richer's observation and Halley's on St Helena in Book III of the *Principia*.

Halley compared the populations of London and Paris, a forerunner of his later work on bills of mortality in Breslau. He told Hooke that he had paced out the dimensions of Paris, 3 English miles south to north from the Observatoire to the Faubourg St Martin and $2\frac{1}{2}$ miles west to east from the Porte St Honoré to St Antoine. The first is a little too great, the latter, roughly from the Madeleine to the Bastille, is about right. In 1680 there were 24 411 burials as compared with less than 20 000 in London. He found births and marriages to be in the same proportion and estimated that each couple would need to have four children to maintain the population.[65]

On 10 May Cassini introduced Halley to the Académie des Sciences, then in the Royal Library in the rue Vivien (now Vivienne) near the present Bibliothèque Nationale:

> Mr Cassini a amené a la Compagnie Mr Hallei qui a fait son rapport de l'observation qu'il a faite dans l'Isle de Ste Helene de Mercure dans le Soleil et qui a fait un nouveau catalogue de presque toutes les principales etoiles australes.[66]
>
> Mr Cassini brought Mr Halley to the meeting, who reported on the observation he made of Mercury in the Sun on the island of St Helena, and who has made a new catalogue of nearly all the principal southern stars.

By 21 May the travellers had moved to Saumur. Tillotson's letter of 2 June to Nelson shows that they lodged with an English lady for it is addressed 'À la Mademoyselle Griggs à Saumur'. She was a friend of the Tillotsons: on 7 November 1681, Tillotson wrote to Nelson, by then in Rome:

> Our friend Mrs Griggs is returned safe with her charge to the unspeakable joy and contentment of Sir Ch. and my Lady. My humble service & thanks to Mr Haley for his Book.

They had an accident on the way to Saumur: in his letter of 2 June, Tillotson wrote:

> I thank God that you receiv'd no harme by the overturning of the Coach to which I suppose the wicked weight of the fat Frier contributed as much, as his sanctity did to your safety. My humble service to Mr Haley & to my good friend Mlle Griggs, whose zeal I extremely commend, but I doubt she hath not pitch'd upon ye most effectual means for ye conversion of France.[67]

How pleasant to have a letter from Tillotson.

The book that Tillotson received from Halley must have been the southern catalogue, most probably in the French translation, for Halley could readily have obtained that in Paris.

The main purpose of Tillotson's letter of 2 June 1681 was to say how relieved he and Nelson's relatives and friends were, that Nelson had not taken up an offer of an office at Court procured for him by Henry Savile and his brother, the Marquess of Halifax. Tillotson waited on the ambassador in London to explain why Nelson had been advised to decline and Tillotson writes of his 'declaring the very great kindness & esteeme he had for you and yore friend Mr Haley'.[68]

Writing to Hooke from Saumur on 19/29 May, Halley set out how they planned to go to Rome:

> ... about the beginning of July or latter end of June, we shall part from hence and go by Rochelle, Bourdeaux, Toulouse and Narbonne, with intention to see the new Canall* which is almost finished, and soe by Montpellier and Avignon into Provence ...[69]

So indeed they did. Halley's itinerary after Saumur is recorded in a note book of Cassini. When he returned to Paris in January 1682 on his way back to London: 'M.Hallei retourna de Rome et me passa la lettre du P. Eschinard' ('Mr Halley has returned from Rome and has given me Father Eschinard's letter'), he left with Cassini a list of places he had passed through, with the dates when he was there and the latitudes as he had observed them. On 15 January 1682 Cassini noted:[70]

---

*This is the Canal du Midi, the great work of Louis XIV, that runs from the Garonne near Bordeaux through Toulouse to the Mediterranean at Narbonne.

M Hallei m'a communiqué cett observation de l'éclipse de lune [Mr Halley has communicated to me this observation of the eclipse of the moon]

1681 August 29   Avenione [Avignon]

2.10.0.	Init. Eclipsus
2.20.12.	tetegit paludem marius
2.22.40.	
2.25.20.	limbus primus nuclei sinni
2.42.50.	cornua perpendicularia
2.48.10.	? ?
2.49.10.	tetegit Aristarchus
2.52.0.	? ?
3.17.45	[end of eclipse]

il me donna aussi les observations suivantes des hauteurs du Pole prise par un quart de cercle de 18 pouces par le Soleil
[he has also given me the following observations of the height of the Pole taken with a quadrant of 18 inches.]

Paris	48.51	Apr 18 1681
Saumur	47.19	May 21
Tholose	43.41	Aug 12
Narbonne	43.15	Aug 18
Montpellier	43.39	Aug 20
Marseille	43.18	Sept 3
Tholen	42.57	Sept 14
Freius	43.19	Sept 21
Gennes	44.16	Sept 27
Legorne	43.27	Oct 7
Florence	43.41	Oct 21
Rome	41.48	Oct 28
Siena	43.14	Oct 23

Halley left very little record of what he did nor of whom he met on his travels through Europe, nor the reasons for his itinerary, although as he was travelling with Nelson it may be supposed that Nelson's interests also influenced their route. Certainly he would have wished to visit Gallet at Avignon. English travellers often visited Saumur and Montpellier because they were the seats of Huguenot academies, the two that remained of the four guaranteed by the Edict of Nantes. That at Sedan had been suppressed just before Halley's visit. Montpellier had a distinguished medical school where Locke studied shortly before Halley's tour and Saumur was known for its wide and liberal studies, especially of law, medicine, and natural philosophy. Locke had also been at Saumur, as had other English radicals, notably Algernon Sidney.[71]

Abraham de Moivre, who later moved to England, was a student at Saumur, and although there is no record of it, Halley may first have met him there. After the Revocation of the Edict of Nantes in 1685 he was detained in the Prieuré de St Martin until he was released in 1688, when he moved to London. There he opened a private school for mathematics and, in 1692, met Halley who introduced him to Newton.[72] It is at least plausible that Halley remembered him from Saumur. He came to know Halley well and his correspondence with Jacques Bernoulli between 1704 and 1714 contains many references to Halley.[73]

Nelson was still at Saumur in June 1681 when Tillotson wrote to him on 2 June about the post at Court which Tillotson and others advised him to decline. The travellers must have spent almost three months in Saumur, for Halley did not observe at Tholose (Toulouse) until 12 August. They thus spent more than seven months altogether in Paris and Saumur and became acquainted with English residents such as the ambassador as well as with distinguished French savants and courtiers. Halley bought books and charts and had the opportunity to become thoroughly familiar with French ideas and developments in astronomy, mathematics, and natural philosophy.

4.5 *On to Rome*

Halley and Nelson moved rather directly to Montpellier and Rome from Saumur, not spending much time in any place on the way. The dates of Halley's observations between Toulouse and Rome allow for little more than direct travel between the towns in the conditions of those days. Toulouse is not far from the Huguenot centre at Montauban and a direct route leads through Huguenot country to Montpellier by way of Narbonne. Nelson may have travelled separately part of the way for he was in Paris in July and is said to have visited Lyon on the way to Rome.[74]

Gallet had observed the transit of Mercury from Avignon when Halley did so from St Helena. He had observed the comet of 1680 in Avignon*, as Halley would have learnt from Cassini. Cassini and Flamsteed both noted Halley's observation of the lunar eclipse of 18/28 August at Avignon. Flamsteed observed it at Greenwich, and derived the difference of longitude between Greenwich and Avignon.[75] Cassini observed the eclipse in Paris, as did William Molyneux in Dublin. Halley's sojourn in Avignon is commemorated in the rue de Halley.[76]

---

*In the first edition of *Principia*, Newton put Gallet's observations in Rome (Prop. XLI, Prob. XXI, Exemplum), but later changed it to Avignon.

The direct way from Marseille to Italy was by sea, and all the places where Halley observed between there and Florence are ports. He cannot have put off much time in sightseeing except possibly in Montpellier, Avignon, and Florence. He would have found English people in Montpellier, in Leghorn, mainly Levant merchants, and in Florence.

The friends intended to spend some time in Rome and in all probability to go on to Venice as most English travellers did, but Halley only reached Rome late in October, and was back in Paris with Cassini on 15 January 1682; it took almost a month to travel from Rome to Paris. In his letter to Hevelius of 5/15 November, Halley explains that he has been travelling almost continuously since May all round France and into Italy, and that he will shortly be leaving for his own country which he hopes to reach about the middle of January.[77] Nelson stayed on in Rome and met Theophile.

Halley and Nelson most probably had a 'sights man' to guide them to the notable monuments. In the previous 200 years popes had built or rebuilt many of the great buildings that now evoke Rome. They had restored the acqueducts and fountains from the dereliction of medieval times to the useful and graceful state they still have today. St Peter's and the obelisk and fountains in Bernini's piazza before it, the Spanish Steps and the fountains there, the fountains of the Naiads and the Marforio, and the Castel S. Angelo, all were much as we see them today. The Roman Forum was in ruins and choked with rubbish. Like Evelyn, Halley could have visited the Vatican Library, the gallery of maps of Italy, the Sistine Chapel.

Halley probably stayed in the Campo Marzio district in the bend of the Tiber opposite the Vatican, possibly at the English College as many English visitors did (Fig. 4.4). There were the remains of Imperial Rome, the Pantheon, the Piazza Navona, the Campo dei Fiori; and great Renaisssance palaces of the Farnese and the Pamphilij. There were hotels known to English visitors. There was the Jesuit Collegio Romano, a centre of astronomical study where the renowned Athanasius Kircher had died the year before. There was the Oratorians' church of Santa Maria in Vallicella (the Chiesa Nuova), with the greatest collection of manuscripts in Rome after the Vatican and with an astronomical observatory.[78] The third centre of natural science was across the Tiber in the residence of Queen Christina of Sweden.

Christina lived in Rome from 1663 in the Palazzo Riario on the via della Lungara. In the next century it was incorporated in the new Palazzo Corsini, where the reconstituted Accademia dei Lincei now has its seat.[*]

---

[*]The first Accademia dei Lincei, founded by Prince Cesi, of which Galileo had been a member, and whose works, especially *Il Saggiatore*, it had published, had come to an end with the death of Cesi in 1630

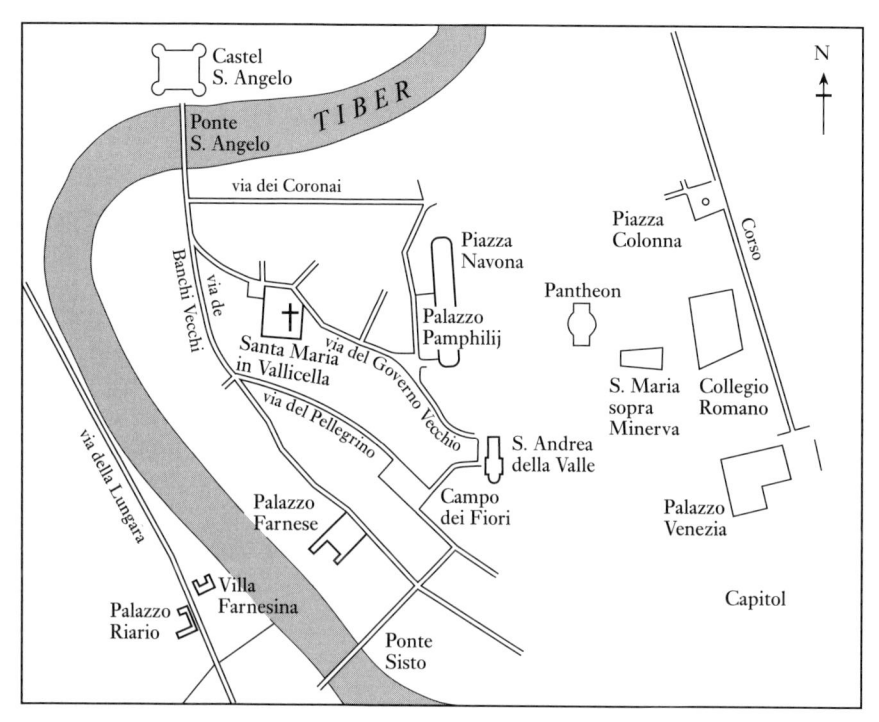

Figure 4.4 Sketch plan of Roma vecchia.

Its gardens later became the present Botanic Garden. Christina brought to Rome an extensive library and collection of manuscripts that are now in the Vatican.[79] She had an outstanding collection of Italian paintings, probably the largest and finest of its day. Her heirs dispersed it after her death, but some idea may be had, even in Britain, of its splendour from the few pictures that have reached Britain, such as the Venus and Cupid of Titian in the Fitzwilliam Museum in Cambridge, and the Venus with a Shell in the National Gallery of Scotland. The National Gallery in London has works by Coerregio, Veronese, and Raphael (a particular favourite of Christina's), and there are others at Hampton Court. The Queen was a munificent patron of music. Corelli and Alessandro Scarlatti worked for her and she maintained a band and singers, men, women, and castrati. She supported two academies, her own Accademia reale and the Accademia fisico-matematico of G.G. Ciampini (1633–1698) established in 1677. Neither was exclusively or even mainly scientific. Many in Rome had great interest in the past of the City. The intense concern for republican and imperial times had faded and early Christianity and the monuments in the earliest churches now occupied people.[80]

G. D. Cassini was in Rome when the comet of 1664 appeared and Christina had observed it with him at her own wish. Later, at Cassini's suggestion, she set up an observatory in the grounds of her palace and there from 1678 a group of astronomers, among them probably, Cellio, Pontio, and Eschinardi, observed the eclipses of the satellites of Jupiter to determine longitudes. Shortly after such an observation on 23 October 1680, they saw the comet of 1680 on 17 November. When Halley arrived in Rome he already knew of those observations, for Cellio had sent them to Cassini while Halley was still in Paris, and Halley had sent them on to Flamsteed.[81]

Queen Christina had offered a prize for predicting the future path of the comet; Cassini, Hevelius, and Bianchini all attempted it. Her interest in comets is an instance of the change of ideas from the Renaissance to the Enlightenment. Millenial and messianic predictions associated with comets that were then current in Protestant circles may have influenced her decision to abdicate from the throne of Sweden and her choice of 1654 to do so. The later comet of 1664 carried similar associations and Cassini thought it might be a return of Tycho's new star of 1572. Cassini and others also thought that the comet of 1680 might bring about a shift in the position of the Earth's pole so as to give France a more northerly position. Christina was knowledgeable about astronomy but was more interested in astrology, which Halley despised, and she practised alchemy in the Palazzo Riario. However, she became more sceptical, in line with the times. By 1680 she doubted the influence of comets on terrestrial affairs.[82]

There are few direct indications of what Halley did in Rome. He measured the latitude as he reported to Cassini, and he recorded the lengths of the Roman and Greek feet as he related in an unpublished paper of 6 May 1696.

> By the way when I was in Rome I measured the Roman and Greek foot at the Campidolio and compared them with the London foot and found the Roman foot $\frac{4}{10}$ of an Inch less than the London, and the Greek $\frac{1}{20}$ of an Inch more, whence it is observable that the Roman foot to the Greek is very near in the proportion of 24 to 25; whence it follows that 8 Greek Stadia of 600 foot each was the same measure as the Mille passus or 5000 foot of the Romans, who made their mile the Octastadion of the Greeks, as I suppose exactly so.[83]

Halley certainly met Eschinardi, for he carried his letter back to Cassini. He would surely have met Cellio and Pontio and discussed their observations of the comet. The Roman observations were not very satisfactory. When the comet was first seen, Ciampini immediately ordered a sextant of five Roman feet radius from Cellio and Pontio, but it was never ready in time, and Pontio had only makeshift instruments at his observatory at Santa Maria in Vallicella.

Halley very probably met the Queen. He, with his European reputation from the southern stars, was just the sort of person whom she would attract to her circle, and Eschinardi and Cellio were members of her academies. He may well have had an introduction from Cassini, and Christina would no doubt have been interested to hear his first-hand account of the observations of the comet of 1680 he made with Cassini. Halley surely met Giovanni Ciampini who later promoted the ideas of Newton, Locke, and Leibniz in Rome. Ciampini was keenly interested in church history and his book, *De sacris aedificiis a Constantino Magno constructis* was the foundation for archaeological studies of early churches in the eighteenth century.[84]

Many books on Roman antiquities and other monuments were among those that went on sale after Halley's death, along with prints of the City, more than about Paris as a city. Many may not have belonged to Halley, but on the other hand they may witness to his antiquarian interests (Appendix 8). There was in particular *Neandrus de mesuris et ponderibus Rom.*, printed in Basel. In 1688 (Chapter 6, Note 58) Halley would consider the proportions of various measures of weight and volume, and in later years he would analyse the Antonine Itineraries knowing the relation of Roman to London measures.[85] Again, when he was surveying the Adriatic coast in 1703, he no doubt had in mind the question of the units used on published maps and charts (Chapter 11).

When Halley wrote to Hevelius in November he deplored the decline of astronomy in Rome and the lack of great artists as there had been in the recent past. He was right on both points. He agreed with Octaviari Pullein, whom Hooke had asked the year before to give some account of Roman science. Pullein mentioned Eschinardi and Ciampini and wrote of Ciampini's academy

> The members of the Society are not over numerous nor over Rich but are very active and vigorous in the prosecution of the Searches and want only the Assistance of some Prince that has Courage enough to spare them money etc.[86]

Galileo had been in the forefront of observational astronomy in his day; half a century later Ciampini, Cellio, Eschinardi, and the others were far behind Hevelius, Flamsteed, Cassini, and Halley himself. Halley must have judged Roman astronomy as inadequate. As for the arts, Christina collected above all paintings and medals of the previous century. Bernini, greatly admired by Christina, had died in 1680, and no contemporary artist could compare with Raphael or Titian. The art of promise was music, but Corelli and Scarlatti, while they astonished their hearers, were still young men with their ways to make. Halley's rather disappointed reaction to Rome seems not unreasonable.

Why Halley returned early from Rome is not known, if indeed he did return prematurely. In his letter of 5/15 November to Hevelius, he had written that he would soon be returning to London where he hoped to be in January.[87] He was married some three months after his return and that may have decided the time he left Rome. He is supposed to have said that it was necessary to protect his patrimony,[88] but according to his own account, that need arose two years later following the death of his father and consequential disputes with his stepmother.[89] For whatever reason, Halley was in Paris on 15 January 1682, when he called on Cassini and gave him his observations of the eclipse and of the latitudes, and the letter from Eschinardi. He was in London on 24 January. Hooke's note of his return shows that he came back through Holland.[90]

4.6 Astronomical observer

Halley lived at Islington after his marriage. There he observed the Moon and other objects regularly from 23 October 1682 to 21 February 1684.[91] His father disappeared on 5 March 1684 (Chapter 5), after which he made only a few sporadic observations at Islington. He made no further astronomical observations until he went to Greenwich as Astronomer Royal—there is no record of any observations at Oxford between 1704 and 1720. Swedenborg, writing from London in 1711, said that in 1698 Halley had sold to the Tsar Peter the Great for £80, the quadrant he had used in St Helena and at Islington in 1683 and 1684.[92] Halley evidently had not resumed astronomical observations after 1684 and saw no prospect of doing so in 1698 (his quadrant was by then more an antique than an instrument). When he went to Greenwich for his final years as an observer he had much better instruments and clocks. Here then is an appropriate point to assess his standing as an observing astronomer at the end of the seventeenth century. The works on which he must be judged are the southern observations at St Helena, the visit to Hevelius, and the observations he made at Islington after his marriage. The southern observations are the most considerable.

The southern campaign gave Halley a great reputation in Europe. He certainly needed gifts of persuasion, persistence, and organisation to carry through his plans, but as discussed in Chapter 3, the results were less valuable for fundamental astrometry than they might have been.[93]

When Halley visited Hevelius he used his small quadrant to measure angular distances between objects. Halley's own letters and the account by Hevelius in *Annus climactericus*, and in his correspondence and the published comparisons, show that both men took great care over the observations. Halley's observations of latitudes during his tour of France and Italy were made with a small quadrant and are of modest precision, as the following comparisons with latitudes currently accepted will show:

	Halley		Accepted		Difference
	deg	min	deg	min	min
Paris	48	51	48	50	+1
Saumur	47	19	47	15	+4
Tholose	43	41	43	37	+4
Narbonne	43	15	43	12	+3
Montpellier	43	39	43	37	+2
Marseille	43	18	43	18	0
Tholon	42	57	43	10	−13
Freius	43	19	43	25	−6
Gennes	44	16	44	24	−8
Legorne	43	27	43	33	−6
Florence	43	41	43	47	−6
Rome	41	48	41	53	−5
Siena	43	14	43	19	−5

A rather evident conclusion is that the quadrant suffered some maladjustment or accident between Marseille and Tholon, but perhaps little better than 5' could be expected from a small instrument which was presumably held in the hand. Those observations are, surprisingly, presumably the first to have been made at some of the sites, since Cassini thought them worthwhile recording.

Much later de L'Isle, who had met Halley in England and obtained from him particulars of his instruments and methods, compared some observations made at St Petersburg after 1730 with Greenwich observations of Flamsteed about 1690 and of Halley about 1732.[94] De L'Isle also made a sketch of Halley's sextant (see Fig. 3.4). His papers contain some comparisons headed by the legend 'Dessein des marques et observations de leur situation et distance pour la verification du Sextant de M. Halley et du Grand Quart du cercle de Paris' ('The design of the marks and observations of their situation and distance for the checking of Mr Halley's sextant and the large quadrant of Paris'). Another set is 'Hauteurs meridiennes du soleil et des étoils fixes observée avec le sextant de M. Halley' ('Meridian heights of the Sun and of the fixed stars observed with Mr Halley's sextant'). Halley's sextant apparently gave results too high by 1' on the upper limb of the Sun. The notes do not say who made the comparisons, but de L'Isle must have taken the sextant to Paris to make the observations.

Halley was not a dedicated observer as Flamsteed was, he had too many other interests to spend a lifetime on astrometry, and so laid himself open to Flamsteed's criticisms. At the same time, he was the first to exploit the possibilities of the new instruments and to fill a serious gap in the knowledge of the heavens. His astrometric work was of a piece with the rest of his career. He saw the possibilities of a new type of investigation of nature and the need for it, and opened up a new field. Then, having shown what could be done, he moved on to something else; he never restricted himself to one subject alone.

The effective collimation length of an instrument, whether with open or telescopic sights, is one factor that affects the consistency of the observations; Hevelius misunderstood it. The graduation of the angular scale, the axis about which the sights rotated, and changes in the mounting, probably caused greater systematic errors.[95] High precision would have been difficult to obtain with even the best mural arcs. Instrumental systematic errors must however have limited the precision of angular observations to about 1' in the seventeenth century.

Clocks had their errors. Halley thought that Hevelius's clocks were poor and he realised that a clock needed a firm support to go well and should not just stand on a table. One second of time corresponds to 15" of longitude or right ascension so that the rate must be uniform to within about 2 s per day if right ascension is to be correct to 30"; that was a demanding requirement although the clocks that Halley had when he became Astronomer Royal seem to have satisfied it.

Atmospheric refraction was well recognised in the seventeenth century, but aberration of starlight due to the motion of the Earth around the Sun had not been discovered, nor had nutation. The new instruments that replaced those of Tycho and Hevelius did greatly improve astrometry, but only in the next century were some major systematic errors understood and overcome.

4.7 A traveller's harvest

When Halley returned to London he would have had a knowledge of contemporary astrometric techniques matched by few. He had direct experience of pre-telescopic methods in the hands of Hevelius. He himself, along with Flamsteed, had developed and used the new telescopic instruments and methods. He had observed with Cassini and others at the new observatory in Paris and had noted that its equipment was nothing special. It turned out that the design of the building was not convenient for meridian observations. English astrometry led the way, and in the hands of Flamsteed, of Halley himself, and of his successor Bradley, continued to do so for many years. France, however, led in geodesy, where Picard had made the first measurement of an arc of a meridian by triangulation.

Halley returned with first-hand knowledge of the continental observations of the comet of 1680, in which he himself had joined in Paris. Thus he was able to pass on to Flamsteed and Newton results that Newton used in his discussion in the *Principia*. He would also have come back knowing of the ideas that comets might return, as put forward by Cassini and by others in France and in the circle of Queen Christina. He himself had attempted to calculate the path of the comet of 1680 and was no doubt aware of the prize that Christina had offered.

There may be a third influence of Halley's European travels. When Christina was young before her abdication, she had a reputation as a religious free-thinker. She was very learned and read the classical philosophers, Plato and Democritus especially, in Greek. She was probably not an original thinker but she knew enough to point out that some of Descartes's ideas, which he thought original, were to be found in Plato and elsewhere. She corresponded with Descartes and he had visited her in Sweden and died there. She also corresponded with Gassendi and appears to have accepted the atomic concepts of Democritus. She was critical of some aspects of orthodox Lutheran positions such as the textual reliability of the scriptures. She knew of speculations in Protestant circles that men existed before Adam and that there might be other created worlds than ours. Whether or not she actually held those beliefs, rumours about them and her evidently sceptical approach, would certainly have given rise to accusations of free-thinking.

Some of Christina's ideas, actual or reputed, correspond to ones that Halley later expressed or was thought to hold, and that gave him the reputation of atheism. That is hardly surprising, for the sceptical approach of Bayle, the atomism of Gassendi, the possibility of other worlds, the age of the world and whether it had an end and a beginning, a critical attitude to the scriptures, all those were advanced by many thinkers in the seventeenth century, and Halley certainly did not have to visit Christina to acquire them. Yet Christina had known Descartes and had corresponded with him and Gassendi. Did Rome help to turn Halley's thoughts towards the rational attitude to religious matters that he later evinced and which others saw as atheism?

5

A wedding and two funerals

> Plots, true or false, are necessary things,
> To raise up commonwealths and ruin kings.
> DRYDEN: *ABSALOM AND ACHITOPHEL*, Pt. 1, 27

5.1 *A City wedding*

EDMOND HALLEY married Mary Tooke at St James's, Duke Place, on 20 April 1682 in the presence of Thomas Crosse, 'first friend'. He had returned to England on 24 January[1] and sometime later he gave Newton an account of his observations of the comet of 1680 (Chapter 4). The letter in which he told Hevelius in the middle of November 1681 that he was preparing to return from Rome does not suggest a sudden decision; he is more concerned with the present state of Rome as compared with its imperial past.[2] Plans for his marriage may have determined his return.[3]

Mary's father, Christopher, was an Inner Temple lawyer, as were her uncles and her paternal grandfather (Appendix 2). Some held or had held high office. Her grandfather and either her uncle, John, or her father had been Auditors of the Court of Wards, which was abolished in 1641. An uncle, Edward, also of the Inner Temple, had been Librarian to the King (possibly Charles II). Tookes held lands in various places. Mary's father had died in 1663 and had left money and copyhold lands to his daughters Margaret, Mary, and Dorothy, in equal shares. Mary also had a legacy from her uncle Edward, the third son of James, which was to lead to a Chancery action.[4] Mary's mother, Margaret, was the daughter of Gilbert Kinder, a mercer of the parish of St Helens, Bishopsgate, just across Bishopsgate from Gresham College and only a step from Winchester Street. Mary's grandmother was buried at the chancel steps in St Helens, where Hooke was buried. An answer to a Chancery action in 1693 shows that Halley's father and Mary's mother had had financial dealings over a mortgage of some plate.[5] Mary was not only a woman of property and related to lawyers with a Court connection; she was also connected with Levant Company merchants for there were Tookes in the Levant factories in Asia Minor.

Mary's relations were clearly well-to-do, prominent in the law and known at Court, and perhaps socially somewhat superior to the Halleys.[6]

Seventeenth-century couples such as Edmond and Mary, from well-to-do families, would not wed for affection alone. Families made arrangements.[7] Edmond at twenty-five had an established reputation and standing in England and in Europe, but he may have come back from Italy when he did because Halleys and Tookes had decided that a marriage between him and Mary would be to the advantage of the families. A little earlier, his cousin, Anne Cawthorne, had married Joseph Chomat, a merchant who had some close relation to the second wife of Halley's father, as stated in the family Chancery actions that followed the father's death in 1684. The date of the father's marriage to his second wife Joane is not known, but if there is any truth in the tale that Halley had to return early to protect his interests, then the three marriages may have been interrelated.

The church of St James's, Duke Place, was near the present church of St Katherine Cree on the east of the City; it stands no longer. Marriages were celebrated in it by special licence without the usual notice of banns because, so the incumbent claimed, the church was a City peculiar. Such 'private' marriages were common among notable City families, who did not like to show themselves to the rabble; St James's was exceptional only in the great number of marriages that the incumbent performed, as many as twelve a day on occasion, no doubt to his considerable profit. Edmond and Mary followed the custom of prominent citizens.[8]

The 'first friend' was the formal witness to the marriage and we must not suppose that he was the only other person in St James's on 20 April. Weddings, as now, were the occasions for great parties; with dinner at midday they went on until well into the evening. The bride and groom would not then depart to the seclusion of some distant hotel. The company would ceremoniously put them to bed and stockings would be flung over shoulders, the groom's by the bridesmaids and the bride's by the groomsman; if any should hit the bride or groom, that was an omen that the bridesmaid or groomsman would be married within the year. Alas we do not know who joined in the hilarity at the Halley wedding. Robert Nelson was still abroad, but his mother and her relatives could have been there. Merchants, lawyers, friends from St Paul's and Oxford, Mary's three sisters and their husbands, fellows of the Royal Society, they could have had quite a party.[9]

Halley and Mary first lived at Islington, a fashionable rural suburb of the City. There Halley set up the instruments that he had taken to St Helena, including his six-foot sextant.[10] It has been said that Halley was in reduced cicumstances when, later, he saw the *Principia* through the press: how then could he maintain a wife, even one with property, before he inherited the father's houses in Winchester Street? He had an allowance from his father of £60 per annum (equivalent to about £25 000 today),[11] and it is likely that some of the real property of the family had been transferred to him as the eldest son and legal heir. Later, in 1684, Edmond and Mary moved to Golden Lion Court, just off Aldersgate Street,[12] and it was there that their surviving children, a son Edmond and two daughters Margaret and Catherine, were born.

In the summer of 1683 Mary and her sisters, Dorothy English and Elizabeth Pearson, together with their respective husbands, started an action in Chancery against Charles Tooke and Francis Bostock Fuller, executors of the estate of Edward Tooke. Edward had left legacies to the four daughters of his brother Christopher (the other was Margaret).[13] Each niece was to have £200, to be paid when they became 21 or were married, if earlier. Edward had died in 1668 and when Mary became 21 in 1679, the sisters had applied to Chancery for payment and a Master made an order for it in 1680. The daughters and their husbands then brought the action to enforce the decision of the Master. Halley is alleged to have said that he would not give the executors more time 'but would prosecute them with rigour'.[14] Subsequently the houses were sold and £200 was paid to Halley for Mary.

5.2 The sky at Islington

The archives of the Royal Greenwich Observatory contain records of observations that Halley made at Islington between 23 October 1682 and 16 June 1684. The list is headed *Observationes coelestis habitae Islingtoni prepe Londinum anno 1682 mense Octobrie E. H.* Halley was observing before his marriage, but he has not recorded where, although some of the records must refer to observations in Paris, and others to Greenwich observations. Most of his observations were of the Moon, many of close approaches (appulses) of the Moon to a star, but also some actual occultations by the Moon, and some distances of the Moon from the Sun. He also observed the planets, particularly Venus.[15] He intended to observe the Moon throughout a saronic cycle, for, as he explained in the Appendix he attached to the third edition of Streete's *Astronomia Carolina,*

> I applied the Leisure I had secured myself about the Year 1683, to observe diligently, as often as the Heavens would permit, the true Place of the Moon.[16]

His Appendix shows that as early as 1682 he had the ideas that the Moon could be used as a clock in determining longitude, which was not novel, and that its position could be predicted from its known place in a preceding saronic cycle, which was apparently new.

He did not get his instruments set up until the autumn but thereafter he observed on very many occasions between October 1682 and February 1684, altogether as he says, two hundred observations in sixteen months. He must have taken advantage of almost every clear night. Thus in 1683 he observed:

in January	12 nights
in February	10 nights
in March	21 nights
in April	7 nights
in May	14 nights
in June	12 nights
in July	12 nights
in August	22 nights
in September	15 nights
in October	17 nights
in November	15 nights
in December	17 nights

Mary's thoughts of this intense nightly activity are not known.

Halley seems to have made no use of those observations, although he did compare some of them with others at corresponding epochs of a later saronic cycle (Chapter 13). Universal gravitation and Newton's attempts at a lunar theory were still in the future and Halley tried to fit his observations to Horrocks's model of a rotating ellipse for the orbit of the Moon about the Earth. He, like Flamsteed, regarded that as the most satisfactory available representation of the Moon's motion.[17] Then in 1684 family matters intervened before he could work on his material, and he put the Moon aside for many years.

Halley published a few papers in *Philosophical Transactions* in his first years at Islington, on topics on which he had spoken to the Royal Society. They included the longitude of Ballasore (1682), the motion of the satellite of Saturn (1683*a*), the magnetic variation (1683*b*), and the tides at Tonkin (1684). The paper on the Earth's magnetic field was his first essay into the theory of the origin of the field. He presented a considerable collection of observations of the variation extending over a century (1580–1680) and from longitude 170° E (New Zealand) to 80° W (Baffin Bay). He probably obtained many of them from the widow of Flamsteed's colleague, Peter Perkins; it was one reason why Flamsteed became so hostile to Halley (Chapter 6). Halley proposed that there were four magnetic poles in the Earth. He left it there in 1683, but later (1692b) would suggest that the Earth had a magnetised shell and core in relative rotation.

There is only one high water at Tonkin in twenty-four hours and the amplitude of the tide is zero every fourteen days. A gravitational theory of the tides was still in the future, but Halley explained the phenomena by a sort of interference, an idea to which Newton would refer and would develop much more thoroughly in the *Principia*. Those papers of his time at Islington were indications of some of Halley's major interests in later years.

As the New Year of 1684 dawned, Halley's standing in Britain and Europe was high, he was known for the achievement of St Helena, he had been elected to the Council of the Royal Society (1683–1685), he moved in influential circles, he was recently married, and he had completed about a year of serious observations of the Moon. His prospects were good and he might surely look forward to 1684 with high hopes.[18]

5.3 Those bloody and barbarous men

The Fell Reaper blighted Halley's plans.

> ... this Design of mine was soon interrupted by unforesen domestick Occasions which obliged me to postpone all other Considerations to that of the Defence of my Patrimony: and since then, my frequent Avocations have not permitted me to resume these thoughts.[19]

He recorded no observations at Islington in March or April of 1684, only few sporadic ones in the rest of the year, and no more astronomical observations until he became Astronomer Royal.

On 5 March 1684 the father was seen for the last time at his home and on 12 April he was found dead near Stroud in Kent, as Flamsteed reported to William Molyneux in Dublin:

> Mr Halley I suppose may return to live in London, his father having been found drowned about a fortnight ago in Rochester river. This mishap will cause him a great deal of perplexed business, but nevertheless I hope to see him oftener than I have done of late.[20]

The circumstances were rehearsed in the *Complaint* of an action in Chancery about the father's estate raised in 1686 by Joane and her second husband, Robert Cleeter, whom she had married in 1685:[21]

> ... the same Edmund Halley [the father] ... did on or about the fifth day of March in the yeare of our Lord One thousand six hundred and Eighty three in the afternoon of the same day goe out of his dwelling house then

being in Wynchester Street aforesaid he being in good health and without being in any quarell with any person or persons whatever that your Orator or Oratrix did know or ever heare but never returned again nor could be heard of after the strictest and most diligent search and enquiry made for him by your Oratrix until the Twelfth day of Aprill the year following or thereabouts when he was found dead in a certain place in or neare the parish of Stroud in the County of Kent ... and by a Coroner's Inquest that sat upon the Corpse after it was found as aforesaid and inquired how he came by his death it was found that he was Murthered by some unknown person or persons.[22]

An anonymous broadsheet, of which only one copy is known, has more details:[23]

A true/ Discovery/ of/ Mr. Edmund Halley/ of London/ Merchant/ Who was found Barbarously Murthered/ at Temple-Farm, near Rochester in Kent.*

Mr Edmund Halley was a gentleman of public Acquaintance and known Reputation through the whole City of London, and had there lived in Winchester Street for many years, and was no less esteemed for his own Merits, as a Person always forward to the Service of his King & Country, than he was respected as a Gentleman of considerable Estate, and plentifully happy in all the Goods of Fortune ... it appears by the Gazet he went from home about the beginning of March; and could not be heard of tho a Reward of a hundred pound was promised to any Person, who should discover him alive or dead.

Wednesday the 5th March, at Evening, was the first time that Mr. Halley was missing, in the Morning he complained that his feet were tender and his Shooes pinched him, upon which a Gentleman standing by who was his nephew, told him, Uncle if you please I will cut the Lineing out of the Toes and that will give you ease, and accordingly he did so, his Wife asked him if he was going out, yes says he, I'll go and walk, but I shall be back in the Evening, when Night came she accordingly expected him, but not returning she was very much concerned at it, and the next day made all possible Enquiry, but after several days not hearing of him, published his Absence in the News Book ... But on Monday last he was found by a River side at Temple-Farm in Strow'd Parish near Rochester on this manner. A poor Boy walking by the Water-side upon some Occasion spied the Body of a Man dead and Stript, with only his Shoes and Stockings on, upon which he presently made a discovery of it to some others, which coming to the knowledge of a Gentleman, who had read the Advertisement in the Gazet, he immediately came up to London and acquainted Mrs Halley with it ...

Mrs Halley who had all along been most passionately troubled at her Husbands absence receiving this Intelligence sent down the Gentleman his Nephew to view the Corps, whether it was the body of Mr Halley or no. It was judged by all appearances to have lain some time in the River, And

*The date '17 April 1684' is added by hand below the title.

afterwards by the motion of the Water driven a shore; the face was so much disfigur'd he could not be known, one of the Eyes was injured, and several parts of the Body, but by what means 'tis not yet discovered. It was concluded by all that he had not been in the River ever since he was missing, for if he had his Body would have been more Corrupted. The Gentleman knew him by his Shooes and Stockings, they being the same Shooes he had cut the Lineing out of, and on one leg he had four Stockings, and on the other three and a Sear-cloth. The Coroner sat upon him, & the Inquest brought him in Murthere'd. On Wednesday Night his Body was brought to his own house in Winchester Street, from whence it will be suddenly buried.

London Printed by E. Mallet 1684

Temple Farm, or Temple Manor, a farmhouse, originally a property of the Knights Templar, stands on the shore of the River Medway just across from Rochester Castle, slightly elevated above tidal mudflats. There must be a strong possibility that the body was thrown overboard from a vessel coming from London.[24]

The father was buried in linen at St Margaret's, Barking, the last known connection of his family with that place (Chapter 2).

It seems a little odd at first sight that Joane asked a nephew to identify the body. The father had three nephews, of whom Francis, the son of his brother William, seems the most likely to have been in London— he later joined Halley in a London property deal. Another possibility, perhaps the strongest, is Joseph Chomat, probably a nephew by marriage, who had been often in the Winchester Street house, as the Chancery actions tell us. If he was a particular friend of Joane, she may well have turned to him rather than to the stepson with whom her relations were evidently cool or worse. Where was Halley himself when the body was found? He apparently was reluctant to advertise his father's disappearance.

The very odd circumstances of the father's disappearance and the state of his body certainly suggested murder, but because he appeared in good health and without enemies, have on occasion led to the idea of suicide. They call to mind the strange affair of the murder of Sir Edmondbury Godfrey, the magistrate, at the time of the Popish Plot. Was commercial rivalry the cause? Or had the father's duties as a Yeoman Warder perhaps led to a vicious attack? Some certainly thought so.

Some English nobles and gentry, Scottish lords, and radical exiles in Holland had been talking about risings against Charles II and James Duke of York in England and Scotland, and there was a parallel plot, the Rye House Plot, to assassinate Charles and James as they returned to London from Newmarket in April 1683. The plot failed and in July of that year the Earl of Essex, the Earl Russell, Algernon Sidney and others of the conspirators in the two plots were arrested and confined in the Tower for alleged high treason. Russell and later Sidney were tried, condemned, and beheaded. Essex was found dead in the Tower on the day Russell went for trial and while he himself was awaiting trial. A jury hastily empanelled found that he had committed suicide by cutting his throat. His motive might have been that of his wife's great-grandfather who had shot himself in similar circumstances to avoid the forfeiture of his family estates that would have followed his conviction. Some at the time, John Locke among them, thought that Essex had been murdered, and some laid the guilt at the door of the Duke of York who, with the King, had decided, most unusually, to visit the Tower that day of Russell's trial. Robert Ferguson, 'The Plotter', an associate of Essex and others, wrote a tract to that effect which was printed in Holland in 1684 in a number of languages and widely distributed around the English Court and Parliament.[25] Most historians have accepted the jury's verdict, but two recent studies of radical activities consider that there is a strong case for murder. Medical opinion does however favour suicide.[26]

A notorious Middle Temple lawyer, Lawrence Braddon, alleged that Essex had been murdered, and found witnesses to support his claim, so he said, with circumstantial evidence. Braddon had close connections with leading Whigs and radicals and was known for his invention of the 'protestant flail', a sort of night-stick with which God-fearing men might defend themselves against murderous attacks by Roman Catholic assailants. He may well have been on the fringe of the Rye House Plot. Hugh Speke, a lawyer, probably a student, of Lincoln's Inn, and the youngest of a family who were certainly close to the conspirators, helped him to gather evidence. For their pains, Braddon and Hugh Speke were tried before the King's Bench on 17 (or 16) February 1684, for spreading rumours derogatory to the Government and for suborning witnesses to make false depositions. Sir George Jeffreys, then Chief Justice of the King's Bench, presided and seems to have been in good form, displaying both a sharp mind in establishing exactly what witnesses were saying and a great capacity for hectoring defendants and witnesses. The main thrust of the trial was to discredit Braddon's witnesses. Unsurprisingly, Braddon and Speke were found guilty. A letter that Speke wrote to a much older acquaintance at Bath, Sir Robert Atkins, who had previously been an active opponent of the Government but now lived privately near Bath, strongly suggests that he and Atkins had been in some clandestine association. Speke claimed to have written under the influence of drink. He seems lucky to have got off as he did, with a fine, and not to have been pursued for a far more serious offence. Jeffreys, however, was after Braddon. Sentence was postponed until April, Braddon was then fined £2000 and Speke

£1000 and both were confined in the King's Bench prison until they paid.[27]

Braddon apparently remained in prison until released after the flight of James II. Then in January 1689 he repeated his allegations. On 31 January some persons were detained for enquiries on his information. On 2 February the House of Lords ordered the detention of the Gentleman Porter of the Tower and of a yeoman warder and set up a committee to investigate the circumstances of the death of the Earl of Essex. The committee was one of a number seeking to fix responsibility for sundry trials, executions, and other happenings of the previous two reigns. Roger Morrice in his *Entring book* gives some details:[28]

> A very little while after the Earle of Essex's death in the Tower it was reported that the person that killed him was troubled in conscience and could have no peace nor rest, and therefore went voluntarily and told two or three persons that he was the man that cut the Earle of Essex's throat, and by whom he was engaged in it, and what money he had for doeing it, and that there were five other persons whome he named privy to his death, or aiding and assisting in it, and that the confession of this person was the ground of what the author of that booke saith in it, to wit, That if there might be a pardon, he would undertake to make proof of that Earle's murder. And this person that said he did actually cut the Earle of Essex's throat was so impatient that he could not without great difficulty be restrained from making an open confession thereof even before the judges in Westminster Hall, but he was told he was sure to lose his great end, which was the bringing to justice those that had committed it, for Braddon was counted a heynous offender for prosecuting that busines, and if he himself should go in now an accuser he would certainly be disbelieved if not hanged, this prevailed with him to suspend his testimony.
>
> This man or witnesse must certainly be Holland, who is credibly said to be a very likely man to confesse or discover it.
>
> On Monday last the 21 instant all his account was newly revived, and it was commonly reported this man came in now to make this confession and acquainted Mr. Braddon therewith.
>
> NOTA. How much of this is true I cannot positively say, but its thus reported by those that have had great opportunities to look into this matter. About the beginning of this weeke five persons were apprehended and committed to the Tower about the death of the Earle of Essex, I no way doubt but one of those five was committed by his own direction, and gave notice where the other foure lay, and how and where they might be seized on. On Wednesday the 23 Mr. Braddon was called in before the Lords it's said by the privity and desire of the Countesse of Essex and Sir Henry Capoll &c to give them a summary account of the reasons why he thought fit that bloud should be further enquired into, and such reasons did he give them that their Lordships appointed a commiteee to examine that matter. The committee met I thinke that afternoon and ordered the witnesses to be brought

thither on Thursday morning, which time the committeee set to meet there againe.

News evidently got quickly around London. On 1 February Hooke reported in his diary that Halley was 'much concerned about seasing of persons about Lord Essex'. The gentleman porter seized by the Lords' order of the 2 February was Major Thomas Hawley, in whose house in the Tower Essex had been confined at the time of his death.[29] Thomas Hawley ranked high in the Tower for he acted on a number of occasions as deputy for the Constable (Sir John Robinson). He had taken part in enquiries that followed the murder of Sir Edmondbury Godfrey.[30]

The Lords' committee took a number of depositions and on 14 February adjourned to the 16th, but on the 15th the House replaced them by a small committee of only three or four lords who continued to take evidence. There is no minute of its proceedings in the records of the House. Parliament was prorogued in February 1690 and there was no final report. Major Hawley was released from detention after four weeks and he must have been cleared of any suspicion for on 11 March 1690 he received a commission as major of the Garrison of the Tower.[31]

After the Lord's commmittee had concluded their sittings, Braddon published a little book in which he set out the depositions that he had collected, together with the evidence that had been given to the coroner's inquisition in the Tower. He claimed that Essex had been murdered and not committed suicide. He repeated the allegations with further detail in 1725. An anonymous tract of 1689 also contains what appear to be the substances of the statements given to the Lords' committee.[32] Braddon and the Tract both allege that some of the soldiers in the Tower were killed or otherwise persecuted because they made unwary remarks about the circumstances of Essex's death.

Why should Halley be concerned? Was there more to it than just the similarity of the major's name to his own, often written Hawley by others? Indeed there may have been. Halley's father, the yeoman warder, was one of those thought to be a victim, as Ferguson explicitly alleged only months after his death. Ferguson identified the father as 'a Warder of the Tower living in Winchester Street, being a person for Reputation and Estate far above that Hawley in whose home the Earl of Essex was prisoner'. He wrote that the father had spoken unwisely of knowledge that he had of the death of the Earl. When 'a great Man', namely the Duke of York, heard of it, he ordered the father's death. It was known that the father was looking for an estate to buy and, so it was said, he was tempted out of London to see some land. Then he was murdered. Ferguson adds that his wife (Joane) was worried that he might be murdered after she had heard him talking about the death of Essex.

The following extract from the Tract of 1689 is more dramatic, but it was probably derived, with ornamentation, from Ferguson, who was closer to the events and to Essex than Braddon or others. It purports to be a conversation about the Lords' depositions by three lawyers, one from Gray's Inn (G), one from Lincoln's Inn (L), and one from the Temple (T), and it might be supposed that Braddon was the Temple lawyer and Speke the one from Lincoln's Inn:

> T: Hawley the warder [i.e. Halley's father], intimately acquainted with Major Hawley (in whose house my lord was murdered) was found dead in the Medways, about April next after my lord's death (having been murdered in a most barbarous manner.) This Hawley was supposed to be taken off to prevent the discovery of what he knew in this matter; for a little while after Mr Hawley had been missing (viz,) about a month before he was found dead, a warder then in the Tower (supposed to be a papist) told Mr A S* (who had long lain under the pressure of the then misgovernment and then there a prisoner, without any evidence to justify the committment): 'That Hawley was ran away for prating somewhat about the Earl of Essex' but how he ran away, a short time discovered.
>
> This Hawley was in Westminster Hall when Mr Braddon was upon his trial, and said 'he much wondered upon what Mr Braddon should stir in this thing, when to his knowledge, Mr Braddon knew nothing.' A gentleman then present, who knew Mr Hawley,[33] looked on this expression as what argued Mr Hawley not a stranger to the matter, wherefor this gentleman immediately said, 'Mr Hawley, if you know Mr Braddon knows nothing of this matter, what must you then know?' Upon which Mr Hawley seemed surprised (having too far expressed himself,) and made no reply.
>
> G: I have been informed by a warder in the Tower that this Mr Hawley the Warder, as soon as he heard the news of the earl's death, immediately declared, 'It was a damned piece of roguery throughout'.
>
> T: This Mr Hawley was very rich, and a warder only to exempt him from parish services but he never waited, unless it were on very solemn occasions; and that very day my lord died he was waiting, and (as declared by several) was one of the warders that attended in person on the Duke of York, whilst he was in the Tower, that morning the earl died.
>
> L: If so, he might well observe the duke's sending the two men to the earl's lodging just before his death, and their return to his highness, as Mr E deposeth; and M and R declared, with several other passages that might to him discover that barbarous murder, and he had cause enough to say 'it was a damned piece of roguery all over.'
>
> G: Good God deliver us from such bloody-minded men!

*Mr A S was presumably Algernon Sidney.

Braddon's account in his own pamphlet of the father's death is substantially the same; his last paragraph is:

> But greater Cruelties than these, some bloody men may be supposed to have used to prevent a discovery of my Lord's Murther; for Mr Hawley who 'KNEW THAT I KNEW NOTHING WITH RELATION TO MY LORD'S DEATH', and his too freely imparting his own Knowledg'd in the matter, is thought to have caused him (not to run away, but) to be murther'd.

Whatever the facts of the father's death, Ferguson and Braddon clearly thought that it was sufficiently close to the death of Essex to clinch their claim that Essex was murdered.

Many of the allegations that Essex was murdered, that men were sent by the Duke of York to do it, and so on, were unsupported gossip and innuendo. They were readily put about and believed at a time when there were real conspiracies against the King and the Duke of York, when the King was a secret pensioner of Louis XIV, and when all factions did their best to suborn justice. None the less the account of the father's reason for being a warder, and of the easy nature of his duties, rings true. Braddon and the author of the Tract, if not the same, may have derived their account from the earlier, almost contemporary pamphlet of Ferguson. All three identify the father by his standing in the City and by the circumstances of his death, matters that would have been common knowledge. The Tract drew explicit parallels with the Godfrey and Overbury murders. Even though Essex had committed suicide, it may be there were those who thought, whether with reason or not, that the father knew something compromising and disposed of him accordingly. He was said to be a close friend of the gentleman-porter; perhaps that compromised him. Was it purely by chance that he happened to be in Westminster Hall at the trial of Braddon and Speke, or had he gone to hear what was said? Did he perhaps indicate to acquaintances that he might make himself scarce? He disappeared about a month after his presence at the Braddon trial but before sentence was delivered in April.

Braddon and Speke were probably in prison between the trial in February and sentence in April 1684, for they had been found guilty, yet Braddon seems to have been well aware of the overt circumstances of the father's death which occurred in that interval; he quite likely had that information from Ferguson, who wrote that he could produce proofs of how and by whom the father was murdered—but never did.

Why should Halley himself have been concerned immediately Major Hawley and a warder were seized and confined in the Tower five years later in 1689, before the Somers tract and Braddon's pamphlet appeared? Ferguson's pamphlet had already linked his father's death with that of Essex. Did he know Braddon? Or Ferguson? Was the Major 'intimately acquainted' with the father, a family friend or perhaps a relative; or did Halley fear that others might think that he himself had shared some knowledge with his father? Had the father confided in him?

The deaths of Essex and the father are likely to remain unsolved mysteries, but some points stand out. At the political level, Essex's death was remarkably convenient and well timed for Charles and James. The evidence against him was probably stronger than that against Russell and Sidney. Russell might well have been acquitted had the news of Essex's suicide not been brought to his trial while it was underway; the Attorney-General most certainly used it to sway the jury. The Crown outrageously bent the case against Sidney. Essex would probably have been convicted had he come to trial but the King was apparently minded to pardon him for his father's sake. Essex's suicide ensured that all three died. That was in James's interests even more than Charles's. Witnesses to the suicide said that Essex asked for a penknife to pare his nails, got a razor because a knife could not be found, and thereupon took the opportunity to slit his throat, and that in circumstances where his cutlery was carefully watched to see that he did not do away with himself. Holland said he cut the earl's throat but that is inconsistent with the medical view; did he perhaps provide Essex with the means for doing so? That fatal morning seems to have been the only recorded occasion on which the King and the Duke spent any time together in the Tower. Why they made an extraordinary excursion is not known. Why was Halley's father in Westminster Hall at the trial of Braddon and Speke—by chance or idle curiosity, or was he anxious to hear what was said? Essex may have been encouraged to commit suicide by the King, or more likely by James; Holland may have been the man who carried the message and the means, and Halley's father, in attendance on the Duke, may have been aware of it. Those who knew or might have known the truth of the matter seemed determined to keep quiet. The widowed countess, the son and heir, and their confidant, Gilbert Burnet, all insisted it was suicide. They gave Braddon no support. The Lords' enquiry ran into the sand. Neither Halley's father, while still alive, nor Halley himself, said anything.

When the father did not return home, Joane offered a reward of £100, to which Halley objected; might Halley have opposed a reward and a search because he knew that the father wanted to lie low after the Braddon trial? Whatever the reason for Halley's concern in 1689, it seems inescapable that he had some particular knowledge of the circumstances of the deaths of Essex and of his father. His close association with the Tower and its denizens and with the Stuart court, hitherto so useful, might now prove awkward.

Joane resisted paying out the reward to the local boy who found the corpse, but eventually had to, £20 going directly to the boy and the remainder to the overseers of the parish for his benefit. Jeffreys, by then Lord Chancellor, adjudicated.[34] Liability for the payment of the reward was one of the issues in the subsequent family Chancery actions.

Many years later, in 1712, Thos. Hearne reported gossip that the father went in fear of his life from Halley, a story that he repeated in 1718.[35] The origin and circumstances of that tale, which was first put about in February 1704 when Halley had just come into the Savilian professorship, are discussed in Chapter 12. There is good reason to be sceptical of the story, which was not among the reasons for the father's disappearance that were reported in current gossip at the time. There may have been some basis of family friction to it, or it may be a more lurid version of some of the gossip in the *Tract*; perhaps the father did indeed go in fear of his life, if not from his son.

There are further possible complications. Apart from Braddon no one seems to have been very anxious to reopen the matter of the death of Essex. That is not perhaps so odd as it might seem. Radicals such as Braddon, Ferguson, and others were not popular with William's government; in Holland they were associated with republicans who opposed William. Nor were the radicals happy with William: they saw that his accession meant the end of republican hopes in England.[36] Those who supported William, such as Bishop Burnet and the new earl of Essex, and many others in the new administration, would have had little inclination to muddy the waters. In any case, Braddon's claims must have looked as dubious to contemporaries as they do today. In those circumstances the fate of Halley's father cannot have seemed to matter.

It is always possible that the father's death was unrelated to that of Essex, but that Ferguson and others found it convenient to make the connection to attack the Duke of York, just as Pepys's clerk Atkins was accused of the Bury murder to attack Pepys and through him the Duke.

Halley's own position could have been difficult if he was close to James, and if Ferguson was right in saying that the father was murdered by command of James. Did Halley know of Ferguson's pamphlet when, three years later (Chapter 6), he presented the first edition of Newton's *Principia* to James, then king? Might his concern in 1689 derive from a risk that an enquiry might call attention to his dependence on James or to possible services to him in the summer of 1688 (Chapter 9)?

There at present the trail ends.

5.4 Such an impertinently litigious Lady[*]

Whatever, if anything, lies behind Hearne's report that the father went in fear of his life from Halley, the Chancery actions of 1686 and other later ones came about because Halley and Joane disagreed over the division of the father's personal estate, for he died intestate.

Personal estate was only part of the assets of a testator, which were commonly of two kinds, real estate which was landed property that would normally descend to legal heirs, and personal estate which was other assets such as money, debts, plate, mortgages, that might be disposed of in a will. Halley's father also had the assets of his business as a monopolist soapboiler. Real property was frequently transferred during the lifetime of the testator—the father received only £5 in the grandfather's will because a full child's portion had already been transferred to him. Similar arrangements may have been made for Halley, especially after his marriage and the death of his brother abroad. The house in Islington was possibly family real estate transferred to Halley on his marriage. The house where the father was living at Shoreditch when Halley was born may have been real estate; there was also a large house in St Giles's, Cripplegate, the parish in which Anne Halley died, on which the father had paid hearth tax. Halley had houses in that parish from about 1684 to the end of his life. There may have been family interests at Barking. None of that putative property appears in any public document and of course in no will of the intestate father. The father had personal property of value £4000 in money, debts owing to him, leasehold houses, and otherwise; it was over those that legal disputes arose when he died intestate. It was the City custom that the widow of a citizen should have a third part of the personal estate, while another third was divided among the children—the heir at law, Halley, the only surviving child after his brother Humphrey had died abroad, would inherit all the real estate. Joane claimed that she was also entitled to the remaining third part of the personal estate, the so-called 'Dead's part', which City custom allowed to be disposed of by will—but the father left no will.

Disagreement between Halley and Joane broke out with the father's disappearance. As soon as the father's death was confirmed Joane sued in the Prerogative Court of Canterbury for the administration of the estate but Halley opposed her and claimed a share of the 'Dead's part' in addition to his 'Child's third'. The Court resolved the matter, with the agreement of Halley and Joane, by granting letters of administration in July 1684 to trustees, Sir John Buckworth nominated by Halley, and

[*]I. Newton

Richard Young Esq. whom Joane nominated. Buckworth and Young, men of influence in the City and the Levant Company, had acted jointly in other matters, for example, in 1672 they had given security for the due performance of his duties by an official of the Levant Company. They now charged Halley with managing the estate under their direction.[37]

Joane's remarriage to Cleeter in 1685 led to further legal activity. Halley agreed with Cleeter on the general disposition of the personal estate whereby, Joane's claims notwithstanding, they divided the property roughly equally. Halley received the houses in Winchester Street with the exception of the one in which Joane continued to live, and money invested in the Coal Office of the City. Joane had the remainder, comprising other houses in the City, the Dog Tavern in Billingsgate, money in the General Lottery, and sundry debts due to the father. Some issues remained unresolved and led to an action in Chancery by Cleeter and Joane against Buckworth, Young, and Halley. It was the first of two groups of suits, those of 1686 and the later ones of 1693 and 1694. There is no record that any ever came to trial or was pursued further than complaint and answer, and they may have been intended as no more than records of the parties' positions. The issues were minor in the context of the overall settlement of the estate. They included a legacy of £100 from the uncle Humphrey II that had been paid to the father for Halley and a further legacy of £100 due to the estate of Halley's deceased brother Humphrey III that Halley was administering; the status of some of Joane's jewellery, and debts that had been assigned to her in the father's lifetime; and the custody of books of account in which the debts owing to the father were recorded. The later actions were prompted by the fact that Halley was expected to go overseas on his magnetic surveys (put off until 1698), that Buckworth had died, that a final account had not been drawn up, and that Young did not want to be left liable for any debts and obligations. There is no reference in the later papers to any court action upon the earlier claims and no action seems to have followed after 1694.

The schedules (Appendix 3), already discussed in more detail in Chapter 2, are for two periods, the first from the earlier action for about one year up to the Chancery claim, and the second for about nine and a half years up to the actions in 1693. There are naturally discrepancies between them but after allowing for exceptional expenses, the total income from the personal estate was between £350 and £400 per annum and Halley's share from the houses in Winchester Street was unlikely to have been less than £150 and may have been as much as £200 per annum, although in the first years he had exceptional expenses.

Halley must have had other assets as the only surviving child of his father. He almost certainly had real property, which, as already mentioned, it may be possible to identify to some extent. The assets that have left no trace are those of the father's business; what happened to the soap boiling after the father's death? Halley or Joane would probably have put someone in to run it, a senior employee perhaps, as was commonly done by the families of City tradesmen in comparable circumstances; but of that there is no record or hint.

There is also no hint in the Chancery claims and answers of any possible link between the father's death and that of the Earl of Essex. In the strict matter of the disposition of the estate it was no doubt irrelevant, but it would also have been tactless to mention it: the Lord Chancellor before whom the issues would have been tried, and who did decide for the poor boy who found the corpse, was Lord Jeffreys who had convicted Braddon and Speke—thin ice indeed if the father's death really did follow from Essex's.

5.5 *Science with distractions*

When Halley married and set up house and observatory at Islington, he had behind him a considerable experience of observing the heavens, at Oxford with Boucher and on his own, at Greenwich with Flamsteed, and on St Helena; and he had met the outstanding observers of his day and seen their methods, Hevelius at Danzig, Cassini, Picard and Römer in Paris, and Gallet at Avignon. Those experiences, the results he harvested from them, and the acquaintances he had made among the leading astronomers of Europe, gave him already at twenty-six a high reputation in Europe and made him a valued member of the Royal Society, such that he would be elected to the Council in 1683. He had good instruments of his own, the one thing he lacked was a mural circle. His financial position, with an allowance from his father, seemed secure. He was known to the King and with an MA from Oxford he had the status of a gentleman.

For two years Halley devoted himself in the agreeable surroundings of Islington to systematic observations of the Moon that he planned to carry on over an eighteen year cycle.[38] It was not to be. The outside world interfered and he did not return to the Moon until he went to Greenwich as Royal Observer thirty-five years after his regular observing at Islington came to its abrupt end.

Halley was abroad for much of the three years from 1679, the years in which the struggle between Charles II and his parliamentary opponents over the powers of Parliament and the exclusion of James, Duke of York, from the successsion was going the way of the radicals. By 1682 Charles's finances had so improved that he could dispense with parliaments and had put in Tories as Lord Mayor and sheriffs in the City. Next year the Rye House Plot was discovered. Those events do not appear in the records of Halley's travels abroad nor in those of his visits to Greenwich, although he must have known some of the actors. He cannot have been oblivious to them when he met Henry Savile in Paris.

Savile leads us at once into high politics and clandestine plotting and intrigue. He was the brother of George Savile, Marquis of Halifax, who held high office under Charles II, James II, and William and Mary. The Saviles were nephews of Anthony Ashley Cooper, the Earl of Shftesbury, to whom Halifax was opposed politically, as he was to Algernon Sidney. Halifax apparently tried to intercede for Sidney and Earl Russell after the Rye House Plot, but without effect, while earlier Henry and Sidney corresponded when Henry was in Paris.[39] Politics and conspiracy changed Halley's course, if his father's death was in fact a consequence, albeit fortuitous, of the death of the Earl of Essex. Had that not occurred he might well have pursued his observations at Islington and not have gone to see Newton in Cambridge in August 1684. He ceased regular observing after the father's death, he did go to see Newton, and the upshot was the writing and publication of the *Philosophiae naturalis principia mathematica*.

Halley's visit to Newton and the writing of the *Principia*, so crucial in the evolution of science, took place in turbulent times. In France the Edict of Nantes was revoked in 1685, with great consequences for the intellectual life of England and for some of Halley's colleagues. On the other side of Europe, Turkish armies were beaten back from Vienna in 1683 and from Buda and Belgrade a little later. Those events bore upon Halley's mission to the Adriatic in the new century. In England Charles II died and James II succeeded to the throne; many radical opponents of the regime, John Locke among them, sought refuge in Holland, and some plotted rebellion. Monmouth's rebellion was crushed, James seemed triumphant. It all collapsed, William invaded, James fled to France. William and Mary claimed the throne. Newton finished the *Principia* and Halley published it just at the time when it must have seemed that James would carry all before him. Those stirring political events, with constitutional results still with us, were also part of the history of science. They certainly affected the timing, if nothing more, of the stirring scientific events of the same years with scientific results that are still with us. The events that summed up and established a revolution in science coincided with those that summed up and established a revolution in English politics. Of all the connections that there were between politics and natural philosophy, none surely was more bizarre than Halley's.

6
Achilles produced[1]

> Il maestro di color che sanno
> [The master of those who know]
>
> DANTE: *INFERNO*, IV, 131

6.1 *The origin of Principia*

HALLEY, along with Flamsteed, was elected to the Council of the Royal Society at the Anniversary Meeting for 1683 and served for two years. At first he regularly attended meetings, when often there were no more than seven or eight councillors out of twenty. He did not attend for the latter part of the 1684–1685 session when family matters and the *Principia* must have taken up much of his time. At meetings of the Society he performed some magnetic experiments relevant to his ideas about the Earth's magnetic field, and he estimated the latitudes of Constantinople, Rhodes, and Alexandria from astronomical observations that had been sent in to the Society.[2]

In the evening of 14 January 1683/4 (os) after the meeting of the Society, Halley fell into conversation with Wren and Hooke about the orbits of the planets about the Sun. He afterwards recalled to Newton,

> ... and this I know to be true, that in January 83/4, I having from the sesquialtera proportion of Kepler, concluded that the centripetall force decreased in the proportion of the squares of the distances reciprocally, came one Wednesday to town, where I met with Sr Christ. Wren and Mr Hook, and falling in discourse about it, Mr Hook affirmed that upon that principle all the Laws of the celestiall motions were to be demonstrated.[3]

In 1673 Huygens had argued that the centrifugal force on a body moving with a uniform velocity, v, in a circular orbit of radius r was v^2/r. Halley had seen that Kepler's third law (the square of the period of a planet is proportional to the cube of the major semi-axis) implied that the forces with which the planets are pressed out from the Sun were inversely proportional to the squares of their mean distances from it.

Wren was the immediate past-President of the Royal Society and Halley was a councillor. Hooke, the Society's curator, was Wren's close colleague in his architectural practice and in rebuilding the City after the Great Fire. Halley had been known to them at least since the days when they all went to Greenwich to view the site of the Royal Observatory in 1675.

All three were familiar with the problem of the orbits of planets. Hooke in 1674 had argued that a body moving in an orbit is continually deflected from its path by an attraction to a centre, although he did not then have the idea of an inverse square law. He would later assert that he had suggested the inverse square law to Newton. Wren had thought for some time that gravity might decrease as the square of the distance and had spoken with Newton about it as early as 1677.[4] When the three talked together after the meeting of the Royal Society, Hooke claimed that he could derive all the laws of celestial mechanics from an inverse square law of force but Halley and Wren accepted that they had not been able to do so. Wren, who doubted Hooke's claim, offered a prize of a book to the value of forty shillings to whomever could produce a demonstration within two months.

About the time that Wren's two months expired, Halley's father disappeared and he became involved with all the consequences, among them managing the father's estate on behalf of the trustees. Those matters would certainly have taken much of his time up to the end of June and no doubt for some time afterwards, for he had to collect rents and other sums due to the estate and to pay debts and expenses. John Eaton, the rector of Sawtry near Alconbury, close to Huntingdon, had owed the father £12. Halley would have been responsible for paying an annuity of £3 per year to Susanna Sandwith at Alconbury, as the will of uncle Humphrey II provided; she had previously had property dealings with the Halleys (Chapter 2). At some time after July 1684 Halley probably had to go to Alconbury, and possibly to his relatives in Peterborough, to settle those matters. He visited Newton in Cambridge in August and the problem of orbits under the inverse square law came up in their discussion.

Why Halley called on Newton is not clear, even though he may have been nearby on family business. No one had suggested asking Newton in January despite Hooke's correspondence with him about orbits some four and a half years before.

Halley had probably met Newton in 1682 after he was back from France. Directly and indirectly, he provided many of the observations of comets that Newton used in Book III of the *Principia*, for Flamsteed's knowledge of many foreign observations came initially from Halley in France. When Halley returned to England he told Newton of his observations of the comet on the road to Paris (Chapter 4). He had observed with Cassini and could have given Newton a first-hand assessment of the observations by Cellio and Pontio (Ponthaeus in the

Principia) in Rome and by Gallet at Avignon. Newton's following note in his manuscript Waste Book may reflect what Richer told Halley when they met in Paris about his pendulum in Cayenne. Knowing of and contributing to Newton's earlier interest in comets, it is hardly surprising that when Halley visited Newton he mentioned Hooke's claim to have solved the general problem of orbits.

Newton later told de Moivre:

> In 1684 Dr Halley came to visit him at Cambridge, after they had been some time together the D^r asked him what he thought the Curve would be that would be described by the Planets supposing the force of attraction towards the Sun to be reciprocal to the square of their distance from it. Sr Isaac replied immediately it would be an Ellipsis, the D^r struck with joy & amazement asked him how he knew it, why saith he, I have calculated it, whereupon Dr Halley asked him for his calculation without any further delay, Sr Isaac looked among his papers but could not find it, but he promised him to renew it, & then send it him ...[5]

Newton always claimed to have done the calculation in 1679–1680 when he and Hooke had corresponded about planetary orbits. He apparently put the matter aside when he became concerned with comets at the end of 1680. He certainly did not lose interest in celestial mechanics in the years prior to Halley's visit of 1684, for he was deeply interested in the orbit of the comet of 1680. Comets brought Newton and Halley together in the first place, and were an essential element in the genesis of the *Principia*.[6]

Newton is known to have had problems repeating the demonstration when he came to look at it again.[7] Then in November Edward Paget brought to London the little article of nine dense pages, *De motu corporum in gyro*, in which Newton showed that an elliptical orbit could arise from an inverse square attraction of gravity. Newton also derived Kepler's second and third laws and the trajectory of a projectile under constant gravity in a resisting medium.[8]

The problem that Newton had solved was crucial to the development of celestial mechanics and dynamics.

In Halley's calculation, which Newton had already done in 1660, orbits are supposed circular, and it has to do with the relation between concentric circular orbits, not the form of an individual orbit. Kepler's third law shows that the attractions on different planets are as the inverse squares of their distances from the Sun. Newton broke new ground when he calculated the form of an individual orbit. In his paper *De motu*, and subsequently in the *Principia*, he showed that Kepler's first two laws, that planetary orbits are ellipses and that the areal velocity is a constant, imply that the attraction to the focus in any one planetary orbit is everywhere as the inverse of the distance from the focus.

Newton's demonstration depends on two results. The first (*Principia*, Book I, Prop. 1) is that the areal velocity (angular momentum) under a force directed to a fixed point is constant. Consequently time may be replaced by area swept out as an independent variable. The second result is an expression for the accceleration in an orbit in the form of a certain geometrical ratio related to the curvature of the path (*Principia*, Book I, Prop. 5).[9] In subsequent propositions Newton relates that expression to well known geometrical properties of conic sections and hence shows that an orbit in the form of any conic section, ellipse, parabola, or hyperbola, would entail an inverse square attraction to a focus (*Principia*, Book I, Props 11–13).

Newton's treatment of the inverse problem—what is the orbit under an inverse square law of attraction?—is far less clear than his treatment of the direct problem—what is the attraction when the orbit is a conic section? In *Principia*, Book I, Prop. 41, he developed a general method for attacking the problem but Johann Bernoulli I considered that he never demonstrated that orbits under an inverse square attraction were necessarily conic sections, and he seized upon that lacuna in his criticism of the *Principia*. An argument can however easily be constructed. The geometrical propositions mentioned above, together with the areal velocity theorem, lead to an expression for the curvature that agrees with that for a conic section. A conic section is therefore a possible orbit under an inverse square law attraction. Newton also showed that any initial conditions would correspond to a conic section, ellipse, parabola, or hyperbola, and hence that whatever the initial conditions, the orbit under an inverse square law to a centre would be a conic section.[10]

With those arguments about orbits, Newton took the first steps to founding celestial mechanics on a dynamical basis.

Halley realised the importance of *De motu* as soon as he received it. After a further visit to Cambridge he reported to the Royal Society on 10 December, when the President, Samuel Pepys, was in the chair:

> Mr Halley gave an account that he had recently seen Mr. Newton at Cambridge, who had shown him a curious treatise, *De motu*; which, upon Mr. Halley's desire, was, he said, promised to be sent to the Society to be entered on their register.
>
> Mr Halley was desired to put Mr Newton in mind of his promise for securing his invention to himself till such time as he could be at leisure to publish it. Mr Paget was desired to join with Mr Halley.[11]

Halley had longer to wait for the new version than he may have expected. In the early months of 1685 Newton was extending *De motu* and developing the principles of dynamics in a rigorous manner. By the end of 1685 he had expanded *De motu* into two volumes known now as *De motu corporum* that Halley read and annotated lightly. In April 1686, Halley gave a 'Discourse on Gravity' in which he reported Newton's 'incomparable *Treatise of motion* almost ready for the *Press*' and summarised what was to become Book I of *Principia*.[12] The first part arrived at the Royal Society on 28 April:

> Dr. Vincent presented to the Society a manuscript treatise entitled *Philosophiae naturalis principia mathematica*, and dedicated to the Society by Mr. Isaac Newton, wherein he gives a mathematical demonstration of the Copernican hypothesis as proposed by Kepler, and makes out all the phaenomena of the celestial motions by the only supposition of a gravitation towards the centre of the sun decreasing as the squares of the distances therefrom reciprocally.
>
> It was ordered that a letter of thanks be written to Mr. Newton; and that the printing of his book be referred to the consideration of the council; and that in the meantime the book be put into the hands of Mr. Halley, to make a report thereof to the council.[13]

The minute was no doubt written by Halley as Clerk, to which position he had recently been appointed. He was probably the only person in the Society in any position to make a report.

Newton took about two years to write the *Principia*, from the late summer of 1684 to the middle of 1686, and the course of his composition is known in great detail.[14] That is not repeated here: this is the story of Halley's part in the birth and epiphany of the *Principia* from early in 1684 to the summer of 1687.

6.2 Halley in 1686

Halley was elected Clerk to the Society on 7 February 1686, one of a number of candidates, Sloane among them. On 9 December 1685 the two honorary Secretaries of the Royal Society, Francis Aston and Tancred Robinson, had resigned from the posts to which they had just been re-elected at the Anniversary meeting of 1685. Halley thought that Aston wanted a greater financial reward for his efforts.[15] Council discussed the position on 16 December and decided by 10 votes to 5 to have another salaried clerk (Hooke had been the last). The Society then elected John Hoskyns and Thomas Gale as Secretaries. Council resumed its discussion of the clerkship on 13 and 27 January. They drew up a list of twelve conditions and five duties: the clerk could not be a Fellow and must have no other employment, he must lodge constantly in Gresham College, he must be a single man without children, write a fair and legible hand and be a master of English, French, and Latin, and knowlegeable in mathematics and experimental philosophy; he was to obey the orders of the President, Council, and officers, to submit all his letters for signature by the Secretaries, to keep the Minutes and a catalogue of gifts. However, the President told the Society that the choice was theirs and that they might dispense with any of the qualifications, 'which were the results of the Council's thoughts'. Plainly the Clerk, as a junior paid officer, not provided for in the statutes, was to be subordinate to the honorary officers (he was to sit hatless at the end of the Council table).[16]

There were four candidates at the meeting of the Society on 27 January 1686. Sloane had 10 votes in the first ballot, Papin, 8, Salisbury, 4, and Halley 16 votes. Salisbury dropped out and in the second ballot Sloane had 9 votes, Papin, 6 and Halley, 23. Halley was thereupon sworn before Council as clerk. He fulfilled few of the Council's conditions, although he did resign his Fellowship and was re-elected only in 1700 after he ceased to be Clerk. Some must have found his appointment unacceptable for on 7 July 1686 the Council asked the Society to confirm Halley despite his marriage, his children, and his residence outwith the Society's premises. Even so, on 10 December a new election was called, and the Society balloted for Halley's continuance. On 16 January 1687 a committee was chosen to inspect the books of the Society that Halley kept and, having done so, reported on 20 February that they were in a very good condition. Only on 17 July 1687 did the Council settle Halley's salary at £50 with a gratuity of £20, all to be paid in copies of the *Historia piscium* (*The history of fishes*), of the nominal value of £1 each.[17]

The *Historia piscium*, left unfinished by Francis Willoughby at his death, had been completed by John Ray. Tancred Robinson had persuaded the Royal Society to publish and pay for it. It had numerous plates and was very expensive. The Society was left with many copies unsold, just when Newton's first draft of the *Principia* arrived.

Within the month of April 1686 Halley described his measurement of the density of mercury to the Society, he spoke on gravity and announced the imminent arrival of Newton's work, he was charged to report on the manuscript when it arrived, he was sued in Chancery when his stepmother and her new husband, Robert Cleeter, started their action against him (Chapter 5). No doubt the Chancery suits were vexatious for Halley, but for us they open a window on to his life just as he was reading the *Principia* critically and seeing it through the press. The estimate of his income from investments as between £150 and £200 a year (Chapter 5) puts a different light on his position as Clerk to the Royal Society from that in which it is sometimes presented. If, as is likely, the house at Islington and that at Golden Lion Court in St Giles', Cripplegate, to which he later moved, were family real estate, Halley was not almost destitute, he was rather well off. With the MA degree from Oxford, he ranked as a gentleman, he had substantial property, and he was well known and respected in London and overseas. Why then did he seek the clerkship? Why, indeed, did Sloane? He also probably did not fulfill most of the Council's conditions. About four years younger than Halley, in 1686 he had recently returned from France where he had taken an MD degree. He was just getting established as an up-and-coming physician in London (later that year he went to Jamaica for eighteen months). Money was probably not important, for Halley's salary of £50 was to be paid in copies of the almost unsaleable *History of fishes*.[18] The Clerkship must have been attractive to two young and able men for the opportunities it opened up, rather than for the modest financial reward it might offer.

Halley's income without the Clerkship was enough to enable him to bear the charge he was about to undertake in seeing *Principia* through the press. He also apparently took financial responsibility for *Philosophical Transactions*, for on 29 November 1686, when the Council decided that there should be a new election for the clerk, it ordered that Halley should be paid £13 10s. in full for 5 dozen of *Philosophical Transactions*, numbers 179–183 inclusive, for the Society's use.[19]

The Society continued Halley as Clerk despite his clear breaches of many of the Council's conditions. He evidently performed his duties efficiently and to the satisfaction of the Society, and with considerable independence. He seems to have carried on the Society's correspondence on his own reponsibility, as he wrote to Cassini, 'l'affair de manager la correspondance ... de la Societe a été mis entre mes mains' and concluded, 'Vous pourrez addresser vos lettres au Secretaire de la Societe Royale a Gresham College a Londres'.[20] Halley appears to have acted in some ways as a modern Executive Secretary, and to have shown more responsibility and initiative than the Council may have envisaged, if the Council's original conditions are any guide.

6.3 *Printing Principia*

Although the Royal Society had at the end of April asked the Council to consider the printing of the *Principia*, the Council had not met by 19 May 1686 and that day the Society, not the Council, resolved

> That Mr Newton's *Philosopiae naturalis principia mathematica* be printed forthwith in quarto in a fair letter; and that a letter be written to him to

signify the Society's resolution, and to desire his opinion as to the print, volume, cuts &c ...[21]

Halley thereupon wrote to Newton, on 22 May, to tell him of the decision:

> S^r
>
> Your Incomparable treatise intituled *Philosopiae naturalis principia mathematica*, was by Dr Vincent presented to the R. Society on the 28th past, and they were so very sensible of the Great Honour you do them by your Dedication, that they immediately ordered you their most hearty thanks, and that a Councill should be summon'd to consider about the printing thereof; but by reason of the Presidents attendance upon the King, and the absence of our Vice-Presidents, whom the good weather had drawn out of Town, there has not since been any Authentick Councill to resolve what to do in the matter: so that on Wednesday last the Society in their meeting, judging that so excellent work ought not to have its publication any longer delayd, resolved to print it at their own charge, in a large Quarto, of a fair letter; and that this their resolution should be signified to you and your opinion therin be desired so that it might be gone about with all speed. I am intrusted to look after the printing it, and will take care that it shall be performed as well as possible, only I would first have your directions in what you should think necessary for the embellishing therof, and particularly whether you think it not better, that the Schemes should be enlarged, which is the opinion of some here; but what you signify as your desire shall be punctually observed.[22]

Halley warned Newton that Hooke was claiming priority in the suggestion of the inverse square law, and was expecting some mention in a preface. Newton's reply, of 7 June, was a moderate account of his previous correspondence with Hooke in 1679 about the path of a body falling to the centre of the Earth and demonstrating the rotation of the Earth. Before Halley responded the Council of the Royal Society had met on 2 June and

> It was ordered, that Mr. Newton's book be printed and that Mr. Halley undertake the business of looking after it, and printing it at his own charge; which he engaged to do.

Subsequently, on 30 June, the Council asked the president (Pepys) to licence the printing of *Principia*, as the Society was authorised to do under the Charter.[23]

People have often remarked upon the Council's decision that he, Halley, should look after the printing and do so at his own charge. Why should the Council place that responsibilty upon a young man, their subordinate Clerk, and how could they expect him to bear the costs? As to 'looking after the printing', Halley was clearly familiar with dealing with printers; his correspondence with Newton shows that he had command of the details of presswork, and as Clerk to the Society he already had experience of seeing issues of *Philosophical Transactions* through the press. The financial responsibility is the one that has attracted general notice and at first sight it seems a great burden to have put upon Halley, although he already had a similar responsibility for the *Philosophical Transactions*. The reasons for the decision will probably never be known but may include the weak financial state of the Society on account of the *Historia piscium*, Halley's evident enthusiasm for *Principia* and determination to see it published, and his own resources. It does seem that it was a matter between the Council and Halley, and that a mean Newton did not impose it on Halley.

No accounts are known to exist for the first edition of *Principia*, so that a close estimate of the cost of printing is not possible, but a rough idea may be obtained from a comparison with the accounts for the second edition, printed at Cambridge under the eye of Richard Bentley. Paper would have been the greatest cost of both. Consequently, it may not be far out to take the costs of the two editions as proportional to the numbers printed. It used to be thought that there were between 300 and 400 copies of the first edition, of which almost 100 were distributed as gifts. More recent estimates however put the print run at 600 or 700. We know that 700 copies of the second edition were printed at a total cost of £117, so that the cost of the first edition would have been between £60 and £120. The price in sheets was 5 shillings, sufficient to have given Halley a modest return.[24] He may, however, not have made a profit, for he seems to have borne the costs of the more expensive presentation copies that were on better paper and bound in leather. He wrote to Newton,

> ... I am satisfied there is no dealing in books without interesting the Booksellers, and I am contented to lett them go halves with me, rather than have your excellent Work smothered by their combinations.[25]

Halley probably did not lose by his very risky venture but did not make much of a profit. He is unlikely to have known this when he undertook the publication, just after he, Buckworth, and Young had delivered their answer to the Chancery action of Cleeter and Joane. *Principia* was certainly an adventure.

Halley's crucial part in the printing and publication of the *Principia* can be followed in great detail in his correspondence with Newton from 18 June 1686 to 16 July 1687, when he told Newton that copies of the complete work were on the way to Cambridge.

On 18 June 1686 he sent a proof for Newton:

> I here send you a proof of the first sheet of your Book, which we think to print on this paper, and in this Character; if you have any objection it shall

be altered: and if you approve it, wee will proceed; and care shall be taken that it shall not be published before the End of Michaelmas term, since you desire it. I hope you will please to bestow the second part, or what remains of this, upon us, as soon as you shall have finished it; for the application of this Mathematical part, to the System of the world; is what will render it acceptable to all Naturalists, as well as Mathematiciens; and much advance the sale of ye book. Pray please to revise this proof and send it me up with your Answer, I have already corrected it, but cannot say I have spied all the faults. when it has passed your eye, I doubt not it will be clear from errata. The printer begs your excuse of the Diphthongs, which are of a Character a little bigger, but he has some a casting of the just size. This sheet being a proof is not so clear as it ought to be; but the letter is new, and I have seen a book of a very fair character, which was the last thing printed from this set of Letter; so I hope the Edition may in that particular be to your satisfaction.[26]

Newton in reply (20 June 1686) explained in some detail why he was not beholden to Hooke and why Hooke's contribution was very modest; he then continued

The proof you sent me I like very well. I designed ye whole to consist of three books, the second was finished last summer being short & only wants transcribing & drawing the cuts fairly. Some new Propositions I have since thought on wch I can as well let alone. The third wants ye Theory of Comets. In Autumn last I spent two months in calculations to no purpose for want of a good method, wch made me afterwards return to ye first Book & enlarge it wth divers Propositions some relating to Comets others to things found out last Winter. The third I now designe to suppress. Philosophie is such an impertinently litigious Lady that a man had as good be engaged in Law suits as have to do with her. I found it so formerly & and now I no sooner come near her again but she gives me warning. The first two books without the third will not so well beare ye title of *Philosophiae naturalis Principia Mathematica* & therefore I had altered it to this *De motu corporum libri duo*: but upon second thoughts I retain the former title. Twill help ye sale of ye book wch I ought not to diminish now tis yours. The Articles are wth ye largest to be called by that name. If you please you may change ye word to *sections*, tho it be not material. In ye first page I have struck out ye words *uti post hac docebitur* as referring to ye third book. Which is all at present from

Your affectionate friend &
humble Servant
IS. Newton.[27]

In a lengthy addendum Newton gave further details of the relationship of his work to Hooke's. Halley did adopt *Sections* for the major divisions of the three Books.

Newton's letter is particularly significant for the history of the *Principia*. He appears discouraged by the problem of comets and no doubt also by his resentment at Hooke's claims to have forestalled him on the inverse square law and its consequences, and reacts with his intention of suppressing the third book. The dispute with Hooke concerns Halley in so far as he tried to compose matters between the two, and to dissuade Newton from suppressing the third book, as his reply to Newton shows. The second point of interest is that Halley is in effect the publisher, as is plain from Newton's phrase, 'now tis yours'. Finally, the sentence about *articles* and *sections* seems to carry the clear implication, as do other indications, that the letters we have are only part of the intercourse between Newton and Halley.

Newton's remark about philosophy being a litigious lady would be apposite if Newton knew of the Cleeters' action, and would indicate a close relation between Newton and Halley.

Cohen has shown more directly that the correspondence we now have is not the whole story. The printer worked from the fair copy (Cohen's M) made by Humphrey Newton, now in the Royal Society. It was written over an extended period, and Halley had the first part and it was in use by the printer at the time of the letters just quoted; Halley did not receive the parts for the second and third books until later. Halley made suggestions and editorial corrections to M identified by Cohen in the Variorum edition of the *Principia*. They are not numerous and are far fewer than those by Newton himself but show that Halley was in closer touch with Newton from the time of the receipt of the copy for the first part than the known letters might indicate. Among the more common of Halley's amendments are corrections to the labelling of figures and improved references to them in the text.

Cohen identified certain of Newton's papers as drafts for M and he further attributed suggestions for their revision to Halley. If that is so Halley would have seen a preliminary version of the *Principia* before Newton separated the first two books. The suggestions he made at the earlier stage are more numerous than those on M, in particular, he gave considerable attention to the enunciation of the laws of motion, and the form in which they now appear is to some extent the result of discussions between him and Newton. Halley also picked up infelicities in the Latin. Cohen's purpose in the very thorough analysis he made of the critical comments, and of Newton's response to them, was to demonstrate that there was a complete but shorter version of the *Principia* prior to M. His argument does not depend on identifying the critic but he put forward strong reasons for believing it could only be Halley.[28]

Halley replied promptly to Newton's letter of 20 June and on 29 June wrote:

> I am heartily sorry, that in this matter, wherein all mankind ought to acknowledge their obligations to you, you should meet with anything that

> should give you disquiet, or that any disgust should make you think of desisting in your pretensions to a Lady, whose favours you have so much reason to boast of. Tis not shee but your Rivalls enviling your happiness that endeavour to disturb your quiet enjoyment, which when you consider, I hope you will see cause to alter your former Resolution of suppressing your third Book, there being nothing which you can have compiled therein, which the learned world will not be concerned to have concealed; These gentlemen of the Society to whom I have communicated it, are very much troubled at it, and that this unlucky business should have happened to give you trouble, having a just sentiment of the Author thereof.

Halley went on to tell Newton of a conversation he had had with Wren about previous discussions of the inverse square law, and then relates his own meeting with Wren and Hooke in January 1684 and his visit to Newton in Cambridge in the August of that year. He suggests that Newton has heard a more unfavourable account of Hooke's claims than actually occurred when Vincent presented Newton's book at the meeting of the Royal Society.

> After the breaking up at that meeting, being adjourned to the Coffee-house, Mr Hook did there endeavour to gain belief that he had some such thing by him, and that he gave you the first hint of this invention, but I found that they were all of opinion, that nothing thereof appearing in print, nor on the Books of the Society, you ought to be considered as the Inventor; and if in truth he knew it before you, he ought not to blame any but himself, for having taken no more care to secure a discovery, which he puts so much Value on. What application he has made in private I know not, but I am sure that the Society have a very great satisfaction in the honour you do them, by your dedication of so worthy a Treatise.
>
> Sir, I must beg you, not to let your resentments run so high, as to deprive us of your third book, wherein the application of your Mathematicall doctrine to the theory of Comets, and severall curious Experiments, which as I guess by what you write, ought to compose it, will undoubtedly render it acceptable to those that will call themselves philosophers without Mathematicks, which are by much the greater number.
>
> Now you approve of the Character and Paper, I will push on the Edition Vigorously. I have sometimes thought of having the Cutts neatly done in Wood so as to stand in the page, with the demonstrations, it will be more convenient, and not much more charge, if it please you to have it so, I will trie how well it can be done, otherwise I will have them in somewhat larger size than you have sent up.[29]

Diagrams were often printed on fold-out sheets that were tipped in to the letterpress when volumes were bound, usually between gatherings. It was inconvenient for the reader. Halley's idea of using woodcuts, so that the diagrams would appear on the same page as the relevant text, was obviously much more convenient.

Newton, replying on 14 July, accepts Halley's suggestions about woodcuts, shows himself somewhat mollified about Hooke, but ends with further justification of his own priority and independence. He refers to Halley's observation that the pendulum swung more slowly at St Helena and the implication that gravity at the equator was reduced by the centrifugal force of the diurnal rotation of the Earth. In fact, the oblateness of the Earth also contributes to the reduction of gravity, as Newton himself was to show in Book III of *Principia*, and his letter suggests that he now has the solution as it appears there. Halley was as yet unaware of the contents of Book III as they would eventually be; Newton was actively engaged on it in the summer of 1686. Then or later Halley persuaded Newton to continue with it, and it does not figure in their correspondence again until March 1687, and again on 16 April when Halley received the copy for it.[30]

Halley and Newton exchanged letters again in October 1686. Halley told Newton of the progress of the printer and raised questions about what is known as Kepler's problem and about a transformation of homogeneous coordinates that converted a parallelogram into a trapezium; Newton gives Halley an expansion in English of a rather terse Latin sentence in the text of *Principia* (Book I, Lemma 22).[31] There are no further known letters from Newton to Halley until 13 February 1687, but in December 1686 Halley, writing to Wallis about the motions of bodies in a resisting medium, had sent him some results of Newton's that he had himself copied out some two years before. Wallis published his own results on the subject shortly afterwards.[32]

Newton's letter to Halley of 13 February 1687 shows that there must have been previous correspondence about sheets of the manuscript that had been lost. Newton also states that he believes he has the solution to the problem of the parallactic inequality of the Moon's motion, the perturbation of the Moon's orbit that depends on the ratio of the distance of the Sun to the distance of the Earth from the Moon. Newton never attained a satisfactory theory of the Moon's motion under the combined attraction of the Earth and the Sun, but he did realise that there were perturbations from which the distance of the Sun could be derived. Halley had already studied the geometrical method for finding the distance of the Sun that depended on observations of a transit of Mercury or Venus in front of the Sun, as he was to discuss in detail in 1691 (see Chapter 8). Newton and Halley exchanged further letters about that, and Wallis's papers and the progress of the press, on 24 February and 1 March, when Newton told Halley he was sending the copy for Book II.[33]

The press was busy in the spring of 1687, for on 7 March Halley told Newton he was employing a second printer to speed things up. No further letter from Newton to Halley is known until their joint work on cometary orbits in 1695 (Chapter 9), but there must have been others, for Halley wrote to Newton about the progress of the press on 14 March and 5 April. On that last occasion he commented on Newton's theory of comets and on his methods for determining their orbits, foreshadowing their studies of 1695.[34] At some time before the printing was finished, Halley composed the eulogy in Latin hexameters that now follows Newton's Preface (see Appendix 6).

Shortly after he received the manuscript of Book III on 4 April, Halley gave Wallis a comprehensive summary of it:

> I have lately been very intent upon the publication of Mr Newtons book, which made me forget my duty in regard of the Societies correspondants; but that book when published will I presume make you a sufficient amends for this neglect. I have now received the last Book of that Treatise wch is entituled de Systemati Mundi; wherein is shown the principle by which all the Celestiall Motions are regulated, together with the reasons of the several inequalities of the Moons Motion and the cause and quantity of the progression of the Apogeon and the retrocession of the Nodes. How he falls in with Mr Hook, and makes the Earth of the shape of a compressed sphaeroid, whose shortest diameter is the Axis, and determines the excess of the radius of the Equator above the semiaxe 17 miles, and from this quantity shows that the retrocession of the Equinoctiall points does necessarily follow, and computes it to be 49″ 58‴ yearly. Then he gives the reason for the tides to be the decrease of Gravity, from the contrary attraction of the Sun and Moon, whereby the Water being less pressed rises where they are verticall, and subsides when they are in the Horizon: he computes the effect of the Moon to be about 6 times as great as the Sunns, and seems to account very well for the severall phaenomena of the Tides. He concludes his book with the theory of Comets, showing that their Orbs are sufficiently near parabolicall; and upon that supposition shows how to find from observation the parabola wherein they move, and gives an example of the Motions of the great Comet of 1680/1, and having stated the Orb from the observations of the Evening Comet, he finds that the Comet that appeared in November in the Morning was the same with that long tailed one; and that his calculus will agree as well with its motion, as the ordinary tables do with the planets. I hope to get the whole finished by Trinity term, but the correction of the press costs me a great deal of time and paines.

That was a resounding thunderclap of a recital. So it must have struck Wallis, to read of all those novel results.

The work was nearly complete by 25 June when Halley wrote again to Wallis:

> I was willing to make you a more valuable present; I mean Mr Newton's book of Mathematicall Philosophie, which is now near finished; but I must entreat your patience yet ten days more, by which time it will be compleat. To hasten

the edition of this book, I have been obliged to attend 2 presses, which has so farr taken me up, that for that reason, and for want of such communications as might be worthy of you, I have so long forborn to write to you, but the book, which needs must please you, will I hope obtain my pardon.[35]

Finally it was finished:

Honoured S^r

I have at length brought your Book to an end, and hope it will please you. The last errata came just in time to be inserted. I will present from you the books you desire to the R. Society, Mr Boyle, M^r. Pagit, Mr Flamsteed and if there be any elce in town that you design to gratifie that way; and I have sent you to bestow on your friends in the University 20 Copies, which I entreat you to accept. In the same parcell you will receive 40 more, w^{ch}, having no acquaintance in Cambridg, I must entreat you to put into the hands of one or more of your ablest Booksellers to dispose of them: I intend the price of them bound in Calves leather and lettered to be 9 shillings here, those I send you I value in Quires at 6:shill to take my money as they are sold, or at 5.sh a price certain for ready or elce at some short time; for I am satisfied that there is no dealing in books without interesting the Booksellers, and I am contented to lett them go halves with me, rather than have your excellent Work smotherd by their combinations. I hope you will not repent you of the pains you have taken in so laudable a Piece, so much to your own and the Nations credit, but rather after you shall have a little diverted your self with other studies, that you will resume those contemplations, wherein you have had so good success, and attempt the perfections of the Lunar Theory, which will be of prodigious use in Navigation, as well as of profound and subtile speculation. Sr I shall be glad to hear that you have received the Books, and to know what further presents you would make in town which shall accordingly be done. You will receive a box from me on Thursday next by the Waggon that parts from hence to morrow.[36]

It was July the fifth (os).

Halley was the publisher, paid the costs and received the income from sales, and paid for presentation copies. One such copy sent to Cassini at the Observatoire in Paris carries the inscription in Halley's hand (Fig. 6.1):

Spectatissimo Domino Dno Joh: Dominico Cassino Grati animi tessaram Humillime offert. Edm. Halley
To the most worthy master Jean Dominique Cassini, presented in great humility as a token of a grateful heart. Edm. Halley.

Another presentation copy inscribed by Halley was for Fatio de Duillier.

For some years afterwards Halley received letters thanking him for presentation copies of *Principia*, from William Molyneux in Dublin and from Reiselius in Wurtemberg among others, while W. Rooke, travel-

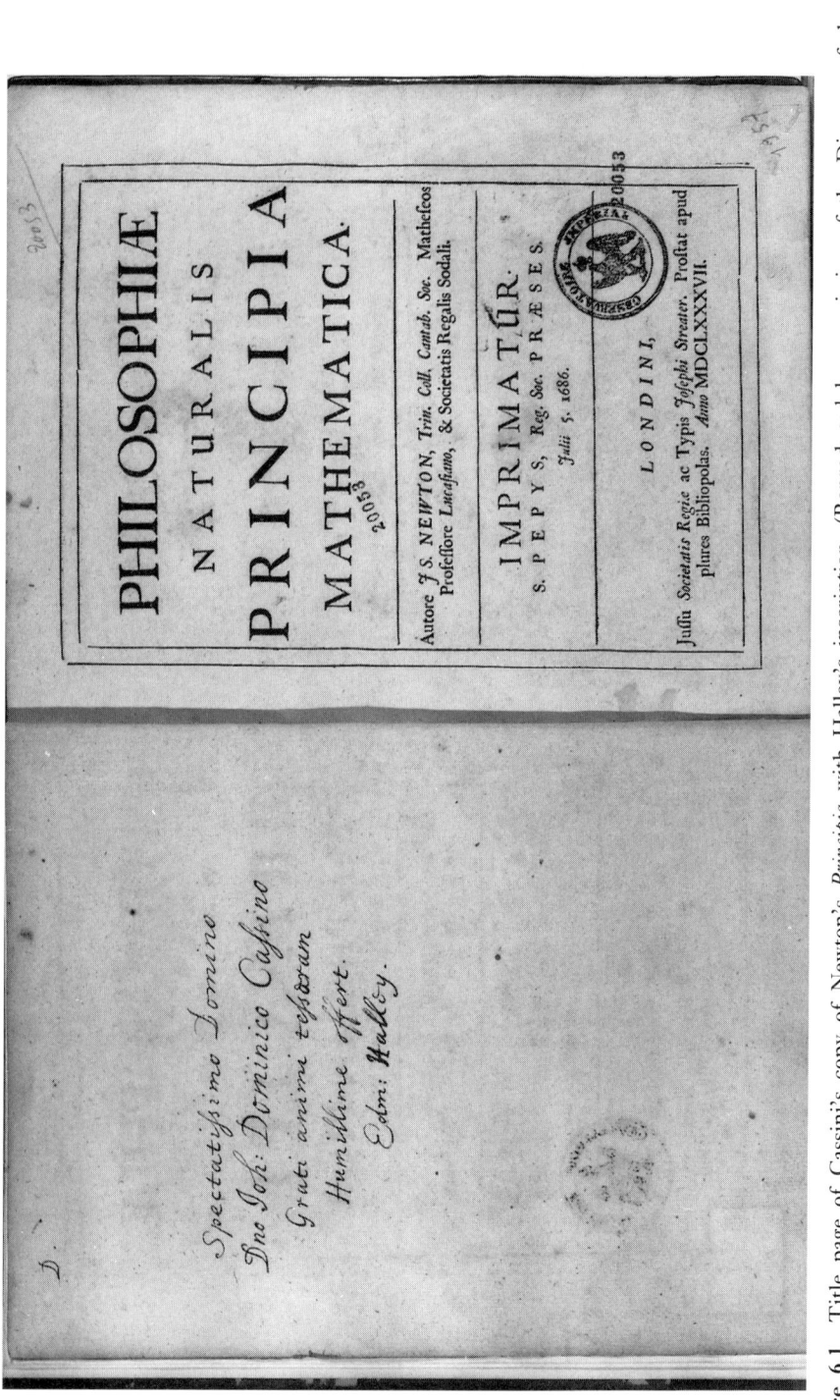

Figure 6.1 Title page of Cassini's *Principia* with Halley's inscription. (Reproduced by permission of the Director of the Observatoire de Paris.)

ling to Florence, carried a copy from Halley to Viviani, who was so impressed that he showed it to the Grand Duke.[37]

There is no known reply by Newton to that last letter from Halley. He did include in his Preface to the *Principia*, dated 8 May 1686, a warm acknowledgement of Halley's part which, in Motte's English translation, reads:

> In the publication of this Work, the most acute and universally learned Mr Edmund Halley not only assisted me with his pains in correcting the press and taking care of the Schemes, but it was to his solicitations that its becoming publick is owing. For when he had obtained of me my demonstrations of the celestial orbits, he continually pressed me to communicate the same to the Royal Society; who afterwards by their kind encouragement and entreaties, engaged me to think of publishing them ...[38]

Halley for his part expressed his admiration and his awe at Newton's achievements in the Latin ode with which he prefaced the book. Dr. N. E. Emerton has made the following English translation:

> Behold the patterned sky by Newton scanned,
> The mass of Heaven and Earth divinely grand,
> The Laws the great All-Father did uphold –
> Creator of the primal things of old –
> To found on them his work of age untold.
>
> To us the secret depth of heaven's revealed;
> The Force that turns the Spheres lies unconcealed.
> The Sun commands to bow before his throne
> The planets circling through the void alone;
> He from the centre holds each one in place
> By strong attraction from the depths of space.
>
> Along what paths the fearsome Comets swerve,
> Without alarm or terror we observe.
> We learn the silver Moon's uneven gait,
> Unreckoned by Astronomers of late;
> Why there is backward movement of the Nodes,
> Why Apogees advance on forward roads.
>
> We find the Moon upon her wandering course
> Propels the Ocean's tides with strongest force;
> The ebbing waves lay down the weed they bear,
> Exposing sandbanks that the sailors fear;
> Then up the beach the breakers' crests uprear.
>
> The minds of ancient Sages were perplexed,
> In vain by hoarse debate the Schools were vexed;

> Such things we can discern with ease today –
> Astronomy has cleared the clouds away.
> No longer doubts and errors make us blind,
> For now the Genius of Newton's mind
> Allows us to ascend towards the Sky
> And penetrate the regions up on high.
>
> Rise, Mortal Man, leave earthly cares behind
> And realise your heavenly powers of Mind
> Which raise you far above the brutish kind.
>
> He who first laid down Law-codes to restrain
> Crimes of Adultery, Murder, Lust for gain;
> Who to walled Towns the wandering races brought
> And Agriculture's blessings to them taught;
> Who first made Wine to soothe away men's woes;
> Who took papyrus, where by Nile it grows,
> And on it sounds by symbols did portray
> So that the eye reads what the Voice did say –
> These added little to the human store
> And brought small profit to a life still poor.
> But we're invited to the Realms on high,
> Allowed to study laws that rule the sky,
> Those hidden laws to which Earth once was blind;
> The World's unchanging order now defined
> Though long unknown, lies open to our mind.
>
> Ye Muses, fed on nectar from above,
> Join me in praising Newton, whom ye love;
> Newton, who opened Truth's sealed casket wide,
> Newton, whose heart Apollo sanctified,
> Whose mind to things Divine did access gain:
> Closer than this no Mortal may attain.[39]

A modern mind may find Halley's *Ode* an uneasy introduction to the most profound work of the age on natural philosophy. His contemporaries would not have seen it so. James I wrote such an ode as a preface to the works of Kepler, and a whole bouquet of Latin verse from around Europe poured out on opening the *Machina coelestis* of Hevelius. Leibniz admired Halley's ode:

> J'ay vû des beaux vers latins que M. Hallé a mis devant les principes de Mons. Neuton, cela me fait connoistre que M. Hallé n'est pas moins habile dans les belles lettres que dans la connoissance de la nature.
>
> I have seen the splendid Latin verses which Mr Halley has put in the front of the principles of Mr Newton; it makes me believe that Mr Halley is no less skillful in the humanities than in the knowledge of nature.

John Conduitt remarked that 'Tycho Brahe made verses upon Copernicus, Kepler upon Tycho. Urania was their muse. Halley upon Sr I went as much beyond the others [as] his subject'.

When Bentley brought out the second edition of *Principia* he 'improved' Halley's verses. Halley was not pleased. According to Conduitt, Newton remarked

> Bentley to show he spared the living no more than the dead altered Halley's verses when he printed the Principia, here could be no error in M.S. – or various lection, and he was reduced to his own peremptous criticism.[40]

The *Ode* illuminates the ideas of the day. The opening lines summarise the main achievements of Book III in the framework of God as Creator of the cosmos. At the same time Halley seems unduly dismissive of morals, civilisation, agriculture, and writing of earlier times. The ending is adulatory, but Halley has his feet on the ground, or rather on the deck, when he writes of the sandbanks that the sailors fear, to which he was about to turn his attention.

People would become greatly occupied in the early eighteenth century with the debate between the Ancients and Moderns, the Battle of the Books. Was the old or the new learning more to be admired and followed? Halley is here firmly on the modern side. People also became deeply devoted to the sublime in literature, art, and music. Handel's oratorios, forty years on, appealed because they were sublime.[41] Halley expresses the sublime by the heightened diction of some of his lines.

6.4 Previews and reviews

The *Principia* did not burst wholly unexpectedly upon the learned world. Halley had foreshadowed it in letters to a number of correspondents. Writing to Johann Christoph Sturm at Altdorf in Bavaria in 1686, he summarised Newton's investigations on gravity and celestial mechanics and on the effect of resistance by fluids. Again, writing to Reiselius on 30 July 1686, he set out Newton's achievements in showing that orbits should be ellipses under the inverse square law of gravitation and in deriving the law of areal velocity. He had given Wallis a full account of the contents of the third book, *De systemate mundi*, in his letter of 9 April 1687. Flamsteed also had heard of Halley's work with *Principia*, for on 4 November 1686 he wrote to Towneley:

> Mr Newtons Treatise of Motion is in the presse. Mr Halley takes care of it and 13 sheets (hee tells me) are wrought of.[42]

The letter shows ('hee tells me') that Flamsteed was on speaking terms with Halley at the end of 1686 (see below, Section 6.5).

Halley presented a copy of the *Principia* on particularly fine paper to James II, with a letter which summarised the theory of the tides. James was an experienced sea captain, who had distinguished himself as an admiral in fleet actions and who would undoubtedly have been interested in tides:

> May it please Your most Excellt.Maty.
>
> I could not have presumed to approach Your Maties. Royall presence with a Book of this Nature, had I not been assured, that when the weighty affaires of Your Governmt. permit it; Your Maty. has frequently shown Your self enclined to favour Mechanicall and Philosophicall discoveries: And I may be bold to say, that if ever Book was so worthy the favourable acceptance of a Prince, this, wherein so many and so great discoveries concerning the constitution of the Visible World are made out, and put past dispute, must needs be gratefull to Your Matie; being especially the labours of a worthy subject of your own, and a member of that Royall Society founded by Your late Royall brother for the advancement of Naturall knowledge, which now flourishes under your Majesties most Gracious Protection.
>
> But being sencible of the little leisure wch. care of the Publick leaves to Princes, I believed it necessary to present with the Book a short Extract of the matters conteined, together with a Specimen thereof, in the genuine Solution of the Cause of the Tides in the Ocean. A thing frequently attempted But till now without success. Whereby Your Matie may judge of the rest of the Performances of the Author.

Halley first summarised Newton's concept of gravity, and how it controlled the motions of the planets, without any need for Cartesian whirlpools. He went on to describe Newton's explanation of the motions of comets and of the behaviour of the Moon. The tides of the oceans occur because, relative to the solid Earth, the Moon attracts the waters on its side of the Earth and repels them on the opposite side, so that on both sides they bulge out into a spheroidal form (Fig. 6.2). As the Earth rotates under the Moon, the bulge moves round the Earth giving a high tide every twelve hours approximately. The Sun also attracts the seas, though to a lesser extent, with the result that the tides are highest, spring tides, when the Sun and the Moon are in line at new and full Moon, while they are least, neap tides, when the Sun and Moon are at right angles in the quarters. The height of the tide is about 0.6 m, more or less according to the position of the Sun relative to the Moon.

That simple scheme, with the bulge of the oceans occurring under or opposite the Moon, only applies to deep oceans far from land. Even so, the oceans would continue to oscillate at their own natural frequency if the tide-raising forces of the Sun and Moon were suddenly removed. Hence the response of the oceans lags behind the forcing attraction of the Sun and Moon, and therefore the spring tides do not occur exactly

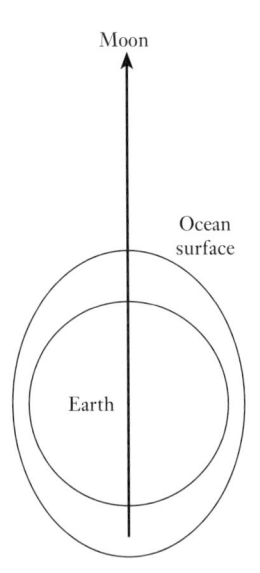

Figure 6.2 The tidal bulge.

at full and new Mooon, nor the neaps at the quarters. The most obvious difference from the simple theory is that the tides are higher, sometimes very much higher, in shallow seas and estuaries than in the open ocean, reaching 7 m in the mouth of the Thames and as much as 10 m in a few bays and estuaries. The volume of water crossing a line supposed drawn across a shallow bay is the same whether it comes from the deep ocean or flows into the bay, and so the change of the depth of water in the bay is much greater than in the deep ocean. Halley explained all that to James and finally gave Newton's reason for the abnormal tide at Tonquin where there is only one tide in twenty-four hours. Two tidal streams flow into Tonquin from the South China sea, and one is delayed by exactly six hours behind the other because of the long and tortuous route it has to follow and the obstructions to the flow. The two semidiurnal tides are therefore very nearly out of phase and cancel out, but the diurnal tides are out of phase by 90° instead of 180° and do not cancel.

Halley ended:

> If by reason of the difficulty of the matter there be anything herein not sufficiently Explained, or if there be any materiall thing observable in the Tides that I have omitted wherein Your Matie. shall desire to be satisfied, I doubt not but that if Your Majesty shall please to suffer me to be admitted to the honour of Your Presence, I may be able to give such an account thereof as may be to Your Majesties full content:
> I am great Sr. Your Maties. most Dutifull & obedient Subject
> EDMOND HALLEY.[43]

What a remarkable letter. If the Royal Society wished to present a copy to the king, why did not Pepys do so, Pepys the President, who had long worked closely with James on naval affairs? Newton himself would have been *persona non grata*, for the previous February he had advised the University of Cambridge to resist the admission of a Benedictine monk, Fr Francis, to the degree of MA, and the imposition of a Roman Catholic (Joshua Basset BD) as Master of Sidney Sussex College by Royal mandate. He was summoned along with the Vice-Chancellor to answer for the University's defiance before the Lord Chancellor Jeffreys. Indeed just about the same time that Halley wrote to the King, James sent a letter, dated 13 July, to Sidney confirming alterations in the statutes that Basset had demanded.[44] It seems bold, if not foolhardy, of Halley to send Newton's book to the King just at that juncture, however interested in tides James might be.

The letter is comparatively sober in an age of adulation: its presentation is unadorned, and it foregoes comparisons with deities, classical or Christian. Furthermore, the content of the letter and the fact that it came from Halley and not Pepys, suggests that Halley was writing for himself as publisher, not for the Royal Society. Why did he give James such a lengthy account of the tides, simplified though it may have been from Newton's discussion? Was it because he was a young man quite remarkably full of his own achievement and insensitive to James's lack of concern? How did he come to be assured that James had 'frequently shown Your self enclined to favour Mechanicall and Philosophicall discoveries'? Was it through a third party, or was it because he was close enough to James to know that James would take a serious interest in a theory of the tides? Halley had been patronised in younger days by Jonas Moore who had been briefly James's tutor, and he was also well known to Pepys who worked closely with James.

Halley's review in *Philosophical Transactions*, in February and March 1688, was the first to appear and he followed it with an *Advertisement* explaining that the delay in the appearance of issues of *Philosophical Transactions* was on account of his seeing *Principia* through the press, and hopes his readers will accept that he had 'been more serviceable to the Commonwealth of Learning' in so doing than in bringing out *Philosophical Transactions*. Few will disagree. Halley's review is a clear, compact account of the three books of *Principia*. He calls attention to a number of particularly significant results, such as Newton's conjecture that the orbits of some comets might be elliptical although indistinguishable from parabolae in the neighbourhood of the Sun; or that air is composed of particles that repel each other; or Newton's pendulum experiments on air resistance, or his investigation of circular motion of fluids and vortices, quite inconsistent with Cartesian notions.

Halley begins:

> This incomparable Author having at length been prevailed upon to appear in Publick, has in this Treatise given a most notable instance of the extent of the powers of the Mind; and has at once shown what are the principles of Natural Philosophy and so far derived from them consequences that he seems to have exhausted this Argument, and left little to be done by those that shall succeed him ...

and concludes:

> ... and it max be justly said, that so many and so valuable Philosophical Truths as are herein discovered and put past Dispute, were never yet owing to the Capacity and Industry of any one Man.

Effusive, but justified surely by the history of the influence of *Principia*.[45]

It is not often that a major work of scholarship receives such a review as Halley's of *Principia*, which shows a grasp of essential points of Newton's achievement in the principles of dynamics, in the scope and elegance of the mathematics, and in the power and versatility of the applications. Halley knew the contents of the *Principia* better than any other living person at the time, but that was not enough; he had also a rare gift for explaining the principles and results of mathematical argument to those who were not mathematicians.[46]

Two French reviews quickly appeared, one in the *Bibliothèque Universelle* (probably by Locke), and the other in the *Journal des Sçavans*. They showed no understanding of the Newtonian method of combining mathematical argument with experiment and observation of the actual world. Locke gave little more than a summary of notable results, while the author of the *Journal des Sçavans* (probably Régis) strongly criticised Newton for giving no physical basis, in the Cartesian sense, to his mathematical constructions. The review in *Acta Eruditorum* was quite a different matter. It may well have been by the editor, Otto Mencke, to whom Halley had sent a presentation copy of *Principia*, and shows a deep understanding of Newton's principles and methods.[47]

That was not the last of Halley's direct involvement with *Principia*. His calculations of orbits of comets (Chapter 8) were surely occasioned by Newton's preparations for a second edition.[48] Newton allowed some of his friends, Halley among them, to annotate their copies of the first edition with the considerable changes that he would make in the second edition. Cohen lists the copies that he could identify, of which two seem to have been Halley's. There is, it would seem, yet a third among the Keynes papers in King's College, Cambridge.[49]

No copies of the first edition were among Halley's books that went on sale after his death, nor any of the Cambridge (1713) printing of the second edition, but a copy of the pirated Amsterdam reprint of 1723 was there (Appendix 8). Bentley presented copies of the second edition to various people and told Newton that he would send him copies so that he could present them personally himself to Halley and de Moivre. He never did. On 28 June 1714 de Moivre wrote to Johann I. Bernoulli to explain why Bernoulli had not had a copy of the second edition: Bentley had kept copies in his own hands so that Bernoulli did not get one, nor did Newton have copies for de Moivre and Halley, who had to buy theirs, as did others to whom Newton had presented his other works. Nicholas I. Bernoulli, Johann's nephew, did however receive a copy, which he acknowledged in 1717 with grateful memories of his reception by Newton and Halley when he visited England.[50]

Very shortly after the second edition was published, Halley had set up in type the tables of the motion of the Moon that were only to appear posthumously almost forty years later. The tables were based on the theory of the Moon's motion as Newton expounded it in the second edition and (Chapter 13) they are the only existing numerical representation of that theory. Halley had known of Newton's development in the tract *The theory of the Moon's motion* of 1702, and was able to complete his tables very soon after Newton finalised the changes he made in the second edition of *Principia*.

6.5 Other pursuits

The *Principia* and legal and family matters were not Halley's only occupations from 1684 to 1687, for in the later years he was responsible, as Clerk, for bringing out the *Philosophical Transactions* although, as he apologised to Wallis, *Principia* was taking precedence. He also carried on the Society's correspondence with people outside London, especially Molyneux in Dublin and Wallis in Oxford, together with Hevelius, Leeuwenhoek, Reiselius in Wirtemberg, and Sturm in Altdorf.

His letters were in part news letters to acquaint his correspondents with the activities of the Royal Society and other matters of general interest. Thus in the latter half of 1686, in letters in which he reopened an apparently interrupted correspondence with Oxford, he told Caswell he had

... seen a great curiosity, viz a Calicoe shirt brought from India, which is wove without a seam all of one peice which I should have thought impossible had I not seen it, it explains the Scripture relation of our Saviours coat which was without seam.

In a letter to Reiselius ten days later, he sends news of Newton's *Principia* with a summary of its contents, he mentions some experiments of Hooke on the barometer and on a level of very long period to determine the vertical, and at the same time refers again to the seamless shirt.[51] In view of the irreligious, indeed atheistic, reputation that Halley acquired, his attitude to the seamless shirt is interesting. Clearly he accepted Jesus Christ as Saviour, but in this matter, as in discussions of evidence for the age of the Earth, he was prepared to test the scriptural account against other evidence.

Other topics of his letters in those years were advance notices of the *Principia*, and an account of the printing of *The history of fishes* at the expense of the Royal Society when he sent a copy to Leeuwenhoek as a present from the Society. He wrote a number of times to Wallis, often about *Principia*, but also about antiquities and ball lighting at Soissons, and he sent him from time to time accounts of proceedings at the Society. He sent instructions to Hayley, a sea captain of the Levant Company, for observing a solar eclipse at Smyrna, and he wote to Valvasar in Ljublana (Laibach) about the geography of Carniola (now Slovenia) and they exchanged verses.[52]

Halley did not observe regularly at Islington afer his father's death but he did, with Hooke, observe occultations of Jupiter by the Moon and published eight other papers in *Philosophical Transactions*, on the tides at Tonquin (Chapter 5), on gravity and projectiles, on the height of the barometer at elevated places, on trade winds, on solid geometry, on evaporation from the sea, on algebraic equations, and on the latitude of Nuremburg. The paper on gravity is notable for its concise summary of the properties of the gravity field of the Earth, as they were known before Newton's theory of the figure of the Earth and of the variation of gravity over it as developed in the third volume of *Principia*, but the bulk of the paper is given over to the theory of projectiles and the principles of gun-laying. The other papers of 1686 and 1687 are discussed in the next chapter.[53]

Halley proposed to measure the length of a degree in England in order to determine the size of the Earth. In 1635 Norwood had obtained the length of a degree of latitude, assuming the Earth to be spherical, from the difference of the latitudes of London and York and the distance between them. He found the difference of latitudes from the altitudes of the noonday Sun on the same day of the year at the two cities, and he derived the distance from pedometer measurements while walking from York to London.[54] His result, equivalent to 10 072 km, or within one per cent of the accepted value, was known to Newton, while the best determination of the size of the Earth when Newton was writing *Principia* had been made by Picard in France (Chapter 4). In 1669 the Royal Society was told by John Wilkins, the bishop of Chester, that Charles II wished that a degree of latitude might similarly be measured in England, and requested the assistance of the Society, but the Society had no money and the King provided none. Somewhat later one Adams asked the Royal Society to support his intended triangulation of the whole country, but nothing came of it. Halley now proposed to find distances by triangulation. On 11 July 1686, when the Council of the Royal Society ordered 'that Samuel Pepys the president be desired to license the book' (*Principia*), they also ordered

> That the treasurer, to encourage the measuring of a degree of the earth, do give to Mr HALLEY fifty pounds or fifty copies of the History of Fishes when he shall have measured a degree to the satisfaction of Sir CHRISTOPHER WREN, the president, and Sir JOHN HOSKYNS.[55]

Halley was encouraged by William Molyneux writing from Dublin on 4 August. He had previously explained to Caswell that £50 would not cover the cost of triangulation, and asked his advice on methods, but later he investigated triangulation for, writing to Wallis on 13 November 1686, he says:

> I attempted about the end of last August to make a survey of a degree of the Earth, by a scale of Triangles; but found severall obstacles that obliged me to desist. The chief whereof were that I found a great and insuperable difficulty to come to the objects that I had seen at a great distance, for the country people I observed could tell me nothing of places above 7 or 8 miles off: And that at about 20 miles North from London, the country is very thick of high Woods, and the hills so near of a hight, that there were no conspicuus objects to be found: so that I saw an absolute necessity of making a preparative perambulation, to find out the eminent Objects so that I might not be at a Loss upon my second attempt. Near London and to the Southwards the Country is open and fit for the purpose, but for 15 miles viz from Ware to Royston there will be great difficulty to continue a scale of Triangles. I should be very glad to know your sentiments about the method of finding out the highest Objects, such as may be certain to see one another; if I cannot do this I must be forced to trust to actuall mensuration, wherein it will be hard to come within a thousandth part of the true distance.[56]

By 'actual mensuration' he evidently meant direct measurement on the ground as Norwood had done. The letter gives a vignette of rural

England. Nothing came of Halley's project and the triangulation of England had to wait until the middle of the next century. After the Jacobite rebellion of 1745 the Ordnance Office began to carry a network of triangles over the whole land between signal stations on prominent hills.

The scale of the great triangulation of Britain hung upon the bases measured at Hounslow Heath and elsewhere and so upon the standards of length used in them. English standards of length derived from the yard of Edward III and a new standard bar, kept in the Tower, had been prepared in the reign of Elizabeth, but the Ordnance Survey had more precise yard standards made for its use. The Royal Society has always recognised the importance of metrology and especially that of precise and consistent standards of measurement. Many various feet, ells, and toises were in use in different countries. The unique English standard yard was superior to them. The possible deficiencies of a metal bar as a standard for scientific purposes were recognised and at the meeting of the Society on 7 June 1681, Wren had suggested that the standard of length might be a part of a degree on the Earth, and on other occasions he and others had proposed the length of a pendulum beating seconds.[57] The meridian quadrant of the Earth was chosen as the basis for the metric system in the nineteeenth century, but the search for a convenient natural standard instead of a metal bar went on until length was defined in the middle of the present century by the time of passage of an electromagnetic signal.

On 28 October 1685 Halley was asked to prepare scales 'showing English, Paris, Roman and other feet most in use' and on 15 July 1688 he reported on the relations of the ounce troy to the ounce avoirdupois, on the weights of water in cubic feet of London and France, and on the ratio of the pound avoirdupois to the French pound; he does not say which French units, of which there were at that time many versions, but no doubt they were the foot and pound of Paris. He also gave equivalents for Roman measures, as he had taken in Rome.[58]

On 20 May 1688 Halley discussed the problem of determining longitude and at the following meeting suggested the use of a pendulum clock. On 15 July, in a paper that foreshadowed one of his most significant discoveries, the acceleration of the Moon, he examined the different forms of the equation of time and argued that inequalities attributed to the Earth's rotation were really in the Moon's motion, and that times determined from the rotation of the Earth relative to the fixed stars were uniform.[59] The same methods that he used were to show later, when more precise observations became available, that the rotation of the Earth was not uniform and that variations of it could be detected both in astronomical observations and through comparisons with quartz clocks when those were developed in the present century.

In summary, during the whole time that he was occupied with *Principia*, Halley was both involved in family and legal matters and was investigating some of the most significant and fundamental questions of physics and astronomy, the bases and standards of measurements among them.

In the same years his friendship and cooperation with Flamsteed turned to sour hostility. Flamsteed had written of the death of Halley's father in 1684 and had hoped that because Halley would be more in London, he would see more of him. On 4 November 1686 when he wrote to Towneley in Lancashire of Halley's progress with *Principia*, he also complained to him and to Molyneux in Dublin of Halley's behaviour. He had four complaints. First, he thought that Halley had filched his ideas about the four magnetic poles of the Earth from Peter Perkins.

> His discourse ... concerning the variation of the needle and the four poles it respects I am more than suspicious he got from Mr Perkins, the master of the mathematical school at Christs Hospital who was very busy upon it when he died ... Mr Halley was frequently with him and had wrought himself into an intimacy with Mr Perkins before his death and never discoursed anything of his 4 poles till some time later I found it published in the transactions.[60]

Perkins (FRS 1679) died in 1680. Letters exchanged in 1700 between Flamsteed and Thomas Perkins, Peter's brother, show that Halley purchased some of Peter's papers from his widow about three weeks after his death.[61] Perkins had collected magnetic observations from all over the world and had discussed his studies with Flamsteed. He had addressed the Royal Society in 1680 but it does not appear that he published anything.[62] Halley almost certainly obtained some of Perkins's collection of observations and he probably used them in his paper of 1683. Perkins was apparently mostly concerned with accidental causes of the variation but also with the relation between dip and variation and he suggested there were six meridians on which the variation vanished. His remarks do not seem to be closely related to Halley's analysis.[63] Halley published his own first paper on the anomalies of the Earth's magnetic field and their westerly drift in 1683. He suggested that they could be accounted for if the Earth had four magnetic poles (Chapter 5), and recognised Perkins's earlier work.[64] There seems nothing out of the way in Halley's doing so but it may have been an early cause of Flamsteed's antipathy. However, Halley's paper of 1683 cannot have been the immediate cause of Flamsteed's disagreement with him, for on 12 May 1684 Flamsteed wrote to Molyneux in Dublin about the death of Halley's father and his hopes to see more of him in London (Chapter 5, Note 20).

Flamsteed's second complaint was to do with the tides at Dublin as compared with those at London. Flamsteed had published some tide tables with a simplistic rule for estimating the times of tides at various ports, and especially the phases of spring and neap tides, from those at London; Halley had argued that matters were more complex. The information about the Dublin tides had come from Molyneux, and in a series of letters to him Flamsteed discussed the observations very thoroughly.[65] At issue was the dynamical interpretation of the differences between them, and Flamsteed considered that Halley had dealt slightingly with his own view. He disagreed with Halley's analysis, he took offence at some comments of Halley and, that recurring theme, he thought Halley had used results that really belonged to others. Halley in 1686 might have seen himself as something of an expert on tides. He had been editing *Principia*, in which Newton had set out the first dynamical theory of tides as driven by the gravitational attractions of the Sun and the Moon, and he had explained it in his letter to James II. Molyneux thought that Flamsteed was making too much of an argument about the interpretation of the observations, and, a perpetual issue with Flamsteed, he advised him to publish some of his results to avoid being misrepresented. Flamsteed ignored that sound advice.

Flamsteed had also heard that people he identified as Halley's associates were putting tales about him round in London. He was told to his face by C.B., a young man whom Eggen identified as Charles Boucher, that he should resign so that Halley might have his post. Boucher, if it was he, must have been back in London from Jamaica in 1683, and he and Halley and a friend of his, M. Crispe the common sergeant, evidently spent some sociable evenings together.

Lastly, Flamsteed complained that he had entertained Halley very hospitably at Greenwich, but when he stayed with Halley overnight, presumably at Islington, 'my entertainment was so rude and homely'. What lies behind that we do not know, but it was at a time when Halley was preoccupied with his disputes with his stepmother as well as being deeply involved with *Principia*. Perhaps in the circumstances he was somewhat offhand.

Up to the summer of 1684 Flamsteed seems to have found Halley acceptable; after 1687 he was beyond the pale, and Halley's approach to magnetic and to tidal studies together with other real or imagined slights seem to have changed Flamsteed's attitude.

Flamsteed's correspondence with Molyneux continued up to April 1687, by which time Molyneux was becoming distinctly more sympathetic to Halley.

Flamsteed's letter to Towneley of 4 November 1686 shows one reason why he and Halley fell out. Their styles of science were quite different: Flamsteed worked sedulously and intently with one main aim throughout his life and regarded everything he did as his personal property, whereas Halley not only worked on many different topics, but one of his common methods was to collect data widely, and therefore from other people, to investigate geographical and historical patterns. So he studied the geomagnetic variation and probably used Perkins's material among others, he studied the trade winds and again made use of observations of others, and then published them openly. Flamsteed readily saw that as thieving.

6.6 *Achievement and preparation*

The three years from the summer of 1684 to the summer of 1687 were the most significant in the history of the physical sciences to this day. After 1687 people would see the physical world in completely new ways. The significance of what Newton had done was widely recognised, by philosophers who knew little mathematics, just as by practising astronomers, in Europe as well as in Britain.

Almost as soon as *Principia* appeared, a few followed its methods. Halley was one of the first, as the next chapters will show. Little more than a decade after its publication, David Gregory wrote a textbook of astronomy expounding its principal results. The problems that the most able mathematicians attacked were to a large extent those set out in *Principia*. The influence of *Principia* in Europe was as great—none could ignore it, even if and perhaps especially if, they disagreed with its metaphysics.

Newton did four things in the *Principia*. First he set out the principles of dynamics as they still are in essence. Secondly, he proposed the concept of universal gravitation, that all bodies attract each other by a force that is proportional to the products of their inertial masses. Thirdly, he showed mathematically that all or almost all the phenomena of celestial mechanics can be accounted for by gravitational forces. Those results were accompanied by a fourth, to a large extent implicit but even more profound, that the secrets of nature are to be teased out by comparing rigorous deductions from theory with the results of careful observation and experiment. In Book III of *Principia* Newton compares his theoretical results with a wide range of the best observations, of the planets and of the satellites around them, of comets, of the Moon, of tides, of the oblate figure of the Earth, of the precession of the Earth.

It sometimes seems that Newton's mechanics has been changed out of all recognition by relativistic ideas, by quantum phenomena, and by the realisation that Newton's own dynamics may have chaotic outcomes. It is not so. Newton applied his laws of motion in a geometry in which space and time were independent. In special relativity we recognise that is not the proper geometry for observations made at a distance and we use one in which space and time are handled together and are not independent; within that geometry the laws of motion are Newton's. Quantum mechanics appears to have a different form from Newton's mechanics, but that appearance dissolves when it is realised that quantum mechanics has the same form as the mechanics of Hamilton and Jacobi which is itself another way of stating Newton's dynamics.

Newton's dynamics was so successful that people came to think that the whole world evolved according to rigorous predictable consequences of differential equations. Yet from early days there were those who recognised that in some circumstances, the equations of dynamics might not have unique solutions. One major question was whether the solar system itself was stable or might eventually collapse into the Sun or fly apart. Many thought that with sufficient care that might be decided, but at the end of the nineteenth century Henri Poincaré showed that the equations of celestial mechanics could have solutions that were not predictable, that the dynamics was, as we say now, chaotic. Newton's dynamics is not wrong, but as with relativity and quantum mechanics, it works differently in conditions wider than those in which Newton and his contemporaries applied it.

The nature of gravitation has attracted powerful theorists and subtle experimenters from Newton's day to this. He himself was loath to speculate in the *Principia* on the nature of gravitation, although he did allow himself some thoughts in the *Scholium generale* and at the end of the *Opticks*, and he performed early experiments with pendulums to test what we now call the Weak Principle of Equivalence.* After 300 years gravitation remains a great challenge to theory and experiment alike. That it works none doubts, but why it works, few would claim to understand. In that we are with Newton's contemporaries such as Huygens and Leibniz.[66]

At the deepest level the way of investigating the physical world is still Newton's, trying to organise empirical results of observation in a logical abstract structure. The changes since Newton's day have been the consequences of extending the scope of observation and finding appropriate mathematics. The principles have proved wide enough and flexible enough to accommodate the progress of technologies of experi-

*The first were by Galileo.

mental and observational physics and of more profound mathematics. Not only that, the methods of the *Principia* are increasingly finding application in biology, as those sciences move from being primarily observational to develop more theoretical structures. If there has been a change since Newton's time it has been the recognition, due in part to the emergence of relativity and quantum mechanics, that theory and observation are not independent. Science is a matter of understanding what is there in nature, empirical observations must always be primary and the form of theory must be consistent with what and how we can observe.[67]

No one would claim that Halley was an equal collaborator with Newton in the evolution of *Principia*, rather the accounts show that he was surprised by each successive development as the new map of the physical world was unrolled. He did however appreciate the significance of each stage as it was revealed to him, for he, like Newton, combined experimental skill with mathematical ability, and, perhaps even more than Newton, he was deeply interested in navigation and the related problems of celestial mechanics. There can be no doubt that without Halley there would have been no *Principia*. Stimulus, critic, sustainer, editor, publisher: bringing forth *Principia* was his greatest contribution to natural knowledge. In the words of Augustus De Morgan,

> But for him in all probability, the work would not have been thought of, nor when thought of written, nor when written printed.[68]

Truly Edmond Halley justly stated his own part when he said that he was the Ulysses that had produced that Achilles. The persuasion of Ulysses secured for the Greeks their hero who overcame Troy. Stimulated by Halley, Newton attacked the citadel of mathematical philosophy. When other Greeks returned to their homes and cities, Ulysses set out upon a voyage that would take him to far-off places and to astonishing adventures. Halley now became a seafarer for months on end for more than fifteen years, in dangerous shoals, in distant icy oceans, upon the bed of the sea. He was to meet astonishing things in nature and the mind.

PART II

Often at sea

(1688–1703)

7
Improving natural knowledge

> ... fatti non foste a viver come bruti
> ma per seguir virtute e canoscenza.
> [... we were not made to live like brutes but to follow virtue and understanding.]
> DANTE: *INFERNO*, XXVI, 119, 120

7.1 *The Royal Society, twenty-five years old*

HALLEY was Clerk to the Royal Society for fourteen years from 1686 to 1700, for the last four of which he was away from London and others stood in for him. In the same years he was often at sea, eventually in his three voyages in *Paramore*. When he returned from the Atlantic he ceased to be Clerk and was re-elected as a Fellow on 30 November 1700.[1] This chapter and the next recount his activities in the Royal Society from 1687 to 1703, and the many various topics that he brought before them. Chapter 9 is about his maritime pursuits in the years before he took command of *Paramore* and leads on to two chapters, 10 and 11, about his voyages in her and his Adriatic mission.

The times were disturbed and almost disastrous for England when Halley became Clerk. In the years just before and for some time after the Dutch invasion of 1688, the English constitution developed into the form that it still has, distinct political parties began to form, and those who did not conform to the Church of England enjoyed some tolerance. It must have seemed very different at the time. The accession of William and Mary removed the danger of arbitrary rule by a Catholic monarch, but not everyone approved or accepted that accession nor the limited religious toleration that went with it. Even as the House of Lords was starting to consider what to do after the departure of James, many sprang to settle old scores and reopen old issues (Chapter 5). Legal toleration made the Church of England more defensive and less tolerant of sceptical or heterodox opinions. The Nine Years War from 1688, in which England at first suffered severe reverses at sea, strained the economy and finances of the country almost to breaking. The great silver recoinage of the late 1690s aggravated the problems; Halley was then deputy controller of the country Mint in Chester. He may have suffered from the great disaster of 1693, the loss of many ships of a

Levant Company fleet off Lagos. The war lasted for much of the time that Halley was Clerk, and it undoubtedly delayed his project to observe the magnetic variation in the Atlantic. It was an anxious decade, a dangerous decade for anyone holding responsible office; in it Halley had some of his most original and influential ideas.

Almost as soon as *Principia* was complete Halley surveyed the Thames approaches. He began to develop the diving bell and other equipment in attempts to salvage a ship off the Sussex coast, from 1690 until he went to the Mint at Chester from 1696 to 1698. From 1693 at the latest and possibly earlier, he was planning his cruises to observe the direction of the magnetic field over the Atlantic. Wittingly or unwittingly he was preparing himself in other ways than begging a ship and crew: he was gaining great experience in working at sea, he was becoming known to persons of influence in nautical matters, so that when he did at last sail off into the Atlantic, he himself was equipped to go and he had the confidence of those who sent him. The sea engaged him throughout the time he was Clerk of the Society and writing his many and varied papers.

The state of the Royal Society a quarter of a century after its foundation was not assured and it seemed at times that it might fade away as had the Accademia dei Lincei, the publisher of Galileo, when Prince Frederico Cesi, its founder, died young. The Society grew out of informal meetings of men interested in natural philosophy at Oxford and in London. During the Commonwealth they were similar to the informal academy of Montmor in Paris at about the same time, or to others in Rome and in Florence. The Society might have been ephemeral but that Charles II owed debts to some of the members for his restoration. Sir Robert Moray and John Evelyn in particular took advantage of their position at the Restoration Court to interest the king and obtain his patronage. The Society came formally into being better supported and with a stronger internal organisation than any contemporary, and it was taken as a model by Colbert when he formed the French Académie four years after the first Charter of the Royal Society.[2]

Little came financially of the initial support of the Court and of the nobility. Some of the most distinguished charter fellows, such as John Wilkins, Sir Robert Moray, and Henry Oldenburg, died in the early years. When Halley became Clerk, nearly twenty-five years after the Charter, the Society had weak finances, many fellows were ejected for not paying their subscriptions, and only one-third were active natural philosophers. Against that the Society had been fortunate to have had distinguished and effective presidents, Lord Brouncker, mathematician, courtier, and a colleague of Pepys as a Commissioner of the

Navy, Sir Joseph Williamson, amateur musician and Secretary of State, Christopher Wren, Sir John Hoskins, a lawyer, Sir Cyril Wyche, a statesman and lawyer, and Samuel Pepys. Brouncker and Wren were true natural philosophers, the others were not. They exemplify the wide range of interests of Fellows of the Society, as courtiers, administrators and politicians, musicians, lawyers, architects, and others.[3]

The Society had begun with an explicit commitment to experiment, as recorded in the minutes of the first preparatory meeting of 1660 November 28: 'a designe of founding a Colledge for the promoting of Physico-Mathematicall Experimentall learning', and Oldenburg described it in 1664 as

> ... a Corporation of a number of Ingenious and knowing persons, by ye name of ye Royall Society of London for improving Naturall knowledge, whose dessein it is, by Observations and experiments to advance ye Contemplation of Nature to Use and Practice.[4]

Experiments were important in the early meetings, but by the time Halley became Clerk they had to a great extent given way to the reading of papers and other communications. Halley may have contributed to that change by his many talks and papers, although he also showed experiments. The *Philosophical Transactions* developed in parallel with the change in the character of the meetings. Oldenburg founded the *Transactions*, the first serial scientific publication with the imprint of a corporation as distinct from an individual, although for many years an individual, first Oldenburg, later Halley, published them on behalf of the Society. When Oldenburg died they became *Philosophical Collections* and languished somewhat, but Robert Plot as Secretary revived them. Halley established them securely, not least because he himself often published in them. *Philosophical Transactions* were not all. The Society, licensed by its Charter to print and publish, had embarked on some important but very costly publishing ventures, especially the beautiful and lavishly illustrated *History of fishes* (Chapter 6).

The three decades from 1672 were crucial for the survival and development of the Society. The first enthusiasm and the first enthusiasts were passing, financial constraints were ever present, and by no means all the fellows had any interest in experimental learning and the improving of natural knowledge. None the less, a core of serious and devoted savants and a succession of powerful and effective presidents ensured that the Society did not succumb to its difficulties, but grew when others sank. Halley, the junior officer and author of original contributions from most varied fields, helped it grow.

7.2 Clerk to the Royal Society

Halley kept the records of the Society's meetings in the minutes and the *Journal Book*, he negotiated on behalf of the Society, with printers for example, and about accommodation in Gresham's College. He continued to receive his salary of £50, although somewhat erratically. He read to the Society many of the papers submitted to it by others and carried on the foreign correspondence. He was supposed to be subordinate to the Secretaries but the tone of his correspondence suggests that he often wrote independently. The Letter Books of the Society show that Halley shared the correspondence with the two Secretaries, Gale and Waller. Halley took astronomical and related matters and Waller biological topics, although the division was not hard and fast.[5] Henri Justel in France was a regular correspondent, sending news of events and discoveries in France over many years, keeping the Society well informed of matters there. Sturm of Altdorf in Germany was another correspondent. Halley announced to Cassini that he was now responsible for the Society's correspondence and through Justel, he resumed a correspondence with Leibniz.[6] He discussed the motions of the satellites of Jupiter with Cassini and others. He received letters from travellers to Asia Minor going to Levant Company factories, whom the Society might ask to make such observations, astronomical, geographical, antiquarian, and others, as they might have opportunity, and to send reports of them. He sent out instructions for observing an eclipse that would be seen at Smyrna. Diplomats going to Vienna sent him accounts of the territories of the Austrian Empire recently conquered from the Ottomans, with rather dismissive comments on the state of cartography: he would see that himself at first-hand in 1703 (Chapter 11).[7]

Halley pursued no great project from 1688 to 1698, no expedition to St Helena, no Atlantic nor Channel cruise, no new *Principia*. He was engaged on administration at the Royal Society and in the Mint at Chester, yet for fertility of ideas and influence upon the thoughts of others, the period stands out in his intellectual life. He is commonly known above all for his studies of comets, but his analysis of the use of transits of Venus across the Sun to establish the size of the solar system was no less influential in the long term (Chapter 8). In those and other of his studies in the decade, he developed the dynamics of the solar system, on the basis that Newton had laid in *Principia*.

He wrote testimonials at the Society's direction. One was for John Marshall who had invented a new way of grinding lenses. Halley and Hooke examined the method and the lenses, and the Society ordered that Halley send a letter to Marshall, which he did on 15 January 1694:

> I Have, by Order of the Royal Society, seen and examined the method used by Mr John Marshall, for grinding Glasses, and do find that he performes the said Work with greater ease and certainty than hath hitherto been practised, by means of an invention which I take to be his own, and new, and whereby he is enabled to make a great many of Optick Glasses at one time, all exactly alike; which having been reported to the Royal Society they were pleased to approve thereof, as an Invention of great Use, and highly to deserve encouragement ...

The letter worried the Company of Spectacle Makers; they complained that the method was not new, nor his, nor of use. The real complaint was that the method infringed on their monopoly. The Society stuck to its opinion, no doubt Halley's opinion, and to its guns.[8]

He wrote another testimonial at the Society's request just after he had ceased to be Clerk and had been re-elected to the fellowship. A Signor Alberghetti had written a book on artillery and made tables for shooting bombs out of long guns. They were calculated on principles laid down by Galileo, Torricelli, and others but neglected air resistance. Halley considered they were suitable for practical gunners setting their guns to batter fortifications at short ranges, but calculated with greater precison than was justified. The inaccuracy of the gun and the indefinite recoil would produce larger errors while at greater distances air resistance could not be neglected.[9]

Halley took on the responsibility for *Philosophical Transactions* as editor and publisher from Number 179 for January and February 1686, which appeared with the dedication:

> To the/ Right Honourable/ JOHN/ Earl of CARBERY/ PRESIDENT/ of the/ Royal Society &c,/ this Sixteenth Volume/ of the/ *Philosophical Transactions*/ is Most Humbly Dedicated/ by Edmond Halley.

It also contained the Advertisement:

> It having been found by Experience that several Curious Persons have been and are desirous to receive some Account of what the Learned part of the World are for the present busied about in the examination of experimental and real Knowledge, and what Discoveries they have made in any part thereof, the Royal Society have therefore thought fit to order that Care be taken for the future, that such Accounts shall be published in these Transactions Monthly as may answer their expectations ... And tho' upon an extraordinary occasion these transactions have for some Months last past been omitted, yet that defect will be soon supplyed by the speedy Publication of what has occurred since December last, and will be for the future continued at least as punctually as heretofore.[10]

That was optimistic. There was further interruption, and in Volume 16, on page 297, immediately after the review of Newton's *Principia* which Halley evidently wrote, the following Advertisement appears:

> Whereas the Publication of these *Transactions* has for some Months last past been interrupted, the Reader is desired to take notice that the care of the Edition of this Book of Mr Newton having lain wholly upon the Publisher (wherein he conceives he hath been more serviceable to the Commonwealth of Learning) and for some other pressing reasons, they could not be got ready in due time; but now they will again be continued as formerly, and come out regularly, either of three sheets, or five with a Cutt, according as Materials shall occur.

Alas for Halley's good intentions: the departure of James II and the advent of William III led to a further interruption as shown by the advertisement in No. 192 Vol. 17 (1691):

> The Publication of these Transactions having for some time past been suspended, chiefly by reason that the unsettled posture of Publick Affairs did divert the thoughts of the Curious towards Matters of more immediate concern than are *Physical* and *Mathematical* Enquiries, such as are for the most part the Subjects we treat of, with the exclusion to many others, wherewith the forein Journalists usually supply their monthly Tracts: These are now to Advertise, that for the future the *Royal Society* has commanded them to be published as formerly, and if possible Monthly. And all lovers of so good a Work are desired to contribute their Discoveries in Art or Nature, addressing them as formerly to Mr *H. Hunt* at *Gresham* College, and they shall be inserted herein, according as the Authors shall direct.

Hunt was the Society's operator. Evidently there was no review procedure. Halley's correspondence, with Wallis for example, shows that authors had control over what was printed and when.

Halley as Clerk was supposed to bring before the Society accounts of experiments, ideas, events, and writings. So he did regularly in the decade from 1686 to 1696; many were his own. Those accounts, on a wide range of topics, are recorded in the *Journal Book* of the Society. Some of his own papers were published more fully in *Philosophical Transactions*, others that survive only in manuscript in the archives of the Royal Society have been printed by McPike (*Halley Correspondence*), who also copied some *Journal Book* entries.[11] Here the paper in *Philosophical Transactions* (if there is one) is quoted, otherwise the unpublished paper, while *Journal Book* entries are given to show when Halley first discussed a subject or when there is no other mention of it. The unpublished manuscripts are often quite rough and were probably notes for a talk.

In the first years of the Royal Society the interests of its members were very wide, from the most abstract pure mathematics and fundamental physical science to medical practice and classical history and archaeology, not omitting accounts and tales of remarkable and strange natural happenings. While away from London at the Chester Mint,

Halley sent many letters about diverse matters: the magnetic variation at Chester, an eclipse of the Moon, the assay of silver, salt springs at Nantwich, a very severe hail storm,[12] of observations on the top of Snowdon to find its height and of another eclipse,[13] of a double rainbow system and of a halo about the Moon,[14] of a Roman altar, and of tales of a birth of a tiny puppy *per anum* by a greyhound dog. The last seems to be another version of a story that was related by Wallis in a letter to Waller slightly earlier. Wallis had it from the chaplain to the bishop of Chester and the dog belonged to one of the canons of the cathedral.[15]

While a schoolboy Halley observed the geomagnetic variation (declination) in 1672; he continued to do so regularly, and wrote most penetrating papers on the origin of the variation and the reason for its change. He laid down a map of the trade winds, and suggested that the Moon was speeding up in its orbit.

Halley continued to investigate topics in optics, physics, and metrology, following up work he had done in the years of the *Principia* (Chapter 5). He studied different forms of thermometer and barometer in addition to his observations on Snowdon.[16] In that way he acquired a reputation as an expert on the barometer and the interpretation of barometric observations. For three years (1686–1688) he edited the almanacks brought out under the authority of the Royal Society.

The topic which then and ever since has been associated with his name was the orbits of comets that he studied in a few months in 1695. The return of 'his' comet of 1682 according to his prediction had a great part in the evolution of the intellectual climate of the Enlightenment (Chapter 8).

While Halley was at Chester, the son of G. D. Cassini (Cassini *le fils*), who had succeeeded his father as astronomer at the Observatoire in Paris, made an extensive tour of the Netherlands and England, participating in the discussions leading up to the Pacification of Ryswick. He observed the lunar eclipse of 28 October 1697 (ns), from which he derived the difference of longitude between Rotterdam and Chester. He must have made use of Halley's observation of the eclipse at Chester. He observed the first satellite of Jupiter in London and obtained a difference of longitude from Paris that differed by less than 14″ from that of his father, as reported by Halley in the *Philosophical Transactions*.[17]

7.3 *Mathematician*

Halley was known in his day as a consummate mathematician, second only to Newton in European eyes, especially after his editions of

Apollonius, Pappus, and Menelaus (Chapter 9). He was no pure mathematician of the calibre of Newton, but he was competent and did publish his results and not hide them.

In 1687 and again in 1694 he set out a method for obtaining numerical solutions to polynomial equations by successive approximations. Mathematicians had addressed that problem for centuries and did not fully resolve it until they developed group theory in the nineteenth century. There are many letters on the topic in the papers of the Earl of Macclesfield thirty years or more before Halley's papers.[18]

The titles of Halley's two papers of 1687 refer to the origin of cubic and quartic equations in the problem of the intersection of the parabola with the circle, as first set out by Descartes. Halley must have been quite pleased with those papers for he holds the diagram of one of them in his portrait by Murray, reproduced as the frontispiece to this book.

The Society evidently appreciated them, for as Waller later wrote to Owen Lloyd in Dublin:

> Mr Halley is daily entertaining us with new Discoveryes and has now shown us a new short and easy method of extracting the Roots of all Equations.[19]

In his third paper of 1694 Halley had generalised his argument to a polynomial equation of any powers.[20]

If the equation is:

$$x^n + a_1 x^{n-1} + a_2 x^{n-2} + \ldots = b,$$

take as the first approximation $x = b^{1/n}$.

For the next approximation put $x^n = b + z^n$.

Halley gives approximate forms for $x = (b + z^n)^{1/n}$ and uses them to reduce the original equation to one in powers of z, making further reductions by the same process until b is sufficiently small.

The method will succeed only if z is small and the convergence of successive approximations is rapid. It finds only one solution close to $b^{1/n}$. Halley's contemporaries knew that multiple roots existed, but did not have general theorems about the number and location of the roots of a polynomial. That led to some obscurities, and Logan in Pennsylvania and Anderson in England questioned Halley's series some years later.[21]

Halley based his work on logarithms on the fact that differences of logarithms are proportional to logarithms of ratios. A letter from Wallis in 1692 to an unidentified correspondent who, from the content, is almost certainly Halley, shows that the relation was well known at that time. Wallis said that whereas there were tables of logarithms, there were none of antilogarithms, or counter logarithms, as he called them. He mentioned that Harriott had begun to calculate tables of antilogarithms and that the work had been continued by Pell, but nothing had been printed. No reply is known to that letter. In 1695 Halley presented a series for the logarithm when the ratio is close to unity and showed how it might be used to construct tables of logarithms to any base with high precision. Following that paper he corresponded with Wallis about logarithms.[22]

Suppose the ratio of two numbers is $(1 + q)$, where q is much less than 1. With a base β, the logarithm λ is given by $\beta^\lambda = 1 + q$. Halley used a series of Newton for the expansion of $(1 + q)^{1/\lambda}$ to derive the series

$$\lambda(\beta - 1) = q - \frac{1}{2}q^2 + \frac{1}{3}q^3 - \cdots$$

An arbitrary constant is implicit in the logarithm and it is usually chosen so that when the base is e, the logarithm of $(1 + q)$ is $(q - \frac{1}{2}q^2 + \frac{1}{3}q^3 - \cdots)$

Newton's series is not complete, for terms proportional to squares and higher powers of $1/\lambda$ are neglected, but Halley's result is the well known series for the natural logarithm when q is small.

Halley also obtained an expression for the antilogarithm, apparently a novel result. Again he used the connection with ratios. If a is a given number and b is an unknown number and if the logarithm of the ratio is λ, that is $\lambda = \log(b/a)$ then Halley showed that

$$b = a + \frac{a\lambda}{1 - \frac{1}{2}\lambda}.$$

Anderson later queried Halley's definition of the logarithm.[23]

Halley used the relation of a logarithm to a ratio as the basis of a geometrical argument to demonstrate that the scale of the *meridian line* is one of logarithmic tangents of the half complements of the latitude.[24] A *rhumb line* or *loxodrome* in navigation is a line making equal angles with meridians (Fig. 7.1), and the *meridian line* was a table of longitudes (at each minute of latitude) on the rhumb line that intersected meridians at $45°$. The theorem that Halley proved had been known to mariners for some

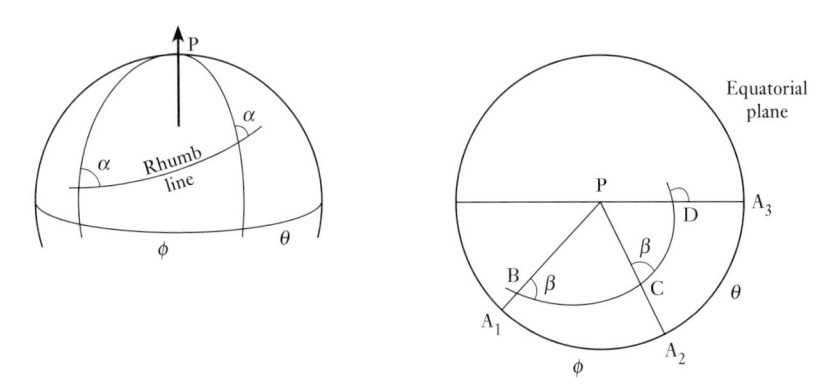

Figure 7.1 The rhumb line and stereographic projection (for explanation of symbols, see text).

time and was important for setting a course of constant bearing at sea. James Gregory had derived the result in 1668, which, wrote Halley,

> ... he did, not without a long train of Consequences and Complications of Proportions, whereby the evidence of the Demonstration is in great measure lost, and the reader wearied before he attain it ...

—a sharp comment that could be made on much other mathematical exposition.

Halley started from the property of a stereographic projection, which he was the first to prove rigorously, that angles between lines in the plane of projection are equal to angles between lines on the projected sphere. Since meridians on a sphere project into radii about the pole in the projection, a rhumb line on the sphere projects into an equiangular spiral that cuts all radii at the same angle.

Let P be the pole and PA_1, PA_2, PA_3, ... be radii of length r, and let the spiral cut them in B, C, D ... Let the arcs A_1A_2, A_2A_3, be ϕ, θ; then the spiral is such that

$$\frac{PC}{r} = a^\phi, \quad \frac{PD}{r} = a^\theta,$$

and so

$$\left|\frac{\theta}{\phi}\right| = \frac{\log(PD/r)}{\log(PC/r)}.$$

Thus intervals of longitude along the meridian line are proportional to the logarithms of the ratios PD/r, PC/r, and so on.

Halley discussed classes of infinite quantities, those corresponding to lines, planes, and solids, and argued that they were incommensurable.[25] He gave an alternative geometrical proof of a result of Caswell and Wallis about the areas of the cycloid and epicycloid, making use of some general results for areas swept out by curves that he obtained by the integral calculus.[26]

Halley's European reputation as a mathematician seems to have followed from his editions of Apollonius and Pappus, and so consideration of his achievements overall is put off to Chapter 12 where those works are described.[27]

7.4 *Geomagnetism and meteorology*

Halley had made experiments on the way in which magnetic force changed with distance and had observed the magnetic variation at places in London from his schooldays.[28] He knew well that the variation changed with time. In 1683 he had assembled many observations of the variation made over about a century from 1580 to 1680 and covering a range of longitude from 170° E (New Zealand) to 80° W (Baffin Bay) (Chapter 5). Many of the observations would have been reported by sea captains, some he may have taken from a book by Athanasius Kircher (1643), and many were probably among the papers that Halley acquired from Peter Perkins's widow, none of which is now to be found. He found that the variation kept its sign over regions of continental extent. He recognised that some of the variation arose from locally magnetised bodies but argued the large-scale effects must originate at depth in the Earth. He identified a generally westerly drift of the variation, and proposed that four magnetic poles were needed to account for it, but did not say how they might arise; his ideas aroused interest in Cassini, among others.[29] At the end of 1691 at three meetings of the Royal Society, he recalled his earlier paper and discussed the possible reason for the change. He published his analysis and ideas in 1692, bringing together a number of observations of the change of the variation in time, among them his own observations in London, as a schoolboy in 1672 and again in 1692, on St Helena in 1677, and Paris in 1681. He noted that the westerly drift is faster in some places than others. The field cannot be just a dipole field and as a first approximation, which Halley recognised it to be, he proposed that it was two superposed dipole fields, that is, there were four poles (the term 'dipole' is an anachronism). He argued that four poles imply two magnetised bodies and suggested that they corresponded to an outer shell of the Earth magnetised in one direction, and a core or nucleus magnetised in a different direction, with the nucleus rotating westerly relative to the shell about the common centre of mass.[30]

Some of Halley's ideas are now out of date; he thought, following Newton, that the Moon was denser, not lighter, than the Earth, and he supposed the shell to be magnetised. We now know that the shell, or mantle as we (most of us) now call it, is not magnetised, but that the core can have more than a pair of poles because the magnetic field in it is generated by electric currents which are more complex than the simple flow that would generate a pure dipole field. We know far more about the internal physics and mechanics of the Earth than Halley did; none the less he showed an understanding of the essential structure of the Earth's magnetisation. Halley not only identified features that any realistic theory of the field must explain, but he also saw that others later with more observations might need to devise more complex schemes.

In his second paper Halley discussed possible motions within the Earth that might give rise to the secular change of the variation. The first paper was written before *Principia*, the second came after it, and in discussing possible motions within the Earth, Halley had to present suggestions that conformed to Newtonian dynamics and celestial mechanics. The only possible motions were rotations that did not change the centre of mass, and he devised his shell and core model accordingly. The concepts of angular momentum and moments of inertia had not then been developed; it is now realised that the motions that produce the main magnetic field cannot significantly alter the angular momentum of the Earth.[31]

Halley's analysis of the magnetic field and its secular change established the features that had to be accounted for by any realistic theory. He himself would later, by his Atlantic cruises, contribute to the more detailed knowlege of the complexity of the field (Chapter 10). There was no significant advance on his theoretical views until the latter half of this century when it was seen that the field must originate in magnetohydrodynamics that satisfy the constraints that Halley recognised. In the intervening two and a half centuries or more sundry unphysical schemes were put forward. One such is that devised some fifty years after Halley's papers by Emanuel Swedenborg, the mystic who was also a mining engineer and mathematician. He described it as depending on 'the fluxion of a perpetually spiral vortex' from which he claimed, by a very lengthy calculation, to derive the variation at any place and time. His calculations for Uppsala were criticised by Celsius, who could not reproduce them, and who set out his views on Halley's contributions in a letter to Swedenborg:

> I believe that Dr Halley has more than any other applied the deepest thought in establishing a theory of the declination of the magnet which the learned have been seeking to improve ever since; yet he does not venture to determine by geometry the situation of the magnetic poles upon the earth and to establish rules for computing the declination. Meanwhile however, he has empirically constructed crooked lines representing the declinations of the magnet on the largest ocean of the world, and he has had the good fortune to see those lines, which were constructed mostly on the basis of the observations taken during his voyages confirmed more and more by later experiments.[32]

Thus a contemporary assessment of Halley's achievements in geomagnetism. It stands after two and a half centuries.

Sailors on the Atlantic passages had long known that to sail from Europe to N. America, even as far north as Virginia, it was best to go south below 30°, take advantage of the easterly winds for the westerly crossing and then sail northwards up the coast. For that reason, the predominantly easterly winds in the tropics were known as Trade Winds. Halley says that he had talks with mariners familiar with India

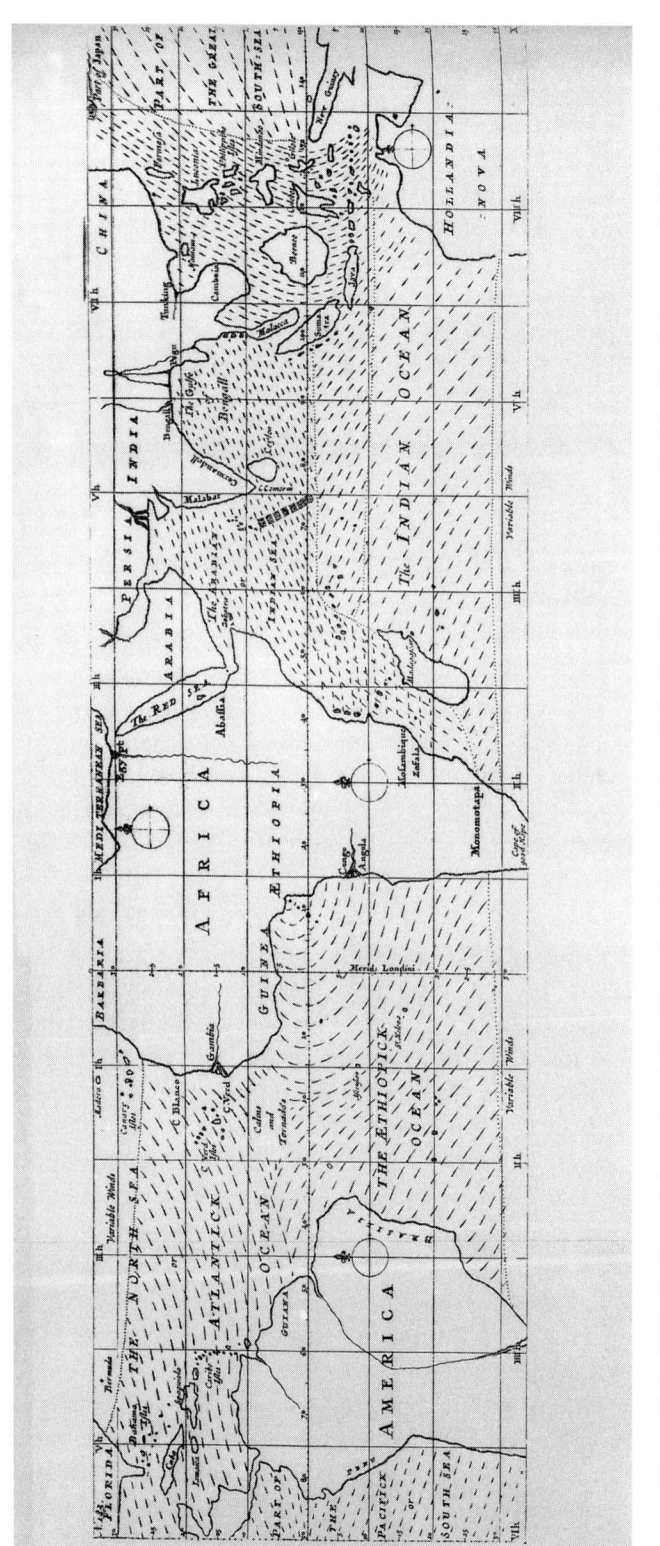

Figure 7.2 Halley's chart of the trade winds. (Reproduced by permission of the President and Council of the Royal Astronomical Society.)

and drew on his own experience in the tropics at St Helena, and so compiled a chart of the general course of the winds in the tropics (Fig. 7.2). It extends in latitude from the mouth of the Nile to the Cape of Good Hope and covers the Atlantic and Indian Oceans. The directions of the winds are shown by short lines covering the seas. As with his studies of the westerly drift of the magnetic variation, the data on which he based his map have not survived. The mariners familiar with India would be captains of the East India Company.[33]

Halley emphasised how useful a map was in showing the world-wide complex structure of the trade winds:

> To help the conception of the reader in a matter of so much difficulty, I believe it is necessary to adjoyn a Schema, showing at one view all the various Tracts and Courses of these Winds; whereby 'tis possible the thing may be better understood, than by any verbal description whatsoever.

There is a concise vindication of cartography.

Halley remarks on the difficulty of sailing southward between Guinea and Brazil on account of the SW winds off the coast of Guinea.

> In this part of the Ocean [St Helena] it has been my fortune to pass a full year, in an employment that obliged me to regard more than ordinary the Weather, and I found the Winds constantly about the South-East, the most usual point SEbE; when it was easterly it generally blew hard, and was gloomy, dark and sometimes rainy weather; if it came to the Southwards it was generally Serene and usually a small gale next to Calme, but this not very common. But I never saw it to the Westwards of the South or Northwards of the East.

Halley also showed that while the winds in the more open oceans blew steadily in the one direction throughout the year those in the northern part of the Indian Ocean, influenced by the land of the subcontinent, blew to the east for half the year and to the west for the other half, the Monsoons. He perceptively identified the West Indies as the source of hurricanes. Two centuries later, Sir Harold Jeffreys showed that hurricanes travelled eastwards from the West Indies and were an essential component in the conservation of the angular momentum of the atmosphere: the conservation is not exact and the small discrepancies contribute to the fluctuations of the rate of spin of the Earth.[34]

Halley argued that the spin of the Earth could not cause the trade winds since gravity would hold the air to the solid Earth. He thought the Sun heated a volume of the atmosphere that travelled round with it and drew the winds after it. Solar heating does cause the circulation of the atmosphere, as that of the oceans, but mainly it drives winds from the hot equator to the cold poles. The winds do not actually blow along meridians because the Coriolis force of the rotating Earth deflects them.[35] A general rotation relative to the Earth would not conserve the angular momentum of the Earth as a whole. The overall structure of the winds is therefore zonal, one way in the tropics, the opposite in mid-latitudes. The surface winds are not the whole story, for a generally poleward flow of heat, together with the overall conservation of angular momentum, is assured by winds in the upper atmosphere. Halley's proposed cause is wrong, but he did realise that land on the borders of the Indian Ocean, or off the Guinea coast, would distort an otherwise uniform flow.

Halley also studied the source of rain. He read a paper to the Royal Society on the rising of water vapour through the air on 25 April 1688, and later reverted to experiments on the evaporation of water.[36] In 1691 he set out the essentials of the circulation of the meteorological water, that it evaporates from the sea, falls upon the land as rain or dew,* and returns to the sea in rivers fed by springs. He later gave an account of a hail storm with very large hailstones at Chester in 1697.[37]

The Royal Society often discussed the height of the barometer and why it varied from place to place and time to time. It was well known that it fell on mountains and Halley had commented on its dependence on winds.[38] At the meeting of the Society on 14 April 1686, Halley and Hooke gave accounts of their measurements of the specific gravity of mercury; Halley weighed a bottle of mercury in water, while Hooke compared the heights of water and mercury that balanced in a siphon.

Halley's results are these:[39]

	lb	oz	gr	
Quart bottle full of mercury	32	00	04	(W_1)
Same in water	28	13	02	(W_2)
Difference	3	02	07	
Weight of bottle in air	2	07	02	(W_3)
Weight of bottle in water	1	08	03.	(W_4)

Halley gives the difference of the first two weights as 3 lb 2 oz 7 gr and, with other figures he gives, that indicates that the weight of the quart bottle full of mercury should have read 32 lb 00 oz 01 gr.

The weight of mercury in air is $[(W_1) - (W_3)]$, or 29 lb 8 oz 7 gr and the weight of water of the same volume is $[(W_1) - (W_2) - (W_3) + (W_4)]$, or 2 lb 4 oz 0gr, whence the specific gravity is 3783/288 or 13.11.

Hooke found that a column of water $79\frac{3}{8}$ inches (2.02 m) high balanced one of mercury $5\frac{5}{8}$ inches (14.9 cm) high, whence the specific gravity of mercury is 635/47 or 13.5. The actual density of mercury is

*On the basis of his experiences at St Helena he thought dew was more important than rain.

13 595 kg m^{-3}, so that both values of the specific gravity were too low, that of Halley seriously so by 4%. It is not clear why that should have been, although he noted that the weights should be checked for an error. The most likely source of error is that the volume of the mercury in the bottle was ill defined, in that the position to which the mercury would have attained in the neck of the bottle would not have been the same as the equivalent position of the water in which the bottle was immersed.

Observations that had previously been described to the Royal Society, in which a flask evacuated and full of air had been weighed, had shown that the specific gravity of air was about 1/800, and Halley calculated that a height of 10 514 inches (267.06 m) of air would correspond to 1 inch (30.5 m) of mercury, consistent with a depression of the barometer by $3\frac{1}{8}$ inches (7.9 cm) at the top of the Puy de Dome (3000 ft/914 m) and by 4 inches (10 cm) at the top of Snowdon (3750 ft/1143 m). On 5 April 1697 Halley wrote from Chester to Sloane to say that he was making up a party to go to Snowdon at Whitsuntide and desired the Royal Society to say what they would have him enquire into there. Whether or not he received any directions, he took with him a barometer and repeated the observations that J. Caswell had made some fifteen years earlier.[40] He reported his results in a letter in the *Philosophical Transactions*. On 2 May on top of the mountain the barometer stood at 26.1 in (66.3 cm) and at Llanberis at the foot of Snowdon that same evening it was 29.4 in (74.7 cm) while at Carnavon at sea-level the next day it was 29.9 in (75.9 cm). Other observations at Llanerch near St Asaph gave nearly constant readings over four days which included Halley's excursion to Snowdon and so Halley took the difference of pressure between sea level and the summit to be 3.8 in (9.7 cm). Caswell had measured the height trigonometrically as 3720 ft (134 m), whence Halley concluded that 1 in (2.54 cm) of mercury was equivalent to 980 ft (299 m) in height, 'which may serve for a Standard 'til a better be obtained on a higher place'. Following that report, W. Derham, the rector of Upminster in Essex, repeated an experiment he had earlier made on the Monument in London, and found 0.1 in (0.25 cm) of mercury to be equivalent to 82 ft (25 m) in height.[41]

The height of Snowdon according to Halley's formula derived from laboratory experiments would have been 3329 ft (1015 m), the modern value is 3559 ft (1085 m). Halley thought Snowdon a 'horrid spot of hills, the like of which I never yet saw'. He found that he could see Ireland, and the Lake District mountains. He enjoyed the trout from the many lakes.

7.5 Physics, metrology, and optics

Thermometers were devised many years before Halley's day. Even so his contemporaries were still very unsure of what was being measured and what meaning could be given to the graduations on the stem of a thermometer. At first thermometers and barometers had been confused in a single instrument but were now distinguished clearly. Halley among others thought that an air thermometer, which had been derived from the primitive combined thermometer and barometer, was the most reproducible and he proposed in effect to graduate thermometers by comparison with the air thermometer.[42] He seems to have had some intuitive perception of Henry's law of the thermal expansion of gases, although that law had not been formulated at the time, unlike Boyle's law; indeed he had to depend at first on a circular argument. An uniform temperature scale was defined many years after by the expansion of a perfect gas.

Halley wrote two short papers on optics: in the one he showed how to calculate the focal length of a thick lens, a substantial theoretical exercise, in the other he discussed the transmission of light through transparent materials.[43]

Halley studied the rainbow and haloes about the Moon.[44] People already understood in Hellenistic times that refraction and reflection of sunlight in raindrops gave rise to the bow, but Antonius de Dominis gave the first realistic account. He had done the experiments on which his explanation was based about 1590 and they first appeared in a book published by Bartolus in 1611. De Dominis was at one time archbishop of Spoleto, but apostasised from the Roman Catholic church. He brought the manuscript of Paola Sarpi's *History of the Council of Trent* to England where it was published, and James I appointed him Dean of Windsor. He became disillusioned with England and went back to Rome to be imprisoned by the Inquisition and to die. After his death his remains were publicly burnt as a heretic in front of the Jesu church in Rome. His fate has been connected with the treatment of Galileo by the Inquisition.[45]

Descartes extended the analysis of de Dominis, but the sequence of colours could not be explained before Newton demonstrated the specific refrangibility of rays producing distinct colours.[46] Halley showed that i, the angular radius of a bow, was given by the extremal of the expression

$$2(i - r) + p(\pi - 2r),$$

where p is the order of the bow. The extremal has to be a maximum if p is odd and a minimum if p is even.

The angular radius of the first-order bow formed in a drop of refractive index μ is then given by

$$\sin i = \left(\frac{4}{3} - \frac{1}{3\mu^2} \right)^{\frac{1}{2}},$$

a result that was described by J. Hermann in a letter to J. J. Scheuchzer. Hermann remarked that although the problem does not appear difficult, none before Halley had solved it, and he pointed out that the result leads to a bow of angle 42° for water of refractive index $\frac{4}{3}$.[47]

In 1691 Halley asked gilders and wire drawers how much gold and silver they used in gilding and drawing silver wire. He estimated the thickness of the gold skin upon gilt wire and determined the maximum possible radius of atoms of gold to be 5.10^{-6} in or 120 nm. The current value for the diameter of an atom of gold is 0.14 nm, one thousand times less. Halley for the first time tried to answer a quantitative question—how big is an atom—by a method: how much gold is there in a given area of gilding? This is in principle the same as much more recent investigations with layers of single molecules.[48] Francis Bacon, Boyle, and Newton among others had atomic theories of matter, which go back to the Greeks, but were speculative and qualitative. Galileo and Gassendi, who traced their metaphysics to Epicurus rather than to Aristotle, promoted the atomic hypothesis, while others, followers of Descartes, attacked it. There was certainly no consensus at the end of the seventeenth century and that is not surprising, for not even the Epicureans had suggested how to detect and examine atoms. The Epicureans and the Cartesians both based their opinions on *a priori* arguments. Halley cut through the philosophical arguments with a physical demonstration of what the greatest size of an atom of gold could be.

7.6 *Demography*

Already in Paris, Halley had compared the areas and populations of London and Paris (Chapter 4) and in 1693, in his papers on the tables of births and funerals in Breslau (Wroclaw), he returned to population questions. Caspar Neumann, a pastor and an ecclesiatical judge in the reformed church in Breslau, had sent the tables to Justel, from whom Halley had them. Nearly a year after his discussion of the tables in *Philosophical Transactions*, and shortly after Justel had died in late 1693, Halley wrote himself to Neumann. He was glad that Neuman was not

Table 7.1 Summary of demographic data from Breslau, 1687–1693

Year	Born	Surviving in year						
		1688	1689	1690	1691	1692	1693	1694
1687	1186	940	884	792	738	708	680	662
1688	1214		956	880	761	714	685	663
1689	1191			954	849	743	690	655
1690	1312				991	903	800	723
1691	1292					988	890	759
1692	1151						857	756
1693								912

displeased by his paper and hoped to hear more of his studies. In return Neumann sent him additional material, summarised in Table 7.1.[49]

The sources of the tables have been described by J. Graetzer. He re-examined the registers of the churches of the Augsburg Confession in Breslau, St Maria Magdalena, St Elisabet, St Bernhardin, and The Eleven Thousand Virgins, and found the mean values to be almost identical with those derived by Halley.[50]

Tables existed for London, but Halley chose Breslau because he considered the London numbers were distorted by immigration, whereas he thought the population of Breslau was steady. He recorded the number of deaths at each age over a period of 5 years, 1687 to 1691, for instance, 1000 up to age 1, 499 from 33 to 34, 252 from 58 to 59, and 20 from 83 to 84. Actuaries have used tables in his form ever since. He gave examples of uses of the table, how to determine the number of men between 18 and 56 able to bear arms, how to determine the chance that someone of age x will live for another y years, how to determine the expectation of life at a given age, and how to calculate the relative price of life insurance.

He calculated the values of life annuities for an interest rate of 6%, showing, for example, that the price of an annuity for a person aged 40 would be 10.57 years' purchase. Those figures may have been based on formulae for compound interest that he published in a chapter in *Sherwin's Tables* of 1706.[51] Beside the fundamental formulae Halley gave very many examples of their application.

Many have commented on how original Halley was in his treatment and how his tables and formulae have been followed in all subsequent actuarial practice,[52] but an important philosophical point seems not to have been made, namely that Halley's procedures depend on the assumption that the statistics of a sample, Breslau from 1687 to 1691, are truly representative of a much larger population, such as Breslau for the next century, or the whole of Silesia. Only if that is so are calculations of annuities based on the statistical frequencies in the Breslau sample justified. The philosophical justification has been argued from the time of Bayes and Laplace until now.

7.7 The student of antiquity

The study of the past gripped many people in Halley's day and if his stay in Rome had not established his own interest, he would have been led to it by contemporaries in England such as Aubrey and Sloane.[53] Like most educated people of his day, Halley had a thorough grounding in the classics in his schooldays and at Oxford, and his whole corpus of writing shows how at home he was with Latin and Greek. Not only was Latin his alternative language for scientific papers and correspondence, but he composed a Latin eulogy in verse for the *Principia* and he commented on and emended Greek and Latin inscriptions from remains in Britain and Palmyra. He also had some knowledge of Arabic, presumably acquired with the help of Hebrew. He interpreted ancient records to determine astronomical data, but his interests were wider, although for the most part he examined the relation between classical records and geography. In this decade of his life he concerned himself especially with three problems: where did Julius Caesar land, how reliable were the Antonine itineraries, and the astronomical information to be gained from historical records. He was a fellow of the Society of Antiquaries of London from 1720 until his death.

At the meeting of the Royal Society on 15 January 1690, Halley was asked to explain his assertion that Julius Caesar had landed from the Downs (between the coast and the Goodwin Sands). He did so at the next meeting on 22 January. He later published his argument.[54] He inferred the year from the fact that the invasion took place in the consulate of Pompey and Crassus, and that the death of Augustus, which fixed the date of that consulate, was established by an eclipse. Caesar's commentary showed that the first two legions landed four days before a full Moon and that they were followed up by ships with the horse at the full Moon, which occurred at midnight. Halley concluded that the full Moon was that of 30 August and therefore that the first landing was on 26 August 55 BC.

Halley considered that the high land reported by Caesar was the cliffs of Dover. Caesar landed on low ground to one side of the cliffs and Halley inferred that it was to the north from the behaviour of the tides.

He knew how the course of the tides in the Straits of Dover was related to the phases of the Moon, and concluded that Caesar must have been carried north of Dover before landing. He thought that the general set of the winds in the Straits, from the SW, confirmed that conclusion. Thus, by interpreting the accounts of Caesar and Dion Cassius, mainly the former, in the light of known astronomical and oceanographic data, he found that Caesar landed on the 26 August 55 BC, from the Downs, in the low land between the North and South Forelands near the present town of Deal.

Sir George Airy reopened the subject. He did not contest the date, but considered that Halley's knowledge of the tides was imperfect (Halley did not himself observe the tides in the Channel until 1701—see Chapter 10) and that Caesar would have landed near Pevensey. A century later the legions came ashore for the definitive conquest close to Halley's suggested site at Richborough where they would build a great port and fort.[55] Halley's discussion, like other comments he made to the Society, shows that he was familiar with the south coast.

On 19 March 1701, between returning from his second Atlantic cruise and beginning his survey of the tides in the Channel, Halley read a paper on the geography of the ancients and moderns.[56] An important source of evidence for the survey methods of the Romans is the Antonine Itineraries, tables of distances along the principal roads of the Roman Empire, Britain included. Halley argued from the agreement between distances in the British part of the Antonine Itineraries and measurements of his day with a measuring wheel devised by Ogilby, that the distances in the Itineraries were actually measured and

> ... not put down by estimate, as is usuall with us, but were an actuall survey, and perhaps there were on the way side Miliary Columns for the information of Travellers, as is certain there were in Italy and some other provinces ...

—and as have since been found in Britain.

Halley observed that although the road surveys were accurate, many latitudes and longitudes assigned to places in the Empire, especially in the East, were wrong. He thought that the Romans did not measure them carefully and that the dimension of the Earth given by Ptolemy was too small (the earlier value of Erastosthenes was better). Thus the longitudes assigned to places east of Rome by contemporary geographers who used Ptolemy's values were too great.

The length of the degree of latitude had been well established by Norwood in England and Picard in France, and Halley and others had measured the dimensions of the Greek and Roman foot as preserved on the Campidoglio in Rome.

I think the measures left us by the Ancients will be very sufficient to shew the great errors committed by most of our Modern Geographers in making the length of the Mediterranean, for an instance, near a quarter part too long, and the lesser Asia and Greece near double what it ought to be in breadth.

He himself estimated some differences of longitude, as of Constantinople, not more than 18° from Rome (in fact about 17°), and Alexandria not more than 27° from Rome—still an overestimate, for it is only slightly east of Constantinople. Observations of Jupiter's satellites at Rome and at London showed Rome to be 13° east of London; the modern value is nearer 12°. Elsewhere he remarked that modern French measurements of longitudes had reduced the length of the Mediterranean from 60° to 40°.[57] Clearly the maps of the eastern Mediterranean in Halley's time were greatly distorted. If such errors still persisted at that time, it is hardly surprising that Columbus had got it wrong in planning his voyages of discovery.

Halley was right to insist upon the accuracy of Roman survey, but it seems that he was not familiar with the texts of the Roman agromensores nor with the later books of the *De architectura* of Vitruvius, even though Vitruvius had been the authority for architecture for fifty years or more in France and England. Had he been familiar with them he would have known how the Romans carried out their land surveys, how they set up milestones along their roads, and in particular, he would have known the description that Vitruvius gives of a hodometer to be used with a four-wheeled carriage on the same principle as Ogilby's wheel.[58] After the number of revolutions of the carriage wheels corresponding to one mile (a thousand paces) a stone would be released by a series of toothed drums and wheels to fall into a box. At the end of the day's journey the number of stones in the box was the number of miles traversed. Vitruvius also describes the adaption of the hodometer to ships, the carriage wheels being replaced by paddle wheels at the side of the ship; that was probably far less accurate than the carriage.

The topics of this chapter may seem random and miscellaneous compared with *Principia*, or the celestial mechanics that follow in Chapter 8, or the seamanship of Chapter 9, but they are not inconsiderable. They occur in the interstices of greater works and we may wonder how Halley found the time or the inspiration for so many of them.

8

Celestial architecture

> ... whose faculties can comprehend
> The wondrous Architecture of the world;
> And measure every wand'ring planet's course ...
> MARLOWE: *TAMBURLAINE THE GREAT*, 1, 872–4

8.1 The paths of heavenly bodies

THE *Principia* changed the purpose of astronomy. Astronomers had hitherto plotted the paths of heavenly bodies with increasing accuracy but had no idea of why the paths should be as they were. Newton showed how all celestial motions could be accounted for on a common basis of universal gravitation. He himself, in Book III of the *Principia*, elucidated many phenomena but left substantial problems for himself and his successors.

Some problems had important practical applications: navigation, for example, needed a detailed knowledge of the Moon's motion. Halley's connection with navigation is evident. He was also deeply concerned with fundamental issues; in particular, whether universal gravitation really did control all motions in the solar system. He made some of the first and some of the most significant new applications of Newton's celestial mechanics.

Nowadays, after three centuries, the *Principia* appears to have changed the entire view of the cosmos. It did not appear so at the time. Various approximations to Kepler's second law of planetary motion were in use, such as that of uniform angular velocity about the empty focus; the publication of the *Principia* certainly did not convince everyone that the law was exact. Newton had shown right at the start of the *Principia* that a force directed to a centre would give a constant areal velocity or angular momentum; for many years natural philosophers in Europe, Huygens, for example, and Leibniz, could not accept the concept of action at a distance. They would not agree that the force on a planet was necessarily directed to the Sun, nor that Kepler's law applied universally. They tried to reproduce the effects of the inverse square law of gravitation by some means of direct contact between bodies, in particular, systems of vortices, schemes that would not necessarily conserve angular momentum. Kepler's second law is not in fact universally exact because attractions by third bodies cause deviations from it, and so a law of attraction to a centre cannot be demonstrated just from very careful observations of the Moon and planets. In the years following the publication of the *Principia*, there were few who understood its principles as thoroughly as did Halley and David and James Gregory. Comets had a special place in the Newtonian system. Vortices might be able to produce the motions of the Moon and the planets in and close to the ecliptic, but the orbits of comets far out of the ecliptic were very hard to accommodate in any vortical theory.

Newton had shown that the paths of two bodies about each other, neglecting all others, would be conic sections about a common focus. There are, however, many bodies in the solar system, and not all the others can be ignored in considering the paths of any two. Four phenomena are especially important: the orbit of the Moon about the Earth in the presence of the Sun, the orbits of Jupiter and Saturn as attracted by each other, the orbits of comets in the presence of the major planets, and the orbit of the Earth about the Sun in the presence of Jupiter and Saturn. Halley as a schoolboy was first excited by astronomy when he realised that the motions of Jupiter and Saturn in his day differed considerably from those at earlier times. The one was early, the other late, as compared with the tables he had. The problem of three bodies, each attracting the others, of which all of those four phenomena are instances, has no exact solutions in general.* It can be dealt with only by approximate methods which it is hoped may converge. Newton mentioned all four problems in greater or lesser detail in the *Principia*, but obtained satisfactory solutions to none of them. Euler, d'Alembert, and Clairaut developed the first effective analytical theories in the middle of the eighteenth century, not long after Halley had died.

The elliptical orbit of an isolated planet about the Sun is commonly defined by the size and eccentricity of the ellipse and the direction of its major axis, that is of perihelion; it is also defined by the inclination of the plane of the orbit to some reference plane, such as that of the ecliptic and the direction of the line of nodes in which it intersects that plane. If the changes in an orbit arising from attractions of other bodies are small, they may be conveniently expressed as changes in the inclination, node, eccentricity, and longitude of perigee (see Appendix 1); that is so for the orbit of the Earth about the Sun under the attraction of Jupiter and Saturn, but is less satisfactory for the motion of the Moon around the Earth (Chapter 13).

---

*One of the soluble problems is the *restricted problem of three bodies*: three bodies move in a plane and the mass of one of them is negligible. None of the above four problems here is of that form.

Comets are unlikely to move in simple elliptical orbits for, as Newton and Halley both realised, between one return of a comet and another the attractions of Jupiter and Saturn could perturb it. The paths close to the Sun can, however, be represented by parabolae (for which the eccentricity is unity). Inclination, node, and perihelion are meaningful parameters and Halley and his contemporaries were concerned to determine them. There is no *a priori* reason why a comet should not move in an hyperbolic orbit about the Sun; it would usually be just as close to a parabolic orbit near the Sun as an ellipse.* The form of an orbit depends on the initial conditions which, like those of the planets, depend on how the solar system and its various constituents came into being. If the path of a comet is appreciably distorted by the attractions of the major planets, it may be difficult to say whether or not it is the same one on a subsequent apparition. Halley supposed that two apparitions were of the same comet if the parabolic elements were similar, but when the perturbations were large it might be difficult to identify a comet on its return.[1]

8.2 Comets

Comets have long been seen as possible harbingers of notable events, mostly disasters. A comet was associated with the assassination of Julius Caesar. Halley's comet marked the death of Agrippa in 12 BC, the defeat of Attila at Chalons in AD 451, and the Norman invasion of England in 1066. The representation of the comet on the Bayeux Tapestry is striking (Fig. 8.1). It has often been proposed that the Star of Bethlehem was a comet, and possibly Halley's (Fig. 8.2).[2]

When Halley was a young man anyone interested in astronomy was also intensely interested in comets. Some of his friends and acquaintances had observed and studied comets with particular care. He included a précis of earlier views on the nature of comets in his summary of his own orbital calculations.[3] Were comets physical objects or were they some sort of optical illusion? Were comets atmospheric phenomena or did they lie beyond the Moon? Tycho Brahe found that the comet of 1577 showed no diurnal parallax and hence must lie beyond the Moon and possibly beyond some of the planets. The nature of the tail was for long a matter of speculation.

Halley himself observed comets carefully to determine their orbits. He surely discussed comets with Hevelius when he was in Danzig, for

---

*Hyperbolic orbits were considered for comets by Newton and Halley but thereafter were generally ignored. Ernest Rutherford took them as models for the scattering of α-particles by atomic nuclei.

206 Edmond Halley

Figure 8.1 Halley's comet in the Bayeux Tapestry. (Reproduced by permission of the President and Council of the Society of Antiquaries of London.)

Figure 8.2 The comet in Giotto's *Adoration of the Magi* in Padua (from a colour transparency in the Royal Astronomical Society and reproduced by permission of the President and Council).

Cometographia was one of Hevelius's best known works. Halley and Nelson saw the spectacular comet of 1680 when they were *en route* to Paris.[4] The same comet had already been seen in November in England, in Saxe Coburg and in Rome before it passed through perihelion in December. Or was it the same comet? Newton at first thought, as did almost everyone else, that the apparitions of 1680–1681 were of different comets. Flamsteed said they were the same, and so they were, although he thought the comet was repelled by the Sun. In his later feud with Newton he never forgot he had been right.[5]

The indirect correspondence between Newton and Flamsteed about the comet of 1680–1681, in which Halley took part, sets out clearly the issues that astronomers faced in the years before the dynamical theory of cometary orbits in the *Principia*. Kepler had thought that comets moved on straight line paths. When a comet with a nearly parabolic orbit is far from the Sun, the path is close to a straight line, so that Kepler's representation was not unreasonable. More careful and precise observations showed that the orbits were concave towards the Sun. Halley made some relevant calculations, as he wrote in a letter to Hooke from Saumur:

> I tryed but without Success to represent the Observations by an equable motion in a right line. I made a theory to hit the first and last and two intermediate Observations, but then the Latitudes differed a little too much and the rest of the Longitudes would not hit right ...[6]

Flamsteed had not observed the comet of 1680–1681 in November when it appeared before sunrise moving towards the Sun, but he predicted that it would appear at a later date moving away from the Sun; he continued to observe it at Greenwich after Halley had left for France.[7] Many other astronomers in Europe and America observed it. Cassini summarised many of the results in a tract he presented to Louis XIV, including the observations that he and Halley made in Paris, and those of Gallet in Avignon, Cellio and Pontio (Ponthaeus) at Rome, and Montanari in Venice. Halley, while still in Europe, sent them to Flamsteed, and subsequently in 1682 discussed some of them with Newton.[8] Newton told Flamsteed of the striking sight of the comet as seen from Trinity College, Cambridge, with its tail over King's College Chapel (Fig. 8.3). Tillotson told Halley of the observations by Hill, the local astronomer in Canterbury (Chapter 4). Halley, like Flamsteed, supposed that there was just the one comet, for in a letter known only from a reply to it by Flamsteed, he appears to have argued that the Sun repelled the comet. Flamsteed, in his reply, said that he had a similar opinion but went further in thinking the Sun and the comet to be magnetised so that the comet changed direction in front of the Sun because

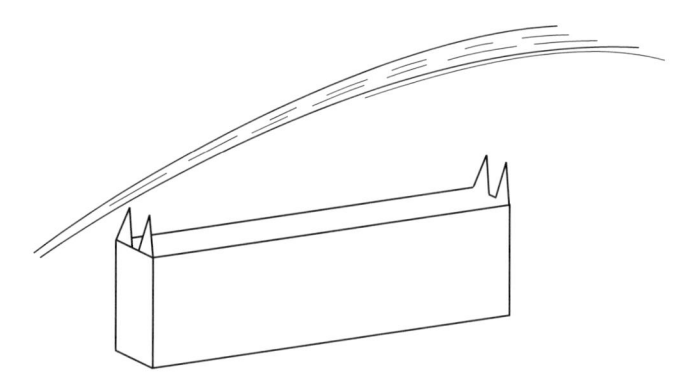

Figure 8.3 The comet over King's College, as sketched by Newton in the *Principia*.

the Sun at first attracted and afterwards repelled it. Newton demolished that idea, but then said that his objections would be removed if the comet went behind the Sun and was always attracted by it.[9] A little later he accepted that the two apparitions of 1680 and 1681 were of the same comet.[10]

Newton may well have had a theory of elliptic orbits by 1681, but not of comets. Cassini had a more radical picture of comets, as Halley reported to Hooke: he thought that they followed circular orbits, and that the comet of 1680–1681 was the same as one observed by Tycho in 1577 and also one of 1664–1665 that had been observed by Newton, by Auzout in France, and by Cassini and Queen Christina in Rome.[11] Ideas that comets returned periodically and moved in elliptical orbits were by no means new and various earlier astronomers had held such ideas, but held them as guesses unsupported by accurate observation or mathematical theory. Halley would surely have been familiar with those notions as speculations; he was to establish them as fact upon sound observation and fundamental theory. He was also probably familiar with the notion that cometary orbits could be parabolae. William Lowther, a friend of Harriot, had apparently suggested as early as 1610 that the paths of comets might be very elongated ovals; Hooke thought the same, and so did Hevelius, whose *Cometographia* was well known in England. Georg Samuel Dörffel of Plauen in Saxony seems the first to have suggested that the orbit was a parabola with its focus in the Sun. He set forth his idea in a pamphlet apparently unknown in England, for he is never mentioned in the correspondence between Newton, Flamsteed, and Halley. Newton may have come to think that comets moved in parabolic orbits about the Sun as he was writing the *Principia*; in Book III he discussed the observations of the comet of 1680 in considerable detail.[12]

Almost all the positions of comets available to Halley when he began his calculations of orbits had come from observations of the distances of

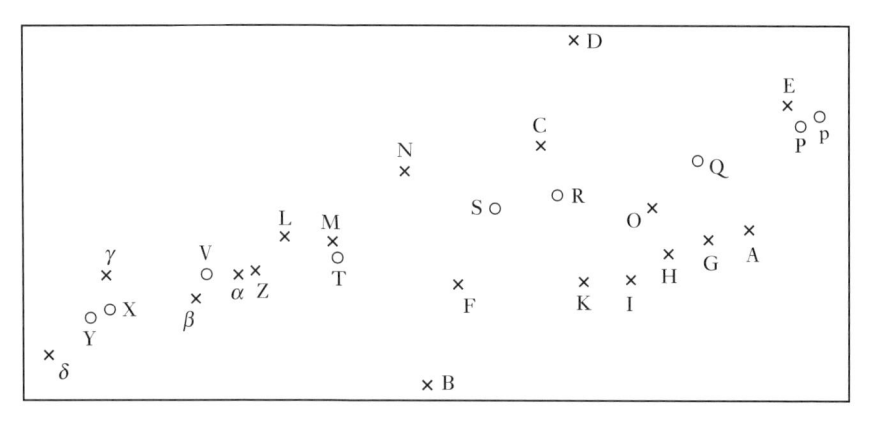

Figure 8.4 A comet in a field of stars, from Newton's *Principia*. (Book III, Prop. 41)
○: positions of comet, P to Y;
×: positions of stars, A to Z and α to β.

a comet from fixed stars. He and his contemporaries measured distances with telescopic sextants. Hevelius had used a sextant with open sights or pinnules (Chapter 4). Newton measured the distance of the comet from known stars with a telescope fitted with a micrometer eyepiece (Fig. 8.4). Many of the older observations were not measurements but estimates of the position of a comet in a field of stars. They were often crude. Very few observers had instruments for the direct determination of right ascension and declination, so it was natural to use ecliptic coordinates, latitude and longitude, derived from distances from fixed stars whether measured or estimated. After Newton had given a dynamical theory for the motion of comets about the Sun, the elements of cometary orbits were naturally referred to the ecliptic.

Halley observed the comets of 1682 and 1683 from Islington using the sextant that he had taken to St Helena.[13] He made no contribution to the theory of comets as it appeared in the first edition of *Principia*. After the publication of *Principia*, Newton moved on to other matters. Halley was occupied with going to sea and with his duties as Clerk to the Royal Society. Only in 1695 did he and Newton discuss the improvement of the treatment of comets in the *Principia*. Newton had become dissatisfied with that in the first edition. Halley visited him in Cambridge in August 1695, and evidently arranged to repeat the calculations.[14]

There were five steps in Newton's study of the orbit of the comet of 1680–1681 in the first edition of the *Principia*. He first showed from the observations that the areal velocity about the Sun was constant, just as for planets and satellites. The force between the Sun and the comet was therefore along the line joining them. Secondly, he had shown that bodies attracted by an inverse square law could follow parabolic orbits.

Thirdly, he showed how to interpolate between observations to obtain positions of the comet at equal intervals of time. That was the first published statement of his general methods of interpolation and extrapolation by finite differences. He then showed how to find a parabolic orbit from just three observations at equal or nearly equal intervals of time using a semi-graphic method (*graphicas*), and finally, again semi-graphically, he checked that other observations would fit.[15]

When Halley offered to undertake calculations for the second edition of *Principia*, he would have been very familiar with the practical problems of observing comets, with the theory of the orbits, and with the methods of computation that Newton had set out at some length in Book III. He now devised a fully numerical way of finding heliocentric positions in a known orbit. He obtained initial values of the elements from three suitable observations by Newton's geometrical method; he calculated positions at the times of other observations from a general table for parabolic orbits; and he then adjusted the elements until he obtained satisfactory agreement with the observations.[16]

The orbits of comets are both simpler and more complex than those of planets. Like planets, they move in planes that contain the Sun, but whereas the orbits of the Moon and planets lie in planes that are all very close to the ecliptic, orbits of comets are inclined to it at arbitrary angles. An observer on the Earth moving around the ecliptic would find it more difficult to determine the plane of the orbit of a comet than that of a planet.

Most comets move in orbits so highly eccentric that they are effectively parabolae near the Sun. All parabolae are geometrically and kinematically similar, with only two orbital parameters to be determined: the direction of perihelion in the plane of the orbit and the distance of the comet from the Sun at perihelion. Halley was therefore able to construct a single table from which the position of any comet in a parabolic orbit could be obtained for any time from perihelion (see Appendix 1).

Halley may have used numerical methods from the start, not graphical ones, to determine the parameters of orbits, for he wrote in a letter to Newton of 21 October 1695:

> I am now become so ready at the finding a Cometts orb by Calculation, that since you have not sent the rulers,* as you wrote me, I think I can make a shift without them.[17]

Halley began with the comet of 1683, which he had himself observed from Islington,[18] and found a good fit to a parabolic orbit. He also obtained a satisfactory result for the comet that Newton had observed in

---

*The rulers were those that Newton had had made for his graphical determinations of the comet of 1680–1681 in the first edition of the *Principia*.

1664. He thought that the comet of 1682 ('Halley's') was probably that of 1531 and 1607.[19] He identified a number of errors in various observations, including some that he thought Hevelius had adjusted to fit a supposed orbit.

Halley was able to fit the observations of the comet of 1680–1681 to a parabolic orbit over the period 21 December to 5 February with errors of little more than 1' in latitude and longitude. The errors reached 3' at earlier and later dates and Halley concluded that the deviations indicated that the orbit was an ellipse of fairly short period. Newton agreed, and made some suggestions for improving the parabolic orbit.

If Halley had only begun his calculations after visiting Newton in August, his results from less than a month's work were substantial,[20] for he would have had to fit them in between working on the salvage of the ship of the Royal African Company sunk off Pagham, and taking up his post at the Chester Mint (Chapter 9). He may have been thinking about cometary orbits before his August visit to Cambridge.[21]

Halley next tackled the comet of 1664. He analysed the observations of Hevelius at Danzig and identified some errors, but then found agreement with a parabolic orbit to within about 2'.

He now turned to the comet of 1682. He wrote to Newton on 7 October to thank him for observations of the comet of 1682,

> ... which next after that of 1664 I will examine, and leave it to your consideration, if it were not the same with that of 1607 ...

However, he clearly realised that attractions by the major planets could alter the period of a comet and hence the time of its return, for he continued:

> I must entreat you to consider how far a Comets motion may be disturbed by the Centers of Saturn and Jupiter, particularly in its ascent from the Sun, and what difference they may cause in the time of the Revolution of a Comett in its so very Elliptick Orb.

To that Newton replied:

> How far a Comets motion may be disturbed by ♃ and ♄ cannot be affirmed without knowing the Orb of ye Comet & times of its passage through ye Orbs of ♃ and ♄. If in its ascent it passes through the Orbe before its heliocentric conjunction with ye Planet, the time will be shortened, if after, it will be lengthened & the decrease or increase may be a day, a week, a month, a year or more; especially if the Orb be very eccentric & ye time of ye revolution long.[22]

Newton never developed a satisfactory theory of the perturbation of comets. The problem rested until Clairaut attacked it for the return of Halley's comet. Meanwhile Halley could not look for exact agreement between the elements of the comets of 1532, 1607, and 1682, and his assertion that they were all the same comet was based on somewhat inspired insight. He wrote to Newton on 21 October:

> I have almost finished the Comet of 1682 and the next you shall know whether that of 1607 were not the same, which I see more and more reason to suspect.

Meanwhile, Newton had written on 17 October: 'I can never thank you sufficiently for this assistance & wish it in my way to serve you as much'.[23]

In his final letter at this time to Newton on the subject of comets, Halley considered that those of 1682 and 1607 were the same, notwithstanding that the inclination of the latter to the ecliptic was less than that of the former. So his collaboration with Newton on the orbits of comets came to a close: the letter itself shows two reasons why it did not continue. He wrote: 'I will waite on you at your lodgings to morrow morning to discourse the other matter of serving you as your Deputy'—that is, at the Chester Mint; the letter must have been written just after Newton learnt of his own appointment to the Mint in March 1696.[24] Newton became deeply involved in the affairs of the Tower Mint and Halley spent two troubled years at Chester. The second reason is in Halley's opening phrase: 'I had waited on you on Saturday, but I was obliged to go on board my friggat'. He had for some time been going to sea in various pursuits (Chapter 9), and in two years' time he would embark upon his Atlantic and Channel cruises and after that set off for the Adriatic (Chapters 10 and 11). Only when he became Savilian Professor of Geometry in Oxford in 1704 did he again find time for comets (Chapter 12).

The correspondence on comets comes to an end with Halley and Newton both accepting that some comets moved in elliptic orbits, and Halley close to asserting that some returned periodically.

Halley spoke of his belief that some comets return periodically to meetings of the Royal Society on 3 June and 1 July 1696. He announced his conclusion that the comets of 1607 and 1682 were the same, moving in an elliptic orbit about the Sun and receding to a distance of 35 astronomical units. He also gave a parabolic orbit for the comet of 1618.[25]

At some time over the next ten years, Halley calculated the orbits of a further twenty comets. He must have completed many of them when he presented a table of elements to the Royal Society on 18 March 1702. Almost as soon as he arrived in Oxford as Savilian Professor, he published them in the *Philosophical Transactions* of 1705 along with the calculations he had done for Newton in 1695, and issued them separately as a broadsheet from the Sheldonian Theatre, Oxford.[26] He summarised the history of the observations of comets and of ideas about them; he showed how, following Newton, the orbits of comets may be determined from three observations; and he explained the construction and use of his general table for the calculation of the coordinates of a comet from the elements of its orbit, from which the position of a comet in its orbit may be found at any time. As examples he gave calculations of the positions of the comets of 1680–1681 and 1683, as observed from Greenwich and from Islington respectively. He collected his calculations of orbits in a table of the elements of 24 comets from 1337 to 1698. He predicted that the comet of 1682 would reappear in 1757 or 1758 and concluded with a few remarks on possible close approaches of comets to the Earth, but left it to others more expert in physics to say what effects there might be.[27]

Halley passed through Hanover on the way to Vienna in 1703. He met Leibniz, who told him of the observations that Kirch had made of the comet of 1680 in early November of that year. He received the observations somewhat later, he reassessed the positions of the comet in relation to new positions of stars obtained by Pound, and Newton included them in his discussion in Book III of the third edition of the *Principia*.[28]

Later, while still at Oxford, Halley compiled a considerable set of astronomical tables that were set up in print shortly before he became Astronomer Royal.[29] He delayed publishing the tables until he could compare observations of the Moon with his tabular values; they never appeared in his lifetime. John Bevis published them posthumously, along with comparisons with Halley's own lunar observations made at Greenwich.[30] The work as a whole is best considered separately in Chapter 13. However, the book does include the final version of *Astronomicae cometicae synopsis*, which Halley worked on while still at Oxford and which was apparently set up in print, in both English and Latin, at the same time as the tables. Halley continued to study comets after he became Astronomer Royal, as shown by a letter to Sloane of November 1722 in which he asks for the loan of a book containing observations of the comet of 1580 by Michael Maestlin, and by another letter to Newton of February 1725 in which he apologises for a mistake in some calculations. The final 1726 version probably incorporates some work done at Greenwich.[31]

Halley had emphasised that the function of the table of the parabolic elements of 24 comets was to detect possible multiple apparitions of comets. He was at pains to say that calculated elements of comets seen in past years were afflicted by considerable errors both of observation and on account of perturbations by Jupiter and Saturn. It was not realistic to expect close correspondence between successive apparitions:

The Principal Use therefore of this Table of the Elements of their Motions and that which indeed inclined me to construct it, is, that whenever a new Comet shall appear, we may be able to know, by comparing together the Elements, whether it be any of those which has appeared before, and consequently to determine the Period and Axis of its Orbit and to foretel its Return.

In the 1726 version, which is essentially the same as those of 1705, Halley goes on to argue that the comets of 1531 and 1607 are the same as that of 1682; he refers to the perturbations of Jupiter and Saturn; he speculates that the comets of 1456 and 1305 were still earlier apparitions of it; and he predicts the return of 'his' comet in 1758. He also suggests that the comet of 1680 may be periodic, with a period of 575 years (a figure that was much in the air when he was in France in 1681). He says he has recently thought how to calculate elliptical orbits; he mentions an idea of Nicolas Fatio de Duillier about finding the solar parallax from that of a comet, should it come close to the Earth; and finally he leaves to others to say what might happen if a comet were to hit the Earth.

The 1752 posthumous version continued quite differently. First (a minor point), he included first differences in his general table for a parabolic orbit. Then he added a completely new section, *Of the motion of comets in elliptic orbits*, which he had written, according to the Preface, before 1718. He discussed elliptic orbits and how they might be calculated, and gave a general table for finding positions in them similar to, but of course more elaborate than, his parabolic table. Finding the position from the time is Kepler's problem (Appendix 1). Halley gave tables of the calculated and observed positions of the comets of 1682, 1607, and 1531 which were based on extensive and critical analyses of the observations that were available to him: at Greenwich for 1682, by Kepler at Prague and by Longomontanus in Denmark for 1607, and by Apian in Germany for 1531. An elliptical orbit fits the 1682 observations to better than 2' 24" in longitude and 2' 48" in latitude, but there are discrepancies of up to 34' for 1607. Not surprisingly, the observations of the apparition of 1531 were crude and there were appreciable errors, especially of latitude, in which Halley thought that Apian had made an error of some degrees. None the less, he considered that the elliptical orbits of 1607 and 1531 were close to that of 1682, and he further considered the still earlier apparitions of 1456 and 1305. He also made a thorough study of the brilliant comet of 1680. For that he had the observations at Saxe Coburg on 4, 5, 6, and 11 November that Kirch had published in his *Novus nuncius coelestis*, as well as the later observations at Greenwich and Paris in which Halley himself took part.[32]

The final version of the *Synopsis* is, accordingly, a more thorough analysis than that of twenty years earlier of the orbits of all comets known to Halley and of the hypothesis that comets are periodic. It is a considerable advance on all previous work on the orbits of comets. It is difficult to know what influence it had as a posthumous publication, but although theory has improved and computing power vastly expanded since Halley's day, the principles of modern studies of the orbits of comets are recognisably still his, and firmly based on Newton's prior analyses in *Principia*.

Halley died in 1742 before he could see his comet return. It was widely appreciated, especially among French savants, that the return would be a powerful confirmation of Newtonian dynamics. A number of people produced predictions upon unreliable empirical bases.[33] De L'Isle, who had visited Halley in London, urged his French colleagues to try to be the first to recover the comet. They made a great effort to calculate the date of return from sound dynamical theory and then to search for it. Halley and Newton had realised the key difficulty: what were the effects of the attraction of Jupiter and Saturn upon the orbit, and thus upon the time of perihelion and the expected positions? In parallel with Euler and d'Alembert, A.-C. Clairaut had made the crucial mathematical advances that enabled the problem to be tackled. He had applied the latest techniques for solving differential equations to the problem of the Moon's motion under the attraction of the Earth and the Sun, an intricate problem of three bodies moving under mutual gravitational attraction. He had thought at first that the discrepancies between Newton's lunar theory and observation showed a failure of the inverse square law of gravitation, but then by careful analysis found that inverse square law forces could account for all the anomalies of the Moon's motion known at that time.[34]

Clairaut had also applied his methods to the orbit of the Earth under the attractions of the other planets, as well as to the mutual perturbations of Jupiter and Saturn, and had developed a theory of the figure of the rotating Earth. He now adapted his lunar theory to the perturbation of a comet by Jupiter and Saturn. That was difficult because approximations that could be made for lunar and planetary theory were not valid for comets. The inclinations to the Sun's orbit are small for the Moon and planets, but arbitrarily large for comets, a difficulty met again in our own times when lunar theory was adapted to the orbits of artificial satellites about the Earth. The specific results for the return of Halley's comet still needed very heavy numerical calculations— d'Alembert abandoned the work when he realised what was involved. The astronomer de LaLande did most of the numerical work in cooperation with Mme Nicole-Reine Étable de la Brière Lepaute (1723–1788), a friend of Clairaut. She seems in particular to have organised the scheme of the calculation. In his account of the recovery of the comet, de LaLande wrote that the immense ensemble of detail would have seemed frightening to him if Madame Lepaute, who had for a long time applied herself successfully to astronomical calculations, had not taken part in the work.[35]

Six weeks after Clairaut announced his prediction to the Académie Royale des Sciences, Messier, de L'Isle's assistant at the Naval Observatory in the Hôtel de Cluny in Paris, found the comet on 21 January 1759; it later turned out that there had been some earlier German observations. The return of the comet virtually on time (perigee was in fact about a month earlier than predicted) was hailed as a powerful affirmation of Newtonian dynamics and of the regularity and predictability of the natural world. De LaLande could say, with some justified self-satisfaction,

> The universe beholds this year the most satisfactory phenomenon ever presented to us by astronomy; an event which unique until this day changes our doubts to certainty and our hypotheses to demonstration.

Needless to say, the acceptance of Newtonian dynamics by the learned did not affect popular notions of disaster and portents, but they came back in different guise. People might no longer see comets as signs of Heaven's grace, or more likely, displeasure, but would fear them as actual causes of disaster. Halley had perhaps contributed to such ideas by his passing remark in the *Synopsis of comets*: 'that if a comet came near the Earth it might cause some remarkable effects'. Some took that speculation as a prediction that disaster would indeed occur, and so feared the apparition of Halley's comet in 1910. Associations with notable events live on. When the European Space Agency launched a spacecraft to pass close by Halley's comet at the apparition of 1986, they called it Giotto, after the painter of the frescoes in the Capella degli Scrovegni at Padua, who showed a comet over the scene of the Adoration of the Magi. Giotto is supposed to have had in mind the apparition of Halley's comet in 1301, shortly before he painted in Padua (Fig. 8.2). Now most recently, we have seen what pieces of a comet can do to Jupiter, and some have speculated that the collapse of the Roman Empire in the West and the onset of the Dark Ages in Europe may have been consequences of a cometary collision with the Earth about AD 530.

When Halley's comet returned in 1986 it launched spacecraft. It led to great efforts in celestial mechanics and in historical astronomy, with attempts to identify sightings of the comet at very distant times. One question is whether all variations of successive orbits can be accounted for by the perturbations of the planets. The intervals between the established apparitions vary greatly, from less than 75 years to more than 79. The attractions of the major planets cause much of the variation, but there seems to be an unaccountable residuum. Non-gravitational forces, such as jets of gas ejected by the comet, have been invoked in explanation. However, as in the past, when non-gravitational forces were suggested to get round difficulties in celestial mechanics and then better mathematics showed that there is no need for any but inverse square law gravitational forces, so it may be again.

About a century ago, Henri Poincaré saw that the differential equations of celestial mechanics might have solutions so sensitively dependent upon the initial conditions that they were, in effect, indeterminate. He himself explained this very clearly and simply for meteorology. His work was the beginning of what we now call chaotic dynamics and although most of the well-known examples occur in fields other than celestial mechanics—in fluid mechanics, for example, or in biology—yet instances are known in celestial mechanics, in particular the orbits of some planetary satellites that have been discovered recently. It is possible that the forces acting on Halley's comet are purely gravitational, but that the motion is to some extent chaotic, generating the residual perturbations.[36]

Chaotic dynamics has reintroduced uncertainty and unpredictability into our study of nature. The Enlightenment was based in part upon the idea that the methods set out by Newton in the *Principia*, and apparently validated by such phenomena as the return of Halley's comet, showed that unambiguous principles determined the behaviour of the natural world. It could be predicted with certainty. What irony if the behaviour of an object that played so large a part in convincing people that the world was governed by predictable mechanical schemes is itself, in some degree, chaotic and unpredictable.

8.3 *The problem of the longitude*

Halley was deeply concerned with the problem of the longitude. He reviewed the possible methods for determining it on 9 May 1688 and concluded that the only generally satisfactory way was to use the right ascension of the Moon as a clock, but he also saw that for that to be useful to seamen, a great improvement in lunar theory was needed.[37] Newton's theory in the *Principia* was not good enough, and well over half a century would pass before Clairaut, d'Alembert, and Euler improved upon it. Halley recognised that the inequalities recurred in successive saronic cycles of eighteen years, and thought that careful observations over a complete cycle would allow empirical tables of the Moon's motion to be constructed. He had had the idea as early as 1682, before it had any theoretical basis in the *Principia*. In his last years, as Astronomer Royal, he did indeed carry out such a programme; meanwhile, lunar right ascensions, though imperfect, were the best way of getting longitude at sea, and Halley developed an instrument for finding the right ascension of the Moon from a nearby star of which the position was known. He placed a scale in the eyepiece of a telescope by which the distance between the Moon and a star could be found when they were close together.[38] He also discussed using lunar and solar eclipses to find longitude.

The satellites of Jupiter were another clock independent of the Earth's rotation. Cassini had made tables of their motions. He had not accepted Römer's explanation of certain anomalies in the times of the eclipses of the satellites as consequences of the finite speed of light, but had postulated an effect that depended on the position of the Earth. Halley considered that to be irrational. Halley printed the tables for London and gave examples of how to use them. He compared the predictions with observations and found good agreement. He remarked that some of the small discrepancies were probably due to the gravitational attraction of the equatorial bulge of Jupiter which is visibly oblate. Newton had shown in *Principia* (Book III, Props 18 and 19) that spinning bodies such as the Earth and Jupiter would be oblate spheroids. He had also shown that the gravitational attraction of the equatorial bulge would cause the nodes of orbits of their satellites to regress (*Principia*, Book I, Prop. 66, Cor. 21). As soon as the first artificial satellite was launched around the Earth, the flattening of the Earth was found by the same principles with far greater reliability than ever before. Halley's is the first application of those important and original theorems of the *Principia*. His posthumous *Tables* contain tables of the satellites of Jupiter.[39]

An account of the determination of the longitudes of French sea ports in a publication from the Academy of Sciences in Paris had stimulated Halley's essay. The eclipses of the first satellite of Jupiter were the timekeeper. Halley considered that, lacking an adequate theory of the Moon, it was then the best method for finding longitude on land. It was impractical at sea because the satellites had to be observed through long telescopes that could only be used under steady conditions on land. Even on land Halley was to find conditions sometimes unsuitable, and it was no doubt for that reason, as well as to make better observations of the Moon over a long period, that he devised a way of using a long telescope in an horizontal position with light from the object directed into it by a plane mirror that was moved in step with the rotation of the Earth. He suggested that the mirror might be turned by clockwork without the intervention of the observer.[40]

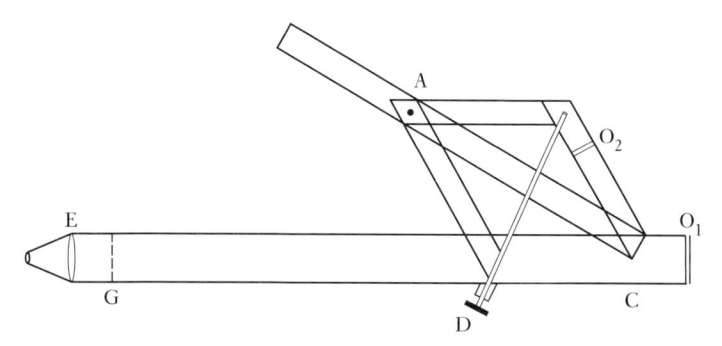

Figure 8.5 Halley's instrument for observing the Sun and the horizon simultaneously. O_1, O_2: objective lenses; D: adjustment screw; A: pivot; E: eyepiece; G: commom focal plane.

Latitude at sea was found from the altitude of the noonday Sun. On land, Halley and others used a quadrant provided with a plumb bob that defined the vertical. That was not practical on a moving ship at sea, and altitudes were taken from the horizon of the sea. Halley devised a telescope in which the horizon was viewed directly but the Sun was viewed by reflection in a rotatable plate placed between the object glass and the eyepiece (Fig. 8.5). This, the forerunner of the modern nautical sextant, had apparently been proposed by Hooke many years before, but Hooke had done nothing about it. Halley had a model made and found it satisfactory.[41]

8.4 How far is the Sun?

Halley had hoped to derive the scale of the solar system, in terms of the radius of the Earth, from his observations of the transit of Mercury from St Helena (Chapter 3). Transits of Mercury occur in April and October in years when the Earth and Mercury pass simultaneously through their respective nodes. The intervals between transits are determined by the ratio of the periods of the Earth and Mercury. Halley's lists were:

April transits: 1615, 1628, 1661, 1674, 1707, 1720, 1753, 1786, 1799.
October transits: 1605, 1618, 1631, 1644, 1651, 1664, 1677, 1690, 1697, 1710, 1723, 1730, 1736, 1743, 1756, 1769, 1782, 1789.

October transits occur more frequently than those in April; the transit that Halley observed from St Helena was the October transit of 1677.

Transits of Venus occur less frequently because its orbital period is longer and it is closer to the Earth at transit. There are two series, in November and in May or June, according to Halley's calculations:

November transits: 918, 1161, 1396, 1631, 1639, 1874, 2109, 2117.
May transits: 1048, 1283, 1518, 1526, 1761, 1769, 1996, 2004.

Some transits occur in pairs at intervals of 8 years; those that led to Cook's voyage were of 1761 and 1769. (They occurred early in June according to the Gregorian calendar, but late in May by the Julian calendar.)

Halley appreciated that observations of Venus would be better for the determination of the distance of the Sun, and on 23 September 1691 he read a paper on their use, but he did not publish his definitive paper on the subject until 1716, after he was in Oxford.[42]

The determination of the distance of the Earth from the Sun by transits of the inferior planets depends upon two geometrical properties. The first is that the projected path of the planet on the Sun, which is not necessarily along a diameter, is almost independent of the position of the observer on the Earth. The duration of a transit does however vary with the path of the planet across the face of the Sun. The second property is that the time of the transit is reduced or increased by the distance travelled by the observer as the Earth rotates on its axis, that is according to the latitude of the observer.

The following account of the method for determining the distance of the Sun is set out algebraically, but no such exposition appeared in any contemporary publication, by Halley or anyone else. Halley and his immediate successors gave numerical values for all the observable quantities, such as the duration of the transit and the difference betwen the durations at different places, as well as for the calculation of the solar parallax. They offered no general discussion of errors.

Suppose that a fictitious planet is mid-way between the Earth and the Sun—Venus is closer to the Earth and Mercury is further off. Suppose that a transit is observed from two sites on the Earth. The paths of the projection of the planet on the Sun will be separated by the difference of the two sites normal to the ecliptic (Fig. 8.6). If one site is in mid-latitudes and one near the equator, that difference might be about 3000 km, or 1/450 of the diameter of the Sun, 4″ as seen from the Earth. Such a displacement of the tracks could not be detected in Halley's day. The difference of times of the transit over the two tracks depends on how far they are from the equator of the Sun. A typical track might be off by about 1/10 of the radius. The difference of times of transit would then be about one thousandth of the duration. Transits last typically for 6 h, so that they would differ

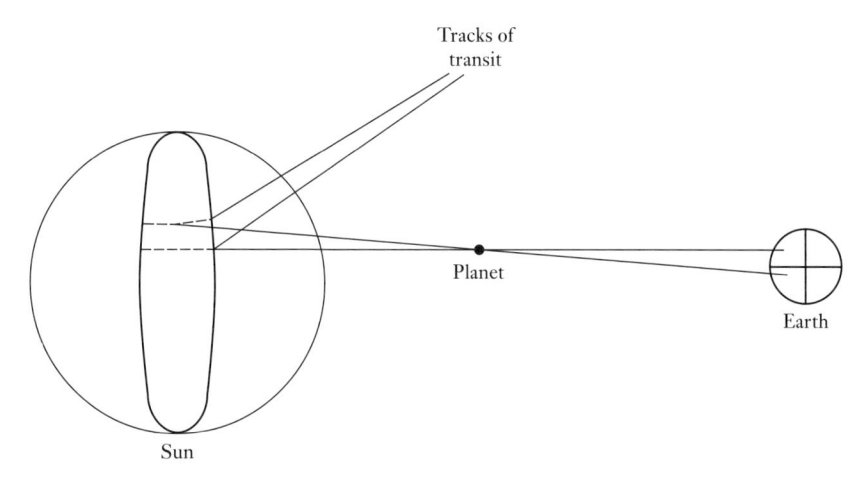

Figure 8.6 Passage of a planet across the Sun.

by about 20 s as seen from the two sites, more for Venus and less for Mercury. Such a difference is comparable to the errors of observation and could easily be allowed for.

The displacement of the track of a transit on the face of the Sun is not the reason for the substantial difference in times as seen from different sites on the Earth: the reason is the spin of the Earth. Let the radius of the Earth be a. A site on the surface of the Earth at latitude λ is displaced from the axis by $a \cos \lambda$. Thus the rotation of the Earth carries sites along the orbit by distances proportional to $\cos \lambda$. If ω is the spin angular velocity of the Earth and T the duration of a transit, the site rotates about the axis by the angle ωT in that time, and the projection of the chord CC' parallel to the orbit, assuming the direction of the Sun to bisect that chord, is

$$2a \cos \lambda \cdot \sin(\omega T/2) \cos I,$$

where I is the inclination of the equator to the ecliptic (Fig. 8.7).

The distance that each observer travels along the Earth's orbit has to be the same in the course of the transit. That distance is the sum of the length of the chord and the distance moved by the centre of the Earth in the duration of the transit, namely $R\Omega_E T$, where R is the radius of the Earth's orbit and Ω_E is its angular velocity in the orbit. Thus for two sites, 1 and 2, at which the transit lasts for times T_1 and T_2,

$$RT_1\Omega_E + a(\omega T_1)\cos \lambda_1 \mathrm{sinc}(\omega T_1/2) \cos I = RT_2\Omega_E + a(\omega T_2)\cos \lambda_2 \mathrm{sinc}(\omega T_2/2) \cos I;$$

here $\mathrm{sinc}\left(\tfrac{1}{2}\omega T\right)$ is $\sin\left(\tfrac{1}{2}\omega T\right)/\left(\tfrac{1}{2}\omega T\right)$.

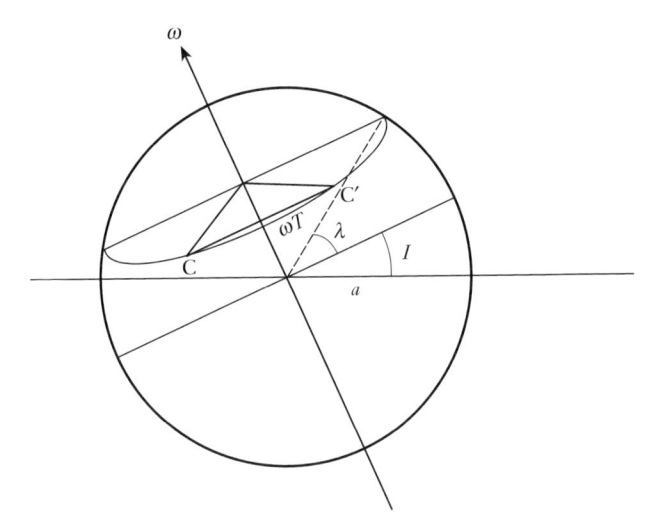

Figure 8.7 Distance travelled by observer on the rotating Earth. ω: spin angular velocity of the Earth; C, C': termini of observation of transit in time T; I: inclination of equator to ecliptic.

The difference between the times T_1 and T_2 is not great and is given very nearly by

$$\frac{\delta T}{T} = \left(\frac{a}{R}\right)\left(\frac{\omega}{\Omega_E}\right) \mathrm{sinc}\left(\tfrac{1}{2}\omega T\right)(\cos\lambda_1 - \cos\lambda_2)\cos I$$

where T is the mean of T_1 and T_2.
Thus

$$\frac{R}{a} = \frac{T}{\delta T}\frac{\omega}{\Omega_E}[\mathrm{sinc}\left(\tfrac{1}{2}\omega T\right)(\cos\lambda_1 - \cos\lambda_2)\cos I]$$

The Earth rotates through 90° in 6 h, so that the factor $\mathrm{sinc}(\tfrac{1}{2}\omega T)$ is about 0.9, as is $\cos I$; $\cos\lambda$ is also about 0.9 for low latitudes. All the factors in the expression are well known, except for the times of transit and their difference. The uncertainty in the estimation of R is therefore determined by the uncertainty of $\delta T/T$.

Suppose that the Earth is reduced to rest (Fig. 8.8) by giving the solar system an angular velocity of $-\Omega_E$. The angular velocity of the planet about the Sun in that frame is $(\Omega_P-\Omega_E)$ and its linear velocity is $R_P(\Omega_P-\Omega_E)$, where R_P is the distance of the planet from the Sun. The linear velocity is then

$$R_P\Omega_E\left[\left(\frac{R}{R_P}\right)^{3/2} - 1\right].$$

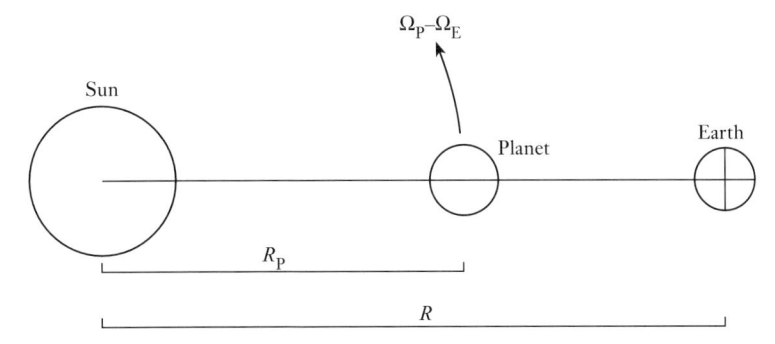

Figure 8.8 Duration of a transit. Ω_P: orbital angular velocity of planet; Ω_E: orbital angular velocity of Earth; R_P: distance of planet from Sun; R: distance of Earth from Sun.

The distance the planet travels as its shadow crosses the Sun is $D_S(R - R_P)/R$, where D_S is the diameter of the Sun, and consequently the duration of a transit is

$$\frac{1}{\Omega_E}\left(\frac{D_s}{R}\right)\left(\frac{R}{R_p}-1\right)\left[\left(\frac{R}{R_p}\right)^{3/2}-1\right]^{-1}.$$

If the projected path of the planet were along a diameter of the Sun, a transit of Mercury would take rather less than 6 h and one of Venus nearly 7 h. The times are less for paths that are along chords. The transit of Mercury that Halley observed from St Helena lasted for 5 h 20 min, and that of Venus that Cook observed from Tahiti in June 1769 lasted for almost 6 h.[43]

The differences of the times from those on a non-rotating Earth are nearly 15 min. The time of a transit as seen from different places might vary by up to 30 s on account of different geometry, but that can be allowed for. If, allowing for errors of observation, the uncertainty of the duration is about 10 s, the relative uncertainty of the distance of the Sun, equal to 10 s in 15 min, would be 1 part in 90.

It might seem that there is no advantage in observing Venus rather than Mercury, but it is because there are transits of Venus in May or June that they are to be preferred. The obvious course would be to observe a transit at sites where the mid-point occurs at midday—that was very nearly so at Tahiti and not far off at St Helena. However, because one series of the transits of Venus occurs in May or June, when the time between sunset and sunrise in high northern latitudes is less than 6 h, sites can be found where the transit begins at sunset and ends at the following sunrise, the mid-point being at midnight. Such sites will differ in longitude by 180° from those where the middle of the transit occurs at midday. At the one site the rotation of the Earth reduces the duration of the transit, at the other it increases it, and so the latitude factor in the difference of times is $(\cos \lambda_1 + \cos \lambda_2)$ instead of $(\cos \lambda_1 - \cos \lambda_2)$. If, as might be typical, $\cos \lambda_1$ is 0.9 and $\cos \lambda_2$ is 0.5, the factor changes from 0.4 to 1.4, an appreciable advantage. Ingress and egress need not be observed from the same site; they may be observed from different sites at the same latitude but displaced in longitude, provided the differences in the longitudes are well known. There was no single high latitude site in the north corresponding to Cook's at Tahiti, but a number of places in Scandinavia and Britain of the same latitude did as well.[*]

Halley had suggested that the transit of 1761 should be observed from Fort Nelson on Hudson's Bay at longitude 90° W (the site in high latitude) and from a place on the Ganges at longitude 90° E (the site in low latitude). The 1761 campaign was to a large extent organised by de L'Isle, and many other sites were occupied in addition to those that Halley had indicated. Bad weather and other difficulties bedevilled the results and a further programme was planned for 1769. The Royal Society took the lead and obtained the support of the Admiralty. James Cook was sent to Tahiti. That was fortuitous. The northerly sites would be in Europe and the low latitude site, 180° away, would have to be in the Pacific. But where? Knowledge of the Pacific was poor, but by good luck, just as Cook was preparing to sail, a ship came into London with news of the discovery of Tahiti at just the right longitude and close to the equator. A voyage to Tahiti would enable Cook to explore other parts of the southern Pacific after he had observed the transit. So Tahiti and New Zealand and round the world it was for Cook and Joseph Banks as they set sail in *Endeavour*. Hawaii would have been a good Pacific site, but Cook did not discover it until 1778.[44]

The solar parallax derived from the 1769 observations was much better than anything that had gone before, and certainly better than just the right order of magnitude. It did, however, suffer from problems that Halley, who had not seen a transit of Venus, seems not to have anticipated. The northern observations, being at sunset and sunrise, were made close to the horizon and were disturbed by the Earth's atmosphere, and the thick Venusian atmosphere rendered all four instants of contact uncertain by some seconds.

Halley's papers on the determination of the size of the solar system from the transits of Venus were important and influential. The method he proposed had, for the first time, sound dynamical, geometrical, and astrometrical bases, and was least subject to errors of observation since

---

[*]November transits, with a short night in the southern hemisphere, would have done just as well, but the sites are not so suitable.

it required measurements of times only and not angles. It was the basis for the size of the solar system, and hence of our own galaxy and the cosmos beyond, until Sir David Gill measured the parallax of Mars at apposition in 1877, using far more satisfactory astrometric procedures than Flamsteed or Cassini had. Observations of the parallax of the minor planet Eros gave still better results, but real improvements in precision only came with direct measurements of the distances of the planets by radar and other electromagnetic means in the mid-twentieth century. The indirect result of Halley's ideas was also far-reaching. James Cook was sent to the South Seas, and after that on his voyages of geographical and botanical and zoological discovery, as the direct outcome of Halley's proposals to observe the transits of 1761 and 1769. Few proposals for observations to be made in the future have had such sequels.

The reason for the efficacy of the method of transits is intriguing. The method would not work nearly so well, if at all, in different geometry. Because the transits last for about six hours there is a range of longitude in which the ingress and egress both occur in the daytime; were the transits to last for twelve hours or more, as they would if the angular diameter of the Sun as seen from the Earth were greater, then the observations would be much more difficult or perhaps impossible. Again, the rate of spin of the Earth happens to give a sufficient difference between the times observed in high and low latitudes. The two factors are connected, for a slower spin would give a longer day and more chance of seeing a transit in daylight. The present geometry of the solar system happens to be particularly favourable for the method. In a similar way, the almost equal angular sizes of the Sun and the Moon enabled astronomers in ancient and more recent times to advance astronomy in ways that would not have been open to them, had solar and lunar eclipses not occurred so fortunately.

8.5 The Moon is speeding up

Halley's third major astronomical discovery in the 1690s was that the Moon might be speeding up. He came to it from his study of the observations of the Arabic astronomer, al-Battānī (Albategnius) (858–929), who lived in Raqqa (Arracta) on the Euphrates and at Antioch, mid-way, as Halley remarked, between the time of Ptolemy and his own day.[45] The works of al-Battānī survived in two printed editions of a Latin translation by Plato Tiburtius, of whom Halley said that it was plain from his translation that he knew neither Arabic nor astronomy. Halley emended the text and drew astronomical conclusions. He discussed the location of Raqqa in a later paper. He showed that al-Battānī's observations of the noonday elevation of the Sun at the summer and winter solstices entailed 36° 1' for the latitude of his site and 23° 35' for the solar obliquity. Ptolemy, following Hipparchos, had taken the obliquity to be 23° 51' 20" (the present value is 23° 26' 21").

Al-Battānī found that the autumnal equinox of AD 882 occurred on 18 September at 13h 15m p.m., while Ptolemy in AD 139 had 25 September at 19h, from which Halley calculated the interval between successive autumnal equinoces to be 365d 5h 46m 24s.

Al-Battānī observed a solar eclipse at Raqqa in AD 891 and one at Antioch in January, AD 901 and a lunar eclipse at Raqqa in AD 883 and another at Antioch in August, AD 901. From those and the equinoctial observations, Halley derived the mean motion of the Moon and the places of its node and apogee in the years AD 881, 882, 883, 891, and 901. The further interpretation of those results depends on knowing the location of Raqqa–Arracta. Halley returned to that in a later paper on Palmyra.

The ancient oasis city of Palmyra (Tadmor in the seventeenth century) lies at the crossing of major trade routes in the Syrian desert. It is about 300 miles NE of Jerusalem, east of the Levant Company factory at Aleppo, and about midway between there and the Euphrates. In its heyday it was a great city which came under Roman domination, and in the ninth century the caliph al-Ma'mūn built an observatory there. Palmyra had declined in Halley's time to a collection of hovels among mighty ruins. Two members of the Levant Company, Timothy Lanoy and Anson Goodyear, visited it from Aleppo in 1678 and 1691, and the Revd William Halifax went there in 1692.[46] Halifax recorded a number of Greek inscriptions on the ruins of Roman buildings and there is a view of the ruins in *Philosophical Transactions*. Halley discussed the inscriptions in a subsequent paper, displaying his classical learning.[47] He summarised the history of Palmyra, so far as it was known from the classical historians, Appian, Pliny the Younger, and others, as well as from coins, and he emended and commented on the records of the inscriptions that were reproduced by Halifax. Halley says that a gentleman who communicated the results to him determined the latitudes of those places in 1680. Halley corrected the accepted value for Aleppo and identified the site of Arracta, where al-Battānī observed, as Racca on the Euphrates, and probably the Nicephorion built by Alexander. Halley hopes that others will observe the latitude there to determine if there is any evidence for a shift of the Earth's axis since the time of al-Battānī, and that they will also observe eclipses at Bagdad, Aleppo, and Alexandria to obtain their longitudes.

For in and near those places were made the Observations whereby the Middle Motions of the Sun and Moon are limited: And I could then pronounce in what Proportion the Moon's Motion does accelerate; which that it does I think I can demonstrate, and shall (God willing) make it one day appear to the Publick.

Halley did not ever set out his estimates, but it has been amply shown since his day that his conjecture was correct and that the Moon does accelerate in its orbit about the Earth. His argument was based in part on the longitudes assigned to Babylon, Antioch, and other places in the East by Ptolemy and others following him, although he observes that they involved some misidentifications of ancient places. Halley thought that the planets were slowed down by a medium in the solar system and Newton seems to have shared that view, for he included a note on Halley's result in the second (but not the third) edition of *Principia*, after remarks on an interplanetary medium.[48]

The angular momentum of the Earth and the Moon together must stay constant, so that as the Earth's rotation slows down, the Moon must speed up in its orbit. The present deceleration of the rotation of the Earth is nowadays found from comparisons with atomic clocks, and the acceleration of the Moon from laser observations and from observations of occultations of stars by the Moon and planets. The behaviour of the Moon in the long term can only be found from historical records of events such as transits of Mercury across the Sun, eclipse observations, and the times of the Sun's passage through the equinoces.[49] Halley used the eclipses and the times of equinoctial passages observed by al-Battānī, and it may well be that it was in that connection that he subsequently showed that the times of summer and winter solstice could be found as accurately as the times of equinoctial passage.[50]

The Moon accelerates in its orbit about the Earth because the attraction of the Sun on the Moon speeds it up, and because tidal friction slows the Earth down. Laplace identified the effect of the Sun, which comes about because the eccentricity of the orbit of the Earth about the Sun is decreasing. His calculations for the consequent acceleration of the Moon agreed very well with the observed value at the time.[51] Then John Couch Adams shewed that Laplace's theory was incomplete. The series that gives the acceleration converges very slowly and Laplace had neglected important terms. The true theoretical value of the solar effect is about half that observed.[52] Adams's paper generated considerable controversy. For a long time there were those who refused to accept his result because they argued that the agreement with observation had proved Laplace correct in his mathematics. In fact Adams was right and the solar attraction is not the only cause of the lunar acceleration. G. I. Taylor and Harold Jeffreys, studying frictional losses from tidal flows in the Irish Sea, demonstrated that tidal friction in shallow seas was most probably the missing cause of the lunar acceleration, as Kant had apparently suggested previously.[53] The agreement between the estimates of tidal friction and the non-solar part of the observed acceleration of the Moon is now close, and an important fact of geophysics.

8.6 Halley's ways

Halley is perhaps best known for his calculations on comets and his study of transits. His investigations of the motion of the Moon are not so familiar. All three have had long-lasting consequences. They demonstrate different aspects of his approach to natural knowledge. The ideas about comets and the use of transits had been current for some while before he worked on them, but until he made specific calculations, neither of them was more than speculation. Halley made them concrete.

By contrast, no one had had the idea that the Moon's mean motion had changed over millenia until Halley arrived at it from his studies of historic observations. It was not the first time he had used historical records in natural science, nor would it be the last. Whether he had the idea of a change in the Moon's motion before he embarked on his analysis, or whether the results suggested it to him, he does not say; it is none the less a combination of careful study of records of the past and controlled imaginative interpretation of them.

Halley's studies of the three topics could not have been made before the dynamical framework of *Principia* had been erected. He came to his other idea, that the position of the Moon could be inferred from its position one saronic interval before, as early as 1682, well before *Principia*. Therefore, prior to a real dynamical basis, Halley extended the known cosmos beyond that of Book III of *Principia*, in space and in time. He showed himself aware of the natural world as a system with an history. The cosmos is not static, its history has to be understood in dynamical terms, and its dynamics have to be elucidated from historical records.

Halley was one of the very few people who took Newton seriously in the last years of the seventeenth century and applied the methods of the *Principia* in practical ways. Then, and for years to come, extensive controversies continued, especially in Europe, about the foundations of natural philosophy. As late as the middle of the eighteenth century, Leibniz and his associates criticised Newton and his followers for concepts considered as irrational, occult and tainted with mysticism, and inclined to atheism. It was both down-to-Earth and imaginative for

Halley to apply the methods of *Principia* to new problems of the cosmos while others argued about the principles of natural philosophy.

Halley adopted the same attitude to atomic theory. Galileo and Gassendi had revived the Epicurean arguments for atoms, which were based on general principles. Halley drew a practical conclusion about the sizes of atoms.

Some of Halley's ideas in the decade after the publication of *Principia* seem almost more like today's science that that of his time. His estimate of atomic dimensions, his suggestion that the orbits of the satellites of Jupiter were perturbed by the equatorial bulge of Jupiter, his proposal that the Moon was accelerating, all are far beyond concepts of his day. Halley, as will appear, was a very practical man, but with an imaginative insight into the physical world of exceptional range.

9

Use and practice of the contemplation of nature

> Quale nell'arzanà de'Viniziani
> ...
> chi ribatte da proda e chi di poppa
> altri fa remi e altri volge sarte
> chi terzeruolo e artimon rintoppa ...
> [as in the arsenal of Venice ... one hammers at the prow, another at the stern, others make oars and others twist ropes ...]
>
> DANTE: *INFERNO*, XXI, 7, 13-15

9.1 *A glorious revolution*

*P*RINCIPIA had been published for a year and a half when William and Mary assumed the throne. Many people settled old scores, and Hooke reported that Halley was much concerned when the House of Lords began to enquire into the allegations of Laurence Braddon about the death of the Earl of Essex.[1] That, perhaps linked to his father's death, was apparently an unwelcome reminder of the past in the new reign. The past may have pursued him in another way. Shortly after William III came to the throne, he is said to have enquired about Halley because he knew he had been favoured by Charles II and James II. William, so it was reported, found that Halley was concerned only to return to his telescope, and was no risk to the new government.[2] There is no record in State Papers and the report is not wholly credible. Why should William have been concerned about Halley? Halley was about thirty-three and although well known to astronomers and in the Royal Society, he enjoyed as yet no reputation in the wider world of affairs. He had been to St Helena, he had published *Principia*, but, unlike Newton, had taken no part in the political events that led up to William's accession. He was a promising man who had just taken on the subordinate clerkship of the Royal Society; hardly, it might be thought, for all his European standing, someone whom William had to see in the first months of his reign. The notion that all Halley wanted to do was to get back to his telescopes rings false. He had not been at his telescopes for almost seven years and, if he did intend to get back to them, he was most effectively diverted in the next ten years.

If William did enquire about Halley, he most likely wanted to know about Halley's sea-going activities. Halley had presented a chart of his survey of the Thames Estuary to the Royal Society on 3 July 1689, and had probably been at sea already in 1688, if not before. The most cursory glance at a map shows how sensitive were the approaches to the River (see Fig. 1.2). Through them flowed the greater part of the trade of England; on them stood the naval bases of Chatham and Harwich; and in them were fought the major battles of the Dutch wars. There lay the bases and anchorages from which the navy of James II would contest the control of the narrow seas when William's expected invasion took place. No other area was of comparable importance to the trade and safety of England and her merchants. Good charts of those dangerous waters were essential; to survey them required very detailed close observations at risk of foul weather.

An incident during the preparations against William's invasion illustrates the deficiencies of the charts. At the end of November 1688 a frigate, the *St Albans* (named perhaps for Nell Gwynn's son), commander Capt. Constable, was ordered from the River to join the main fleet. She took a Trinity House pilot on board for the passage to the Gunfleet anchorage. She was not out of the Thames when the Admiralty learned that the battle fleet had sailed for the Downs and they ordered Constable to follow. The pilot objected that he would only go as far as the Gunfleet for he could not safely take the ship further. Trinity House could not find another suitable pilot and Constable was ordered to take one up at Harwich. That also was not possible and the original pilot continued, under protest. He ran *St Albans* aground on North Sand Head, the northernmost point of the Goodwins, so her captain reported to the Admiralty Board:

> My Lords and Gent[n]
> I send you thanks for the Pilott w[ch] was sent me, and whom I desired you to change, having found him incapable, w[ch] you refused, he hath since run the ship upon the North Sand head, where she strucke thrice ...

and added particulars of the difficulties of getting her off and the loss of ship's boats and other items.

Halley later reported to the Royal Society on a project for a light on the northern tip of the Goodwin Sands, but the projector kept his methods secret.[3]

Halley could hardly have been surveying in 1688 without the consent of James II and Pepys; they may have sponsored him. William would surely have wanted to know what he had done, and who had supported it. Would Halley serve the new administration indifferently, as so many of James's officials did? William, if he did enquire about Halley, must have been sufficiently satisfied that Halley was able to show his chart to the Royal Society in the summer of 1689.

9.2 Hally a sayling

Halley could hardly have surveyed the Thames approaches without experience at sea; how he came by it is unknown. His comments in 1688 about Winchelsea on 13 June, and shells in the cliffs at Harwich on 1 August, may indicate that he was working in and around the mouth of the Thames in 1688. Even earlier he had spoken at the Society on 9 May 1688 of how the forts at Sheerness were unhealthy, whereas a guardship four miles offshore remained healthy, and two weeks later he mentioned the thick dirty froth on the sea after a long calm. He may have been surveying the coast in the spring or early summer of 1688 and doing so with the knowledge, if not the cooperation, of the Royal Navy.[4] He may have been at sea before 1687. The *Ode* that he put before the *Principia* (see Appendix 6) has three lines that stand out starkly from the rest:

> ... dum fractis fluctibus Ulvam
> Deserit ac nautis suspecta nudat arenas;
> Alternis vicibus suprema ad litora pulsans.

While the sedge is laid down by slack waves, and sands mistrusted by sailors are exposed; in turn high waves beat on the shores.

They suggest practical experience.

On 6 March 1689 Halley described a diving bell and a means of carrying fresh air down to it. Shortly afterwards Hooke noted in his diary 'Hally a sayling' on 22 March and on 3 April, 'Hally Returned'. Halley may then have been trying out his bell, or he may have been surveying the Thames, for on 3 July he presented his chart of the mouth of the Thames to the Society, saying that he had corrected many great faults in all charts hitherto published. Again, on 24 July 1689, he mentioned the lack of trees on Foulness and Canvey islands at the mouth of the Thames, indicating that he had been working there earlier. Later in 1693, he surveyed the Sussex coast between Selsey and Arundel, evidently in connection with diving on the *Guynie* frigate.[5]

No one is ever said to have seen the chart of the Thames after Halley presented it to the Royal Society and it might be supposed to have disappeared: it almost certainly survives in the chart of the English Channel that he published after his tidal cruise of 1701 (Chapter 10, Plate X). Parts of that chart have been published in a number of essays about Halley, but some omit the coast to the north of the Channel proper. The chart in fact extends as far north as Cromer and as far east as the Frisian Islands on the Dutch coast, while the log of the Channel cruise shows that in 1701 Halley only went south of Dover and west down the Channel. The surveys of the mouth of the Thames, the coast of East Anglia, and the coasts of the Low Countries must have been done at some earlier time. They may not have been by Halley and he may have incorporated someone else's work without acknowledgement, but that seems unlikely, for Halley was at pains to emphasise in 1689 that he had corrected many errors in previous charts of the Thames. He would hardly have published material that he thought to be inferior to his own. The most likely conclusion is that the chart of the mouth of the Thames and of the Low Countries is what Halley had done by the summer of 1689.

If the 1701 chart does incorporate the 1689 survey, Halley's claim seems to be justified. The only published hydrographic surveys of the coasts of Britain in 1688 were those in Sellers's *English pilot* of 1671.[6] The book is a compilation of the published work of others, much of it Dutch, and it was widely recognised to be inadequate. Pepys was especially critical.[7] Sellers's fold-out draft of the coast from the South Foreland to Orfordness is the one original chart. The coast and the coastal waters up to the north of Norfolk had been surveyed by Sir Jonas Moore in 1661, and the soundings in Sellers's chart had been provided by two Elder Brethren of Trinity House, Capt. Gilbert Crane and Capt. Tho. Browne.

So unsatisfactory were the charts of the waters around the British Isles in 1689 that, at the end of June, William's new Commissioners of the Admiralty had to call on Sellers to provide them with whatever maps and charts he had available.[8] When Pepys left at the end of 1688 he took all his papers, including perhaps maps and charts, and there was nothing left in the Admiralty on which they could plan operations against the French in the Nine Years War that had just started. It may have been about that time that William enquired about Halley.

Captain Greenville Collins published a chart of the Thames approaches and the coast of East Anglia in his *Great Britain's coastal pilot* of 1693. There are fewer soundings than on Halley's published chart of 1701, and far less detail north of Southwold. Collins had obtained a warrant from Charles II as his hydrographer in 1681, he had been given command of a cutter to survey the coasts and ports of Great Britain, and at Pepys's instance had been admitted as a Younger Brother of Trinity House. He did not survey the approaches to the Thames and said he took his chart of that area from the surveys of

Trinity House. It is almost certainly based on the survey of Moore and the hydrography of Crane and Browne. In 1688, the year after Collins had completed his surveys, there was no English chart of the Thames approaches based on surveys later than 1671. Even Collins's work did not meet with Pepys's complete approval[9] and it would not be surprising if in 1688 Pepys, then Secretary to the Admiralty, was anxious to have a more up-to-date survey of the approaches to the River. When, in 1723, a second edition of Collins's book was called for, the charts were all reprinted, apparently from the old plates, and the only change was that Halley's tidal chart of the English Channel, going as far north as Cromer, was included.

The publication of Collins's surveys apparently met obstruction. The Court Minutes of Trinity House record that on 21 February 1689 Collins was asked to attend next Saturday 'to state what stops the finishing & publishing of his draughts of the sea coasts that they may give him assistance in the removal thereof'.[10] What the stops were and what the outcome of the Court's minute are not recorded, but Collins did not publish his surveys until 1693. In the first half of 1689, in command of the *Mary* yacht, he was engaged on a wide range of duties, from transferring crews between men-of-war to carrying very senior army officers and diplomats and their families between England and Holland. He was not surveying.

While Collins appears in Admiralty records in 1689, Halley does not. Hooke reported him as '*sayling*' in the spring of 1689 and he presented his Thames chart to the Royal Society in the summer. That is all. There is a curious circumstance of a dog that did not bark. In the late summer and early autumn of 1688, when the invasion of William was expected and Pepys was very active in getting out the battle fleet, an embargo was imposed on all shipping from English ports, with a few general exceptions such as the colliers bringing coals to London. The Privy Council also from time to time exempted specific ships, but if Halley was indeed surveying then, no ship that might have carried him is listed. After William was in control and expecting war with France, a general embargo was reimposed in the spring of 1689 and similar exceptions were made, but Halley again was not among them. If he was surveying privately, he would have had to have had an exemption from the general embargoes and a protection against the pressing of his crew; none is recorded. He is unlikely to have been working from a naval vessel, for all the small craft, such as yachts and ketches, were fully engaged on ferrying, intelligence, and minor convoy duties at a time when the fleet expected to engage the French, and did so at the battle of Bantry Bay in May. Halley's activities from midsummer 1688 to midsummer 1689 may have been secret then and remain mysterious now.[11]

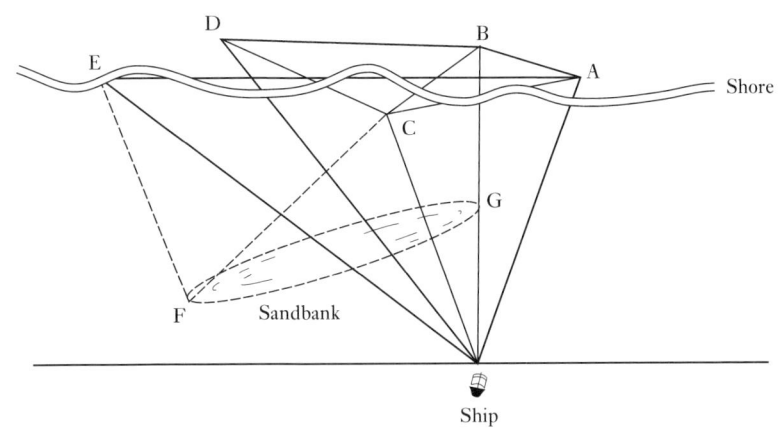

Use and practice of the contemplation of nature 235

Figure 9.1 Halley's method of resection (for description, see text).

Halley probably determined his positions offshore by the methods he later described to Southwell.[12] He made two applications of the method of resections, one to obtain positions offshore from marks on a coast, and the other to obtain positions on an inaccessible coast from ships or boats at sea (Fig. 9.1). In the first, the true bearings of various prominent landmarks were to be taken relative to the rising or setting Sun, so avoiding the vagaries of the direction of the magnetic compass. The position of a ship or boat was then found from the true bearings of some of the marks, as observed from it, and Halley particularly mentioned that in that way the position of a vessel making soundings may be found. Two vessels were needed for the second application, using them as stations to obtain positions of marks on the inaccessible shore. Other positions at sea were then to be found from those marks, as in the first application. He suggested that a scale could be found from the time taken by the sound of a gun to travel between two places, which he put at 15 seconds for a marine league. The principle of resections was not novel, although the author often given the credit for it, Laurent Pothenot, did not publish his solution until 1692. Snel (1580–1621) had applied it much earlier in his Dutch meridian chain of triangulation, published in 1617. Halley was familiar with the principles of triangulation, but whether he knew Snel's book and his method of resection in the late 1680s cannot be stated.[13]

Halley's Channel chart of 1702 also has small charts of the waters around the Isle of Wight and of the entrance to Plymouth Harbour, the new naval base at the west of the Channel that was built up after the accession of William, when France became the enemy. They were included on the plate printed in the second edition of Collins's *Pilot*. Halley spent some days in and around the approaches to Portsmouth at the end of his Channel cruise in 1701 at the request of the Commissioner of Portsmouth Dockyard (see Chapter 10), which suggests that the earlier chart of that area in Collins's *Pilot* was unreliable. Altogether Halley surveyed the approaches to the Thames, the coasts of the Low Countries, parts of the French coast, the Sussex coast by Pagham near Chichester, and the Isle of Wight and Plymouth, as well as the Adriatic harbours of Trieste and Buccari in 1703. All had strategic significance to a greater or less extent. Many compare well with modern Admiralty charts.

9.3 Diving and salvage

In the first days of April 1691 the *Guynie*, a frigate of the Royal African Company, foundered off the Sussex coast near Pagham. The Court of Assistants of the Company learnt of it on 6 April by a letter from the captain, William Chantrell.[14] The chairman of the Court was the Sub Governor, Sir Gabriel Roberts, the uncle of Robert Nelson (the King was the Governor), and Abraham Hill was the Deputy Governor. Halley would have been well known to both of them.

Guynie had appeared in the records of the Royal African Company at the end of her previous voyage to England, when she arrived at Bristol late in May 1689. She then carried gold.[15] Later, the Court agreed to buy *Guynie*—she must up to then have been chartered to them—and eventually they gave Chantrell the command.[16] The minutes of the Court of Assistants record the arrangements and instructions for *Guynie*'s next voyage to the Gold Coast, on which she eventually sailed about the end of January 1690.

On 27 February 1691 the Court, having read a letter from Chantrell of 23 February from Falmouth, asked that the Sub Governor (Roberts), the Deputy Governor (Hill), and Sir Benjamin Bathurst (an Assistant) should wait on the Lords of the Admiralty to obtain a man-of-war to convoy *Guynie* from Falmouth. They were successful, and next day they told Chantrell of the convoy.[17] Nothing in Admiralty papers shows why the Court asked for a convoy, nor why their request was granted so speedily. England was at war with France and the French had defeated the English fleet at Beachy Head the year before: lone ships would be at considerable risk from French cruisers and possibly from the victorious French main fleet. If *Guynie* carried a valuable cargo, the Admiralty might have been sufficiently concerned to provide an escort.

Guynie did carry a valuable cargo. The *Black book* of the Royal African Company of England has a summary of a letter from three merchants on the Cape Coast (the Gold Coast). Their main purpose was to explain that Chantrell carried a considerable quantity of gold (that they had not seen), which he was taking to England on behalf of several Portuguese merchants and others, to be delivered to the RAC. They also sent a bill of lading for 184 elephants' teeth (tusks, probably) of twenty five hundred [weight] gross. Gold and ivory together would have been very valuable and, besides, the gold was entrusted to the ship by foreign traders. No wonder they had a convoy up the Channel.[18]

In the Court's correspondence, *Guynie* is called a frigate, and so Halley later wrote of her. A few other ships of the Company were frigates and it seems that they, like *Guynie*, carried especially valuable cargoes, not bulky ones like blacks (slaves) and cloths. Frigates were fast, and relatively heavily armed for their size. They ranked as fifth or sixth rates and carried thirty to fifty guns. They could evade the heavier ships of a line of battle and fight off most pirates or privateers. A frigate sailing fast on her own would be the best way of sending valuable cargoes; equivalent ships were used in the same way up to the Second World War. Halley's reference to going aboard 'my frigatt' in a later letter to Newton confused the editors of Newton's correspondence.[19]

Just over a month after arranging for the convoy, the Court had the letter from Chantrell at Chichester, reporting the loss of *Guynie*. They sent four of their officers to Chichester to see what might be recovered. On 8 April the Court gave directions for salvage of the ship and her cargo, and ordered a guard of soldiers to be mounted to prevent looting.[20]

On 22 June 1691 Halley wrote to Hill from Pagham:

> Hon. Sir,
> I got down hither this morning by times, and went on board, in order to have gone down and set our people to work; but it was captain Chanterell's advice, that our five-inch hawser, which had scarce been five times used, was so far worn, being exceedingly burnt with overtarring, that he thought it unsafe, and therefore desired he might have a new one somewhat larger, of about thirty fathoms; the casks likewise prove not so well as expected. It is the opinion of all who have seen our ropes, that they are the most tarred of any they ever saw, and I am willing to believe it is done for the advantage of the maker, rather than out of any design to baffle and defeat our business. We shall with all diligence prosecute the affair and I hope now in a short time to give you a good account of your ship. This business requiring my assistance, when an affair of great consequence to myself calls me to London, viz, looking after the Astronomy-Professor's place in Oxford, I humbly beg of you to intercede for me with the Archbishop Dr. Tillotson, to defer the election for some short time, 'till I have done here, if it be but for a fortnight: but it must be done with expedition lest it be too late to speak.[21]

The Oxford election did not in the event take place until December 1691.

Evidently in the two months from April to June, Halley had been asked to examine the wreck and had built or acquired a diving bell to do so.

The Royal Society discussed diving bells in its very early years, and they had been used in work on the great mole at Tangier with which Jonas Moore had been involved. Halley first mentioned his ideas in a paper he read to the Royal Society on 6 March 1689 about a method of walking under water. He described how pearl divers and others can remain under water for several minutes, and how they had been recently employed in the West Indies to look for treasure in a Spanish wreck. The diving bell was used to refresh the divers without their returning to the surface, but they still had to hold their breath to work. Halley proposed to make a bell that could be moved about under water so that men could work while inside it. It had to be weighed down to keep it on the sea bed and he suggested making the weights in the form of heavy wheels so that the bell could be pushed around by those inside it. He pointed out that, as the air inside was compressed by the pressure of the water, the water would rise to an inconvenient height in the bell and the weight of the bell would increase so that it could not be moved. He therefore thought of bringing air down into the bell in casks. A bell of this sort nowhere figures in his subsequent papers and he probably never made one. He did use casks to bring down air into a bell, as he explained to the Society on 6 May 1691. A week later 'he shewed the Method he intended to use in raising the ship'. Shortly thereafter he was at Pagham and writing to Abraham Hill, with a mention of his casks.[22] Then on 30 June the Court of the RAC asked their Committee of Shipping to send down Robert Nicholas, 'to forward the business about the Wreck'.[23]

Halley was not on his own. On 15 September 1691, a Secretary of State gave instructions to prepare a bill for a grant to Sir Steven Evans, Francis Thyssen, John Holland, and Edmund Halley for the use of an engine for conveying air into a diving vessel, whereby they could maintain several persons at the same time to live and work safely at any depth for many hours together.[24] That was a patent. It joined together Halley, the imaginative engineer; two members of the Royal African Company, Sir Steven Evans and Francis Thyssen; and a City financier, John Holland, possibly a Mercer, and the founder of the Bank of Scotland in 1696.

Although Halley became involved with *Guynie* almost immediately after the wreck, no doubt through Roberts and Hill, it must at first have been only informally, for only on 9 February 1692 did the Court of Assistants ask six of their number to discuss the method proposed by Halley and others for saving the goods on the frigate. They had also to meet with the Assurers to make the best terms they could in the Company's interest.[25]

Those were troubled times for the Company. Nine months later, the Treasurer Williamson resigned with his accounts in disarray and £4000 missing. Besides their regular business, the Court now had to appoint a small committee to act as Treasurer *pro tem* and to pursue legal action against Williamson, all at the same time as Halley was working on *Guynie*.

Halley and the Company took time to agree their arrangements. On 25 April 1693 the Court referred the proposal of Halley and partners to the Committee of Shipping 'and any other of the Court that please to be present to treate and conclude with them for the recovery of what was sunke with the Guynie frigatt', and nine more months elapsed before the Committee of Law was asked to act on articles of agreement.[26] Halley continued to work on *Guynie* almost until he went to the Mint at Chester, as an undated letter of his to Newton shows. That letter must have been written after March 1696 when Montague offered Newton the post of Warden of the Mint (see Note 19). Halley may not have undertaken the actual salvage operations himself, for on 9 April 1695, the Court of Assistants considered 'Articles of Agreement with Thomas Toupion, Samuel Pargiter & Benj. Graves for taking up the Elephants Tooth sunk with the Guynie frigatt 21 March last'.[27] *Guynie* did not sink on 21 March 1695; perhaps the three men recovered the tooth that day. Halley may have formed a joint-stock company to exploit his developments, since J. Houghton, in his weekly paper of 20 July 1694, had an article on the value of joint-stocks and a review of the prospects of some of them. He mentioned Greenland fishing, Guinea, Hudson's Bay, New Jersey, Pennsylvania and tapestry, and a *diving* stock, and said of diving, 'If Mr Halley should succeed, of which (were the wars at an end and the seas secure) he seems very sure ... it would be very considerable'.[28]

Houghton published his weekly papers with the support of a number of fellows of the Royal Society and often gave reports of meetings of the Society. He reported Halley's experiments on the evaporation and the discharge of rivers and on the identity of Sal Gemme (rock salt) and common salt (see Chapter 7).[29] He used unpublished estimates of the areas of England and Wales, and the separate counties that Halley got from weighing a cardboard map of the country and comparing it with the weight of a circle cut from the centre of the map. He reported Halley's comment on the fertility of sheep and the prevalence of sheep rot on St Helena, and his remark that the Earth was the shape of a Holland cheese. He had weights of bread in Troy and Avoirdupois pounds from him, and a summary of the properties of diamond as compared with glass.[30] Halley evidently kept Houghton informed on various matters and Houghton's commendation of Halley's diving project was probably well founded.

Halley and the Court did not hurry their negotiations, perhaps because it was not clear how well Halley's novel ideas might succeed. Court minutes show that he was working on the wreck from May 1691 to April 1695, but the letter to Newton quoted earlier indicates that he was still involved a year later.[31] There is nothing to show if anything was ever raised from the wreck, whether Halley and his partners profited from it, nor how the RAC concluded the business.

Things may not have gone smoothly. An undated query of Pepys in his Naval Minutes suggests a serious interruption: 'Mr Halley's having his vessel taken from him by a privateer when he was at work in diving upon a wrack somewhere upon the coast'.[32] The seas were never safe in those days. No doubt it was when diving at Pagham that Halley surveyed the West Sussex coast between Selsey and Arundel, for he produced his chart of that coast at the Royal Society on 15 November 1693.[33]

Halley described his diving operations in considerable detail to the Royal Society, so that we know what his problems were and how he overcame them. He printed no account until more than twenty years afterwards, but spoke to the Society on 6 March 1689 and then in 1691 on 6 May and 12 August.[34] He produced a paper on 26 August and followed it with further reports on 23 and 30 September, 7, 21, and 28 October, and 4 and 11 November. He is best known now for his development of the diving bell, but in those accounts in unpublished papers in the Royal Society, he also describes a diving suit and the use of blocks and tackle in raising great loads.

At the meeting of 12 August 1691, Halley told the Society that his diving experiments had succeeded, and he was asked to provide an account in writing, which he did on 26 August, saying that he was responding to requests for an account of his work on the 'Guiney frigat'.[35] He described his bell and how he replenished it with air. The bell seems to have been quite small, 5 ft (1.5 m) in diameter at the bottom where it was open, 3 ft (90 cm) in diameter at the top where it had a thick glass window and a small cock for letting out foul air, and 5 ft deep. He placed a bench about a foot from the bottom for men to sit on, and says that he kept three men under water at 10 fm (18 m) for an hour and three quarters. It sounds very crowded, but the bell was not for working in but for men to rest after working in a diving suit outside. Nothing suggested that the bell had wheels.

As the bell was lowered, the air in it was compressed and the water rose inside. Halley overcame that, as also the provision of fresh air for long periods of immersion, by bringing air down to the bell in casks. He strengthened the casks with iron bindings, weighted them with lead so that they should sink to the bell, and provided a cock at the top and a bung hole at the bottom. The cask was lowered with the open bung hole downwards and when it was in the bell, with the bung hole still under water, the upper cock was opened and the air in the cask was driven into the bell by the pressure of the water. He spoke of the effect of the increased pressure on men's ears and how the pain was alleviated by oil of sweet almonds.

Halley apparently worked himself at a depth of 10 fm (18 m) and anticipated no difficulty in going to 20 fm (36 m) by the same methods. He described the greenish light from the sea around and the pale cherry-coloured light from above through the window, phenomena for which Newton accounted in the *Opticks*.

> Of this kind is an experiment lately related to me by Mr Halley, who in diving deep into the Sea in a diving Vessel, found in a clear Sunshine Day, that when he was sunk many Fathoms deep into the Water the upper part of his Hand on which the Sun shone directly through the Water and through a small Glass Window in the Vessel appeared of a red Colour, like that of a Damask Rose, and the Water below and the under part of his Hand illuminated by light reflected from the Water below look'd green. For thence it may be gathered, that the Sea-Water reflects back the violet and blue-making Rays[*] most easily, and lets the red-making Rays pass most freely and copiously to great Depths. For thereby the Sun's direct light at all great Depths, by reason of the predominating red-making Rays, must appear red; and the greater the Depth is, the fuller and intenser must that red be. And at such Depths as the violet-making rays scarce penetrate unto, the blue-making, the green-making, and yellow-making Rays, being reflected from below more copiously than the red-making ones, must compound a green.[36]

In his second paper of 1691, read on 23 September, the same day on which he read his paper on the transit of Venus, Halley described a further substantial development.[37] He had evidently tried various unsatisfactory ways of enabling men to work outside the diving bell, of which no report survives; now he set out his idea for what was effectively the diving suit as still used today. Halley claimed priority for his invention:

> A Man having a suite of Leather fitted to his body, with a cap of Maintenance such as I have formerly described, capable to hold 5 or 6

---

[*]Newton's language—*blue-making, red-making* rays—is much more precise than common usage today—*blue rays, red rays*: the rays are not blue or red, they produce blue or red sensations in our eyes.

gallons, must be perfectly inclosed so that the water may as little as posible soack in upon him, must have a pipe coming from the Diving Bell to his Capp, to bring him Air, which must be returned by another pipe, which must go *from* the cap of Maintenance, to a small receptacle of air placed above the Diving bell into which it is to return the Air, that has been breathed; whilest the other brings it to the man.

Halley explained that, by that arrangement, air was forced into the cap from the bell because the pressure in the bell was greater than that in the receptacle above it by the height of the water column between them; the rate of flow of air through the cap can be regulated by the aperture of the pipe or by a cock. Halley was also able to maintain a candle alight under water, and in that way he expected to 'maintain a man and his light in any reasonable depth under water to walk up and down about any business he could do in a shipp were shee above water'. That seems unduly optimistic.

On 7 October he showed the society the pipe he used. It was made with a small wire on which a tape was sewn and covered with several folds of gut drawn over it. They were completely airtight when dried, yet the pipe was very flexible.[38]

On 21 October Halley described how he could go down to the diving bell and return from it, and work on the bottom undisturbed by the waves at the surface.[39] He wore thick flannel or woollen garments against the cold under his diving suit, and by weighting himself down with a quilted girdle of lead shot, could descend to the bell which remained always on the bottom, while to return to the surface, it was only necessary to cast off the girdle. Previously, it seems, it was necessary to raise the bell to the surface to change the men, and that required great effort and many hands and, so he implies, would be difficult in rough weather. Halley also says that he has thought of raising sunken ships by attaching to them large vessels full of water and then sending down casks of air, as he sent down to a bell, to drive out the water. He suggests it should be possible to raise a ship in one lift rather than tide by tide, as was then done.

Working on a sunken vessel involved much raising and lowering of great loads. On three occasions Halley spoke about blocks and tackle, and their use to multiply the force applied by the number of turns of the ropes going over them, as in St Paul's for raising large blocks of stone.[40] It is a little odd that the advantage of multiple pulleys was not already known—it was the sort of technical device that would be familiar to practical seamen. Ropes and block and tackle were common to architecture, seamanship, and quarrying. Halley's mention of his great rope in his letter to Abraham Hill recalls the personal care that Michelangelo took over his ropes more than a century and a half earlier when he was quarrying marble north of Pisa for San Lorenzo in Florence. Michelangelo transhipped columns and blocks of marble with shear legs and block and tackle, just as Leonardo da Vinci had drawn, and no doubt much as Halley used on his frigate. Indeed those methods go back to Vitruvius (Book X). Halley must have used them later at Oxford, when he was asked to look after the moving of the Arundel Marbles from a wall around the Sheldonian Theatre into a gallery of Bodley's Library (Chapter 12). It seems surprising that he needed to describe the multiplicative effect of pulleys to the Royal Society.[41]

In 1693 Halley made just one further report on diving to the Royal Society, on a form of manometer for showing the depth of a diving bell below the surface. Later that same year he showed his chart of the west coast of Sussex to the Society.[42] Until sometime in 1695 or 1696 he was putting his methods to use on *Guynie* off Pagham, while at the same time bringing a very wide range of results and observations to the Royal Society. Whether his diving brought him any commercial advantage cannot be said, but apart from the diving bell itself and the diving helmet, he made significant developments in lifting gear and also in the science of hydrodynamics. He took great risks in trying out those new schemes himself and working under water. Even more important, perhaps, he had gained further experience of practical seamanship that would serve him well in his Atlantic and Channel cruises.

9.4 *Hydrodynamics*

After he had overcome most of the difficulties of maintaining men in a bell for appreciable times, Halley found that the flow of running water greatly hampered work under water. (He also noticed that waves on the surface died away at depth.) He discussed the force exerted on a man by a stream of water and how to measure it, and that led him to a way of calculating the force of wind on a sail and its relation to the force encountered by a ship moving through the water, and then to the size of wing necessary to keep a bird up in flight.

Earlier in the same year, he had done some experiments on the relation between the pressure and velocity of a fluid. Newton had discussed the force of a moving fluid upon a solid body, and the rate at which a fluid would flow out of a hole in a vessel, at the end of Book II of *Principia* (Prop. 37). His analysis was far from complete, and only in the second and third editions did he give results equivalent to the statement that the pressure exerted by a fluid of density ρ flowing with velocity v on a body immersed in it is proportional to ρv^2; Halley's experiments on water flowing out of a vessel were made four years after the first edition of *Principia*, and long before the second. He determined the constant of proportionality, showing that the pressure, p, at a depth z, was equal to $g\rho z$ and to $\frac{1}{2}\rho v^2$, where g is the value of gravity. That is effectively Bernoulli's equation, although the equation as such was not known at the time. Halley used his result to estimate the force on a bullet fired from a gun and the height to which it would be projected by explosion of a charge of gunpowder.[43]

Water falling from a height of 16 ft would acquire a velocity of 32 ft s^{-1}. A velocity of 4 ft s^{-1}, taken by Halley as the strongest tidal flow in the Thames, would be acquired by falling through $\frac{1}{4}$ ft or 3 in. Halley supposed the pressure on a plane surface perpendicular to the flow to be the static pressure $g\rho z$. Then the force of a stream flowing at 4 ft s^{-1}, on a board 2 ft square would be the weight of water standing 3 in deep on the board, namely the weight of one cubic foot of water, or $62\frac{1}{2}$ lb. He recognised that the force on a cylinder would be less.

He argued that, for a man to stand up against the stream, the moment of the horizontal force upon him had to be less than that of his own mass about his feet. He did not recognise that the weight of the man is reduced by buoyancy, but he did realise that a man had to carry extra weights for stability.

In his second paper Halley extended his argument to the force exerted by air on a sail, and showed that if the velocity of a stream of air was the same as that of a stream of water, then the forces on sail and ship would be the same if the areas were in the inverse ratios of the square roots of the densities of air and water. The density of air is 1/850 times that of water and thus the ratio of areas would be 1/29.

In both papers, Halley ignored the form of the flow around the obstacles and the compressibility of the fluid. It would be some time before those issues were formulated, much less resolved, and quite powerful mathematical methods are needed for proper solutions. Furthermore, only if a sail were stretched at right angles to the air flow would Halley's estimate be even approximately correct, for in most cases the air flowing over the sail bows it out so that it becomes an aerofoil with the pressure reduced over the front surface where the air is moving faster.

Finally, in a third paper, Halley estimated the speed with which a bird's wing must beat. He supposed that it must beat down fully extended at the same speed as that of an upward stream of air that would keep the bird aloft on extended wings. He considered a pigeon, 'a bird of Quick flight', and estimated the area of the two wings fully extended to be 70 in^2. He found the weight to be $8\frac{8}{10}$ oz Troy, so that each square foot should support 18 oz. With the same velocity of water as of the air, each square foot would support 29 × 18 oz or $43\frac{1}{2}$ lb Troy, equivalent to a depth of water of about 8 in. That, by the previous rules, would give a velocity of $6\frac{1}{2}$ ft s^{-1} for the stream of air and so, by Halley's argument, for the down beat of the pigeon's wing. Halley then went on to estimate that, with the same velocity, the wings to support a man would have to be 12 ft (3.7 m) long, with a span of about 25 ft (7.6 m). That was not just science fiction, for people were already attempting to glide or fly with bird-like wings.[44]

Halley's calculations of forces exerted by fluids appear to be original, and the relation he assumed between pressure and velocity predated both Newton's more complete discussion in the later editions of *Principia* and the work of Bernoulli. Halley seems to have confined his discussions to talks at the Royal Society, so that these, his early contributions, have escaped notice.

Bird flight is much more complicated than Halley thought. His idea that the bird is supported, like a helicopter, by the downward flow of the air would apply, if at all, only to a bird hovering without horizontal movement. A pigeon is not a hovering bird and, as with most birds, the upward force on it must come mainly from the wings acting as aerofoils. Halley's calculations might apply to hawks and humming-birds, but even then there are complications.

A typical humming-bird weighing about 3 g would be kept aloft by a pressure of air of some 25 N m^2. That would be exerted by an air stream at 7 m s^{-1}. The corresponding pressure and air speed to sustain a sparrow hawk weighing 0.1 kg would be 16 N m^2 and 5.5 m s^{-1}. The amplitude of the wing beat of each bird, according to Halley's model, would have to be considerably greater than the length of a wing. Neither bird, it seems, sustains itself aloft by simple up and down motions of the wings: the wings act in much more complex ways.

9.5 *Astronomy at Oxford: an unsuccessful candidate*

In 1691 the Savilian Professorship of Astronomy at Oxford became vacant by the resignation of Edward Bernard, and Halley declared himself a candidate.

The electors to the Savilian Chairs at Oxford were the Archbishop of Canterbury, the Lord Chancellor of England (or the Lord Keeper of the Great Seal), the Chancellor of the University, the Bishop of London, the Principal Secretary of State, the Chief Justice of the

King's Bench, the Chief Justice of the Common Pleas, the Chief Baron of the Exchequer, and the Dean of the Court of the Arches. Tillotson was the Archbishop. He had aroused the ire of nonjurors and high churchmen when he had been enthroned in Canterbury after the nonjuring Sanford had been ejected. His fellow elector Henry Compton, the Bishop of London, who had aspired to Canterbury himself, was one who was displeased. Compton had been tutor to James's daughter, Mary, and his position was complex, for he was one of the Seven Bishops tried and acquitted for sedition, and he alone upon the bench of bishops had signed the invitation to William of Orange sent to him by persons of influence before his invasion.[45]

Bernard had been talking of resigning for some years, and as early as 1678 he indicated to Flamsteed that he might resign in his favour. Flamsteed, however, was told that he could not hope for the succession because he was not an Oxford man (he was of Jesus College, Cambridge) and that Halley (at that time still in St Helena) might reasonably expect it. Flamsteed wrote to Bernard saying he was content with his position at Greenwich, gave him news of Halley and commented

> He is very ingenious, as I found when he talked with me: and his friends being wealthy, you may expect that advantage by a resignation to him, which it is scarce in my power to afford you.

Bernard did not resign then.[46]

On 22 June 1691, when Halley wrote to Abraham Hill from Pagham about the progress of his work on *Guynie*, he had asked for the election to be deferred if possible.[47]

Hill was then not only on the Court of the Royal African Company, Treasurer of the Royal Society, and a Commisssioner of Trade, but also an official of the Archbishop of Canterbury, on which account no doubt Halley wrote to him. In fact the election did not take place until December, for Bernard only resigned in November (see Note 55).

Halley's paper on Noah's flood, of a much later date, which appeared to date various geological phenomena before the Mosaic Creation,[48] might have led some people to conclude that he believed that the world was eternal, a belief incompatible with the account in Genesis. There is nothing similar in his earlier writings. He argued in 1692, because the planets seemed to be slowing down, according to records of ancient eclipses, 'that the Motions being retarded must necesarily conclude a finall period and that the eternity of the World was hence to be demonstrated impossible',[49] contrary to what he was supposed to assert. Many people at that time speculated that the world was eternal, and some that there might be other worlds. Halley denied the first notion and seems not to have entertained the second.

Halley obtained a certificate from the Royal Society:

> It was ordered, that the Society doe give a recommendatory Letter to Mr Halley signifying their opinion of his abilities to perform the Office of Professor of Astronomy in Oxford now vacant, as likewise to testifye, what he has done for the advancement of the said Science, and that Dr Gale be desired to draw up the Testimoniall.[50]

Halley was not elected. According to the usual account, he fell foul of Richard Bentley, who questioned him on behalf of Edward Stillingfleet, the Bishop of Worcester. There are different versions of the story (Bishop Hough, Hearne, Whiston), but most are second-hand and late. Fatio de Duillier later told Huygens that he had hoped to get the position himself, and said that Halley was excluded on account of religious opinions imputed to him; that seems to be the only contemporary reference to Halley's views.[51] David Gregory, for whom Newton wrote a certificate,[52] was elected. The general reliability of tales about people's religion may be judged by one about Gregory,

> ... a Scot, a stranger came severall times to a Coffee House wch Dr Halley used, and often asked the man after him. But the Dr not happening to come, the man enquired after his pressing business. Why Sr (says he) I would fain see the man that has less religion than Dr Gregory.[53]

Gregory had been ejected from his Edinburgh chair when the Presbyterians took control of the Scottish Kirk after 1689, because he was a firm Episcopalian.

Who entered the caveat against Halley? It must have been Flamsteed, for he himself said it in a note he made on a letter from Wallis of 28 December 1698:

> Dr Gregory is a freind of Mr Halleys tho he was his competitor but I perceive by this transaction he is no freind of mine tho I showed him more freindship than he could reasonably expect on yt occasion & Mr Halley as much enmity but he thinkes Mr Halley has an Interest in Mr Newton & therefore is become his freind, & takes the same course Hawley did to ingratiate himself with him whose favour may be of use to him with Mr Montague.[54]

Flamsteed may have entered the caveat against Halley, but the age of the world may not have been the only reason for it, nor the one that weighed most with the electors. Halley's views about the age of the world were not exceptional, and his published attitude to the relation between the Biblical account of creation and the natural sciences was not so different from that of latitudinarians, Tillotson and Stillingfleet in particular. It seems unlikely that they would have objected to Halley on such theological grounds, especially as ten years earlier Tillotson had thought highly of Halley.

Religion, at least in those terms, may not have been the main reason for the choice of Gregory. How were the two candidates seen by the electors in 1691? Gregory was an able mathematician, he had studied Newton's *Principia* very thoroughly since it had appeared, and he had Newton's support. He had already been a professor in Edinburgh. The bishop-electors would have been sympathetic to him as having been ejected for holding to the episcopalian church. Halley had St Helena to his credit, although Flamsteed, if asked, would no doubt have repeated his criticisms of his methods. Halley had observed for about a year at Islington and then instigated and published the *Principia*. Nowadays we, like him, see the *Principia* as a great achievement, but then it is doubtful if any of the electors would have understood its implications and what Halley had done. Halley and Gregory were perhaps the only two people in Britain who appreciated it, and Gregory had Newton's commendation. After *Principia*, Halley would have been seen to have abandoned astronomy for sea-going and other activities. He had no academic experience and if questions were raised of his religious position, it would seem less sure than Gregory's.

Seven electors signed the certificate of Gregory's election. One was Tillotson. Compton signed. The others were the Duke of Ormonde, Chancellor of Oxford; the Earl of Nottingham, a Secretary of State; and the three judges, Sir John Holt, Chief Justice of the King's Bench, Sir Henry Pollexfen, Chief Justice of the Common Pleas, and Sir Robert Atkins, Chief Baron of the Exchequer.[55] The Earl of Nottingham, Daniel Finch, had been a friend of Charles II, but supported William in the last days of James II and became William's Secretary of State. Nottingham and Holt were again electors when Halley was elected as successor to John Wallis in 1704 (Chapter 12).

The electors were above all a political body. Few persons were more committed by their offices and their pasts to the government of the day. All would adhere to the settlement of 1689 and would be very suspicious of anyone associated with the previous regime. Newton, who had given his support to Gregory, had opposed James II and had been a member of the Convention that seated William and Mary on the throne. Halley's political position, on the other hand, was equivocal at best, for he had been strongly supported by the courts of Charles II and James II and was associated with nonjurors such as Pepys, Robert Nelson, and Sir Anthony Deane, the shipwright. He would scarcely be seen as the administration's safe man in Oxford.

Westfall and Funk have argued that Halley was never in need of, nor asked for, Newton's patronage, on this occasion nor any other, and rather discount the idea that Newton failed Halley by not discharging the debt due for all the work on *Principia*.[56] Newton and Halley did move in different circles and Halley never seems to have called on Newton for support for anything that he really wanted to do. The debt for the *Principia* seems very great to us after three centuries, but perhaps we are more in Halley's debt than Newton was. The *Principia* may not have bulked so large immediately after publication—and Newton had already made acknowledgement in the Preface.

The electors' decision may have turned out for the best. Halley went on to his voyages of scientific discovery and the surveys on the Adriatic, while Gregory brought out his *Principles of Physical and Geometrical Astronomy* (1702), an exposition of astronomy on Newtonian principles. After Halley was elected to the chair of geometry in 1704 he contributed his *Synopsis of the Astronomy of Comets* to Gregory's book, he collaborated with Gregory on the edition of the *Conics* of Apollonius, and after Gregory died in 1708, he took over the preparation of Flamsteed's material for the *Historia coelestis*. What might not have come out of Oxford, had Gregory lived?

9.6 The Mint at Chester

Silver coins, hammered out by hand until 1695, had smooth rims that miscreants could clip. Clipping paid when the market price of silver exceeded the face value of the coins. Over time, and notably after 1695, the coins in circulation came to contain far less silver than their nominal value. The Government was forced to recoin the silver currency. A committee which included Locke and Newton recommended recoinage at the existing standards of weight and fineness, following Locke, as against William Lowndes of the Treasury who argued for devaluation. The Government adopted the committee's advice. Great inflation followed the recoinage, in part because the old coins, however badly clipped, were accepted at face value for melting down. Clippers had a licence to 'coin' money for a short while around 1696 and cost the Treasury the heavy loss of about £2.3m.

A new mechanical process was introduced for the recoinage. A powerful press stamped blanks at almost one a second and a secret tool edged them with either a milled edge or an inscription. Clipping could be detected.

The act for the recoinage set up temporary country mints at Bristol, Chester, Exeter, Norwich, and York. People far from London did not then have to go there to bring in their old coins. The country mints hardly reduced the burden upon the Tower Mint, which produced well over three quarters of all coin: the Chester Mint coined silver worth

£318 377 in the three years 1696, 1697, and 1698, and the Tower coined £5 030 677 in the same years. The government closed the country mints in 1698 after the recoinage; Chester was closed in June. The accounts took much longer to settle: the Treasury did not finally accept them overall until January 1704.

Newton was appointed Warden of the Mint in 1696 after the decision to recoin had been taken, and much the greatest part of the work was done in his first years. His responsibilities were primarily legal and formal. The Warden was a magistrate and had jurisdiction in all issues that concerned the Mint and its officers and servants, except for suits of the Crown. He was the most senior officer but not the highest paid; that was the Master-Worker who was in charge of the actual operations. The Master, Thomas Neale, was an adventurer in more ways than one. His work at the Tower was in effect performed by his able deputy, Thomas Hall. Newton effectively assumed direction of the recoinage. The third senior officer was the Comptroller who checked the bullion accounts and the accounts of coins minted and saw to their safe custody, all on behalf of the King; in 1696 he was James Hoare. The three seniors were supposed to keep a check on each other, while jointly they were a legal corporation. Their divided and ambiguous responsibilities led to persistent serious problems. The country mints were formally branches of the Tower Mint and their principal officers were deputies of those in the Tower and appointed by them, so that the accounts of all the country mints were the responsibility of Neale. It was a clumsy arrangement at best. Neale seems to have set his deputies against Newton's and generated great ill-will and squabbling in the country mints. Some of the deputies were incompetent, some were dishonest. Neale's deputy at Norwich ended in prison for large discrepancies in his accounts. Such was the state of affairs into which Halley was plunged when he accepted Newton's offer to be his deputy at Chester.[57]

A post for Halley at Chester was first mentioned late in 1695 or early in 1696 when, in an undated letter assigned to that period, he says that he will be calling on Newton 'to discourse the other matter of serving you as your deputy'.[58] Halley was not formally Newton's deputy; he was appointed as deputy to the Comptroller. He had not received his appointment by 8 June 1696 when Newton and Hall reported to the Treasury the progress made in establishing the five country mints. The Tower was making the machinery and instruments, and five deputy wardens and five deputy masters had been appointed and were active. 'But the Comptroller refuses to appoint any Deputies'.[59] Halley was eventually appointed, presumably at Newton's insistence. Hoare died shortly after and was succeeded by Mason and Thomas Molyneux acting jointly. Halley corresponded with Molyneux as well as with Newton about the affairs of Chester.

Why did Halley go to Chester Mint? Was Newton, now in a position to do so, at last repaying the debt that he owed for *Principia*? Possibly, but it has been argued cogently that Newton and Halley moved in such different circles that patronage of one for the other had no meaning nor possibility (see Note 56). Would Halley have seen Chester as such a great thing? He knew that it would be for two years only and it was not what he was interested in doing. He was, however, in something of a limbo. Six years had passed since he had first suggested cruising in the Alantic to study the magnetic field. *Paramore* had been built for him, then the project was in abeyance, possibly because England and France were still at war. Was this a chance for Halley to do Newton a service while waiting to go sailing? Might he have been glad to have the salary (£90) if he had suffered loss in the Lagos disaster of 1693?

Halley had spoken for Newton in bringing his book before James II in 1687, but that was nine years ago and James had gone. William III may have been suspicious of Halley when he arrived in England. Newton would be in favour as one who had opposed James and supported William in the Convention of 1689. Newton seems to have seen things in some such way, for when he wrote to Halley on 11 February 1697 about an army engineer post, he said that it might make him better known to the King and so pave the way for something better;[60] he would hardly have written that in 1687, but the wheel of politics may have brought round a change in the positions of Halley and Newton.

Some twelve years later, Newton was asked to advise on the operations of the Mint at Edinburgh, to bring them into line with the procedures of the Tower after the Act of Union of Scotland and England. He asked David Gregory to do this. Gregory, Savilian Professor of Astronomy and with a medical practice in London, would hardly have needed the money; Newton no doubt chose the person he thought best fitted for the responsibility; perhaps he chose Halley for Chester for a similar reason.[61]

Halley had certainly been at Chester for some while prior to October 1696, for on 12 October he wrote to Sloane in London to excuse himself for not having sent the Journal Book of the Royal Society to him—

> ... but a great glutt of business, at the first opening of our Mint, requiring a constant attendance, has hindred me for some time, so that I must begg your excuse at present.[62]

The business probably had to do with finding buildings and installing machinery, for on 10 October 1696 the officers of the Mint at Chester, like officers at the other country mints, were ordered to ask the Mayor of Chester for the services of a surveyor who could check the building of the Mint there to ensure that the Treasury was not being overcharged.[63]

Halley was at first hopeful of good progress:

> Wee proceed lightly in our business of Coining and should soon have done, were we not limited to small mony, only shillings and sixpences.[64]

Things did not turn out so easily. According to letters to Sloane of 2 and 25 November, a considerable amount of clipped money had come in, but the supply had dried up as people waited to see a forthcoming act of Parliament about the terms of exchange; meanwhile the coinage went on apace.[65]

Halley apparently ran into some trouble about the casting of bullion and ingots, and on 28 November he wrote to Newton to explain and justify his procedures.[66] Was it sufficient for the final casting to be of the standard quality, or should certain Tower prescriptions be followed for the materials, scrap, new bullion, and so on, put into the melt? Complaints about Halley's actions may have gone higher, for on 15 March 1697 the Lords of the Treasury declared, 'My Lords will have Mr Haley continued in the Mint at Chester'.[67]

Internal disputes afflicted the Chester Mint throughout its existence. In the early summer of 1697, Halley had suspected malpractices among the tellers. On 21 June 1697 Newton wrote to Halley that tellers were to be sent from the Tower to Chester to inspect the work of the tellers there and to turn out those who proved unsatisfactory.[68] On 25 October 1697, in a letter to Sloane, Halley wrote that he had hoped to have attended the Royal Society before then but the business of the Mint was not yet finished:

> In the mean time my heart is with you and I long to be delivered from the uneasiness I suffer here by ill company in my business, which at best is but drudgery, but as we are in perpetual feuds is intollerable.[69]

More than a year of quarrels lay behind that *crie de coeur*.

Clarke, who was Neale's deputy as Master at Chester, took himself off to London for months at a time, leaving the duties of his office to Halley and the deputy Warden, Weddall.[70] Two of the clerks at Chester, Bowles and Edward Lewis, were generally insolent to Halley and Weddall, wearing hats in their presence and carrying swords. In August 1697 the three senior officers in the Tower reprimanded the Chester officers severely. They were concerned at the quarrels; they insisted that the melting and coinage were to go on, 'for the Mint will not allow of the drawing of Swords & assaulting any, nor ought such language Wee hear has been, *be used any more among you*'. That may or may not have stopped the drawing of swords, but an inkstand was later thrown at Weddall.[71] The clerks also and others, so it was said, performed their duties irregularly. Neale in London supported Lewis, as did Clarke, and on one occasion Clarke, taking offence at something no one else had heard, went to borrow a sword from Bowles to waylay Weddall as he went home. He did not do so, but later sent a challenge to Weddall which Weddall accepted:

> He appeared, however on the ground, before the hour, with his man and horses, and staid not after it, by which means they fought not, and I demonstrated the folly of such decisions that went no further.[72]

Lewis and Clarke raised various charges against Halley and Weddall, including that about the composition of the melt, and that Halley showed preference to individuals in the purchase of silver, and they also accused Weddall of treasonable words.[73] Accusations of treasonable words were a serious matter when they might lead to conviction for high treason.

Even as early as the end of 1696 Halley wanted to resign, and Newton was looking out for another post for him: on 11 February 1697 he wrote to him about the position of mathematics master to engineer officers in the army. Halley was cool about it, for he thought that any army post might not continue after the end of the Nine Years' War, which indeed ended in September 1697 with the Peace of Ryswick (see Note 60). So Halley stayed on at Chester and did get something better, far, far better, in *Paramore*.

Matters came to a head early in August 1697 when Clarke went up to London to lay his complaints against Halley and Weddall.

> If need be, I begg you would interpose your protection, until we can be informed of any sort of accusation, and that we may be heard before we are in any case judged. I hope your potent friend Mr Montague [the Chancellor of the Exchequer who had arranged Newton's appointment to the Mint] will not forgett me if their should be any occasion, but I am conscious to myself of no transgression, so I doubt not to acquit myself of any imputations their malice can invent ...[74]

Lewis was also accused of misbehaviour, and on 9 November 1697 the Lords of the Treasury ordered 'Ed Lewis, clerk to master worker at Chester to attend Lords forthwith to answer accusations'.[75] Halley, meanwhile, had evidently decided to resign at last, but before doing so wanted to present his complaints against Lewis:

> The Parliament having this day voted the Continuance of all the Country Mints, I should be very unwilling to leave Lewis and Clarke to enterprett

my resignation to be any other than a voluntary cession, as they will most certainly do, unless I prosecute, as I have already begun, the undue preferences by them made. Abbom Grays affaire I value not, as being what I hope may be justifiable on many accounts, should the Lords believe me consenting to it, but the Mint at Chester I assure you cannot subsist as it ought, whilst Lewis governs Clark as he does, and Mr Neale supports both. Wherfore I begg that Lewis may appear face to face with me, before the Lords, there to answer to his throwing the standish at Mr Weddell, the giving undue preference to Pulford, and some other accusations of that nature I am prepared to lay before their Lordships. I came to town purposely to charge that proud insolent fellow, whom I humbly begg you to believe the principall Author of all the disturbance we have had at our Mint, whom if you please to see removed, all will be easie; and on that condition I am prepared to submit to all you shall prescribe me. Nevertheless, as I have often wrote you, I would urge you to nothing, but what your great prudence shall think proper, since it is to your particular favour I owe this post which is my chiefest ambition to maintain worthily; and next to that to Approve myself in all things.[76]

Newton evidently was of Halley's mind. On 2 February 1698 Neale, Newton, Mason, and Molyneux, the senior officers of the Mint, attended the Lords of the Treasury. Newton alleged that Lewis was insolent, quarrelsome, and mischievous. Neale defended Lewis and said on his behalf that he understood his business well, but admitted that he was quarrelsome if provoked. He also said that the other country mints were much worse. The Lords decided that Neale should dismiss Lewis forthwith and a warrant of the same day directed him to do so.[77]

So, early in 1698, Halley's time at Chester came to its end. He had carried out his duties and seen the Mint produce a modest amount of coin, he had troubles with colleagues and had been vindicated, while retaining the respect and confidence of persons in authority. He had also found time for scientific and archaeological observations (Chapter 7). He was now about to embark on greater enterprises in which he would meet afflictions like those he suffered at Chester; no doubt he would recall his time there and draw on that experience in handling men.

9.7 *Looking forward*

In the ten years from 1688 to 1698, Halley had put his mind to a wide range of topics in natural philosophy, some imposed on him by his duties as Clerk to the Royal Society, others the outcome of his own interests. Of the latter, his studies of the orbits of comets and his analysis of the use of the transits of Venus had far the most influence on his successors. They have shown the way forward and set agenda for astronomers from his day until now. Methods of observation have become vastly more powerful, as has theory with the assistance of computers; however, the underlying problems remain essentially as he defined them—to establish the orbits of comets and hence say something about whence they come, to measure the distance of the Sun and hence the scale of the solar system, galaxy, and cosmos.

Those ten years foreshadow the next five. Halley's experience at sea as a hydrographic surveyor, and in developing methods for working on the sea-bed and raising sunken vessels, extended his practical background of seamanship which would enable him to propose and to carry out his voyages for the observation of the magnetic variation over the Atlantic and the tides in the Channel. He first proposed the magnetic voyages in 1693, but on account of delays he did not sail until 1698; then he opened a new window on the natural world.

10

Far seas and new prospects

> Heureux qui comme Ulysse a fait un beau voyage.
> [Happy he who like Ulysses has made a good voyage.]
>
> DU BELLAY: *Sonnets*

10.1 *Setting out*

ON Thursday the twentieth of October 1698, a small gale blowing from WSW, his Majesty's pink, *Paramore*, Captain Edmond Halley in command, weighed anchor at the Navy Yard at Deptford and stood down the River to Gravesend. Halley had begun the cruise he had proposed five years before. His orders were to improve navigation by observing the magnetic variation over the Atlantic Ocean and to discover, if possible, lands to the south in the Atlantic. Nearly two years would pass before he returned *Paramore* to the Master of Attendance at Deptford, and a further year in which he took her to sea again to survey the tides in the Channel. Almost three years to the day from first leaving Deptford in 1698, Halley finally delivered *Paramore* into the custody of the Master of Attendance there. In three years at sea he discovered no new lands, he did improve navigation, and he opened up the Earth to physical investigation. The effects of his voyages have continued to this day.

The mariner seeking his position and true course at sea met with three great problems at the end of the seventeenth century. The first, discussed in Chapters 3 and 8, was the longitude. The second was the very poor idea of the currents on the deep oceans, first studied by Athanasius Kircher.[1] The third problem was complex tides near land. The effect of currents is seen in Halley's course westward across the Atlantic to the coast of Brazil on his first cruise. From his speed through the water, as given by his log, he estimated the longitude of Paraiba (Joana Pessoa),* where he made his landfall on the east coast of Brazil, to be about 25° W of London, whereas his astronomical measurements gave 36° W. The true value is about 35° W. He made an

---

*The modern names of places, where they differ significantly from those usd by Halley, are given in Table 10.1, as well as the longitudes he measured astronomically along with the modern values.

Table 10.1 Names and longitudes of Atlantic ports

Voyage	Name Halley	Modern	Longitude (all W) Halley ° ′	Modern ° ′
2	Abrothos	Abrolhos		
2	Anglois	English		
1	Anguilla	Anguilla	62 50	63 05
1	Antegoa	Antigua	61 27	61 49
1	Barbadoes	Barbados	59 05	59 00–12
1	Bermudas	Bermuda		
2	Bon Esperance	Good Hope (Cape)		
1	Bonavista	Boa Vista		
2	Bona Vista		23 00	23 00
1	Carlisle Bay	Carlisle Bay	59 05	59 37
2	Causon	Cawsand		
1	Dello (Cape)	Cabadelo	36 00	34 50
1	Desseada	Désirade	60 30	61 03
1,2	Fernando Loronho	Fernando de Noronha	34 00	32 25
2	Ferro	Hierro	19 00	18 00
1	Fonchiall	Funchal	16 45	16 55
2	Frio	Frio	43 40	42 00
2	Grande	Grande		
1	Guadalupe	Guadaloupe		
1	Ilhas Desertas	Deserta Grande		
1	Madera	Madeira	16 45	17 00
1	Martinica	Martinique	60 20	60 30
2	Martin Vaz	Martin Vaz	27 21	28 52
1,2	May	Maio	22 00	23 13
1	Monteserrat	Montserrat	61 47	62 14
1	Nevis	Nevis	62 10	62 35
1	Paraiba	João Pessoa	36 00	34 53
2	Palma	La Palma		
1,2	Pernambuco	Recife	35 30	34 52
1	Praya	Praia	22 30	23 30
2	La Praya		23 30	22 30
1	Redondo	Redonda	61 55	62 19
2	Rio Janeiro	Rio de Janeiro	44 45	43 17
1	Saba	Saba	62 55	63 26
1	St Bartholomey	St Barthélemy	62 35	62 50
1	St Christophers	St Kitts	62 25	62 45
2	St Christophers		62 08	
2	St Helena	St Helena	6 30	5 43
1	St Iago	San Iago	22 40	22 40
1	St Martins	Sint Maarten	62 50	63 05
2	St Thomas	Sao Tome		
1	Sall	Sal	22 00	22 55
2	Sall		23 00	
1	Sant Eustachia	Sint Eustatius	62 40	63 00
1	Scilley	Scilly	3 32	

Table 10.1 Continued

Voyage	Name		Longitude (all W)	
	Halley	Modern	Halley ° ′	Modern ° ′
2	Sphere	Spear	17 00	16 35
2	Teneriff	Tenerife	54 00	52 51
2	Toads Cove	Tors Cove	18 46	12 30
2	Tristan da Cunha	Tristan da Cunha	–19 20	

Names and longitudes have been taken from Appendices A and B of *Paramore*. Some names have been omitted where the modern name is obvious and Halley did not observe the longitudes.

error of about $\frac{1}{2}$ degree per day from the time he crossed the equator near the coast of Africa. He was being carried westward by the equatorial currents and his error implies a current of no more than 1 kn or 2 km^{-1} h. His error was the greater because he was becalmed for days on end. His measurements of latitude told him that he was being carried northwards by the current, but in default of astronomical measurements of longitude he could not detect the westerly flow.

The North Atlantic is effectively an enclosed basin, and on account of the Coriolis force the ocean currents run clockwise round it. The Gulf Stream carries warm water from the West Indies to the north-west of Europe and then down the west coast of Europe and Africa until it forms the North Equatorial Current flowing north-westerly from the Gulf of Guinea to the West Indies. The circulation in the South Atlantic is generally anticlockwise, with a south equatorial flow from Africa to South America, but the ocean is open to the circum-Antarctic ocean to the south, where a current flows right round Antarctica and couples the circulations in the south Atlantic, the Indian Ocean, and the south Pacific.

Tides hardly affect navigation in the open oceans but they dominate sailing close to land, as Halley's Channel cruise would show.

The uncertainty of navigation in those days can hardly be overstated, and led to some notorious disasters. One was the loss of Sir Cloudesley Shovell and his fleet on the Scillies in 1707, through precisely the errors of navigation against which Halley had warned.[2] We may guess that Halley felt that loss acutely, for he had met the Admiral a number of times. He had sought his help or protection as a flag officer and Shovell had held the court martial on Halley's lieutenant after the first Atlantic cruise. He commanded in the Mediterranean when Halley was on the Adriatic coast (Chapter 11), and it is very likely that Halley's charts of Adriatic harbours were those passed to Shovell when he went into the Mediterranean the next year.

Halley's experiences with *Paramore* show some of the defects of wooden square-rigged ships: leaks, problems of ballast, difficulties of sailing close to the wind, the frailty of masts and yards in sudden gales. Finding fresh water on long cruises was always uncertain, while everyone suspected any single ship to be a pirate. All plans were at the mercy of the winds. Westerlies in the Channel delayed Halley in sailing for Madeira in 1698. Persistent winds in 1703 prevented Sir Cloudesley Shovell's fleet from leaving the Channel until too late in the campaigning season to carry out the plans for naval action in the Mediterranean that Halley's surveys on the Adriatic were to support.

Reactions to Halley as a sea captain exemplify a problem of the Royal Navy that Pepys took very seriously and that was a live matter for decades after he had left office. There were two sorts of senior officer in the Restoration Navy, 'tarpaulins', experienced seamen who had begun as able seamen and were eventually entrusted with command, like James Cook a century later who started out as a merchant master; and gentlemen commanders who had often come to the sea after serving on land. Tarpaulins were apt to be despised by the gentlemen as rough fellows, while tarpaulins thought little of the gentlemen's seamanship. Aristocratic commanders also despised the Navy Board officials, and one reason for Pepys's dismissal in 1689 was the hostility of Russell and other admirals.

Commanding at sea was in some ways not so different from commanding on land. Tactical and strategic ability, and the command of men, were needed in both. There was one crucial difference. No ability to command could be effective at sea without a profound experience of seamanship, which determined everything that could be done, which imposed limits, but also opened great opportunities to the bold, instinctive captain. Commanders, whether merchant or royal, were out of touch with their boards or admirals or Secretaries of State for weeks or months or years on end. In these days of almost instantaneous transmission of orders and information, it is difficult to realise how much the conduct of the affairs of the country was in the hands of remote, isolated, independent individuals. Captains far from home had to be able to understand the grand strategy of the Admiralty and plan their actions accordingly. They had to be diplomats, for whether in command of a fleet or of a single ship, they represented the monarch abroad. Halley, when in Austria, enjoyed the respect due to his rank as captain and colonel and as his monarch's representative (Chapter 11). Commanders had also to manage the logistics of their command, no small matter in distant waters.

Halley, a civilian, was neither a tarpaulin nor a gentleman commander. When Flamsteed castigated him as drinking brandy, smoking, and swearing as a sea captain, he no doubt had the landsman's idea of him as a tarpaulin; when Halley's lieutenant despised him for his ignorance of nautical terms, he saw him as a gentleman commander. What Halley was will appear in this chapter and the next.

Halley and Benjamin Middleton proposed in January 1693 a voyage to observe the magnetic variation.[3] The Royal Society supported them strongly. Queen Mary, with William III abroad, ordered a ship to be built, fitted out, and manned at Deptford. *Paramore* was launched in April 1694 and preparations went ahead. Then, in August 1696, orders came to lay her up at Deptford. Two years later, when he had left Chester, and the Nine Years' War was over, Halley received his commission as Master and Commander of *Paramore*. He had his sailing orders on 15 October 1698 and dropped down the Thames on the twentieth. The weather and the poor state of the ship kept him in English waters until 30 November, when he took his departure for Madeira. His cruise was common knowledge and figured in many private and public reports, as for example, in a letter of James Gregory, the brother of David, to the Revd Colin Campbell in Edinburgh on 29 May 1699:

> Mr Hally has gott a ship from the government, in which he has sett sail to goe round the globe on new discoverys, and the rectifying of geography.[4]

Official papers, now mostly in the Public Record Office, record in detail the progress of the proposal, its adoption by the Admiralty, and its administration by the Navy. Halley's logs of his cruises are in the British Library and have been published by the Hakluyt Society,[5] together with the related correspondence and official papers. Halley corresponded with Josiah Burchett, who had risen from being Pepys's servant to succeed him as Secretary to the Admiralty. Halley published his chart of the magnetic variation over the Atlantic in 1701. Other editions followed, almost all after his death. The chart of the tides in the Channel was published in 1702, a second edition in 1708, and others posthumously.

10.2 *Preparations*

How could Halley have had the confidence to propose that he should command a small naval vessel voyaging over remote seas that, if not unknown, were certainly only poorly known? How could those who supported him, who built a ship for him, who commissioned him as her commander, be sure that they were not squandering resources and putting men's lives at risk? Nothing suggests that Halley had any experience in command, but he must certainly have had considerable knowledge of the sea and seamanship. On his passages to and from St Helena he cannot fail to have studied how a great merchantman was sailed. In the years before his proposal, he had developed the diving bell and diving suit. On the *Guynie* frigate (Chapter 9) he gained experience of working inshore, anchoring in tidal streams, handling equipment over the side: activities that are still difficult today from powered ships and powered tackle—how much more difficult from sailing ships with manual labour. He had made at least two hydrographic surveys, of the Thames and of the coast of west Sussex, again tasks that involved handling a ship and boats close to shore. Navy officials surely knew of those activities and may have commissioned them.

Benjamin Middleton, elected FRS in 1687, was the son of Colonel Thomas Middleton, a Commissioner of the Navy, who had managed the dockyards of Portsmouth and Chatham effectively as a colleague and friend of Pepys. He was also a friend of Sir John Buckworth, who was the trustee of his will. He had died in 1672. Benjamin was a member of Gray's Inn (1670) and a fellow of Emmanuel College, Cambridge, and like many other Fellows of the Royal Society, he contributed to the fund for building the church of King Charles the Martyr in Tunbridge Wells.[6] *Paramore* was stated to have been built for *Col* Middleton, but the colonel was then dead and the father and son must have been confused. Middleton must have been seen as the financier of the project, for he undertook to go on the voyage and to victual and man the ship at his own cost. The Society supported the proposal and Middleton's petition went before the Queen, Mary II. The Commissioners of the Admiralty wrote to the Navy Board on 12 July 1693:

Gentlemen

A petition has been lately presented to this Board by Benjamin Middleton Esq[r] Wherein he proposes, that he will together with M[r] Edmond Halley, undertake a Voyage, wherein he proposes to incompass the whole Globe from East to West, in order to the describeing and laying downe in their true Positions, Such Coasts, Ports and Islands, as the Weather will permitt, to some of which possibly an Advantageous Trade may be found. And also to endeavour to gett full information of the Nature of the Variation of the Compasse over the whole Earth, as Likewise to experiment what may be expected from the Severall Methods proposed for discovering the Longitude at Sea. And praying that he may be furnished with a Vessell of about Eighty Tuns Burthen fitted out for the said Voyage, & maintained therein at their Ma[ts] Charge, excepting only for Victualls and Wages, which he will be at the Charge of; Which Petition having been laid before the Queen Her Ma[ty] is graciously pleased to incourage the said undertakeing. And in pursuance of her Ma^ts pleasure Signified therein to this Board, We do hereby desire and direct you forthwith to cause a Vessel of about Eighty Tuns Burthen to be set up and built in their Ma^ts Yard at Deptford as soon as may be, and that Mr Middleton be consulted with about the conveniencies to be made in her for Men and Provisions, and that when she is built She be fitted out to Sea, and furnished with Boatswains and Carpenters stores for the intended Voyage, & delivered by Inventory to the said Mr Middleton to be returned by him when the Service proposed shall be over.[7]

The project was far-reaching, for a voyage round the world no less, with some of the same purposes as Cook's voyages sixty years later. Halley only ever cruised in the Atlantic and discovered no new lands, but he did discover the longitudes of a number of places by 'severall methods'. He must have hoped to complete his original plan, for when he met Leibniz in Hanover in the summer of 1703 he told him that he intended to go to the Pacific. But he went instead to the Savilian Chair at Oxford.[8]

The Master Shipwright at Deptford sent particulars of the ship to the Navy Board in October 1693, and on 1 April 1694 the Admiralty ordered that the new pink be launched and named *Paramour** and entered on the Navy List in that name.[9]

Fitting out evidently took some time, and not until 11 January 1696 was *Paramore* searched (examined, that is), and the ballast checked. Then on 4 June 1696, the Admiralty issued

> Commission without Instructions for Mr Edmund Halley to be Master and Commandr of his Mat Pink the Paramour, Dated ye 4th June 96.[10]

Middleton now disappears from the record. The boatswain, gunner, and carpenter had their warrants on the same day as Halley's commission, and Halley gave security for the payment of the wages of the crew. It was agreed that, although Halley was to pay the wages, the men would be borne on the King's payroll, presumably so that they should be subject to naval discipline. In fact Sir John Hoskyns of Harwood in Hereford, who was then a vice-president of the Royal Society, stood security for £600, as Halley reported to the Navy Board on 19 June 1696.[11]

All was now ready for Halley to sail, but on 15 August 1696 the Admiralty ordered the Navy Board to lay up *Paramore* in wet dock at Deptford; and there she stayed until 16 March 1698.[12] The Admiralty gave no reason for the order, and it has usually been supposed, and indeed seems most reasonable, that it was because Halley went as

---

*Although spelt *Paramour* in the official order, her name usually seems to have been written *Paramore*.

Newton's deputy to the Chester Mint. At the same time, England was still at war with France, and a lone, small, slow ship in distant waters would have been gravely at risk.

The war at sea must have affected the timing and progress of Halley's project. The Admiralty issued their orders to the Navy Board to build a pink on almost the very day that Londoners learned of the great disaster to the Levant fleet off Lagos, in which Halley's own trade may have suffered.[13] In the following years Louis XIV had no money to maintain a large fleet and England and her allies came to rule the seas, yet the French put out many small ships that attacked commerce. Clearly the Admiralty thought that, war or no, Halley's project was well worth while, but prudently deferred it until peace came.

The Peace of Ryswick was signed in October 1697, but the visit of Peter the Great of Russia may have caused further delay. *Paramore* was rigged and brought afloat in March 1698 to be employed 'as the Tsar should desire'.[14] He was visiting western countries to study up-to-date practice in various fields, ship-building and seamanship included. He arrived in England on 2 January 1698. The government let him see the procedures of Deptford and gave him opportunities to go to sea in various ships, among them *Paramore* and the fleet destined for the Mediterranean. The King put him up in Sayes Court, John Evelyn's house near Deptford, then rented by Admiral Benbow. Sayes Court still exists in Deptford and the Tsar's stay is recalled by Czar St nearby. The Tsar was there from 6 February to 21 April, in which time he and his entourage managed to do damage to the house and gardens and Benbow's possessions that Sir Christopher Wren assessed at nearly £400.[15] Halley is said to have dined with the Tsar on occasion, and may possibly have gone sailing with him, but there is no definite evidence of that. Swedenborg, on a later visit to England, reported that Halley had sold the Tsar the quadrant he used on St Helena and in Islington for £80. By then it would have been more an antique than an example of the latest technology.[16] If there be any foundation for those stories, they imply that Halley was already known to the Admiralty or the Court as someone expert in affairs of the sea.

Arrangements for Halley's cruise were set in train as soon as Peter the Great left. *Paramore* was found to be a poor sailer, perhaps shown up while she was used by the Tsar, and orders were sent to Deptford on 3 July 1698 for the necessary work to be done on her. She was referred to as 'designed on a Particular Service', a phrase taken to mean that in all respects Halley was treated as a naval captain. In the next instructions from the Admiralty, for sheathing and fitting the vessel, for supplying stores and victuals, and allowing a crew of twenty men, a voyage to the East Indies is still envisaged, but *Paramore* is now evidently a royal ship that is being lent to Halley. In the explanation on his published chart he said that his voyage was made 'at the Publick Charge'. Subsequent orders deal with the supply of eight guns, with an imprest account for Halley's expenses, and the appointment of the surgeon, George Alfrey, for whom Halley had asked, as well as a commission for Edward Harrison as mate and lieutenant.

Halley was not happy with the other officers he had been given and had asked for a naval lieutenant to keep discipline, but it did not turn out like that. Harrison was to give Halley a great deal of trouble. He seems to have been a merchant officer who had transferred to the Royal Navy, and after his voyage with Halley he went back to the merchant service. Not long before Halley's first voyage, Harrison had written a little book on methods of finding longitude.[17] He said he was writing for 'tarporlins', or tars. He opted for the lunar method as the best. His longest chapter is on the magnetic variation with a discussion of the drift of the magnetic poles: like Perkins and Halley, he supposed there were four poles. Harrison had served, he says, on six ships of the Royal Navy as well as on merchantmen sailing to the East. He had a poor opinion of the equipment and methods for navigation upon all of them. Scientifically and technically the book is poor, second-hand, and ill-digested. It is shot through with aspersions on mathematicians, the manner is aggressive, and it shows a deep inferiority complex. Halley had reviewed it unfavourably, but seems not to have realised that its author was to be his lieutenant. The situation was ripe for trouble; it occurred.

As mate and lieutenant Harrison would have been responsible for navigating and sailing the pink and, according to naval instructions, he would have kept a journal. After the first Atlantic voyage, while Halley was at sea in the second voyage, Flamsteed made extracts from Harrison's journal and also had a letter from him with an abstract of some of his observations. Comparisons of Flamsteed's extracts with Halley's log show that Harrison and Halley made independent observations, with good agreement on latitudes and magnetic variations but poor correspondence in longitudes.[18]

Halley attended upon the Lords of the Admiralty and received his commission as master and commander on 19 August 1698. Halley's formal status is not entirely clear. According to the List of Sea Officers, he was commissioned as commander in 1696 and as captain in 1698, but the commission that Halley had in 1698 does not differ from that of 1696. In those days the commissioned ranks below rear-admiral were captain and lieutenant. Halley was always subsequently referred to or addressed as 'Captain' and that was presumably his rank. His appointment was as commander of *Paramore*.[19] His expedition was in part private before 1696 but in 1698 it was a Royal Navy undertaking. Certainly in his reports to the Admiralty and in his relations to flag officers at sea, Halley acted as a regular captain. When in 1703 he was sent to the Adriatic (Chapter 11) he was addressed as colonel by officials of the Empire, consonant with the fact that naval captains usually had the honorary rank of colonel in the army. After 1703 he continued to be known as Captain Halley and ranked as a post-captain, that is an unemployed captain who had held a post.

Halley had his orders and instructions on 15 October. He was not now to sail round the world, but to cruise in the Atlantic. The course he followed in his two voyages is very much as laid down in those instructions, which are so similar to the original petition that he in effect wrote them in large part:

> You are to make the best of your way to the Southward of the Equator, and there to observe on the East Coast of South America, and the West Coast of Affrica, the variations of the Compasse, with all the accuracy you can, as also the true Scituation both in Longitude and Latitude of the Ports where you arrive.
>
> You are likewise to make the like observations at as many of the Islands in the Seas between the aforesaid coasts as you can (without too much deviation) bring into your course; and if the Season of the Yeare permit, you are to stand soe far into the South, till you discover the Coast of the Terra Incognita, supposed to lye between Magelan's Streights and the Cape of Good Hope, which Coast you are carefully to lay down in its true position.
>
> In your return you are to visit the English West India Plantations, or as many of them as conveniently you may, and in them to make such observations as may contribute to lay them downe truely in their Geographicall Scituatuion.[20]

The routes that Halley followed and the observations that he made conformed very closely to his instructions, except that, on account of difficulties with his first set of officers, he returned to London prematurely. He went further south than anyone before him in search of the *Terra Incognita*, but without success, and it was only put to rest by Cook sixty years later

Paramore was small, only 80 tons burthen and some 60 ft long, carrying three masts and with a crew of twenty (Fig. 10.1).[21] Pinks were generally slow and not good sailers. They were capacious, an advantage on a long voyage. Of shallow draught, just under 10 ft for *Paramore*, they could sail close inshore.* The fleet with which William of Orange invaded England in 1688 included sixty pinks, no doubt to transport stores and men.

---
*The tonnage, length, and complement of *Paramore* were very close to those of the *Mathew*, John Cabot's ship in which he sailed to Newfoundland 200 years before Halley.

Figure 10.1 A pink setting sail. (Reproduced by permission of the Director of the National Maritime Museum)

A sailing ship can move against the direction of the wind because the sail acts as an aerofoil, the greater speed of the wind over the front of the bowed sail reducing the pressure and pulling it forward. If that were all, the ship would move sideways rather than forwards, but a well designed ship has considerable resistance to sideways motion and little to forward motion. The modern racing yacht has those characteristics in high degree. Ships of Halley's day were far less efficient. The square rig was poor aerodynamically, and the form of a ship that had to carry many heavy guns or much cargo made for a high forward resistance. *Paramore* seems to have been particularly defective. Her shallow draught gave poor lateral resistance, while her broad beam gave high forward resistance. Halley noted that she went more to leeward than most other ships. Barnacles and other creatures that grew on her sides after she had been at sea for some time further increased the forward resistance. Halley took every occasion to remove barnacles. Pinks were lightly armed, *Paramore* carrying six three-pounder guns and two small guns on pivots. She was the only ship ever to have borne that name in the Royal Navy.

There is no list of Halley's instruments, but the observations that he made indicate what some of them might have been. Deptford issued him with two magnetic compasses, but apparently nothing else, although a log line would have been a standard issue. Presumably the other instruments that he took were his own.

Log lines had knots on them to show the length paid out and they had two uses. With a lead weight at the end they sounded the depth of water under a vessel, both for surveying and for pilotage in shallow waters. Halley would have been well used to that from his surveys of the Thames and the coast of West Sussex. With a float at the end the line was cast into the water behind the ship and the length paid out in a given time, such as half a minute, timed with a sand glass, gave the speed through the water.

Halley observed the altitude of the Sun at noon whenever possible. Noon was when the altitude of the Sun attained its greatest value and the noon altitude gave the latitude. If Swedenborg was right that Halley sold his 2 ft quadrant to the Tsar, he could not have had it with him, but he would surely have had some quadrant. He may also have used an instrument that he described to the Royal Society on 23 March 1692, by which the Sun and the horizon could be seen together in the same telescope (Chapter 8). Observation of latitude at sea, though straightforward, was impossible if the Sun was obscured by cloud and not easy when the ship was rolling or pitching. Determination of local time was not straightforward. It involved interpolation between successive sights of the noon Sun, and there was no mechanical clock that would do that on board ship until John Harrison developed his chronometer more than half a century later.

The great difficulty was finding longitude. Halley gives daily values of longitude in his log, presumably at local noon, but few of them were astronomical observations. His practice at sea was to observe the latitude whenever possible and to determine the distance he had travelled each twenty four hours from his speed through the water as given by the log line. He also observed the direction of his course.

Suppose that in 24 h he travelled $\delta\lambda$ in latitude, and that his estimated distance made good was a at an angle θ to the meridian. The three quantities must satisfy the relation

$$a \cos \theta = \rho \delta \lambda,$$

where ρ is the radius of curvature of the Earth in the meridian.

No doubt that relation would not be satisfied exactly, and Halley does not say how he reconciled the measurements. The distance made good was probably the least reliable of the quantities—Halley had doubts about his log line and he made no allowance for currents.

The change of longitude, $\delta\phi$, in the 24 h is given by

$$(\rho \cos \Lambda) \delta\phi = a \sin \theta,$$

where Λ is the latitude of the ship's position.

With a equal to $\rho \delta \lambda \sec \theta$, as above, we obtain

$$\delta\phi = \delta\lambda \tan \theta \sec \Lambda.$$

This result is subject to errors in the bearing θ, which, so far as observations go, may be taken to be random. The systematic effects of currents vitiate it. The calculation assumes that the distance made good and the bearing are relative to a set of coordinates fixed in the Earth, but they are relative to coordinates moving with the ocean currents. The adopted values of a and θ are therefore both in error. The effects of currents are particularly noticeable in the first cruise sailing from Africa to Paraiba and on to the West Indies, with a course throughout in the North Equatorial current.

Halley realised that there were errors in the distances he estimated from his speed through the water and thought that was because his log line was too short and that he did not allow for its distortion by the pitching of the ship in high seas. That may have been so, but currents of the general circulation must have caused a large part of his errors of longitude.

Halley did determine a few longitudes astronomically. He had discussed the problem of longitude in a paper to the Royal Society ten years before, and concluded that there could be no certainty until the lunar theory was perfected (Chapter 8).[22] He used, or attempted to use, two astronomical clocks. The position of the Moon was one, although the irregularities of its motion limited the precision. Halley observed the Moon in two ways, by eclipses, and by the difference of its right ascension from a nearby star, no doubt using the device he had previously described to the Royal Society.[23] Newton later showed the Society an improvement to Halley's instrument but commented that, despite its faults, it gave better longitudes than previous methods. The report of Newton's talk suggests that Halley had observed the Moon at sea. The eclipse observations were not successful and Halley's longitudes depended mostly on measured distances from stars.[24] The eclipses of the satellites of Jupiter were his other celestial clock. They are easy to observe through a telescope on land and give quite precise instants of time. Halley found that it was much more difficult to observe them from a moving ship at sea. In those days, before the days of achromatic lenses, refracting telescopes had to be very long, often 20 ft (6 m) or more, to reduce spherical and chromatic aberration. Long telescopes were unwieldy on land; they were unmanageable on a ship at sea. Halley carried a long telescope for observations of the Moon and of the Jovian satellites and may have provided it with another device he had described to the Royal Society, a reflecting plate that was turned as the Earth rotated to cast the light from a celestial object into a fixed telescope lying on the ground.[25]

Halley already knew something of places such as Barbados and Bermuda, and perhaps some of the Carribean islands, where he had not himself been but which were familiar to Londoners. He would no doubt have had with him the most up-to-date charts of the Atlantic. He would have taken tables of the declination of the Sun for calculating latitude. He would have had star catalogues, tables of the position of the Moon, and tables of the motions of the satellites of Jupiter for finding longitude. Streete had published the most up-to-date tables in his *Astronomia Carolina*. London stationers sold most of Halley's needs.[26] He also had his own catalogue of southern stars.

The log of the first voyage lists latitudes and longitudes but does not say how they were found. The log of the second voyage is more informative and, if we assume that Halley followed the same practice on both voyages, then he found his latitudes from observations of the Sun at noon, and his longitudes at sea from a combination of dead reckoning and latitude, as described above, but also occasionally from lunar distances. Unless there is some particular note in the log, that is how the daily positions should be understood.

Halley had two azimuth compasses to observe the magnetic variation. He usually took the angle between the magnetic needle and the Sun at sunrise or sunset, or preferably both. The direction of the geographical meridian is that of the Sun at noon, but on a rolling ship at sea the azimuth of the Sun at that instant is difficult to determine. At sunrise and sunset the Sun is on the horizon and angles between it and the compass needle are less difficult to observe. The azimuths of the Sun at sunrise and sunset could be had from tables or direct observation. Flamsteed recorded a comment by Harrison that on 18 December, when they were at Madeira,

> I could not well observe in ye Road by reason of ye Rowling and quick motion no doubt of ship and complained to me in defense that by reason of ye smallness of ye ship it rowld and moved so fast y^t he could not observe so exactly as they use aboard ye bigger and steadier East India Merchant men.[27]

Halley had a thermometer and a barometer on his second voyage and made regular daily observations with them. He himself had contributed to the development of both (Chapter 7); Hooke had invented the marine barometer.[28] Instrument makers and chart publishers in London sold thermometers and barometers. The readings indicate that the barometer was graduated in inches. The scale of the thermometer is less obvious. Zero may have been at the freezing point of water and the readings in the tropics suggest that one degree was about 0.3°C.

10.3 The Atlantic

Halley's log has a day-by-day record of his last days in English waters in 1698, and his passage to the Cape Verde Islands. *Paramore* anchored

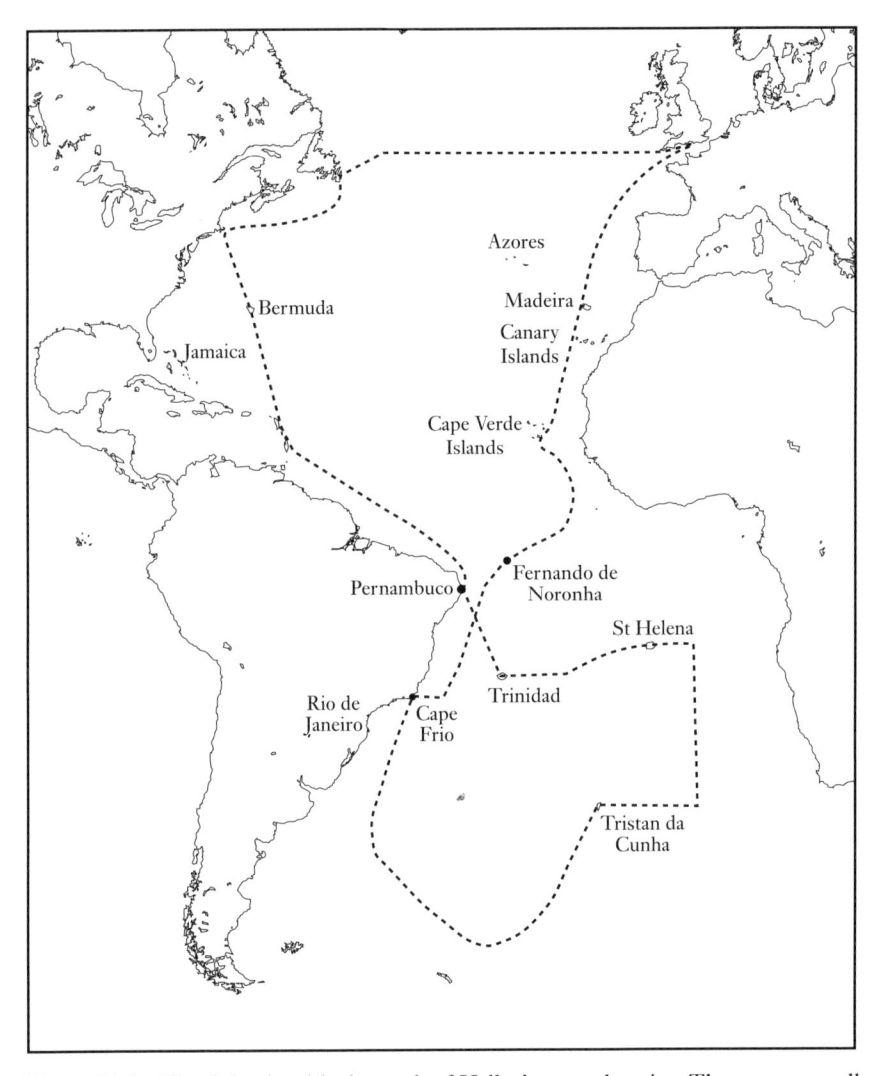

Figure 10.2 The Atlantic with the track of Halley's second cruise. The names are all modern, with the exception of Pernambuco (Recife).

at Gravesend while Halley returned to London to settle his accounts at the Victualling Office. Adverse winds then delayed his leaving the River until 25 October. By the time he made Margate, he had found defects in the pink. They were off the Isle of Wight in a northerly gale on 30 October and saw it covered with snow—winters were more severe then than now. Halley had surveyed the area and knew where he was, but he comments on his crew 'Some that ought to have known it better, tooke it for Portland'.[29] He found *Paramore* very leaky in a gale and the sand ballast was unsatisfactory. As the ship rolled, the pumps brought up sand with water, which damaged them. Accordingly, the next day

Halley went into Portland Road. He observed the magnetic variation, and wrote to the Admiralty asking that the ship should be examined, either at Portsmouth or at Plymouth as the wind would let him sail. With a SW gale he went to the Isle of Wight but could not get into Portsmouth because of the strong tide. Gales prevented their going in on 4 and 5 November, but on the 4th they fired five guns and hung out colours for the birthday of William III. They did the same on the 5th for 'Powder Treason' (Guy Fawkes's Day), and saw a French merchantman go ashore in the gale.[30]

At last able to enter Portsmouth harbour on 6 November, they found orders to examine the ship and change the ballast to shingle. It was done by the 16th. Halley took the opportunity to measure the variation at Portsmouth. With the wind strong from the W and SW they could not get out of harbour until 22 November, when they joined a squadron under Admiral Benbow at St Helens, the most easterly point of the Isle of Wight, 'Saluting him with 5 pieces and he returned me as many'. That was a great compliment to Halley, who was not a regular captain and was in command of a small ship; it is another indication that Halley and his accomplishments were well known in the Navy. 'I sent my Lieutenant with my respects to him and to entreat him to take care of us'. Halley was concerned to have protection when sailing in seas infested with pirates, especially the Sallymen from the island of Sall in the Cape Verde group. The following day his small boat was carried away in a storm and he had to get another from Portsmouth. At last on 29 November the wind changed to the NE and Benbow signalled for sailing. They were off Start Point on the 30th and Halley reckoned his departure from latitude 49°50′ N, longitude 04°00′ W of London.[31]

Halley was still Clerk to the Royal Society, but before he finally left English waters he must have arranged for a deputy, for on 8 February 1699 the Council was told that he left Dr Arbuthnott in his place. The Council already knew that Halley was going to be away, but they did not accept his choice of deputy. They appointed Israel Jones, but he did not last long, for he also went exploring, to the Barbary coast of Africa. Arbuthnott later became very influential as a Fellow of the Society (Chapter 12).[32]

The difficulties Halley had encountered in five weeks from Deptford, from adverse winds and tides and from the deficiencies revealed on the first voyage in a new ship, were far from exceptional in those days, and sometimes had serious consequences for commercial and naval plans.

Throughout the first two weeks of December, in company with Admiral Benbow's squadron, Halley sailed steadily southwards to a landfall at Madeira, having made three observations of the variation of the compass en route. Benbow continued to the West Indies where he was to show the flag and chase pirates, but Halley remained to get wine. He was delayed by surf that prevented his getting his wines off but was able to sail on 21 December for the island of Sall. He reached there on 31 December, having observed the variation once on the way. Surf prevented his going ashore on Sall, so they continued through the archipelago to the bay of Praya on the Island of May, where they were able to take on water and wood 'at a very extravagant price'. On passage from Sall on 31 December they ran into a streak of water that was quite turbid, yellow and muddy, due to great numbers of yellow globules which some took to be the spawn of fish, while Halley thought them small squid. Others have conjectured they were tiny jellyfish.[33]

On 6 January 1699 Halley stood out to sea, intending to make for the island of Trinidada off the coast of Brazil, but the winds stayed very light so that sometimes they made less than 20 miles in the day. Halley obtained six measurements of the variation, ranging from about 1° W to 1° 50′ E. Their slow progress meant that water was running short and Halley restricted the crew to three pints each per day. One day they met a great swarm of mosquitoes. Early in February, Halley decided to make for the nearer island of Fernando Loronho, much farther north, to look for water, and on the 17th he was able to get an astronomical determination of longitude from an occultation which gave him a position to the east of the island. Previously he had found, presumably from measurements of latitude, that they were farther north than dead reckoning indicated—they were already in the North Equatorial Current.[34]

Halley now had trouble with his boatswain. He had written to Burchett before leaving the Channel of his concern at the weakness of his officers.[35] As they approached Fernando Loronho he found the boatswain setting a different course from that he had ordered, such that they would have missed the island and, so Halley thought, would have stopped him achieving his aims. However, they did anchor in the lee of the island the following day (18 February). It offered little in the way of provision and no fresh water, but they were able to get wood, and scrub the ship, and attend to the shrouds and masts. Halley includes a profile and plan of the island in his log and he made a note of the phases and amplitudes of the tidal currents. With no water to be had, they set sail for the coast of Brazil and made their landfall on 26 February. That same afternoon they entered the river Paraiba (João Pessoa) and obtained water.[36]

They remained at Paraiba until 12 March, having taken a long time to replenish their water because of leaky casks. The delay allowed Halley to obtain an astronomical longitude. He intended to use the eclipse of the first satellite of Jupiter, but the planet was too high in the

Plate I Sir Joseph Williamson (Sir Godfrey Kneller). (Reproduced by permission of the President and Council of the Royal Society.)

Plate II Henry Oldenburg (Jan van Cleef). (Reproduced by permission of the President and Council of the Royal Society.)

Plate III John Wallis (Derard Soest). (Reproduced by permission of the President and Council of the Royal Society.)

Plate IV John Flamsteed. (Reproduced by permission of the President and Council of the Royal Society.)

Plate V J.-D Cassini. (Reproduced by permission of the Director of the Observatoire de Paris.)

Plate VI Sir Isaac Newton (Charles Jervas). (Reproduced by permission of the President and Council of the Royal Society.)

Plate VII Sir Hans Sloane (Sir Godfrey Kneller). (Reproduced by permission of the President and Council of the Royal Society.)

Plate VIII Halley in the uniform of a captain RN (Sir Godfrey Kneller). (Reproduced by permission of the Director of the National Maritime Museum.)

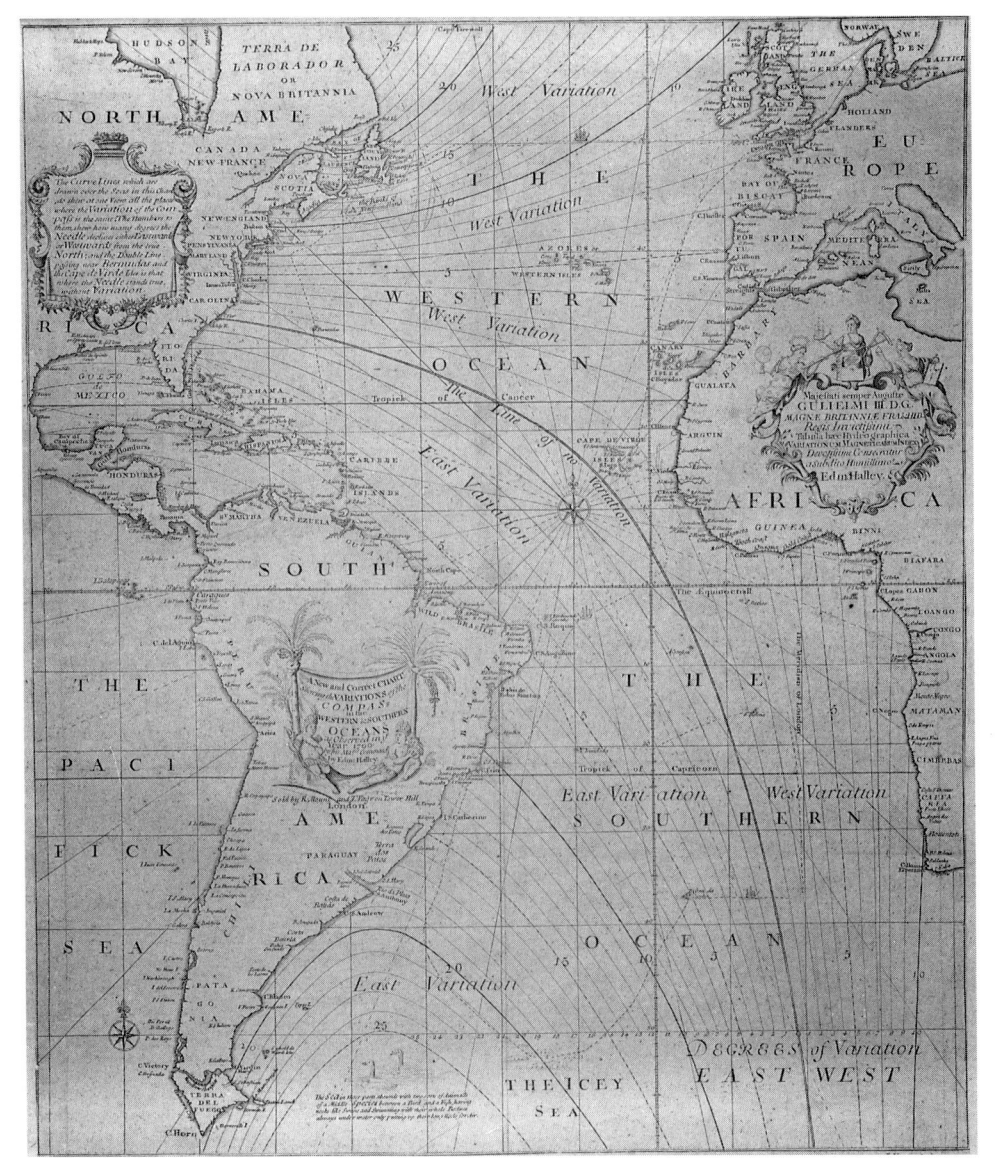

Plate IX Halley's chart of the magnetic variation over the Atlantic. (Reproduced by permission of the President and Council of the Royal Astronomical Society.)

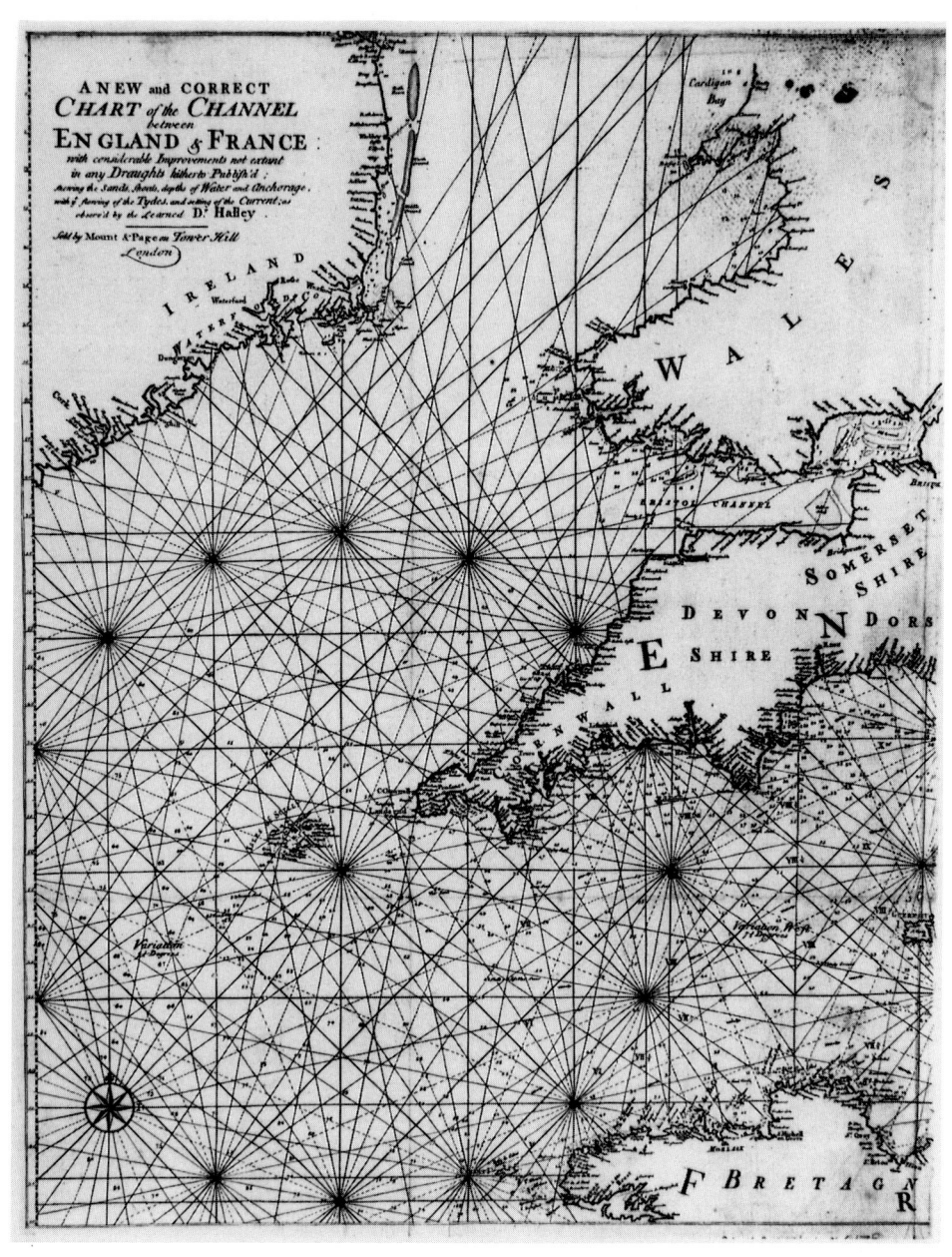

Plate X Halley's tidal chart of the English Channel. (Reproduced by permission of the President and Council of the Royal Geographical Society.)

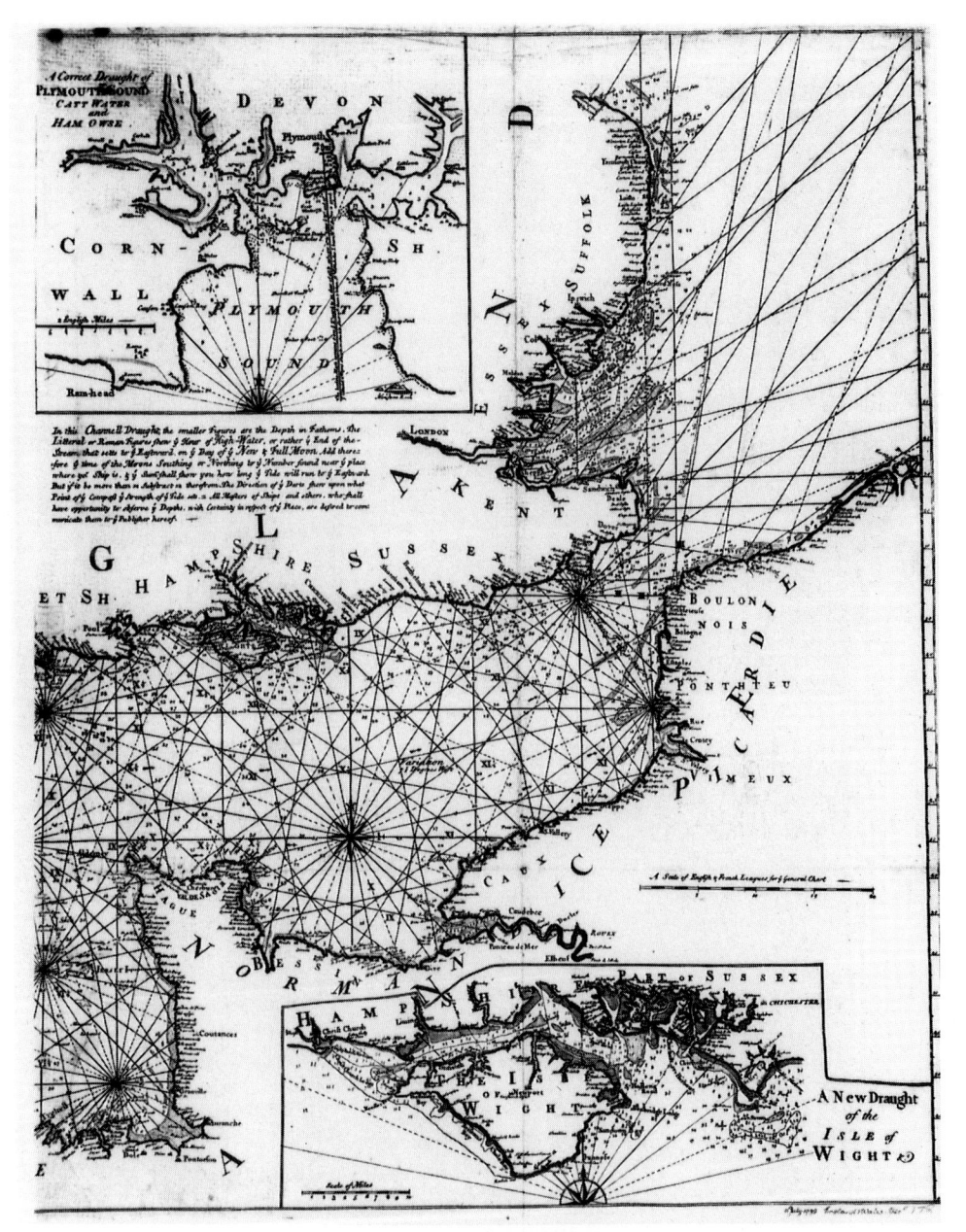

Plate XI John Arbuthnott (Robinson). (Reproduced by permission of the Director of the Scottish National Portrait Galley.) **Plate XII** George Stepney (Sir Godfrey Kneller). (Reproduced by permission of the Director of the National Portrait Gallery.)

Plate XIII G. W. Leibniz (artist unknown). (Reproduced by permission of the President and Council of the Royal Society.) **Plate XIV** Halley in a fresco commemorating the transit of Venus of 1882 at the Observatoire de Paris (E.-L. Dupain). Halley is in the lower left-hand medal of the pair at the bottom right of the picture. (Reproduced by permission of the Director of the Observatoire de Paris.)

Plate XV Halley at Oxford, before 1713 (Thomas Murray). (Reproduced by permission of Bodley's librarian.)

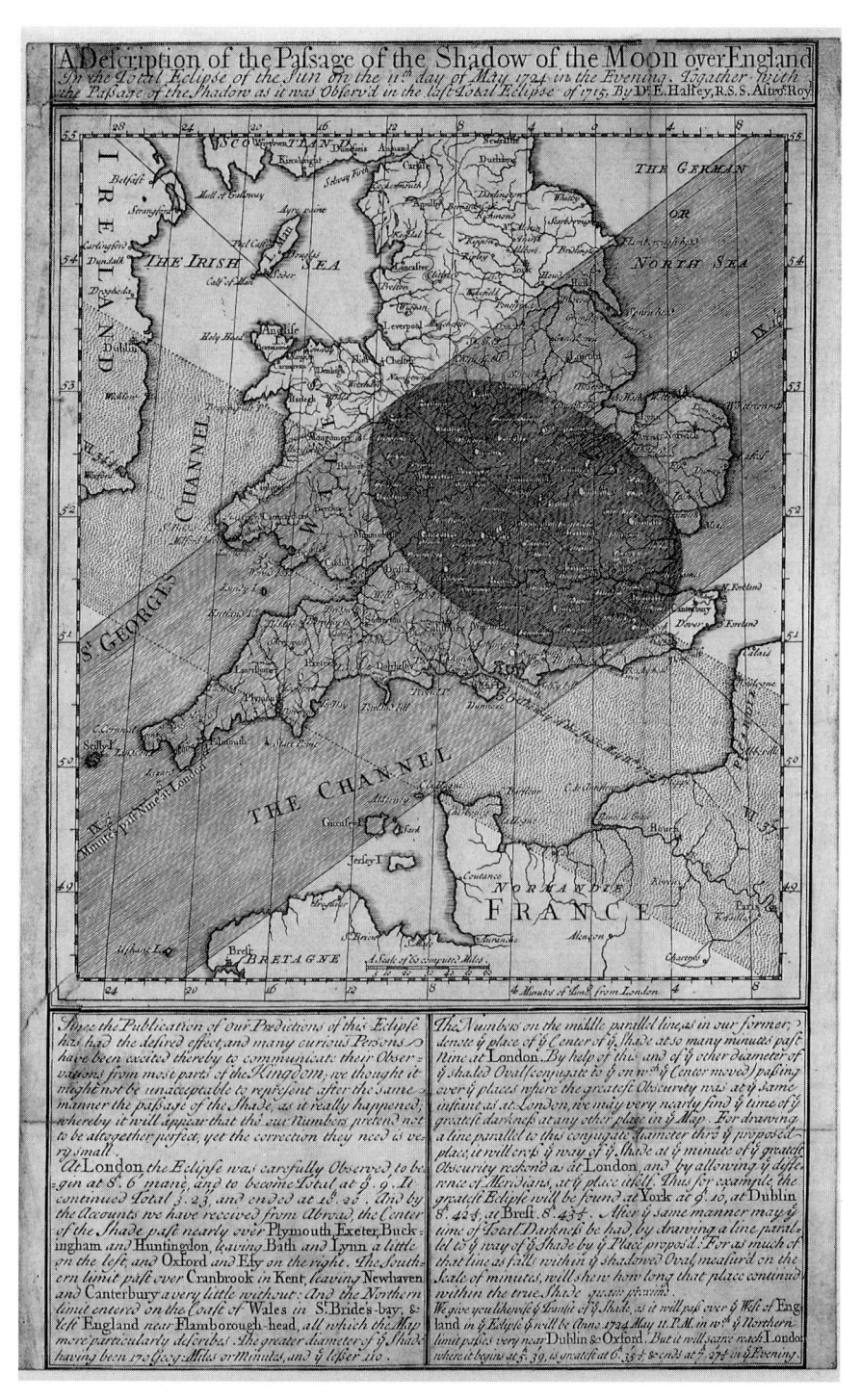

Plate XVI The solar eclipse of 1715. (Reproduced by permission of the President and Council of the Royal Astronomical Society.)

Plate XVII Halley in 1722 (Richard Phillips). (Reproduced by permission of the Director of the National Portrait Gallery.)

Plate XVIII Halley aged 80 (Michael Dahl). (Reproduced by permission of the President and Council of the Royal Society.)

sky for him to observe it without a better support than that which he had for his long telescope. He did observe the end of an eclipse of the Moon on 5 March, while previously, as they were approaching the coast, he had obtained the right ascension of the Moon from a stellar distance. He deduced from those two observations that his longitude was about 36° W of London (actually 35° W), whereas the longitude he derived from latitude and dead reckoning was about 26° W. As previously mentioned, the discrepancy could have been due to the North Equatorial Current. On the other hand, he had observed a lunar position at sea on 17 February when close to Fernando Loronho, which is not far from the coast of Brazil, and so determined his position as being close to that island. On the face of it, Halley made an error of about 100 miles in 220, according to both his own astronomical determinations of longitude and the charts he carried with him, too much to ascribe to currents between Fernando Loronho and the mainland. It is also too much for an error in his log line, as he later supposed, and the cause is obscure. If Halley, who knew what he was about, could be so wrong, what might other less expert navigators not do? Such errors of position raise questions about the magnetic chart that Halley produced, as discussed in the next section.[37]

Halley observed the variation at Paraiba and wished to survey the estuary but was forbidden by the Portuguese authorities. *Paramore* left Paraiba on 12 March and sailed southward for five days, but covered only 3° of latitude in the face of northerly currents. Halley now decided to abandon the South Atlantic. Winter was approaching in the south, his ship was foul and not sailing well, and he was having difficulties with his officers who were evidently unhappy at the prospect of going further south. They could have wintered in Brazil but because of the attitude of his officers, 'uneasy and refractory', he decided to go to Barbados to exchange them if he found a flag officer there. On the morning of 16 March, Halley turned to the north and then north westerly. After sixteen days' uneventful passage along the north-east coast of South America, during which he took six observations of the magnetic variation, they sighted Barbados. Here he found that his lieutenant was steering a different course from that he had ordered to get into Barbados:

> ... he persisted in this Course, which was contrary to my orders given overnight, and to all sense and reason, till I came upon Deck; when he was so farr from excusing it, that he pretended to justifie it; not without reflecting language.[38]

Halley attempted to observe eclipses of Jupiter's satellites on Barbados. He was defeated by clouds on one occasion, and troubled by winds shaking his telescope on another, but he did make an observation. Using Cassini's tables of the satellites, he derived a longitude of 59° 50′ W, very close to the true value, and about 3° more than he obtained from dead reckoning. That again showed the effect of the North Equatorial Current along the north coast of South America. He also observed the magnetic variation and noted the phases of the tides, as he had done previously at a few other places. They went on to Antigua where they made some repairs and got wood, and to St Christopher's where they took in water. Halley found he would have to go back to London to exchange his officers, as he was determined to do after the difficulties he had had with them. On 9 May between Dog Island and Turpentine Tree Island, he took his departure in lat 18° 20′ N, long 62° 50′ W from London.[39]

Nothing of note seems to have occurred during May. The weather was variable, Halley made nine measurements of the variation and one lunar observation, presumably of longitude, on 23 May. They missed Bermuda. June likewise appears unexceptional in the log, good distances sailed each day, one lunar observation giving a longitude 25 leagues more westerly than dead reckoning and eight observations of the variation. On 20 June they made the Scilly Islands and on the 23rd got into Plymouth Sound. They were in the Downs on the 28th. Halley waited on Sir Cloudesley Shovell there for orders and, having none, obtained his leave to go to London.

Halley went to London to follow up a letter he had sent to Burchett from Plymouth to explain why he had returned early from his cruise. He was too late to go south that season, but also something happened on 5 June that is not recorded in the log:

> But a further motive to hasten my return was the unreasonable carriage of my Mate and Lieutenant, who, perhaps because I have not the whole Sea Directory so perfect as he, has for a long time made it his business to represent me, to the whole Shipps company, as a person wholy unqualified for the command their Lopps have given me, and declaring he was sent on board here because their Lopps knew my insufficiency. Your Honour knows that my dislike of my Warrant Officers made me Petition their Lopps that my Mate might have the Commission of Lieutentant, thereby the better to keep them in obedience, but with a quite contrary effect it has only served to animate him to attempt upon my Authority, and in order thereto to side with the said officers against me. On the fifth of this month he was pleased so grosly to affront me as to tell me before my Officers and Seamen on Deck, and afterwards owned it under his hand, that I was not only uncapable to take charge of the Pink, but even of a Longboat; upon which I desired him to keep his Cabbin for that night, and for the future I would take the charge of the Shipp my self, to shew him his mistake; and accordingly I have watched in his steed ever since, and brought the Shipp well home from near the banks of

Newfound Land, without the least assistance from him. The many abuses of this nature I have received from him, has very sensibly toucht me, and made my voyage very displeasing and uneasy to me, nor can I imagine the cause of it, having endeavoured all I could to oblige him, but in vain. I take it that he envys me my command and conveniencies on board, disdaining to be under one that has not served in the fleet as long as himself, but however it be I am sure their Lopps will think this intollerable usage, from one who ought to be as my right hand, and by his example my Warrant Officers have not used me much Better; so that if I may hope to proceed again I must entreat their Lopps to give me others in their room.[40]

Harrison's opinion of Halley, as reported by Halley, agrees very well with the attitudes Harrison showed in his book.

Not only is there no mention of those events in Halley's log, Harrison apparently did not record them in his journal, or at least, Flamsteed did not copy them. Flamsteed's copy does, however, show that Harrison continued to observe; he recorded latitudes and variation on 9 June, and latitude, longitude, and variation on 15 June.

The upshot was that a court martial convened on 3 July in *Swiftsure* in the Downs, with Sir Cloudesley Shovell as president and three other admirals and eight captains as members, to enquire into the behaviour of Lt Harrison. The Admiralty orders were to try Harrison, but the Court enquired into the behaviour of the warrant officers as well. They concluded that there had been no actual disobedience to Halley's commands,

> ... tho there may have been some grumbling among them as there is generally in Small Vessels under such Circumstances & and therefore ye Court does Accquitt ye Sd Lt Harrison & the other Officers of his Majties Pink ye Paramour of this Matter givng them a Severe reprimand for ye Same.

Narcissus Luttrell mentioned the case.[41]

Shovell was generally considered the most able admiral of his day; he clearly understood the pressures on men in small ships. Harrison's behaviour was far from unusual in the Navy of the time; flagrant disregard of orders, high and insulting words, even, or especially, among officers of the superior ranks, were among the undisciplined manners that Pepys tried to control. The events on *Paramore* were small beer by comparison with those at almost the same time on *Roebuck*, in which William Dampier explored New Holland (Australia). Dampier's lieutenant, Fisher, similarly regarded him as an incompetent amateur, and Dampier clapped him in irons until they got to Bahia in Brazil, where he was transferred to a Portuguese prison and left there for four months. At the court martial after they were all back in England, of which again Sir Cloudesley Shovell was a member, Dampier was fined the whole of his wages for three years.⁴² It is said that the Admiralty never again put a civilian in command of a royal ship after their experiences with Halley, but Dampier's outrageous behaviour may have swayed them more: at the time they continued to have confidence in Halley.

Halley was not happy with what he evidently considered unduly mild treatment of his officers, and only now realised that Harrison was the author of the book that he had reviewed severely four years earlier. London gossip, reported by Narcissus Luttrell, was that Halley's crew had wanted to turn pirate. The Board of Admiralty must have been satisfied with Halley, for they soon acceded to his request for a second commission, as Luttrell again related. On 21 July they instructed the Navy Board to have the pink surveyed and either propose improvements to her or suggest another ship that Halley might use. The Board made their report on the 27th, detailing some work to be done on *Paramore*, and so it was ordered.⁴³

While the work was being done, Halley gave some account of his first cruise to the Royal Society, with a chart of the variation and improved positions of places in Brazil.⁴⁴ He received his new commission as Master and Commander on 23 August 1699 and his instructions on 12 September. Having dealt with some manning problems, he left Deptford on the 16th. He had, as before, suggested the content of his instructions, which differed little from his first set but were slightly more specific about the discovery of lands between the Straits of Magellan and the Cape of Good Hope and between 50° and 55° S.⁴⁵

Halley was in the Downs on the 27th, and wrote briefly to Burchett. He passed Portland on the 29th and from then on took his latitude whenever the Sun was visible at noon. The log of the second voyage is much more detailed than the first, in particular it has daily thermometer and barometer readings and notes good determinations of latitude.

They had the company of the *Falconbird*, a ship of the Royal African Company named after a prominent member of the Company, as far as Madeira. They did not call there, despite having to forego wine, because the surf was too high, and then on 14 October, the boy Manly White fell overboard and was drowned, for although the ship was put about and an oar thrown overboard, they could not see him in heavy seas. He was the only member of any of Halley's crews that he ever lost, and it seems that it affected him deeply. Two days later they were in the Canaries and a week after that at St Jago in the Cape Verde archipelago. Halley observed the latitude on most days and measured the variation nine times. He wrote to Burchett from St Jago to report his good progress and to say that his ship's company was well and that he was pleased with his new officers. They had had mixed weather, but generally strong and favourable winds from the north, and had made good time. He met an English merchantman, took on water and provisions, and prepared for the next stage of his voyage.[46]

They sailed first to the SE and then to the SW, and crossed the equator in longitude 23° W of London on 16 November 1699, steering SW for Fernando Loronho, but Halley was beginning to fear he might miss the island on account of the currents that were bearing them northwards. They began to see sea-birds, a Man-o-War or frigate bird and a tropic bird among others, showing that they were getting close to the coast of Brazil. Halley observed the latitude every day and the variation on most days, and on 5 December he observed a lunar position and so obtained his longitude. On the 9th they 'all smelt a very Fragrant Smell of Flowers which the Wind brot of the Land; and Severall Butterflies flew on board'. Halley decided they were south of their intended landfall of Cape Frio as a consequence of southerly currents, but in fact they made their landfall at Cape St Thomas which is to the NE of Cape Frio. They passed the island of Cape Frio on the 13th, saw the Sugar Loaf at the entrance to Rio Janeiro harbour and on the 14th went in.[47]

There is a gap in the log until 29 December when they sailed from Rio. Halley apparently wrote to Burchett from Rio, but his letter has not appeared and possibly never reached London.[48] They set their course from Rio generally south by west, with fair weather and fresh gales and the occasional storm until they were in latitude 40° S, the Roaring Forties, when Halley decided that if they were to go further south before the season was too advanced, they should now go to the SE.

They had a severe thunderstorm the following day, 18 January 1700, and the temperature fell. They were leaving the warm southerly current that had carried them down the coast of South America and were entering the cold currents around Antarctica. The wind grew stronger and the temperature was dropping. The colour of the sea changed to green and then to white, they met drifts of seaweed and saw many sea-birds, all apparently signs of land. Between 24 and 25 January, as they were in latitude 49° S or thereabouts, with strong winds, it became much colder, only slightly above freezing in Halley's cabin. Haze veiled the Sun. On the 27th, in latitude 50° 45′ S, fog enveloped them. They kept seeing penguins and Halley thought they must be near land. Later they saw what were probably bottle-nosed whales. The fog thinned on the 31st to give seven hours of sunshine, but despite that the temperature remained below freezing. Halley got good observations of latitude and variation. He began to think of going north, for he was concerned that if the wind should change to the east, it would be difficult to make the Cape of Good Hope.[49]

On the last day of January 1700, in the afternoon, they were in latitude 52° 20′ S and longitude 35° 13′ W of London. Three great islands rose ahead, too late in the day to explore them, flat on top, covered with snow and falling to the sea in tall cliffs. Next day the fog cleared briefly. They saw that the islands were all of ice. Halley thought they were about 200 ft (61 m) high. They could not sound the bottom in 140 fm (256 m). Halley knew that only one eighth of the ice was above water and thought it must be grounded—in fact the water is much deeper and the ice would have been floating. They had sailed into the region where the pack ice comes furthest north around the Antarctic continent, driven up by a cold current. The Straits of Magellan and the Falkland Islands to the west in about the same latitude are generally free of pack ice. Halley was roughly due north of the bay named after him where a research station was set up for the International Geophysical Year. He turned back when close to South Georgia Island, which was discovered and named by Cook in his second voyage. Cook was about a fortnight earlier than Halley in those parts and he was able to go about $1\frac{1}{2}°$ further south, but his account of the conditions and of the birds he saw is very like that of Halley.[50]

For three days they were in great danger. They could see little more than a furlong in fog. They moved slowly through the water, a mass of ice loomed up ahead, they tacked to avoid it, another with loose ice floated nearby. A third iceberg appeared but with a wind and a clear sea they could avoid it. Halley and his people were very conscious of their danger in a lone ship. On 5 February in latitude 50° S they saw the last of the ice and fog but it remained very cold. Halley says nothing about his rigging but ice accumulated on Cook's masts and rigging in the same southern seas, which made the sheets difficult to handle and the ship dangerously unstable. Cook had snow, Halley did not, but the persistent fog soaked their linen, their clothes and papers, even in the cabins.[51]

Clear of the ice, Halley set a course generally to the north and east for the Cape of Good Hope. A week's sailing took them to Tristan da Cunha and as they came close they saw many albatrosses. They did not go ashore but set course for the Cape. In the cold Benguela current the weather improved, the fogs cleared away, but the temperature fell and the seas were high. The sky was often overcast, preventing noonday sightings of the Sun. There was a furious storm on 26 February. About six in the morning, a great sea broke over the starboard quarter. The pink nearly capsized. Fortunately the water ran clear through the gunwales and 'it pleased God she wrighted again'. Water got into the bread room and flour, bread, and cheese were ruined and had to be cast overboard. Southerly gales for another day drove them north of the

Cape of Good Hope. Water was running low. To save time Halley went straight to St Helena. In the first days of March they had strong favourable winds and Halley could observe the noon Sun every day and the variation most days. They saw white birds on 10 March, possibly terns with which Halley would have been familiar from the year he spent on St Helena. They saw the island on the 11th and on the 12th anchored in Chapell (James) Valley. Halley wrote to Burchett from St Helena and he left on 30 March, having got provisions and some rather unsatisfactory water.[52]

There is little in the record for the next fourteen days other than courses set, winds, and daily observations of latitude on every day and variation on most days. On 14 April they sighted rocks off Martin Vaz and on the 15th came to the island of Trinidada. They found good water and replaced that from St Helena, and put some goats, pigs, and guinea fowl ashore to breed. Halley took possession of the island in the King's name and rowed all round it and surveyed it.[53] They left on 20 April and for the next eight days sailed generally westward, making daily observations of latitude and six of the variation. On the 28th they made their landfall on the coast of Brazil and the following day entered Pernambuco (Recife).

The Portuguese governor allowed them to purchase wine and other commodities and assured them that there was still peace in Europe—the War of Spanish Succession, generally expected, had not yet broken out. Then they had difficulties. On 30 April a Mr Hardwick, who presented himself as the English consul, suspected them of being a pirate, despite Halley's commission and instructions, but after examining two of the seamen he seemed satisfied and allowed them to load their purchases. The next day he arrested Halley while he searched the ship. He found no signs of piracy and apologised. The following day the pilot would not take them out without the governor's order, and when Halley went to obtain it he learnt that Hardwick was no consul but the agent of the Royal African Company and of the owners of a ship, *Hannibal*, that had been taken by a pirate.[54]

They sailed from Pernambuco on 4 May and made Barbados on the 21st; for the most part they had favourable winds and good weather, though with some storms. Halley observed the latitude on most days and the variation on nine, being prevented on other days by cloud or haze at sunrise and sunset. He noted that the currents along the north coast of South America had taken them north and west of their supposed positions. The governor of Barbados, Ralph Grey, advised them to leave as soon as possible because sickness was rife on the island. They left next day after getting water, but too late, for when they sailed, Halley fell ill. They put into St Christopher's on 28 May for him to recover and took the opportunity to take in water and overhaul rigging. They sailed on 5 June to Anguilla and stayed there until the 10th, laying in wood and provisions. They then left for Bermuda. Nothing remarkable occurred, said Halley, and so just listed his daily course, position, and variation in a table. They stayed in harbour at Bermuda from 20 June until 11 July, overhauling the ship to prepare for the passage to England. Halley made astronomical observations for the longitude. He wrote to Burchett with a cheerful account of their progress and state, and of the affair of Hardwick at Pernambuco. The mate had an offer of better employment that he wished to accept and because there was a Bermuda master looking for a passage to England, Halley took him on instead. They sailed for Newfoundland and England on 11 July. The passage to Newfoundland was uneventful for the most part, with almost daily observations of latitude and variation, but Halley stood off from Nantucket on account of a gale and shallowing water.[55]

They made their landfall on the French side of Newfoundland by Cape Anglois (English) on 31 July in fog, and were fortunate to fall in with some French ships that warned them of the risk of being driven ashore. Further along the coast of Newfoundland, they met a fleet of English fishing boats on 2 August, who fled, believing Halley to be a pirate. When they entered the harbour of Toad's Cove, a Bideford fisherman, Humphrey Bryant, fired four or five shots through their rigging, but damaged nothing. Halley, ever conscious of his commission in the Royal Navy, anchored and sent his boat to bring Bryant on board, but when he heard of the pirate, let the matter rest.[56]

They took on water and provisions and sailed for England on 7 August. They had fog for many days and not until the 17th could Halley get an observation of latitude, but thereafter he observed the latitude and variation almost every day until they made the Scillies on 25 August and entered Plymouth Sound on the 27th. There were no orders for him, so he continued up Channel to anchor off Dungeness, where he found orders to proceed to Long Reach to deliver his guns and to Deptford to be laid up. That they did on 8 and 9 September, and brought to a close the most eventful two years of Halley's life and a remarkable episode in scientific oceanography.[57]

Halley was very gratified that he had brought back healthy crews from both voyages, and his achievement has been compared with that of Cook; they both sailed in days when knowledge of deficiency diseases was almost non-existent. There is a substantial difference between Halley's voyages and Cook's. Halley, sailing for the most part close to land and often calling at established ports, never spent very long at sea, usually about two weeks or less. Whenever he called at a port or otherwise made land, he took the opportunity to take on fresh provisions, to overhaul and clean the ship, and to allow the crew on shore. Their longest time at sea by far was the six weeks in the south from Rio to Tristan da Cunha, but they had spent two weeks in Rio beforehand. Somewhat less than another month took them to St Helena for a stay of three weeks. The next longest time at sea on the second voyage was the three weeks from Toad's Bay to Plymouth. The sea times on the first voyage were all relatively short, except for the final part to England which took about six weeks. *Paramore* carried flour and they baked their own bread or bought it on shore. They had large quantites of cheese. They had beer from England and obtained wine when possible. They bought or hunted birds and other game and they had fishing gear.

10.4 *The magnetic chart*

The Royal Society had had news of the second cruise on 10 July 1700 when they had read a letter from the surgeon, George Alfrey, with some account of where they had been and especially the islands of ice: Alfrey said they had been in 52° S, 34° W of London. Halley himself produced a chart of the variation and spoke of the islands of ice at the meeting on 30 October 1700. From time to time he mentioned sundry observations of plants, birds, fishes, and animals at other meetings of the Society.[58]

Halley published his observations of the variation as a chart of the north and south Atlantic with isogonic lines, that is, lines joining points of equal variation. He presented his results very quickly and showed his chart to the Royal Society less than two months after he got back to Deptford. He must have reduced most of his data while at sea. Later, on 4 June 1701, he presented a copy of his published chart to the Royal Society, who ordered that it should hang on their walls. It has disappeared.[59]

The chart is a handsome example of the skills of the engraver and printer (see Plate IX). It is drawn on the Mercator projection, and has the conventional compass roses with radiating lines as on a Portolan chart. The map of the coasts of the Atlantic takes account of Halley's observations of position and is an improvement on anything that had gone before. Halley stated in his explanation on the chart that it showed the results of his cruise of 1700, and indeed it plots the track of the second voyage but not that of the first. However, there are two sets of observations from the first cruise, from Madeira over to Brazil and from Newfoundland to England, which are not duplicated in the second cruise and which must have been used in drawing the isogones in the north Atlantic.

How reliable were Halley's results? The uncertainty of the measured variation would have depended on whether azimuths were observed at both sunrise and sunset, for on many days only one was possible. The general uncertainty seems to be within 10′ and perhaps less. The accuracy of positions is more in question, for there were sometimes substantial errors in longitudes at sea. Some, perhaps most, were found by dead reckoning from differences in latitude, but the discussion at the Royal Society on 16 August 1699, when Newton suggested an improvement to an instrument of Halley's, implied that a few at least were found from lunar distances. If the errors were for the most part due to unknown currents, they would give systematic distortions of scale in the E–W direction.[60] The latitudes, on the other hand, seem to be free of systematic errors. Most observations were made along north–south coasts. Since lines of equal variation in the north Atlantic run predominantly east–west, the general disposition of the lines of equal variation in the north Atlantic is unlikely to be seriously wrong, even though Halley's longitudes at sea may be unreliable. In the south, on the other hand, the isogones run more nearly north and south, and consequently are probably less reliable. The point is important, for Halley's were the first organised geomagnetic measurements over a large area, and as such continue to have a considerable influence on models of changes of the magnetic field in the long term and, in particular, of its westerly drift.[61]

The chart was in some respects as remarkable as the voyages. No one had ever published such a chart before; afterwards it became the standard way of showing the results of almost all geographical observations. Halley's isogonic lines are often said to be the first instance of a representation of observations by isolines or isarithms,* but according to Athanasius Kircher, Christofor Borri, a Jesuit, had represented magnetic observations in a limited area with parallel lines in 1630. A Peter Bruinss had drawn a chart with lines joining equal depths of water in 1584, and another was made by Pierre Ancelin in 1697. None of those charts was printed or published. Kircher had died the year before Halley was in Rome, so they never met, and although there were four works of Kircher in the sale of Halley's books (Appendix 8), his *Magnes* was not one of them. Halley did, however, know of Kircher's work on magnetism, which was in some ways very original, and so might have have heard of Borri's representation.[62] Halley's was, none the less, the

---

*The term *isogone*, for a line of equal variation, was coined by the Norwegian astronomer, Christopher Hansteen (1784–1873), in about 1820.

first chart to be printed and published. Thrower thinks that Halley was unaware of the earlier charts and considers that his claim stands:

> What is here properly New is the Curve-Lines drawn over several Seas, to show the Degrees of the Variation of the Magneticall Needle or Sea Compass: which are design'd according to what I my self found in the Western and Southern Oceans, in a Voyage I purposely made at the Publick Charge in the Year of our Lord 1700.[63]

Halley's chart was certainly the first to be printed and published with an isarithmic representation. For that reason Thrower sees it as one of the most important charts in the history of cartography. That seems to have been recognised by Halley's contemporaries, who called isolines *Halleyan* or *Halleian* lines.

The chart is a great advance in the way we represent the natural world, and hence of our understanding of it. Later, in 1702, Halley published a chart of the whole world with isogones over other seas for which he had some information from journals of recent voyages of ships in the 'India-Seas', especially the Indian ocean, the East Indies, and the west coast of Australia. There is nothing over the Pacific for he noted on the chart, 'I durst not presume to describe the like Curves in the South Sea wanting accounts thereof'. It shows the Falkland Islands by name for the first time. He had prepared the world chart by 5 February 1701, when he showed it to the Royal Society. It was republished in several states, some with revisions and some abroad, until the end of the eighteenth century. It was widely used at sea.[64]

Halley was by no means the first nor the only seaman to observe the variation at sea. The pages in which Flamsteed records Harrison's observations are followed by others with extracts from journals of voyages to India and elsewhere going back fifty years and more. Halley certainly knew of those earlier records, but they were casual observations made from ships on regular trade routes, and not for the purpose of a systematic study of the magnetic field. He was unaware of many observations that had been made in the Pacific and Indian oceans by Dutch, Spanish, and Portuguese captains. The Spanish records are probably in the Archive of the Indies in Seville, but neither they nor the Portuguese records have ever been investigated. The Dutch material was available to W. van Bemmelen in the Observatory of the Dutch East India Company in Batavia, of which he eventually became Director. He worked on the records over many years, correcting positions and misidentifications of places where the declination was observed, and so was able to produce isogonic charts for every fiftieth year from AD 1500 onwards, including 1700. The main difference between Halley's world chart and van Bemmelen's for 1700 is that van Bemmelen was able to draw realistic isogonic lines over much of the Pacific. Chapman, in 1941, compared the two charts in common regions and found that they agreed to within 1° in the declination almost everywhere. While many discrepancies probably arise from differences in position, especially at sea, a comparison of Halley's values with those Harrison made at more or less the same time suggests that errors in the magnetic observations could reach 1°, although for the most part they were well within a degree.[65]

A chart of the variation helps the navigator to determine his true course from a magnetic course, but it might also be used to help to find position in favourable circumstances. If the isogones run regularly north and south and are closely spaced, as they are off the Cape of Good Hope, then, as Halley pointed out in the explanation of his charts, a measurement of the variation gives an estimate of the longitude. If the lines run east and west, as in the north of the Atlantic, or if they curve round, as off the southern coasts of south America, they are no help.

The chart, a great feat of observation and mapmaking, is underlain by a fundamental implicit assumption about the nature of the natural world. Halley's observations lie almost all on the margins of the two basins of the north and south Atlantic,[66] yet he drew isogones right across their interiors. He implied or accepted that it was reasonable to infer the unknown state of the interior from the known states on the margins; he relied on nature being rational and knowable, exactly as Boyle had said. Isarithms which show what there is between observations are otherwise nonsense.

10.5 Tides

Halley must have given up any idea of returning as Clerk to the Society after his second Atlantic cruise, for at the Anniversary Meeting on 30 November 1700 he was re-elected to the fellowship, and shortly afterwards he was elected one of the auditors of accounts, as he was again the next year.[67] He had observed some tides near land on his Atlantic voyages, he was well aware of the problem that tides presented for navigators in the Channel, and he had got involved in a dispute with Flamsteed about tidal predictions (Chapter 13). He must have been thinking about a tidal survey in the Channel during the winter of 1700–1701, if not earlier, and he evidently discussed his ideas with the Admiralty, for only three days after he submitted his formal proposal on 23 April 1701, the Admiralty gave orders for *Paramore* to be cleaned, fitted out, manned and victualled, and supplied with extra boats, cables, and anchors. Guns were ordered and Halley received his commission the same day. He had some problems getting men and officers but on 12 June he had his sailing instructions, which he had in effect written for himself on 11 June; his suggestions were:

> You are to use all possible diligence in observing the Course of the Tides in the Channell of England as well in the mid sea as on both Shores, and to inform your self of the precise times of High and Low Water; of the sett and strength of the Flood and Ebb and how many feet it flows in as many places as may suffice to describe the whole. And where there are irregular and half Tides to be more than ordinarily curious in observing them. You are likewise to take the true barings of the principall head lands of the English Coast one from another, and to continue the meridian as often as conveniently may from side to side of the Channell in order to lay down both coasts truly against one another.[68]

The final instruction, to improve the surveys of the coasts, is mentioned in a single sheet that appeared in 1701, *An advertisement necessary to be observed in the navigation up and down the Channel of England, communicated by a Fellow of the Royal Society*, and printed by the printers to the Society. It is usually supposed to be by Halley. An earlier version had appeared in the *Philosophical Transactions* of 1700. The main point of the paper is to warn of the dangers of ignoring the change of the variation of the compass, which had gone from easterly to westerly in 1657. It also mentions the risk of striking on the Caskett-Rocks and adds

> For by the late curious Survey of the Coast of France compared with what had previously been done for our own (though perhaps not so exactly) it appears that the Course from the land of Beachy or Dungyness to the Caskett-Rocks is but West 26 Degrees Southerly.[69]

Halley does not record a survey of the French coast in his log of the Channel cruise but the clear implication from that passage must be that he made one.

It has often been suspected that Halley was also asked to collect information about French waters that might be useful in the coming war, and it is possible that the addition that the Admiralty made to the instructions he proposed has given rise to that conjecture:

> And in case dureing your being employed on this Service, any other Matters may Occur unto you the observing and Publishing whereof may tend towards the Security of the Navigation of the Subjects of his Maj^{tie} or other Princes trading into the Channell you are to be very careful in takeing notice thereof.[70]

That, however, seems to envisage the results of Halley's surveys becoming public knowledge. On the other hand, there is a much earlier remark in Hooke's *Diary* for 24 March 1693, 'Halley alleged spys', which may refer to some espionage activity at the time he was working on *Guynie* off Pagham (Chapter 9), or it may refer back to his Thames survey.[71] Halley, as a naval captain, was of course under Admiralty orders.

Halley had given an account of Newton's explanation of the tides when he presented a copy of *Principia* to James II (Chapter 6). The amplitude of the equilibrium tide, as it is called, is about 600 mm and only observable with delicate modern instruments measuring pressure on the sea-bed. It is but a small disturbance of the surface of the deep ocean, but drives large currents in shallow waters. The volume of water flowing in and out of a shallow sea has to match that in the adjacent deep ocean and consequently the currents are much greater: at the western end of the English Channel the tidal heights and currents are far greater than in the middle of the Atlantic. That tide drives a wave up the Channel which, on the whole, increases in amplitude as the Channel gets narrower, so that the further east one goes the higher the tide and the later is high water, since it takes time for the tide to travel along the Channel. At the Straits of Dover the tidal wave coming from the west meets one coming down the North Sea and the two interfere. Another complication arises from tidal flows around islands, as in Spithead and Solent, which makes pilotage out of Southampton and Portsmouth rather tricky. There are circular flows around some of the Channel Islands so that a strong swimmer may go right round an island in phase with the tide. Although tidal streams in the Channel are complex, Halley supposed that the flow at any one point should vary in a regular way with the position of the Moon, and he related the times of high water to the direction of the Moon.

Tides are not so simply related to the Moon. The Moon is not stationary in the sky but is moving round the Earth and there is a diurnal as well as a semidiurnal tide. The solar tides, also diurnal and semidiurnal, and about one-fifth of the lunar, cannot be neglected. Components of the tides have different phases relative to the Moon at any given place and the net flow does not correlate uniformly with the position of the Moon. Halley expected that the flow would vanish at high water, as it would if there were just one single tidal component. He found it was not always so. Not only are there many components with different phases, but friction on the sea bed damps the flow and distorts the relation between high and low water and flow. When Halley surveyed the tides in the Channel, understanding of those matters was still very rudimentary and it is hardly surprising that he related everything to the position of the Moon, for there was nothing else he could do.

Halley left Deptford on 14 June and by the 20th was off the South Foreland. His log for that day gives a good idea of how he worked.

> ... a little before Sunrise I ankered under the pitch of ye South foreland, about a mile from shore, the middle between the two Light houses baring NNW, here I saw the turn of the Tide both ways, and observed that it ran to the Westward till a SE moon, and to the Eastward which is commonly calld the Flood till a SW or IIIh Moon, but that it was high Water by the Shore with a S$\frac{2}{3}$E ☽ or XI$\frac{1}{2}$, the set of the Flood between ENE and NEbE. The tide being done about 11h, I weighed with an easy gale at NEbN and at 5h I came to an anker off of Dungeness a small mile from the Beach the Light house baring NbE. Here I rode till the next day morning when I found it was high water by the Shore a little before eight, that is with a SbE moon; but that the Eastern tide held out to a SWbS or II$\frac{1}{2}$ Moon.[72]

Halley was working his crew very hard again, not with the hazards of sailing in unknown and dangerous seas, and frequent cleaning and overhauling of the ship, but with the physical labour of continually anchoring and weighing, and in addition, often putting out a boat and rowing round to follow a tide.

Next day he sounded over a bank called the Riprapps and observed the tides. On the 22nd he made over to the French coast and the following days sailed down to Boulogne and Dieppe, Cap d'Antifer, and the mouth of the Seine, anchoring, sounding, and observing the tidal flow. On 28 June they anchored in the Channel and the following day were back in English waters by Selsey Bill. They were off Spithead on 2 July and found Sir Cloudesley Shovell in *Triumph*. Halley spent the next week around the Isle of Wight, where the tides are complex, and returned to the French coast to anchor off Querqueville near Cherbourg. For the next ten days they worked around the Channel Islands in company with a local pilot and on 20 July went into Portland, where they were delayed for three days by a gale that prevented their leaving harbour. On the 23rd they went back to Alderney but returned to an anchor in Cowes Road to ride out a gale. They weighed to sail down the Channel on 2 August, anchored in Plymouth Sound on the 4th, where Halley measured the latitude and variation, and then sailed to the newly built Eddystone Lighthouse. Here he anchored and went ashore and observed the tide. He was fortunate to see the lighthouse: two years later the great storm of November 1703 destroyed it. Halley later made further measurements of latitude and variation at Plymouth. When they left on the 10th they ran into a gale that stopped them weathering the Lizard and they had to go into Fowey until the following day.[73]

The next two days they anchored in mid-Channel and observed the tides there, and then made over to Ushant. Halley had intended to take his instruments ashore to observe the variation at Ushant but the sea was too rough. They went back to the English side and ran into a thunderstorm; Halley went ashore at Pendennis Castle to observe the variation. Gales continued—they had to anchor, and saw a fleet under Sir George Rooke shelter from the winds in Torbay. On 22 August they were forced into Dartmouth where they stayed until the 28th, when they were able to sail for Sept Isles off the coast of Brittany. They cruised off Brittany and among the Channel Islands for the next few days and then, because the ship had become foul, they cleaned her at the pier of St Aubin on Jersey. They spent a further day in the Channel Islands before returning to Spithead where they took on provisions. Over the next two weeks they surveyed the various shoals and rocks around Spithead and the entrance to Portsmouth Harbour, and observed the tides. Finally on 28 September they stood up the Channel. They came into Gallions Reach on 8 October and delivered up their guns, they were at Deptford on the 10th when Halley delivered the pink into the custody of the Master of Attendance, and she was paid off at Broad Street on the 16th.[74]

Halley kept the Royal Society up to date with his progress. On 18 June 1701 the Vice-President, Sir John Hoskins, told the Society that Halley had gone 'to make nice observations on the Tides and Currents in the Channel for the Improvement of navigation', and on 30 July the Society received a letter from Halley at Guernsey with an account of fair weather and good progress. He showed his new chart of the Channel on 12 November.[75]

Halley published his tidal results in a chart:

> A New and Correct Chart of the Channel between England & France with considerable Improvements not extant in any Draughts hithero Publish'd; showing the Sands, shoals, depths of Water and Anchorage, with ye flowing of the Tydes, and setting of the Current; as observ'd by the Learned Dr Halley. [see Plate X]

The new features which give it its value are the Roman figures that show the hour of high water, or rather the end of the stream that sets to the eastward, on the day of the new and full Moon, together with 'darts' that show by their directions the direction of the tidal streams, although not the velocities. Halley gives a rule for finding the time of high water at other times than new and full Moon. Proudman suggested that Halley's remarks on the Chart were to lead some to the incorrect conclusion that high water would coincide with the end of the flood tide. Halley himself, however, was under no such misapprehension, for on a number of occasions in his log he recorded different times for high water by the shore and the end of the flood just offshore. In the middle

Table 10.2 Tidal flows in the English Channel from Halley and the Admiralty *Atlas*. The table shows times of end of easterly flow at points on the north (1), in the middle (2), and on the south (3) of transverse lines across the Channel (hours on days of full and new Moon).

Traverse	Halley 1	Halley 2	Halley 3	Admiralty *Atlas* 1	Admiralty *Atlas* 2	Admiralty *Atlas* 3
Lizard/Ushant	5	7	5	8	8	6
Start Pt./Isle de Bréhat	$8\frac{1}{2}$	$8\frac{1}{4}$	$7\frac{1}{2}$	$9\frac{1}{2}$	$8\frac{1}{2}$	$6\frac{1}{2}$
Portland Bill/Cap de la Hague	10	10	$9\frac{1}{2}$	10	$10\frac{1}{2}$	$10\frac{1}{4}$
St Catherine's Pt/Cap Barfleur	11	11	11	$10\frac{1}{2}$	11	10
Selsea Bill/le Havre	$10\frac{1}{2}$	$10\frac{1}{2}$	9	$9\frac{1}{2}$	$11\frac{1}{2}$	$10\frac{1}{2}$
Beachy Head/Dieppe	11	12	11	11	12	12
Dover/Calais*	15			15		

*There are no middle and south values for Dover–Calais because Proudman adjusted the times to agree at Dover.[77]

of the Channel he could record only the time when the flood ended, for he measured only the flow of the water and not the height of the tide.[76]

Proudman compared Halley's times for the end of the easterly flow with those given on the *Tidal Stream Atlas of the English Channel*, a British Admiralty publication, as shown in Table 10.2. Proudman selected seven transverse lines across the Channel from the Lizard in the west to the Straits of Dover in the east, and for each of them listed Halley's times and those from the *Atlas* for points at the north, in the middle, and at the south of the line. Generally speaking they agree very closely, and the disagreement may often be the result of the complexity of the tidal streams in the Channel, so that if observations were not made in exactly the same places, they would be discrepant. The *Atlas* is based on far more detailed surveys than Halley undertook and it is hardly to be expected that the two charts would agree in detail.[77]

Proudman emphasised three points: that Halley's survey antedates any others by a century, that his chart antedates any others by a century and a half, and that his times and general picture of the tidal streams agree well with recent more detailed surveys.[78]

The chart includes material that was not the result of the cruise of 1701, for it extends as far north as Cromer, and in 1701 Halley did not go north of the North Foreland. No tidal features are shown to the north, but the main channels and sandbanks are shown with many soundings. They may be taken from the chart of the approaches to the Thames that Halley presented to the Royal Society in 1689 (Chapter 7).

But there is more, for soundings are shown not only off the coast of France, which Halley could have taken in 1701, but also off the coasts of the Low Countries, where, according to his log, he did not go in 1701. That part of the chart also may be derived from a survey of those waters that he made in the sensitive years of 1688–1689.

After his Atlantic cruises Halley drew up the anonymous paper, *An advertisement necessary for all navigators bound up the Channel of England*, on the errors that can be committed in approaching the Channel from the Atlantic. When, after his Channel cruise, he published that notice separately, he added a section on the effects of the magnetic variation and faulty charts in sailing from the English coast by Beachy Head and Dungeness towards the Casquett Rocks (see Note 69). The implication is that he made a survey of the French coast in his Channel cruise.

Halley was far in advance of his times in his tidal survey and the chart he drew from it. A century would pass before there was any similar survey, and a century and a half before any tidal chart of any other sea was published. Only in 1845 and 1851 did F. W. Beechey (d. 1856) give charts for the English Channel and the Irish Sea.[79]

10.6 *At the publick charge*

Halley's voyages were notable achievements in scientific oceanography and geomagnetism, on which others would build in the future. Their importance for the study of the Earth's magnetic field and for cartography has already been discussed, as has the philosophical significance of drawing isarithms. They were well appreciated in his own day, for example by Swedenborg.[80]

There is yet a further way in which Halley's voyages foreshadow modern science. Prior to his cruises, science was a matter for someone by himself in a laboratory or collecting fishes or botanical specimens; it was pursued by individuals almost casually. The one exception was astronomy. Tycho Brahe and Hevelius had devoted large resources to the extended study of the heavens over many years. They had been followed by the French and English monarchs when they founded the first national scientific institutes for Cassini and Flamsteed. Halley's cruises were another step to the arrangements for modern science. For the first time, a scientific study was planned and carried out like a naval or commercial operation. A public authority, the Navy, provided the means and support, and plainly Halley planned his cruises in advance to take account and advantage of known ports and other facilities. The descendant of the observatory is the national or international laboratory, set up to make observations continuously over a long period, while the descendant of Halley's voyages is the space research project, expensive, dependent on public support, having to be planned with great care, requiring secure logistical support.

The seagoing career of Captain Edmond Halley was ended. In his three years in command he had made the first large-scale systematic survey of the magnetic field of the Earth and the first systematic survey of an extensive and complex tidal system. His scientific achievements were recognised as such but the value of his work to the navy and navigation was also recognised, for William III ordered that he should receive, in addition to his pay as a naval captain, a reward of £200 for his magnetic surveys, and Queen Anne commanded the same sum for his tidal surveys.[81] He became known to influential people in the Navy, to Burchett, to Benbow and Cloudesley Shovell, and probably also to Rooke, and almost certainly to Prince George of Denmark, the consort of Queen Anne, who was a member of the Board of Admiralty. These connections undoubtedly led to his next public service. Even while his tidal chart was being engraved, the War of Spanish Succession, so long expected, at last burst out. Strategic plans of the widest scope were being devised between London and Vienna. They required the Royal Navy to operate in waters of which the Admiralty was ill-informed. Who better than Halley to inform the Admiralty?

11

Upon the Dalmatian shore

> This is Illyria, Lady
>
> SHAKESPEARE: *TWELFTH NIGHT*, Act 1

11.1 *War over Europe*

In the late autumn of 1702 Queen Anne gave Edmond Halley an urgent command:

> Instructions for Our tr. & Welb.
> Edmund Halley Esqr. Given at Our Court of St James's the 4 day of November 1702 in the first year of Our Reign.
>
> Having though fit to make choice of you as a person in whose ability integrity and zeale for Our Service we much confide, to go and view the Ports and Havens belonging to the Emperor of Germany on the Adriatick Sea:
>
> You shall immediately upon receipt hereof repair to the Emperors Court at Vienna, or such other Place where he may be, and there apply your Selfe to Our tr. and Welb Stepney Esqr, Our Envoy Extra[ry] or in his absence to his Secretary ...[1]

England was again at war with France. When the Peace of Ryswick was signed in 1697, the Spanish king, Charles II, was expected to die very soon. The parties to the Peace had agreed that the Bavarian prince Joseph Ferdinand would succeed him. Joseph died in 1699. Charles, still alive, thereupon recognised Philip, the grandson of Louis XIV, as his successor. England and her allies, the Netherlands and the Holy Roman Empire, could not accept that aggrandisement of Louis. They proposed the Hapsburg Archduke Charles as Charles III of Spain. The War of Spanish Succession broke out in 1702. In it the Duke of Marlborough fought the notable battles of Blenheim, Ramilles, Oudenarde, and Malplaquet, his campaigns made possible by the political and administrative support of Queen Anne and William Godolphin. Naval operations accompanied war on land. They have not gained so much popular attention but were in the long term at least as effective. The Earl of Nottingham, once again a Secretary of State (the other was Sir Charles Hedges), directed them in the early years. English and

Dutch combined fleets, commanded by Sir George Rooke and Sir Cloudesley Shovell, rather by chance captured Gibraltar, and later acquired a base at Port Mahon in Minorca. So they came to dominate the Mediterranean.[2] The principal purpose of naval operations was to support campaigns in Spain on behalf of the Archduke Charles, but fleets also assisted campaigns in Italy. Allied troops besieged Toulon (ineffectually) with powerful support from the English–Dutch fleet.[3] On the other side of Italy, Venice would not allow the Imperial forces under Prince Eugene to pass through her *Terrafirma* to engage the French and Spanish armies in Milanese territory. Consequently the Imperial administration considered moving troops by sea into Italy to bypass the dominions of Venice.

Three powers controlled the eastern shores at the head of the Adriatic (Fig. 11.1). The Doge of Venice bore the ancient title of Duke of Dalmatia, but of her once extensive Dalmatian territories, the Serenissima retained only the west of the peninsula of Istria. Lands of the Holy Roman Empire lay to the north and east, administered from Graz as part of Steiermark (Styria). The Imperial city of Trieste, the ancient Roman port of Tergestum, stood immediately north of Istria. It was tiny, not at all the great port it became much later in the

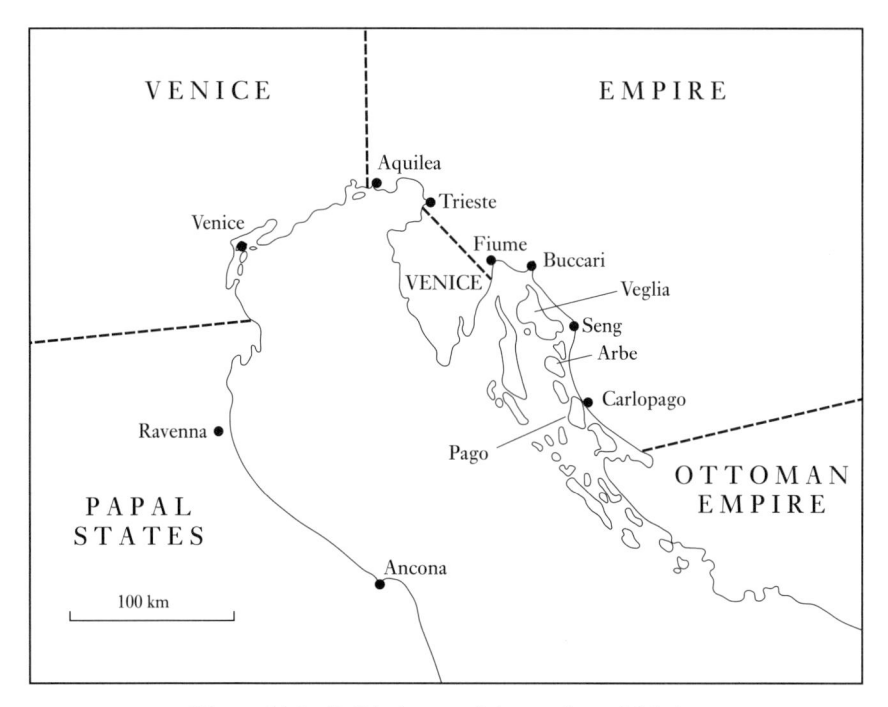

Figure 11.1 Political map of the northern Adriatic.

eighteenth century. A narrow strip of Imperial territory reached down the coast as far as Zadar. The Ottoman Empire extended to that coastal strip and inland up to an east–west line roughly in the latitude of Fiume (Rijeka).* Troops might be shipped from the Imperial shores to Italian ports south of Venetian lands, but Trieste was only a minor port and roads to Graz and major cities of the Empire were poor.

M. le Comte de Forbin commanded the French light naval squadrons in the Adriatic; he would later give English shipping a great deal of trouble in the North Sea. He closed the northern Adriatic and frustrated the movement of troops across to Italy. Forbin bombarded Trieste, ineffectually according to a sardonic couplet:[4]

> Forbinus nuper Tergestum fulmine terret
> Si damnum queris, porcus et gallus erit.
> [Forbin recently threatened Trieste with a thunderbolt;
> what was the damage you ask, a pig and a cock.]

French ships burnt Aquilea, the once great port and military base of the Roman empire at the head of the Adriatic. That escapade prompted Halley to comment ironically in a letter to George Stepney, the English ambassador to the Imperial court in Vienna, that

> It will serve rarely to embellish the Paris Gazette that 300 Francois should be able to carry by assault, plunder and burn a citty once capable to hould out a siege against the Emperour Maximine;* but they will hardly tell how little damage they have done in it, that once famous citty being then scarce other than a heap of rubbish, and a few inconsiderable cottages.[5]

Stepney was a former fellow of Trinity College, Cambridge, a minor poet and a member of the Kit-Kat Club, and a most experienced diplomat. He was a fellow of the Royal Society and Halley may already have known him (see Plate XII).

An allied fleet in the Adriatic might have ended the harassment. None did in the first years of the war, for the combined English and Dutch fleets could enter the Mediterranean only for a few summer months, and scarcely that if the weather in the Channel prevented their leaving England early in the season. Stepney put to the administration in London the value of a squadron that could remain in the Adriatic throughout the winter. In 1701 he recommended that some experienced seamen be sent to Imperial territory to identify a suitable port as a base. The government rejected his proposal, but when he repeated it in 1702,

---

*Names of places in Slovenia and Croatia are given in the contemporary form but are initially identified by the present-day name also.
*C. J. Verus, AD 238.

decided that harbours in Imperial territory should be surveyed. They asked the Lord High Admiral's Council to propose suitable seamen.[6] The Council proposed Halley. Halley had been at sea, off and on, for almost fifteen years. He had been a good husband of the Admiralty's moneys, he had obeyed his instructions, he had brought home a ship and crew in good shape from distant waters. He had demonstrated that he was a most able sea captain, that he knew more about navigation than anyone else. The hydrographic surveys he had made in earlier years were just such as would be needed to assess foreign harbours. The Board of Admiralty were clearly confident in him as a sea captain and as a representative of the monarch abroad; and he had attracted, or confirmed, the good impression of the Queen. He was the obvious person to send. As a post-captain in the Royal Navy, he was under orders. He ranked as a colonel, and Imperial officials in Austria saw him as the personal representative of the Queen.

Halley's survey illuminates the diplomacy and strategy of the English government, and the state of surveying and map making at that time. Halley appears as a military engineer, as a manager of men, and as a diplomat.

11.2 The record of Halley's survey

Halley's mission was supposed to be secret but it was known in London almost as soon as the government had decided on it. Many references by Narcissus Luttrell show that his progress was well known in London. The Dutch newspapers mentioned it, to the intense annoyance of Stepney who thought that the Venetians would obstruct Halley.[7] Subsequent published accounts have been misleading to a greater or lesser extent.[8] Halley was under the Queen's orders.[9] He wrote to Stepney once a week and their correspondence, and that between Halley and the Secretaries of State in London, is in the State Papers, Foreign, in the Public Record Office and in the Stepney papers in the British Library. Together with other correspondence preserved in public and private archives in England, Austria, and Venice, they record Halley's itinerary and activities in considerable detail.

The Adriatic coast was part of Inner Austria, administered from Graz by the Innerösterreichischen Hofkammer that reported directly to the Emperor, independently of officials in Vienna. The archives of the Hofkammer in Graz have letters from the Emperor introducing Halley (the Emperor's signatures still show traces of the gold sand with which they were dusted), and letters and reports from officials on the coast who assisted Halley. Some have useful accounts of Halley's activities.

The most informative accounts of Halley's surveys are in Venice. The Republic was indeed very interested in Halley's expedition, about which their ambassador in London, Alvise Mocenigo III, had alerted the Senate. The Senate instructed the ambassador in Vienna, Francesco Loredan, to spare no pains in finding out what Halley was doing.[10] Loredan was able to copy the report that Halley made to the Emperor and although he could not copy the maps that Halley prepared, he saw them and sent a description to the Senate. When, later in 1703, Halley returned to the coast to inspect the progress of fortifications at Buccari, Dolfin, who had replaced Loredan, sent further detailed reports of Halley's activities to the Senate.* Thus it is that some of the fullest information about Halley's doings rests in the Archivio di Stato in Venice, a small memorial to the efficiency of the diplomatic and secret service of the Serenissima.

Halley drew up three maps or charts,

> ... a generall plan of the coast of Istria, a particular one of the Bay of Buccari and a third of the Entrance into the Bay with a description in writing of what Batteries are to be raised ...

He presented a set to the Emperor in Vienna and at least one set was placed in the Admiralty in London, whence it was probably delivered to Sir Cloudesley Shovell, who was to command in the Mediterranean.[11] None appears to have survived: would that Loredan had obtained the copies that he sought. More or less contemporary maps of the region do survive in various archives, in Vienna, in Graz, in the Archivio di Stato and in the Biblioteca Marciana in Venice, and among State Papers in London. They give some idea of the state of map making in the Empire at that time.

11.3 Maps of Dalmatia

Halley thought that the maps of the Dalmatian coast that were available to him in Vienna were seriously defective and he apparently made some corrections to them. Loredan considered that he had added nothing new, and that would agree with the impression that Venetian maps of the Dalmatian coast were much more reliable than those in Graz or Vienna. That is not surprising. The peninsula of Istria between Trieste and Fiume was Venetian. The coasts further south, which had long

---

*Mocenigo was Doge from 1722–1732, and Loredan and Dolfin were both of dogal families.

been Venetian dominions, were far from Vienna, and the hinterland had been, and still was in part, under Ottoman rule. A collection of maps of Istria affords a good idea of the mapping of the region at the end of the seventeenth century.[12] Trieste is more or less correctly placed in every map from the Roman *Tabula Peutingeriana* onwards. Map-makers were much less sure of the harbour of Buccari (Bakar), Halley's choice for a possible naval base, shown most misleadingly in a map of 1657 by Nicholas Sanson, the distinguished French cartographer. Two Venetian maps of Coronelli of 1687 (*Ristretto della Dalmazia*) and of 1688 (*Golfo di Venezia*) are far more reliable. Maps published in Germany and Holland gave a very poor picture of the Adriatic coast. The published maps of Trieste were also unsound. The first reliable depictions of Trieste are in a view and a map of 1730 by Matthias Weiss, an Imperial engineer who constructed a road from Carlstadt (Karlovac in Croatia) to Buccari.

Ivan Klobucaric, at one time prior of the Augustinian house in Fiume, made a complete survey of Inner Austria. His manuscript maps and views, which remain unpublished, are in the Steiermärkisches Landesarchiv in Graz. They include a quite accurate sketch map and water-colour view of the harbour of Buccari. Halley did not say that he saw any maps in Graz. There are well-drawn manuscript maps in the Archivio di Stato in Venice, but they would not have been known to Halley. In summary, it seems that the maps that Halley could have consulted would have given him only a rough and rather inaccurate idea of the coasts and ports of Dalmatia south of Fiume.[13]

11.4 *Halley surveys Trieste and Buccari*

The Queen's letter to Halley of 4 November 1702 gave him detailed instructions—they read as though he helped to draw them up:

> You shall immediately upon receipt hereof repair to the Emperors Court at Vienna, or such other Place where he may be, and there apply your Selfe to Our tr. and Welb Stepney Esqr, Our Envoy Extra[ry] or in his absence to his Secretary, whom We have directed to be assisting to You in obtaining such Commission or Orders as shall be sufficient to enable you to perform the Service, whereupon you are sent, and herein We cannot Suppose you will meet with any difficulty, the Emperor's Minister in Our Court having assured Us of his Masters readinesse to give all the assistance herein that can be desired.
>
> Having received your dispatches from the Emperor you shall make what haste you can to the said severall Ports, and, take exact Plans of each of

them, making such Observations and Remarks as you think may any way conduce to Our Service.

You shall particularly sound the depth of water in each Port, what Rocks or Sands are in each Port, and in the Entrance into it, what number of Ships can ride there securely against Wind and Weather, what fortifications can be made for the defence of them against an enemy, and in what manner and with what charge the same may be made.

What conveniencys there are or may be made there for careening, cleaning and repairing any of Our Ships.

You are to inform your Selfe of the rates of provision, and the several species, that may be got there for the use of Our Seamen, and what material may be got there for fitting, cleaning and careening Our Ships.

You are to go from hence by the Hague, and there communicate your instructions to the Earle of Marlborough, if he be there and to the Pensioner in case they or either of them should have any direction to give you in this matter, which you are to transmit to one of Our Principall Secretarys of State for Our further pleasure therein.

You are to use all diligence in your journey to Vienna and thence to the Adriatick and returne and in the execution of these Our Instructions, since it is of great importance to Our Service, that we have as speedy an Account of these matters as is possible.

You shall correspond with one of Our Principall Secretarys of State, informing him from time to time of any thing materiall that shall occurre, of your proceedings herein; and you shall observe and follow such further Directions as you shall receive from Us or one of Our said Principall Secretarys of State.

A. R.

Halley would have had, in the course of his various inshore surveys and of his Atlantic and Channel cruise, much experience in all the duties laid upon him. He was to be paid £200 for his expenses.

The Earl of Nottingham gave him a letter of introduction to George Stepney:

Mr Stepney
or in his absence to his Secretary
Recommending Mr Halley

Whitehall Novem 4th 1702

The bearer hereof Mr Halley [written above, 'Hawley' deleted] is the person whom her Maj^y has ordered to view the Emperor's Ports in the Adriatick Sea that we may know what to depend upon, when her Maj^y shall send a Squadron thither. His Passe takes no notice of him but as of a private Gentleman, And not as employed in any Service for her Maj^y but he has letters from the Count Wratislau to the Ministers at Vienna, to acquaint them with his Errand, and to procure for him such Passes and Orders from the Emperor to his Officers in those places, as that he may be assisted by them with Boats and otherwise as shall be necesary for him to performe this Service, which being intended for the interest of the Emperor, I presume he will meet with no difficulty at Vienna but on the contrary be furnished with such authority as may prevent or overcome all difficultys which he might otherwise meet with elsewhere. I must desire you to direct and assist him in the speediest and most effectuall manner.

I am your and &
Nottingham[14]

Count Wratislaw was the Emperor's ambassador to the Court of St James.

Halley and servants arrived in the Hague on 12 December 1702. Marlborough was not there but Halley met the Pensionary (Heinsius), as he reported on 15 December in a letter to Nottingham. The Pensionary was not enthusiastic; he evidently suspected that one of the purposes behind the English suggestion was to obtain commercial advantages over the Dutch in North Africa. Halley obtained equipment for travel in the severe cold of the German winter and set out by way of Leipzig and Prague for Vienna, where he arrived on 10 January 1703, after hard weather. He went a roundabout way to avoid Bavaria, where the Elector held to the French interest. While waiting in Vienna for passes and letters of introduction he studied the descriptions and plans of ports that were available there. He concluded that the harbour of Buccari, just south of Fiume, was the only one likely to be able to receive English men-of-war.[15]

Halley went from Vienna to Trieste. The Emperor wrote to the Hofkammer in Graz on 13 January, instructing them to assist him. Graz is on the direct road from Vienna to Trieste through Laibach (Ljubljana), so that Halley must have stopped there. He arrived in Trieste on 1 February. He has left no account of his journey. Leibniz, who went from Vienna by Graz to Trieste *en route* for Venice thirteen years before at the same time of the year, had been delayed leaving Vienna and was concerned that the roads over the mountains would become muddy when the snow melted.

As soon as he arrived in Trieste, Halley immediately surveyed the port, assisted by Count Vito di Strasoldo, the governor of the city, and Count Johann Ernst von Heberstein, the military governor.* He sent Stepney a brief report in his letter of 4 February, brief because he did not find Trieste suitable as a base. Strasoldo's report to the Hofkammer in Graz was much more detailed.[16]

---

*The Strasoldo family stll flourish in Trieste and near Tolmezzo, and there is a village of their name near Palmanova; the Heberstein family were magnates in Styria and other parts of Austria.

Halley had the company of a Sig. Andreassi from Trieste onwards. Andreassi, an Italian, had been designated as the commander, should an Imperial fleet be formed in the Adriatic. They arrived in Buccari on 10 February and Halley reported to Stepney that, as he had supposed, the harbour of Buccari would be a very suitable base. He and Andreassi established positions of batteries to defend the entrance. Halley went on to Seng on 15 February and after a visit to Carlopago (see Fig. 11.1) he returned to Fiume and thence, after calling at Graz, to Vienna.[17]

Halley was at Vienna on 27 February. He prepared his maps and reports, and presented copies to the Emperor at an audience on 10 March. A week later, on 17 March, the Emperor gave Halley a diamond ring worth £60. The ring attracted widespread attention and figures in all accounts of the visit. Halley immediately left for London with his report and charts.[18] Nottingham's sailing instructions to Sir Cloudesley Shovell of 7 May 1703 imply that the charts were given to him to take into the Mediterranean; a later letter of Stepney indicates that they were in the Admiralty in 1706.[19] If Shovell kept the charts, they may have been among his papers that were recovered from the wreck of his squadron in 1707 and taken to a Cornish port, but were then carelessly thrown away by a passing boy.[20] There is no trace today of the London or Vienna copies of the charts but the report, in the form of a *Memoriall* to the Emperor, survives in a copy in Stepney's files and in one made by Loredan in the Archivio di Stato in Venice. Loredan's description is all we know about the charts.[21]

Trieste (Fig. 11.2) was built around the castle, the cathedral of San Giusto, and the old harbour. Roman remains, the theatre, some arches, and an aqueduct outside the city, can still be seen. A mole, the Zucco, may have been Roman. Trieste stands on a bay that was much more extensive than it is now, for the railway station and the quays and the moles of the old free port stand on land reclaimed from it. The bay is quite open to the south-east and ships riding in it would have no protection against a hostile fleet, which was why Halley said little about his survey. Strasoldo's account is fuller, but it is not easy to relate it to the present topography because of the changes between then and now.

Halley wrote to Stepney from Trieste on 4 February 1702/3

Extract of letter from Capt Halley.

Orders are come both to the Count Strasoldi Governour of the Castle, and to the Count Heberstein Military Governour here, to furnish all that I shall desire for the service; and I suppose the same is gone to the other Ports. I have been oft at sea and have surveyed this bay, which is a little too open to be made defensible for shipping against brave fellows. Here are the remains of an ancient Mole which seems once to have had a Fort on a native rock at the head of it; since converted into a church, which being again

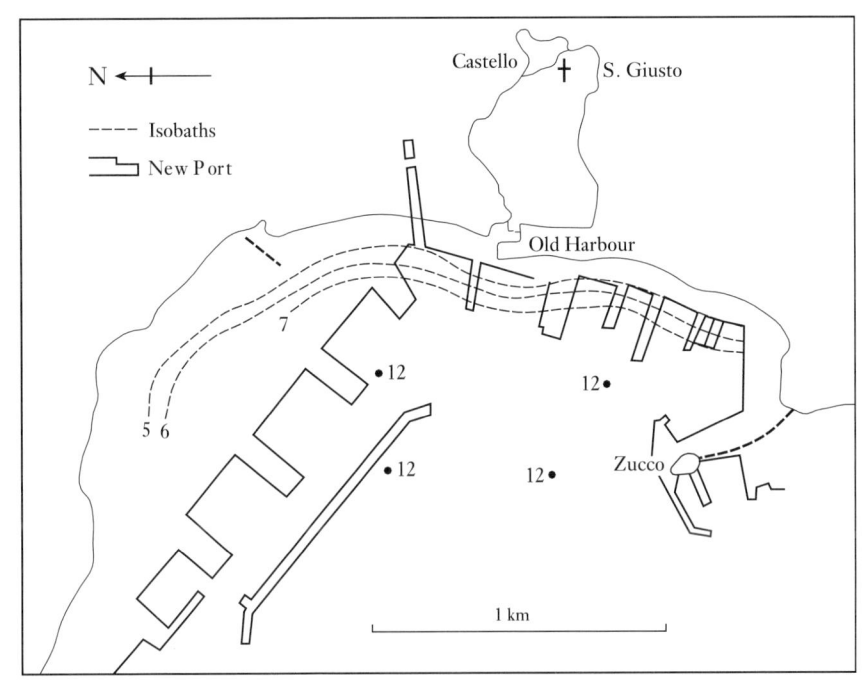

Figure 11.2 Trieste in 1703 with depths in paces as measured by Halley.

made a fort would contribute to the security of the port, it running a great distance into the sea, only it is sommewhat more than an English mile over to the other side, where there is lately erected a battery with 5 peices of Cannon. As to the rest there is water enough for any shipps, and the good Anker ground of Oaze and very clear of Rocks; only the road is open for 4 or five points of the Compass, where shipps would ride hard if it should blow: and this is all I shall be able to say of the Bay of Trieste to morrow I think to set forward for Fiume.[22]

Strasoldo begins by saying that he had given Halley every assistance and entertained him 'with all possible civility and courtesy', and that he had given him information about materials for ships and provisions for seamen and how they might be obtained. He continues (my translation from the Italian original):

Thirdly I had him provided with barques and called on the oldest and most experienced seamen of those seas, and accompanied him myself to sound the depths of the water where it would be of use to ships in the port in front of this city and its fortress and, having drawn back by a distance of about 50 geometric paces from the land, the measure of the water was 4 paces, at a distance of 70 to 80 paces the water was found to be about 5 paces, and at a distance of about 100 paces, the water was found to be about 7 paces deep, except near the island of the Zucco which is about 700 to 800 paces distant from the city where the water was found to be about 8 paces deep. The island of Zucco is found to have a circumference of just more than 50 geometric paces with firm foundations; it is established on rocks where it is judged not only useful and profitable but necessary that a little fort should be erected and because the repair of the mole, or way that leads from the island to the point of the Campo Marzo on dry land, a distance of 200 paces, can not only make the place secure against winds but also will protect from the attacks of enemies both the port and the city itself.

The expense of such construction is judged not to be great for the base of the little island is almost all rock and dry and very easy to build on, and the way or mole, from that island to dry land of the Campo Marzo is of similar quality for a large part of its foundation was constructed at other times in the past.

From that island of the Zucco, we went directly to the side called the Musiella which is opposite to it, the distance one from the other being about one thousand and 300 paces, that forming the mouth of the port, and in sounding the depth of water over that distance it was found everywhere to be 12 paces except near the said point of the Musiella where at 80 paces it was reduced to 5 paces, that point of the Musiella being indeed about 1000 paces from the city.

In all this circuit of the port, which amounted in all to about three Italian miles or 3000 paces, the bottom was found good and secure to hold firmly vessels at anchor besides which they can move closer to the land to tie up with their lines and ropes.[23]

Halley presumably measured depths with a lead line in the conventional manner, and most likely found positions by resection of bearings from known places on land, as he had described in his letter to Southwell (Chapter 9).

Halley surveyed the harbour of Buccari (Bakar) either in the few days between leaving Trieste and arriving in Fiume on 10 February, or after his return to Fiume from Seng and Carlopago on 17 February. In some ways the harbour is little changed from Halley's day (Fig. 11.3). It is an almost landlocked bay surrounded by bare limestone hills, with two small townships at opposite ends; Buccari is the larger while Buccarizza (Bakarac) was and is very small. Porto Re (Kraljevica), another little town, stands to one side of a narrow strait that leads from the bay to the sea. Tuna fishing was the main activity until the latter nineteenth century, but then a coal handling installation was built on the bay between the entrance and Buccari, and later an oil terminal was constructed opposite Porto Re where, in addition, there has been a small ship repair yard. Yet the topography can have changed little, and the depth of the bay and the general configuration of the entrance must be much as they were when Halley was there. Thus it is possible to make rather direct comparisons between Halley's survey and the current Admiralty chart.

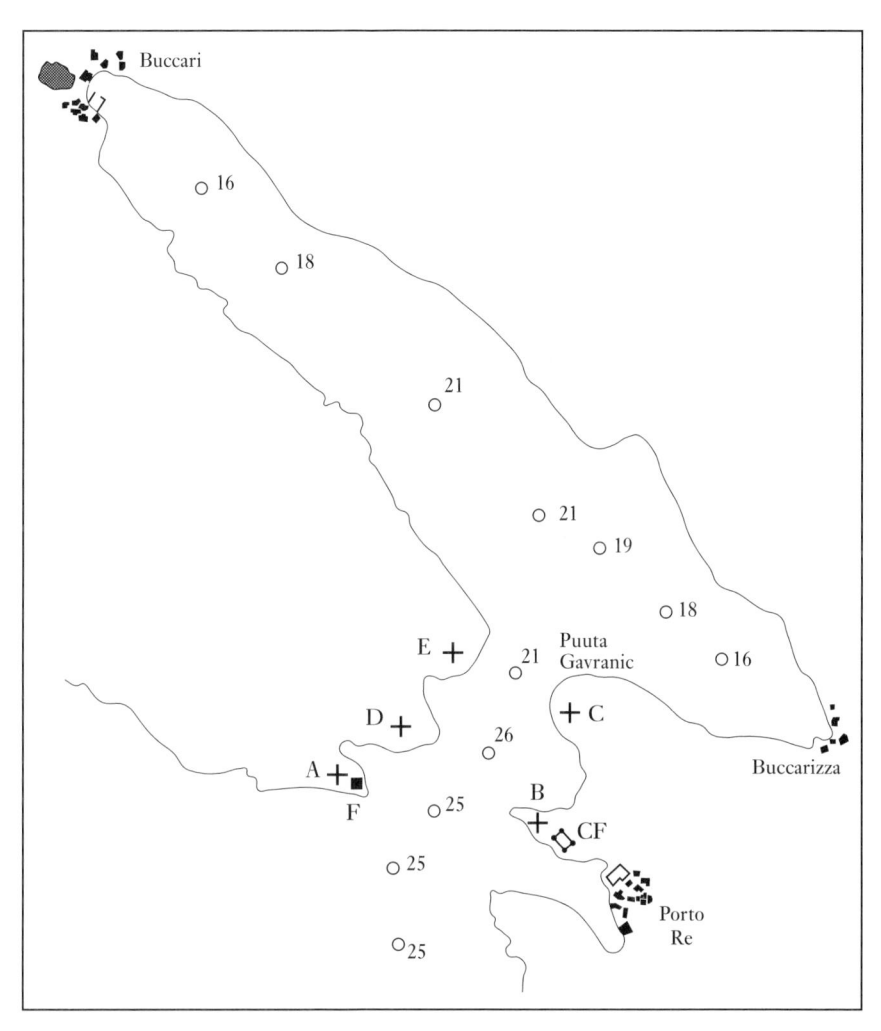

Figure 11.3 The harbour of Buccari with depths in paces measured by Halley and sites of batteries (A, B, C, D, E) and pre-existing forts (F, CF).

Buccari is not open to the Adriatic but is sheltered by the islands of Cherso (Kres), Veglia (Krk), and Arbe (Fig. 11.1). The channel between the islands and the mainland is the Canal di Morlacco.

Halley described the results of his survey in the *Memoriall* that he presented to the Emperor on returning to Vienna. The copy made for the archives of the English embassy in Vienna is in the Public Record Office.[24] Loredan acquired a copy clandestinely that is now in the Archivio di Stato in Venice. He could not copy Halley's maps, but he did see them and described them to the Senate.[25] Halley's own descriptions of his results and Loredan's reports to the Senate in Venice give a reasonable idea of Halley's achievement, even though the maps themselves have vanished.

The harbour of Buccari is about 4.5 km long and about 1 km wide and the entrance is some 400 m wide at the narrowest point. Halley states in his *Memoriall* that the bay is about three Italian miles long and that the mouth is about 200 passi wide and Loredan agrees with the latter figure, as he saw it on Halley's map of the bay. Loredan reported to the Senate (my translation):

> By keen and prudent persistence I have at last succeeded in obtaining a copy of the report that Captain Hall presented to His Imperial Majesty following his observations of the Austrian coast. I am now sending it to inform the State as I would have wished I might have done with the maps, which I have had an opportunity to see but of which I have not so far been able to obtain copies. However the latter do not contain essentially new observations of any significance, other than the improved ones of the Port of Buccari. That occupies a separate chart on which is to be seen clearly delineated the entrance or mouth, about 200 paces across. The depth of the water in it is shown as 25 and there are in outline two batteries on each side to defend it. In the interior the large inlet extends towards Buccari and will be about two miles long, continuing on the other side towards Buccarizza for about one mile, with numbers everywhere showing the depth of the water, 20, 18, 16 paces, which decrease in proportion as they approach the land.[26]

Loredan said he suspected that the talk of military preparations might be a cover for commercial plans detrimental to the Venetian State, but later he accepted that he had found no evidence for that. He wrote apparently of two maps, one of the Dalmatian coast and one of the whole harbour of Buccari. Stepney's correspondence shows that Halley also produced a plan of the mouth of the harbour with proposed sites of batteries, and Loredan seems to have known something of it.

Halley's measurements as recorded by Loredan may be compared with the modern chart. The *passo* used by Halley was probably the Venetian passo of about 1.74 m, so that the Italian *mile* of 1000 passi would be 1.74 km. The comparisons with the modern chart are then (all distances in metres):

	Halley	Admiralty chart
Total length of bay	5220	4600
Width of entrance	350	400
Entrance to Buccari	3500	3200
Entrance to Buccarizza	1700	1400
Greatest width of harbour	580	1000

A scale error of about 13 percent, which might have arisen from a confusion of the local scales of length, would account for some of the differences, but the widths of the entrance and of the harbour are too small. The proportions do, however, agree with those of some more or less contemporary maps, from one of which Halley may have found the positions of his soundings by resection and may have derived the chart of what was evidently a rather hurried survey.

The depths of water, on the other hand, agree well with the modern chart if the passo is 1.74 m, for that is only slightly less than the fathom of 1.83 m.[27] The figures quoted by Loredan, 20 to 16 paces in the harbour and 25 in the mouth, are then effectively identical with the modern values. Halley did not give any depths in his own letters because they would be on the maps he left in Vienna as well as on those he took back to London.

Halley's letters show that he had the help of Sig. Andreassi and that, instructed by the Queen to establish 'what fortifications can be made for the defence of (ports) against an enemy, and in what manner and with what charge the same may be made', they laid out the sites of four batteries at the entrance to the harbour, two on either side. He set them down on the charts that he delivered to the Emperor and agreed them with him. Loredan reported that he had seen the sites marked on Halley's chart. Halley does not say how many batteries he had indicated but he did say that he had provided for sixty cannon, forty heavy and twenty lighter, to be able to mount a defence against a fleet passing rapidly through the entrance under a strong wind from the sea. He argued that sufficient guns should be provided for many shots to be fired while some guns were reloading, and he also argued for heavy guns because lighter ones would not do enough damage.

The surveys of Trieste and Buccari took most of Halley's time, but he also quickly viewed the coast down to Carlopago and he reported on the supplies of stores that might be obtained in the neighbourhood. He wrote to Nottingham on 25 February from Graz:

My last gave your L^{opp} an account of my departure from Vienna for the Adriatick, since when I have had the good fortune to have weather so favourable that I have been able to visit the whole Sea coast subject to his Imperiall Ma^{ty}: and to consider all its ports. Among them I cannot find that there is any one fitting to receive and protect a fleet of Shipps against any enemy except singly Buccari, but that is one of the finest I ever yet saw, being large enough for any squadron, good ankerground, and deep as you please: tis a bason near a league in length and about a third of a mile in breadth, and the entrance thereto is but too hundred paces wide and may be fortified with batterys so as to make it impracticable for any shipps to pass without leave. I have designed the manner and place of those Batterys which I conceive may be built for a small charge, the materialls being upon the place, and tis believed the Emperour will forthwith order it to be gone about. As to Navall stores here are at present none, but very good masts of all moderate sizes grow in the Mountains near the sea side above Zeng; hemp may be raised in what quantity shall be desired in Croatia, from whence and Hungary may be drawn any quantity of beef and corn desired; Beef at a penny English per Lib, but good wheat delivered at the sea side will cost about three shillings ye Bushell, and peas 3^{s}6^d. Pork is scarce here, and very little butter and cheese and that bad, but in lieu thereof oyle grows both at Fiume and Trieste and is both good and cheap. Oatmeale is not used nor known here, but in lieu thereof they use rice brought from Lombardy. Wine of the growth of these parts is also reasonable.[28]

The modern visitor will recognise Halley's description of the splendid harbour of Buccari; olives no longer grow near Trieste and the wine has probably improved since his day. Halley's report on supplies follows closely the instructions for victuallers laid down by the Admiralty in the Standing Instructions for various officers, in particular, for the Navy Board, Yard Officers and Agents in 1691. It is very likely that Pepys drew up those instructions.[29]

Halley left Vienna to return to London on 17 March after his audience with the Emperor the week before. He took with him the Emperor's ring and the high opinions of all whom he had met, as Stepney reported to Nottingham that same day:

Mr Halley being now ready to return to England, I cannot but accompany him with assurances to your Lord^p that he has performed what he was sent about entirely to the satisfaction of this Court and I hope that your Lord^p will herein be satisfied with the discoveries he has made.[30]

Halley was back in London about the middle of April. The government reacted promptly to his report and, even before he was in London, Hedges had written to Stepney (on 27 March) to ensure that 'Buccari be fortified for the security of our ships according to Mr Halley's proposal with all expedition'.[31] So Stepney went to the Emperor 'hand in hand' with the ambassador of the Netherlands, Count de Bruyinx, to present a memorandum on the conduct of the war, which included a passage about Buccari. Having rehearsed the common interest of the Empire in enabling an English fleet to operate in the Adriatic, they conveyed the request of the London ministers that the Emperor might order that all necessary provisions and stores should be provided on the Adriatic shore, and also

... to fortify the port of Buccari so that it may be in a fit state to receive the fleet securely, according to the plans that Captain Halley has had the honour to present to your Imperial Majesty.[32]

On 13 April Hedges wrote for a report on what was happening and on 30 April he told Stepney that Halley would soon be back on the Adriatic. The government clearly intended that Buccari should be fortified in time to receive the fleet that they planned to send to the Mediterranean that same summer under the command of Sir Cloudesley Shovell. Shovell was instructed to report on the fortification of Buccari and the provision of supplies, and shortly after Halley's arrival in London, was given what can only have been Halley's charts.[33] At the same time there were difficulties in despatching the fleet, for the Dutch ships which were to form part of it did not arrive; not until September did the fleet sail from England, too late in the season to do anything.[34]

The English ministers suspected, rightly, that not much was being done at Buccari, despite the assurances that Stepney had received in Vienna.[35] So they sent Halley back. Orders to provide him with funds were given on 19 May and he left London on 22 June, to arrive in Vienna on 23 July, all to the great comfort of Stepney:

It will be an ease to me to have Capt. Halley in these parts to look after the Fortifications and Provisions which are things I do not understand and I should have been unwilling to rely on the assurances these Ministers can give me, that all is in readiness, wherein they are apt to deceive both themselves and others.[36]

On the way to Vienna for the second time, Halley called at Hanover, and perhaps Osnabruck, and met Leibniz, who later wrote of his visit in a letter to Johann Bernoulli. Leibniz was pleased to hear that Newton was bringing out his work on colour and wished that he might produce his new theory of the Moon that Halley had praised. Leibniz said that Halley was now (15 January 1704) in England, having returned from the Adriatic. At the time of his visit to Leibniz, Halley seems to have envisaged a Pacific cruise to perfect his magnetic observations. Leibniz writes of Halley as that praiseworthy, tireless student of useful knowledge. Leibniz told Halley of the observations of the comet of 1680 that had been made by Kirch.

Halley is said to have met the Elector, already designated as George I. Many English people did. Anne dissuaded George from visiting England in her lifetime because of Jacobite reaction, but had it seems no objection to visits to Hanover. Halley would have had the particular reason for passing through Hanover of informing the Elector of what he was doing on the Adriatic, with which George might become directly involved on his accession.[37]

11.5 *Fortifying Buccari*

The Venetian ambassador in Vienna, now Daniele Dolfin, reported Halley's return to the Adriatic:

> Captain Halley has already arrived and has left for Buccari preceeded by two engineers who go to review and progress the operations already arranged for the security of that port.[38]

No instructions for Halley's second visit can be found, but it is easy to surmise what they would have been: to report, much as Shovell had been instructed to do, on the progress of fortification and provisioning at Buccari. He was very soon to discover that, as London had suspected, very little had been done, and he vigorously set about hurrying things on.

When Halley reached Vienna on 23 July he was presented to the Emperor, who sent orders to Graz that he should be assisted. At the coast, he met the Imperial engineer, Rauschendorf.[39]

Halley wrote to Stepney on 5 August from Fiume:

> I gott not to Fiume before last night, and this morning went to Buccari to see what was there done. I found only the first Battery on the left hand going in (being for twelve guns) in some measure finished; and one other that has been begun for some weeks is but little advanced, there having been some stopp putt to the work. The Ingeneer Rauschendorff went with me to Buccari and showed me what he had done, and I find him very ready to hear reason: and the Commissioners who are here have promised me that all thats wanting shall without delay be perfected; but what they determine to do and in what time it may be done, I shall be better able to judge when I see the result of a view all parties concerned are to have upon the place tomorrow morning.[40]

The commissioners who were supposed to oversee the work and to pay for it were J. A. Endtres, an official of the Hofkammer at Graz, and Count Herberstein who had assisted Halley at Trieste.[41]

Halley's reactions when he saw the state of affairs at Buccari were not favourable. He reported to Nottingham that just the one battery had been built and that the foundations for a second by Porto Re had been cleared and that some materials had been assembled, but no stones had been laid, and continued:

> I find that upon supposition that ye fleet would not come this year they had suspended ye work, tho otherwise absolutely necessary for the defence of so good a port. Since I have been here I have endeavoured to convince those who have the conduct of the work of ye necessity of dispatching what they have in hand, and how little they correspond with her Ma^{tys} intentions to assist their master with her Naval force, upon the sole condition of being

safe in port. They have promised the utmost diligence to retrieve ye lost time, but being well paid by the day while the work is in hand, I shall be glad to see them prefer their Masters interest to their own.[42]

To Stepney on 11 August he wrote yet more pungently:

Wee have here three Comiss^{rs} for carrying on the affair of building the batterys, whose pay being great they intend shall continue; and as far as I can find their allowances being about 40 Guilders per diem, and about treble what they have expended on the Batteries themselves, they pretend that not believing that fleet would come, they saw no reason to press the dispatch; though I am told that with the Emperours commands they have actually received 50,000 Gld which would do all the work and buy the cannon necessary: They now promise that all the rest of the designed batteries shall be forthwith gone about, but all our last weeks work makes so little show that I have no great hopes of it. I find by all their discourse, that they think any thing may serve, and are for doing it a juste prix; but I have given them to understand that the charge of our Fleet is such as in no measure corresponds with their good husbandry.[43]

A week later Halley could tell Stepney that work had been pressed ahead on the battery by Porto Re with some 150 workmen, with the prospect that everything might be completed before the winter set in. On 1 September he reported to Nottingham that the stonework of the Porto Re battery was finished and that good progress had been made with the remaining two, and consequently he hopes he may return before the winter.

Meanwhile, another problem, which had been in the background for some while, came to the fore: how to provide the heavy cannon for the batteries. Halley had told Nottingham that he thought they would have to be sent from England, but Nottingham was strongly opposed to that. The Austrians, on the other hand, expected that they would be able to take cannon from the fleet when it arrived, but Halley had provided for sixty cannon in four batteries, the whole armament of a third rate ship of the line, and such a depletion of the fleet would plainly be unacceptable to the Admiralty. The Imperial authorities had made trials of casting guns in the neighbourhood, but without success, and although their field artillery was good, as the wars with the Ottoman Empire had shown, it is likely they did not have the experience to cast heavy ships' cannon. Furthermore, the roads to the coast were poor, and it would have been difficult to bring heavy guns overland to Buccari. The impasse was not resolved when Halley left and there is no record of guns ever having been placed in the batteries. Nor did the fleet ever come, for the capture of Gibraltar and Port Mahon gave the

English–Dutch fleets winter ports and control of the Mediterranean, so that in subsequent years an Adriatic base was unnecessary.

11.6 Sites and construction of Halley's works

Although Halley's own charts do not survive, it is possible to locate the sites of the batteries with some assurance, making use of his own accounts and of some later records which seem to refer to his works. The probable sites of Halley's masonry works, labelled A to D, are marked on Fig. 11.3. The evidence for the positions comprises Halley's reports, the account by Loredan of what he saw and learnt of Halley's maps, later maps and views in archives in Vienna, and an account of a visit by a nineteenth-century topographer.

In his letters Halley writes of four batteries of masonry construction, but a note on one letter in the archives of the English embassy in Vienna says, 'there are 4 batteries of stone for 58 great guns and one of earth for 8 more'. The charts probably did not show the site proposed for the earthen battery, for Loredan reported 'e comparriscono abozzate due Batterie per parte per diffenderla', that is, the sketch shows two batteries on each side for defence.[44] Halley apparently provided for 66 cannon. He does not explicitly describe the sites because they were set down on the chart he had given to the Emperor, but letters from his second visit refer to particular batteries. When he first returned to Buccari one battery had been built, 'on the left hand side going in'.[45] The second battery to have been built was by Porto Re, and there was another on the same side 'which commands the entrance of the Harbour completely, and will be of 20 Guns'. The two batteries on the Porto Re side are marked B and C on the plan.

In 1730, as part of the preparation for building a new road (Via Carolina) from Trieste by Buccari to Carlstadt (Karlovac, south east of Zagreb), Matthia Antonio Weiss assembled a collection of maps and plans that is now in the Hofkammerarchiv in Vienna. A plan of the harbour of Buccari is a version of one in another Viennese archive, the Kriegsarchiv, and is associated in the Weiss collection with a plan of the entrance and a pen and wash sketch of Porto Re and parts of the entrance to the harbour. The dates of the plans and the sketch are important, for Porto Re is said to have been fortified under the Emperor Charles VI, about 1729, twenty-five years or so after Halley's activities. Weiss's plans were assembled at about the same time. However, the catalogue of the map collection in the Kriegsarchiv gives the date of the large plan as 1726, before any Caroline construction. A further map in the Kriegsarchiv, by

Upon the Dalmatian shore 311

Figure 11.4 (a) Fortifications in a pen and wash sketch in the Weisskarte von Buccari of the entrance to the harbour of Buccari in about 1730. (Reproduced by permission of the Director of the Finanzarchiv und Hofkammerarchiv, Vienna.)

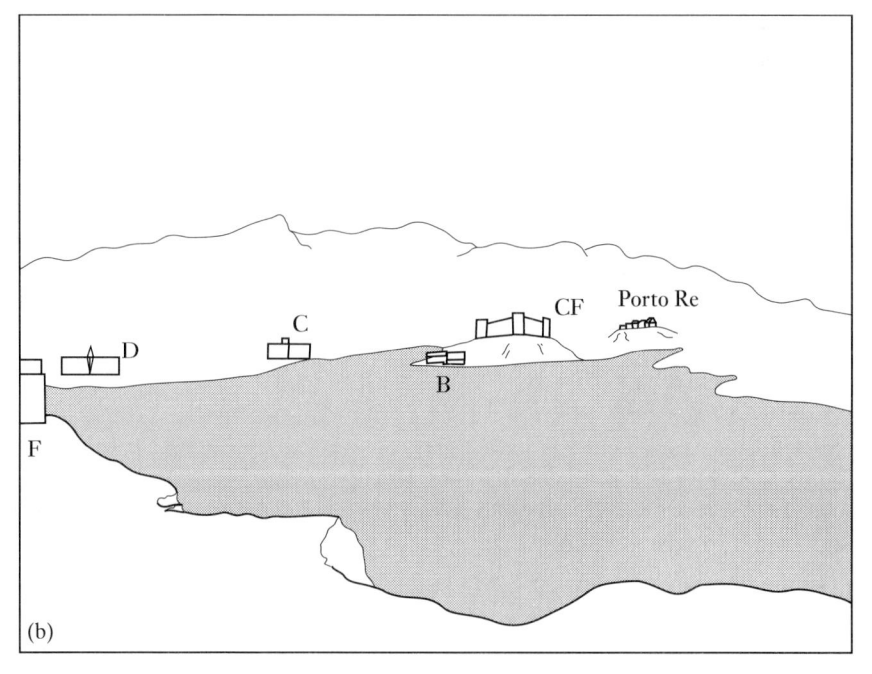

Figure 11.4 (b) Diagram showing positions of batteries B, C, and D on pen and wash sketch (labelling as in Fig. 11.3).

Antonio de Vernada, of 1733, labels batteries at points A, C, and D, and has a symbol that might indicate a battery at B but without a legend. That map was produced in connection with 'new works' at Porto Re, perhaps the fortifications of Charles VI, and shows the Via Carolina with the spur to Buccarizza which appears also on the Weiss map. On none of those maps is there any indication of the fifth earthwork battery (E?). The pen and wash sketch in the Weiss collection (Fig. 11.4) is drawn from a point near A, where there was a sixteenth-century fortress, F, and it shows batteries at points B, C, and D, of which that at C is the dominant one, a substantial work which could certainly have mounted 20 guns and which commands the whole entrance.[46]

The Hauptmann at Buccari arranged for one hundred and fifty corveé labourers to prepare the sites of the batteries. The work had to be finished before the autumn so that the men, local peasants, could harvest the grapes. The stonework required skilled masons and carpenters who were paid from money brought by the commissioners; one of Halley's complaints was that the commissioners were very reluctant to spend money on the wages of the craftsmen. The batteries probably consisted of stone retaining walls filled in with earth on which were wooden platforms for the guns. The walls were no doubt protected by earth banks on the outside. Nothing of the batteries now remains. The possible sites at A and D are covered by modern oil tanks, there are later houses at B, and a road around Punta Gavranic crosses the site of C.

There were some remains of fortifications when the Archduke Louis Salvator of the House of Habsburg-Lothringen explored Buccari late in the nineteeenth century. He described the topography, fauna, and flora of the harbour of Buccari in the form of a guided walk clockwise right round the harbour, illustrated with a map and sketches. The map is based on Austrian surveys from 1821 to 1867, it carries a scale of Wiener Klafter and gives depths of water in Wiener Fuss. It shows remains of two batteries at sites C and D.[47]

Salvator writes of the battery at site D:

> ... one can see a fallen shore battery which once served to defend the harbour entrance and Porto Re, which it commands. It lies on a small ledge of teeth-like rocks and has three sides, of which only the two longer, that form a blunt angle to the south-west, show gun embrasures. The side that one first reaches has five gun positions, the second, turned more towards the harbour entrance, has nine.

A battery with fourteen guns would be consistent with Halley's description at this place.

Salvator saw what may have been ruins of a battery at site B and something more substantial at C, where he again saw fourteen embrasures in ruins covered with mesembryanthemum, with stones fallen into the water.[48]

The Weiss sketch shows this battery somewhat above the shore, whereas Salvator speaks of the waves washing over its ruins, but they might have fallen down into the sea. The location in the Weiss sketch is consistent with the function of the batteries, which would be to disable, not to sink, men-of-war. The most effective fire would dismast ships and would come from batteries at but not above deck height. The Weiss delineation of the battery at site D accords with Salvator's description of the ruins and with Halley's comments.

The works shown by Weiss and those seen by Salvator could have been the somewhat later Caroline works (of which there is only slight documentary evidence) rather than Halley's. Batteries on Halley's sites certainly existed in the 1730s and their ruins were described by Salvator in the nineteenth century. If not Halley's construction, they undoubtedly owe their positions to him, and the Weiss sketch may give a good impression of what Halley's simple works were like.

11.7 Consequences

The ostensible purpose of Halley's expeditions was never fulfilled. No allied fleet entered the Adriatic to make use of Halley's surveys and fortifications. Sir Cloudesley Shovell's squadron of 1703 was considerably delayed in home waters and could spend only a few months in the Mediterranean before it withdrew for the winter. Next year English forces captured Gibraltar rather by chance, they destroyed the French fleet in harbour at Toulon, and they captured Port Mahon in 1706. Thenceforth the allied fleets dominated the Mediterranean and there was no further need for an Adriatic base, although Stepney did raise the question again in 1706, when he referred to Halley's plans and report being in the Admiralty.[49]

The change in the naval situation in the Mediterranean was not the only reason why nothing came of plans for an Adriatic base. The supply of guns for Halley's batteries was not resolved. The Austrians wanted to take them out of English ships; the English Government refused to weaken their force in that way. Even if Austrian gunfounders could have cast the large cannon that Halley called for, it would have been difficult to bring them to Buccari. The foundries were at Salzburg, the roads from there across the Alps were dreadful.

It was not practical at short notice to base a squadron at Buccari, or indeed anywhere on the Imperial coast of Dalmatia. The naval shipyards of England and France, at Deptford, Chatham, Brest, or Toulon, were the greatest industrial establishments of early modern times, even as the Arsenal at Venice had been in Dante's day. The construction and repair of the ships themselves had behind them auxilliary activities, gun-founding, rope works, and forestry for the ships and their masts. Halley reported that timber for masts was to be obtained in the country behind Buccari, as was hemp for rope, but the facilities for working them were not there. Buccari, Buccarizza, and Porto Re are tiny today, and they were even smaller in the eighteenth century with only twenty houses at Buccarizza, so Halley reported. They could never have accommodated the crews of a sizeable squadron nor the workmen to maintain it. The crews alone of ten third-rate ships of the line would number at least three thousand men to be housed throughout the winter. Those places had none of the timber stores, rope yards, magazines for munitions, the warehouses and slaughterhouses for food and other provisions. Much of the correspondence of the Hofkammer in Graz was concerned with negotiations with a contractor about provisions, so that in June, Count Bruenes, the president of the Hofkammer in Graz, was concerned about the possibility that no fleet might come to Buccari because 'a sufficient quantity of victuals are provided and that their loss would be great if it should so happen that the Fleet should not appear in the Gulph'.[50] But the fleet did not come.

Halley's work apparently had no influence on later maps. His maps were kept securely in Vienna, so it seems, and twenty years after his visit maps of the region of Buccari were no better than when he went there. Maps shown at an exhibition in Trieste in 1982 include one of Trieste of 1720 that is quite unreliable, and the earliest plans of Trieste that are reasonably accurate are those of Weiss and Pallavicini, both about 1734. They do not show depths of water in the harbour and appear to owe nothing to Halley.[51] Similarly the plans of the harbour of Buccari of the same period are unreliable. One, of about 1720, shows some depths of water and what may be batteries on the shore, but is quite erroneous. The first printed map of any accuracy is that of Florjancic, published in Ljubljana (Laibach) in 1744. There is in the archives of the English consulate in Venice, among the papers for 1714, a set of undated plans of Dalmatian ports. It is a miscellaneous collection with different scales, and legends in different languages.[52] A plan of Buccari resembles one in the Kartensammlung of the Kriegsarchiv in Vienna and shows fortifications 'that might be built' in about the places where Halley laid them out. The surveyor is not identified but could have been Oberingenieur Hans Friedrich von Hollstein; did he have some knowledge of Halley's work?

Halley's missions tell us something about surveying, map making, and fortification at the beginning of the eighteenth century. The techniques of surveying such as Halley practised were capable of producing useful maps and charts but the maps that were published at that time were often very poor and not based on direct survey. The batteries themselves were built quickly when forcefully driven, showing that quite simple constructions were very effective if provided with heavy guns. The missions also tell us something about Halley. There is no comparable set of continuous letters from him and to him at any other period of his life and they are most informative. We see him carrying out his mission speedily and competently, gaining the admiration and respect of all he met, while he himself had a realistic and sardonic view of the capacities and intentions of the Imperial officials with whom he was in contact. When he returned and saw how little had been accomplished, he ceased to be just the expert adviser and became the site engineer, driving on the commissioners and the workmen to finish the construction before the onset of winter. Only then did he leave.

There is another sideline to the missions. Halley came to the Adriatic at a turning point in the affairs of the allies generally and of Mediterranean and naval matters in particular. Mediterranean affairs were an integral part of the Blenheim campaign and they turned out as they did because the sea power of Venice was no longer dominant there. The depredations of Forbin in the Adriatic, the bombardment of Trieste, and the burning of English ships by the French in the heart of Venetian waters in the lagoon at Malamocco, would not have happened in the heyday of Venice. Now the northern powers, and especially England, would control the Mediterranean for two hundred and fifty years. Halley's instructions and activities, and the reaction to them in Austria and Venice, show something of why and how that came about.

11.8 *Surveyor and cartographer*

When Halley left the Adriatic coast for the last time, he came to the end of a period of fifteen years during which he had made hydrographic surveys of the Thames estuary, of the Channel coasts and ports of some Atlantic islands, and of Adriatic ports. He would not go to sea again. This is the place, then, to look at his achievements as a surveyor and in particular at the accuracy of his charts.

Halley had described how he obtained positions in his Channel surveys in his letter to Southwell that was apparently written in the course of that cruise (Chapter 9).[53] Halley's methods, determining position by trigonometry and depths by lead line, would be effectively unchanged until echo sounding and radar came into use.

Halley probably never made surveys of the land in the way he described. He had maps of the coasts for his principal undertakings, the surveys of the Thames approaches and of Trieste and Buccari. For the first he probably had the map made by Sir Jonas Moore and Dutch charts of the Thames approaches which were available before his survey. He found the existing charts very defective. He surveyed Trieste and Buccari so quickly that he probably did not have time to make a map of the coast, but may have relied on the faulty maps available in the Empire, as the imperfections of his chart of Buccari perhaps show. His soundings are, however, his own. The hydrography of the harbours of Trieste and Buccari seems to have been unknown before his surveys. His charts of the approaches to Portsmouth and Southampton and of the harbour of Plymouth also probably depend on some existing land survey; according to his own logs his time in the Solent was all spent at sea and at Plymouth he only observed the latitude on land.[54] It is thus likely that Halley sounded the depths of water in the respective harbours at places that he located by resection from landmarks on available maps. His surveys were brief, a day or two in the harbour of Trieste, little more at Buccari, and a few spring or summer weeks in the Thames approaches. He can hardly have had the time to make a proper survey of the coasts from which he took his positions at sea.

Halley's survey of the harbour of Buccari compares well with the modern chart, the depths agreeing for the most part to within 1 fm. Discrepancies would arise from errors in position as well as from errors in sounding, but the floor of the main harbour of Buccari is rather flat and errors of position would have little effect in it. The only place where the depth changes rapidly is in the entrance where there are unlikely to be great errors of position. In essence then, Buccari shows that Halley could measure depths to 1 fm (1.83 m).

The tidal range at Buccari is small and would not seriously affect the datum for depths. The great range in English waters, 7 m or 3 fm at the mouth of the Thames, would affect the measured depth. In addition, the floor of the sea is far from flat in the Thames approaches and the Solent, so that the observed depths would depend greatly upon the location of soundings and therefore on the accuracy of maps and of the resection procedure. For both reasons, tides and uneven floor, it might be expected that Halley's surveys in English waters would show greater discrepancies from modern charts. His printed chart of the Thames approaches is on too small a scale to make a close comparison but, such as it is, it agrees well. In particular there seems to be no systematic difference in depths, implying that Halley used something close to a low water datum. The general delineation of the major banks and channels is also close to the modern chart, which probably indicates that his positions were both more accurate and also more numerous than in earlier surveys. In summary, Halley's surveys were an advance on previous hydrographic surveys.

Halley surveyed a few Atlantic islands in the course of his Atlantic cruises but the results survive only as sketches in his logs, and add nothing to the assessment of his methods. He also sounded some French waters, but apparently rather hastily.[55]

Halley's great contribution to cartography was his introduction of isarithms, as has been recognised from his day to this.[56]

11.9 Sea captain and engineer

Halley went to Austria with considerable experience of seamanship, of practical navigation and survey, and of command and management. The logs of *Paramore* show us that he was a skilful navigator, and we know that he was concerned with navigation all his life and thought deeply about it. He was not a deep water sailor only, for in the confined waters of the English Channel he showed himself adept at inshore survey. He surveyed the harbours of Trieste and Buccari and produced a bathymetric chart of Buccari that compares well against modern productions, even though he can have spent little more than two or three days there. Halley was more than a surveyor or navigator, for his proposals for the forts at Buccari call upon the thought he had already given to problems of gunnery.[57]

Halley and James Cook share a characteristic that all great sailors have: care for their crews and ability to control and get the best out of men. Both had very healthy crews for their day, Halley quite remarkably so, and he took his crew with him in what must have seemed to them pointless wanderings about the Atlantic, into the dangerous waters of the south, and upon what was certainly the very hard work of continual anchoring and measuring currents in the tidal survey of the Channel. He displayed the same ability at Buccari when he had to overcome the inertia of both the official commissioners and the corveé labourers to get the batteries constructed. Halley was a masterful man and his sardonic reports from Buccari display his impatience of slow procedures on the part of the Imperial administrators and the venality of the commissioners. He was sent on his second mission by the Queen's ministers to report on progress, but he did more, for when he saw that progress was slow, he effectively took over the management of the site and ensured that the building work was completed before the grape harvest took the labourers away.

Halley's attainments as an engineer were considerable. Not the least of his qualifications when he went to the Adriatic was that he had worked on diving and salvage over a number of years at Pagham in harsh and dangerous conditions. He had experience of the force of currents and of handling heavy loads. At Buccari he laid out batteries and ensured they were built. Subsequently, in Oxford, he handled the move of some heavy marbles for the Delegates of the Clarendon Press (Chapter 12). Engineering is far from Halley's main claim upon our attention, but it is not negligible.

Halley attracted golden opinions for his surveys in Dalmatia, both because he was skilful and also, as Nottingham wrote, on account of 'his zeal in the public service'. Stepney had warm praise for his responsibility and abilities and for the high regard in which he was held in Vienna. He was trusted by the Queen and Nottingham to represent them with the Pensioner in Holland and the Emperor and his officials in Austria, and to visit the Elector in Hanover. Halley was a great navigator and surveyor and he carried through difficult projects, while keeping on good terms with most people. Energetic and forceful, attractive and dependable, clear-sighted in the failings of others, he was for the most part admired and respected by all with whom he had to deal. In great matters of European strategy Halley's surveys were inconsequential, but they and the character he displayed had great consequences for him.

In the late September of 1703 Halley turned inland from Trieste upon the imperial road to Graz and Vienna. He still hoped for a Pacific expedition. It was not to be. Oxford called him. He was leaving the sea behind him for the last time. He was bound, although as yet he did not know it, from the confusion and fury of the seas to the confusion and fury of the academic world. All Europe would see him as a brilliant mathematician and outstanding classical scholar.

PART III

Scholar and sage

(1704–1742)

12

In the Savilian Chair

> Very nice sort of place, Oxford, I should think,
> for people that like that sort of place
>
> SHAW: *MAN AND SUPERMAN*, Act II

12.1 *The Electors meet*

ON 6 November 1703, Viscount Hatton, the head of the Hatton-Finch family, wrote to the Earl of Nottingham, in a letter otherwise concerned with family interests in Guernsey:

> I am so much obliged to Dr Keith of North[ampt?]on for his care of me in my late sickness that I cannot but recommend his brother, at his request, for the 'mathematique' Profesorship at Oxford, for which he is a candidate. He is, I know, very capable of it, 'but if Mr Halley be thought of, or aim at it, he acquiesces and would by no means oppose it'.[1]

John Wallis (1616–1703, see Plate III) died on 28 October 1703. He had been elected Savilian Professor of Geometry at Oxford in 1649 under the Commonwealth. Later he became Keeper of the Archives of the University. He was one of the founders of the Royal Society. He was a leading mathematician and well known to Halley, for they had corresponded since Halley was an undergraduate. Within days of his death, candidates for the professorship were being mentioned.[2] Halley must have declared his interest as soon as he was back from the Adriatic, for in the middle of December Flamsteed wrote to Abraham Sharp: 'Dr Wallis is dead—Mr Halley expects his place—who now talks, swears and drinks brandy like a sea captain'.[3]

Halley was still abroad when Nottingham replied to Hatton on 20 November:

> I can't say y^t Mr Halley thinks of ye Mathematick professor's place but everyone who has a vote in the election thinks of him; and am very glad y^r Lop^s recomendation of another is accompany^d with this condition, if Mr Halley does not pretend to it, for I have seen his zeal in the publick service (on which he is now abroad) the which added to his extraordinary skill above all his competitors obligd me to be very forward in promoting him to this place.[4]

Nottingham was better placed than most to know Halley's capacities. He would probably have known of his survey of the Thames approaches at the time of William III's invasion, he would have been acquainted with his career as a naval captain before he was chosen to go to the Adriatic. To him had Halley reported the results of his surveys on the Adriatic and the progress of his fortifications. He knew the praise lavished on Halley for his execution of the Queen's commands (Chapter 11). Nottingham would have been confident that, if Halley were elected to the Savilian Chair, the administration would have its man in Oxford, whatever the aspersions cast by Flamsteed and others and whatever his associations with nonjurors. Nottingham was not just a powerful canvasser, he himself was one of those with 'a vote in the election', he was a statutory elector.

Of the nine electors to the Savilian Chairs (Chapter 9), six were present on 8 January 1704 and signed the certificate.[5] Tenison, who had succeeded Tillotson as Archbishop of Canterbury, was, like him, a latitudinarian, but there is no indication that Halley's religious views were an issue. Anne was more of a Tory than Mary or William, and her administration was less suspicious of those who had been lukewarm in their acceptance of William and Mary. It has been suggested that Newton gave his support to Halley at this election but there is no indication at all that Halley asked for it or that Newton proferred it; in any case, as already observed, Newton was never a patron of Halley, their circles of influence were quite distinct. Halley was elected because he had given effective and tactful support to the administration in a matter that directly involved him with persons at the highest levels, the Queen and Secretaries of State, Marlborough, the English ambassador in Vienna, the Emperor and Prince Eugen, the Grand Pensionary, and the future George I.

Halley's election was entered in the Acts of the Consistory of Canterbury, he came to Oxford at the end of February, and was admitted into the Congregation on 7 March.[6] He delivered his inaugural lecture on 24 May 1704. Hearne noted:

> Mr Hally made his Inaugural Speech on Wednesday May 24, which very much pleased the Generality of the University. After some Complements to the University, he proceeded to the Original and Progress of Geometry, and gave an account of the most celebrated of the Ancient and Modern Geometricians. Of those of our English Nation he spoke in particular of Sr Henry Savil; but his greatest enconiums were upon Dr Wallis and Mr Newton, especially the latter, whom he styled his Numen etc. Nor could he pass by Dr Gregory, whom he propos'd as an example in his Lectures; but not a word all the while of Dr Bernard ...[7]

while David Gregory wrote:

> ... he is to publish Ptolemys Geography; a Treatise concerning all the Comets that have appeared, a treatise concerning the Variation of the Needle, with an account of his voyage towards the South Pole; The Contraction of the Equations of the Sixth Dimension or Power. The treatise of Comets is to be Astronomia Cometica & to contain 60 or 70 sheets in folio; but one sheet with the elements of them all is to be printed just now.[8]

Halley never proceeded with Ptolemy. His references to Savile and Wallis foreshadowed his edition of Apollonius, which he began in collaboration with Gregory and for which he undoubtedly made use of editions by Wallis. He probably had material collected by Savile as well as some preliminary matter from Bernard. Bernard, Gregory's predecessor as Professor of Astronomy, more a classical scholar than a mathematician or astronomer, had done much to develop both the Bodleian Library and the University Press in Oxford.

The Savilian Professorships were founded in 1619 by Sir Henry Savile, successively Greek tutor to Queen Elizabeth, Warden of Merton, and Provost of Eton. Savile had travelled in Italy as a young man and had collected Greek manuscripts which were the basis of the library associated with the professorships—they are now in Bodley's Library. He had lectured in Oxford in the 1570s on astronomy and especially on the ideas of Copernicus, and later he edited the works of St John Chrysostom. His foundations conspicuously advanced the mathematical sciences in Oxford, for the first professors were leading scholars. The first professor of geometry was Henry Briggs, who devised logarithms to the base 10. He lectured on Euclid, the conics of Apollonius, the mechanics of Archimedes and practical applications. Wallis brought out editions of Greek mathematicians, wrote an history of English mathematics, and made many deep original contributions. Christopher Wren was the professor of astronomy for twelve years from 1661. David Gregory, who wrote the first textbook of astronomy on Newtonian principles,[9] was Halley's colleague from 1704 until he died in 1708. The professors' rooms were in the New Schools quadrangle over against Bodley's Library.[10]

Four estates that Sir Henry had conveyed to the University provided the stipends of the professors. Little Hays was in Essex, Norlands in the Isle of Oxney in Kent, Morton Hindmarsh (Moreton-in-the-Marsh) in Gloucestershire, and Purston was in Northamptonshire. Halley visited all four in 1704 shortly after he arrived in Oxford. He went by himself to Little Hays on 20 April and to Norlands on 22 April, and with Gregory to Purston on 5 June and to Morton Hindmarsh on 6 June.[11]

He recorded the rents as they were at the time and as they had been in the past and he noted any particular features of the estates, especially delapidations and necessary repairs. He made a plan of Morton Hindmarsh. He remarked on Purston, 'This estate (as all the rest) seems to be ill used by the Tenants for not having Leases and often another farm or land of their own near at hand'. That comment, and other remarks in the notes, suggest that the estates had not been regularly visited before Halley and Gregory did so (although Wallis had visited) and that they were not in good condition. Similar comments might be made about many college estates for which the titles, the conditions of tenure, and the payment of rents were often all unclear,[12] but college bursars endeavoured to visit their estates more frequently than the Savilian Professors. There may have been some uncertainty about whether it was the responsibility of the University or the professors individually to see to the Savilian estates.

After repairs and certain taxes, the maximum income available for the stipends in Halley's day could not have been more than £300, and perhaps less, depending on taxes and repairs. Thus the income of each professor would have been about £150 per annum, not more than Halley's income from the Winchester Street houses in London. The professors did not always receive their full entitlement from the University and shortly before he became Astronomer Royal, Halley had over £1000 owing to him.[13]

Each professor was to lecture publicly twice every week in term and to spend an hour in his lodgings in more informal teaching of those who might call on him. The professor of astronomy was to lecture on the whole of Ptolemy, together with optics, gnomonics, geography, and navigation, but excluding nativities and judicial astrology without exception. The professor of geometry was to expound the thirteeen books of Euclid's Elements, the Conics of Apollonius, and all the books of Archimedes, and to leave in the University archives his notes and observations. He was also to teach and expound speculative and practical arithmetic, land surveying, mechanics, and canonics or music. It is unlikely that Halley lectured on music, for the Heather Professorship of Music had been founded before he assumed his chair, although the holder seems at that time to have been more a Kapellmeister to the University than a musicologist. Wallis had presented a candidate for the degree of MusD, at a time when the music chair had not been filled.[14]

12.2 A life in Oxford

Little is known of Halley's personal life in Oxford. As Savilian Professor he occupied a house in New College Lane, at the western end close to the Sheldonian Theatre. It had belonged to Wallis and after his death his son gave it to the University for the use of the Savilian Professors, so that they might spend more time in Oxford and less in London, according to Flamsteed.[15] Halley built an observatory on it. He had a house in London in Bridgewater Street in the parish of St Giles's, Cripplegate; letters he wrote from London to Arthur Charlett and John Hudson show that, like Gregory, who practised medicine in London, he was often away from Oxford.[16]

Hearne, the antiquarian and Assistant Librarian at Bodley's, retails a certain amount of gossip about Halley. In many brief notes he gave glimpses of Halley's day-to-day activities with Gregory and others, his social round, and his stories of the wider world of London politics. Hearne respected Halley's abilities, he thought most Oxford lectures were poor with very few exceptions, notably those of Halley, 'a great ornament'. As a staunch nonjuror and Jacobite, he found Halley too uncertain in politics and lax in religion. Hearne embodies in some respects the Oxford to which Halley returned. William and Mary came to the throne in part because many who followed Tory principles and held High Church beliefs had deserted James II in the face of what they saw as his attacks on the Church of England. Those High Church Tories had blocked William's plans for religious comprehension and considered that the Toleration Acts went too far in providing for dissenting consciences. Some, and not only High Church adherents, became nonjurors, unable in conscience to break the oaths they had previously taken to James by swearing new oaths to William and Mary; they were excluded from office, especially ecclesiastical office. Many heads of colleges were Tories and High Churchmen. Hearne was more extreme: a Jacobite, he looked for the return of the Old Pretender ('James III'). Oxford Jacobites accepted Anne grudgingly as a High Church Tory Stuart, in contrast to her sister Mary, but could not stomach George I (the 'Duke of Brunswick' to Hearne). Hearne suffered for his loyalties. The most senior men in Oxford, Tory and High Church though they might be, could not afford to offend the Government by allowing such as Hearne to remain in office; he was ejected from his position as second Librarian at Bodley's and not allowed into the Library. That coloured his opinions of people.

Jonas Proast was another prominent Oxford High Churchman. He wrote pamphlets saying that religious conformity should be enforced by the civil magistracy; they called forth John Locke's three *Letters on toleration*. Proast had been an undergraduate at Queen's a few years before Halley. He became chaplain of All Souls, but was ejected for his behaviour after James II had imposed John Fitch as Warden by royal mandate. Proast was not immediately reinstated when William and Mary arrived, for he was opposed to the dominant latitudinarian party on the bench of bishops led by the archbishops, first Tillotson and then Tenison, and by the bishop of Salisbury, Gilbert Burnet. Proast was eventually reinstated and later became archdeacon of Berkshire.[17]

Many of Hearne's stories are hearsay or otherwise unreliable, for example Gregory's memoranda show that Hearne's more scurrilous references to him are baseless. Some of Hearne's references to Halley are likewise no doubt coloured by the rather extreme politics and churchmanship of Hearne and his friends, extreme, that is, relative to London, and should be read in that light.

In October 1712 Hearne reported that

Mr Proast said on February 26th 1704 that Dr Milne said that Mr Hally's father had told him that he went in fear of his life from his son Mr Hally. Mr Proast s^d y^t a friend of Mr Hally's told him y^t Mr Hally believed a God & y^t was all, & y^r was a story that Mr Hally went to Dr Stillingfleet & y^t he told him y^t he believed a God and y^t was all.

Later, on 25 September 1718,

A Gent. who is very honest, lately told me that he thought what is related in one of these Volumes, viz that Dr. Hally's father went in fear of his life from his own Son, the said Halley, is true. I think that Dr. Halley's Father was drownede.

Another of Hearne's stories, for 13 November 1713, concerns Hevelius:

The picture of Dr Edmund Halley (Savilian Professor of Geometry) done exactly like him by Mr Tho. Murray, who gave it, is lately placed in the Gallery of the Bodlejan Library. It hangs by Hevelius whom Dr Halley when he was young, had visited at Dantzick, and for that reason, as well as his skill in Astronomy, Hevelius hath mentioned him in one of his books. And some Persons say that he is very justly placed by Hevelius, because he made him (as they give out) a Cuckold by lying with his Wife when he was at Dantzick, the said Hevelius having a very pretty Woman to his Wife, who had a very great kindness for Mr Halley and was (it seems) observed often to be familiar with him. But this story I am apt to think is false.

Again on 22 January 1718, Hearne noted that Newton is a great Whig and so is Halley, though pretending to be a Tory: 'In short, Dr Halley has little or no religion'. Hearne wrote that Halley defended taking all manner of oaths; he took the oaths publicly as soon as George I came over, and bragged of it in coffee houses.[18]

On 11 February that same year, conversing with Hearne about the hereditary right of kings, 'for my part, says the D^r, I am for the King in Possession. If I am protected I am content I am sure we pay dear enough for our Protection & why should we not have the Benefit of it'. 'The king in possession' seems to have been a phrase used to justify their actions by many who accepted William and Mary and George I. Proast had sworn allegiance to William and Mary 'for the necessity of government'.

Halley was by no means the only person that Proast, Hearne, and others libelled as having no religion; John Locke, though a constant member of the Church of England, was so charged, and so was David Gregory.[19] Hearne's remarks about Halley's irreligion and trimming seem little more than the common coin of polemical gossip, and not to be taken as evidence of actual belief, any more than similar stories about Locke impugn his adherence to the Church of England, his friendship with its bishops, and his influence in ecclesiastical matters. Gregory had been the only Episcopalian professor in Edinburgh and was ejected for it. Hearne likewise castigated Flamsteed, writing of his republican principles and his snivelling temper.

There is another more particular possible reason why Hearne's tales after about 1710 should be treated with reserve. It was about that time that Newton and others asked Halley, to Flamsteed's fury, to prepare Flamsteed's *Historia coelestis* for the press. Thus, on 30 October 1711, Halley told Hearne that he had finished printing Flamsteed's observations in folio, quite against Flamsteed's consent, after Flamsteed had desisted: 'Mr Flamsteed is very angry with him but D^r H is regardless of his anger and his reflexions'.[20] The 'some persons' who recounted the story of Halley's seduction of Elizabeth Hevelius almost certainly included Flamsteed, who was more closely in contact with Halley at the time of the visit to Hevelius than anyone else. He had earlier done his best to get Halley rejected for the astronomy chair (Chapter 9) and now, after more than thirty years, may have resurrected or invented the tale. Is it coincidence that Hearne hears that story, and others such as that about Halley's father or Halley being irreligious, just at the time of the terrible quarrel?

Halley as a professor would have participated in University ceremonies. Evelyn described an earlier one in 1669 in which Wallis took part. There was a speech by the Terrae Filius, the University clown, of which Evelyn disapproved for its personal abuse rather than wit, then

> After this ribauldry, The Proctors made their Speeches: Then began the Musick Act, Vocal and Instrumental, above in the Balustred Corridore,

opposite to the Vice-Chancelors seate: Then Dr Wallis the Mathematical Professor made his Oration, and created one Doctor of Musique, according to the usual Ceremonies, of Gowne (which was white Damask) Cap; Ring, kiss etc ...[21]

It would have been on account of the provision in the Savilian statues that required the Professor of Geometry to lecture on music that Wallis presented and created a doctor of music before there was a professor of music.

Four days later Evelyn received the honorary degree of Doctor of Laws, in part at least for bringing the Arundel Marbles to Oxford. Two years earlier he had persuaded Henry Howard, the grandson of the Earl of Arundel, Earl Marshal of England, to donate the Marbles to Oxford. At his visit in 1669 he saw that they had been built into the wall around the Theatre, where people were beginning to scratch graffiti.[22]

On 19 November 1714 Hearne noted

The East Wall of the Theater Yard being to be pulled down, the Marmora are to be removed & are accordingly ordered by the Vice-Chanceller to be removed, to another Place, and Dr Halley to take care of the Matter.[23]

On 3 October 1715 at the meeting of the Delegates of the Press,

Mr Vice Chancell^r proposed to gratifie Dr. Hally for his care at ye removing ye Statues and Monuments from ye Theatre yard & placing them in Order. Agreed to give him 30 gns ...[24]

The next minute explained the reason for the 30 guineas: the architect Hawkesmore was to be given £100 for his designs and oversight of the construction of a new Printing House. When John Fell, the Dean of Christ Church and subsequently Bishop of Oxford, established the University Press after the Restoration, he persuaded Archbishop Sheldon, the donor of the Sheldonian Theatre, to ask that his building be used as a printing house when it was not required for ceremonies. Thus the imprint of Oxford books of Halley's day is *The Sheldonian Theatre*. The joint use of the Theatre was inconvenient; the Delegates wanted their own premises. The large profits from Clarendon's *History of the Great Rebellion*, together with the sale of stones from Osney Abbey,[25] enabled them to build the Clarendon Building next to the Sheldonian Theatre in Broad Street. The Press moved there about 1713. Statues and monuments were in the way, principally the Arundel Marbles in the wall around the Theatre. Halley, with his experience of moving heavy objects in his diving projects (Chapter 9), was well placed to help the Delegates. He moved the Marbles to the Schools Gallery above the rooms of the Savilian professors. They remained there until transferred to the new Ashmolean Museum in the nineteenth century.

Halley received the degree of Doctor of Civil Laws in 1710, upon the recommendation of the Chancellor, the Duke of Ormonde:

Mr Vice Chancell^r & Gentlemen

    Mr Edmund Hally having been Master of Arts near thirty years and often employed by her Majesty and her Predecessors in the Service at Sea in the remotest parts of the World to the Great Satisfaction of the Lords of ye Admiralty and others of the first Quality in the nation; And now being Your Professor of Geometry and well known to be a Person of great knowledge not only in that Science but in most of ye other parts of Learning, I have thought fitt in consideration of his great merit and the Service he hath done to the publick both at home and abroad to recommend it to You that you would conferr the Degree of D^r of Civil Laws on him without fees & not doubting your concurrence herewith I am Mr Vice Chancell^r & Gentlemen

                                Y^r affectionate friend & Servt
                                           Ormonde

London July 13 1710

The letter was read to Convocation on 18 July with a number of other recommendations from Ormonde for higher degrees.[26] It might seem that it was just one among many. Not so. The others, like many that Ormonde wrote as Chancellor, were brief and generally recited that he was supporting a request from the applicant, often a medical man. The qualifications of the candidate that were cited were mostly formal. The recommendation for Halley differs in three ways: there is no statement that Halley asked for the degree, the grounds for it are given in some detail, and fees are to be remitted. It was an honorary degree for notable public service: when Handel came to Oxford in 1733 and put on concerts in the Sheldonian and elsewhere, he declined the doctorate of music that the University offered him, because, it was said, he would not pay the fee.[27]

It may be that Halley's publication of the *Conics* of Apollonius, which took place in 1710 (see below), prompted Ormonde's recommendation, but it was not advanced as the ground for the degree, except perhaps implicitly in the phrase, 'well known to be a Person of great knowledge not only in that Science but in most of ye other parts of Learning'. Halley's service in the public interest was the ostensible reason, and Ormonde's words raise questions. They seem to imply more occasions of service than those of which we know. Queen Anne, so far as is known, employed him only once, in 1703, and not 'in the remotest parts of the World'. The Atlantic cruises did indeed go to the remotest parts of the world, as Hearne once noted: 'he went as near the Southern Pole as any man yet did'.[28] While Halley first proposed the cruises, it could be said that in the end, holding a commission in the Royal Navy, he was 'employed' by William III. The survey of the tides in the

Channel was also his idea, but he might have been thought of as 'employed', for he held a naval commission. Do 'her Predecessors' just mean William and Mary, or might they include James II or even Charles II? Who were the 'others of the first Quality in the Nation'?—one would have been the Earl of Nottingham but there may be some so far unidentified, Prince George perhaps.

Halley never received a knighthood as Newton did, yet Ormonde's letter shows that Halley did at least as much, and probably more, in the public service than Newton, the Mint notwithstanding. Why did Halley become Doctor and not Sir Edmond, while Newton remained Master of Arts but became Sir Isaac? Newton's knighthood was conferred in 1705 at the request of the Earl of Halifax, to bolster Newton's chances of election to Parliament that year in Halifax's interest.[29] It had little to do with natural philosophy or public service but much to do with politics; that may have been so also with Halley's doctorate, although no one could doubt that it was justified on academic grounds as well.

Hearne's Oxford may seem to have been dominated by Tory and High Church resentment, Jacobitism, bitter verbal infighting, and elaborate ceremonial; it also nurtured splendid learning and scholarship in the years up to and just after 1700. When the Savilian Professors were Wallis and Gregory, Halley, Caswell, and Kiell, the mathematical sciences were in good hands. The Sedleian Professorship of Natural Philosophy had been founded a century before the Jacksonian Profesorship at Cambridge, and chemistry was taught in the Ashmolean Museum. Oriental studies were vigorous and so were patristics and Anglo-Saxon language, literature, and history. Many in Oxford were interested in antiquities. Thus Halley maintained, against Hearne, that Silchester (to the south of Oxford) was the Roman Calleva Atrebatum. No doubt he based his identification, which is correct, on his earlier study of the Antonine Itineraries in which Calleva Atrebatum is listed. Halley was also of a party that went to view the Roman mosaic pavement at Stunsfield north of Oxford, and he maintained his interest in Roman inscriptions. In 1720, just as he was going to Greenwich, he became a Fellow of the Society of Antiquaries.[30]

All that scholarly efflorescence in Oxford was based on and associated with Bodley's Library as a major public repository of manuscripts, and on a consistent policy of the Delegates of the Press to publish scholarly editions. The two sides of Oxford, the devices and personal animosities of politics and religion, and the individual pursuit of scholarship and its support in others, were often combined in one person. Arthur Charlett was a notable instance of the complete Oxford man, although exceptional as a latitudinarian in religion. He was a consummate Oxford politician, Master of University College, sometime Vice-Chancellor, a Delegate of the Press and influential in the affairs of Bodley's Library, and a gifted amateur architect and musician. Halley found him helpful when he first arrived in Oxford and in his business with the Press. Henry Aldrich, Dean of Christ Church, was another with whom Halley had much to do. He was a leading High Churchman and Tory in Oxford and nationally in Convocation, also sometime Vice-Chancellor, and a considerable scholar who strongly urged on the programme of publishing editions of the Greek mathematicians, and who acquired for Oxford some of the manuscripts that Halley used in his editions; he was another more than competent musician. Bodley's forceful librarian, John Hudson, helped Halley with reading proofs. Resentful, domineering, devious in political and religious affairs some of Halley's contemporaries and colleagues may have been, and some undoubtedly were idle and some dissolute, but many also were driven by a real desire for knowledge and the advancement of learning and scholarship. That Oxford might seem a strange place to us today, and it may have seemed so to Halley returning after thirty years, thirty years in which he had attained a European reputation in mathematical science, in which he had wide experience at sea, and had served the government in various ways. He had discussed seamanship with Peter the Great and strategic matters with the Grand Pensionary of the Netherlands and the Holy Roman Emperor, he had pulled back from disaster among the ice and fogs of the remote southern seas: must he not have regarded some of the preoccupations of colleagues who had never left Oxford with detached amusement?

Halley's reputation in England and in Europe attracted visitors to Oxford. Swedenborg, whose geomagnetic interests have already been noticed, apparently had a courteous welcome from Halley (Chapter 10). The Italian astronomer, Monsignor Francesco Bianchini, who had successfully repeated Newton's optical experiments in Rome, visited England in 1713. He met Newton on three occasions, attended the Royal Society, came to know Keill and Flamsteed, and had a friendly reception in Oxford. Nicholas Bernoulli greatly appreciated meeting Halley.[31]

Ministers and Parliament called on Halley for expert advice. In 1712 he and Newton jointly were asked by the Earl of Oxford (Harley) to give their opinion on a navigational instrument devised by one Coward—who did not produce the goods.[32] In June 1714 the important meeting of a committee of the House of Commons took place to hear the proposal of Ditton and Whiston for a method for finding longitude, and Halley was called on to give his opinion (Chapter 13). Halley was also asked for testimonials, such as he and Newton gave for Willm. Jones to teach navigation at Christ's Hospital—Jones was not successful.[33]

12.3 The Royal Society again

Three prominent fellows of the Society died in 1703, Wallis, Samuel Pepys, who had accorded the Society's *Imprimatur* to Newton's *Principia*, and Robert Hooke. It is said that after Hooke's death Newton was no longer unwilling to publish his *Opticks*, nor reluctant to be elected President of the Royal Society, for on both accounts he is supposed to have feared attacks from Hooke.[34] Newton, as President, involved Halley in two unsavoury disputes, that with Flamsteed over the publication of Flamsteed's life's work at Greenwich, discussed in Chapter 13, and that with Leibniz over who had first developed the infinitesimal calculus. Newton presented his *Opticks* to the Royal Society and, as with *Principia*, they passed it to Halley for a report.[35] There is no record of Halley's report nor of the publishing history of the *Opticks*, which Halley with others had for some years before been urging Newton to publish. Halley may not have been further involved for, unlike *Principia*, the book was already complete. Halley was elected Secretary of the Royal Society at the Anniversary Meeting of 1713 in succession to Sir Hans Sloane who had served for twenty years. Thereupon he took over the publication of *Philosophical Transactions* and Volumes 29 and 30 appeared under his name.

Halley's part in the priority controversy over the infinitesimal calculus was rather marginal. Briefly, Newton devised his method of fluxions in the late 1660s, and Leibniz his equivalent calculus with a different and better notation, the one we now use, about ten years later. Leibniz published; Newton did not until forced to give some account of his work in 1702. At first the dispute was muted but then Newton came to believe, partly through injudicious remarks by Leibniz, partly through the promptings of Keill, that Leibniz had based his work on results of Newton that he had seen on a visit to London after Newton's discovery but before his own. It was not so, but the accusation stuck, and to justify himself Newton had the Royal Society set up a committee to review the papers and his correspondence with Leibniz. Halley was a member and was Secretary when the Society took that decision, but Keill ostensibly took the lead. Newton's case was set out, and his results from before 1669 were revealed for the first time, in the *Commercium epistolicum* compiled in 1712. Most of the composition was Newton's.[36] Leibniz replied with his *Charta volans* of 1713, also anonymously, and Newton then commented on the *Commercium epistolicum*, again anonymously. Halley was certainly consulted in the matter, especially by Keill, but it does not seem he was deeply involved, rather he acted as a postman between Newton and Keill.[37] Various protagonists of Leibniz joined in, especially Jean Bernoulli who had a poor opinion of Keill but retained his respect for Halley and Newton. When Leibniz died in 1716 Newton persisted for a while but the affair eventually faded out.

Throughout the period of the controversy, continental mathematicians kept up a civil and even friendly correspondence with the English, and Jean Bernoulli in particular wrote regularly to de Moivre. In 1712 Jean Bernoulli's nephew, Nicholas, visited London and was received by de Moivre who introduced him to Newton and Halley; Jean Bernoulli was very appreciative of the courtesy that his nephew was shown in England.[38]

12.4 The Greek geometers

The Oxford University Press effectively came into existence in 1662 when the University established a Delegacy to run it. Archbishop Laud had intended to set up an academic press before the Civil Wars and the project was resumed at the Restoration by John Fell. Fell brought in three others to join him as Delegates, Sir Joseph Williamson, who died in 1701 before Halley came back to Oxford, Sir Leoline Jenkin, the judge and the Secretary of State who committed Essex to the Tower, and Thomas Yate, Principal of Brasenose College. They would print two classes of book, Delegates' books that they themselves commissioned and published, and authors' books which they printed at the expense of the authors. In 1672 Fell set out a programme of works that he thought the Press should publish as Delegates' books, although he did not designate them as such, and it included, besides Bible texts and editions of patristic writings,

> The ancient Mathematicians, Greek and latin, in one and twenty Volumes; part not yet Extant, the rest collated with MS, perfected from the Arabick versions, when the originals are lost, with their Scholia & comments; & all illustrated with Annotations, if this proposal shall be thought too vast, we shall enter upon such authors, & publish them in such formes as shall be desired.[39]

The project was similar to that of the Maurists in Paris, albeit much more modest. The Oxford Press was well placed to do it: Bodley's was the first public library in England founded for scholarship and research and its resources were not to be matched elsewhere in England for long after Halley's death. In the seventeenth century Oxford, and especially the Library, had acquired numerous Greek and Latin manuscripts. Archbishop Laud had donated many, and the Earl of Pembroke had given the Barocci Collection from Venice, with over 300 Greek manuscripts. Collections of oriental manuscripts had been bought in the 1690s. No doubt there was little demand for teaching in Arabic, but the professors in the late seventeenth and early eighteenth centuries were mostly considerable scholars. Furthermore, Oxford had men such as Arthur Charlett and John Hudson, outstanding scholars and forceful and effective administrators and university politicians. They strongly supported the publication of the Greek mathematicians and similar projects. Both helped Halley in preparing his editions of Apollonius and in other ways.[40]

Fell set out his plan just as Halley was coming up as an undergraduate. Edward Bernard, then Savilian Professor of Astronomy, drew up a synopsis of Greek, Latin, and Arabic texts that should be included and himself received a grant of £30 or £40 from the Delegates to copy parts of an Arabic manuscript of the *Conics* of Apollonius of Perga at Leyden—that of Golius described below. Bernard proposed an edition of Euclid in fifteen volumes; it did not proceed but later, in 1703, Gregory published an edition of Euclid in which he used the material assembled by Bernard and gave a Latin version as well as the original Greek.[41] Well before that Wallis had prepared editions of Ptolemy, Porphyrius, Brennius, Archimedes, Eutocius, Aristarchus, and Pappus, and had begun to edit Apollonius when he died. Halley therefore inherited an unfinished project when he came to the Savilian Chair and John Aldrich, who had succeeded Fell as Dean of Christ Church, persuaded him to collaborate with Gregory on the works of Apollonius.[42]

Apollonius of Perga, known as 'The great Geometer' to his contemporaries and his Greek and Arabic successors, was worthy of Halley's attention. Little is known of his life. He was born at Perga in Pamphylia in Asia Minor about 240 BC. He visited Pergamum and Ephesus and composed the *Conics* about 200 BC when he was living in Alexandria. The *Conics* was his great work and, apart from an Arabic translation of *De sectione*, a lesser work, his other books are known only through references by Pappus of Alexandria (*fl. c.* AD 320) and through an Arabic epitome.[43] The *Synagoge* or *Collection* of Pappus was an encyclopaedia of the whole of Greek mathematics up to his day, and included a synopsis of the *Conics*. Pappus not only summarised the works of his predecessors, he also gave lemmas that were needed to fill in steps in arguments that the original authors had taken for granted. The lemmas for the *Conics* helped Halley to reconstitute its lost final book. The original intention was that Gregory should edit the first four books of the *Conics*, for which there is a Greek text, and that Halley should do the later books which survive only in an Arabic translation. Gregory died in 1708 before they got very far and the edition of the *Conics* as published is essentially Halley's.

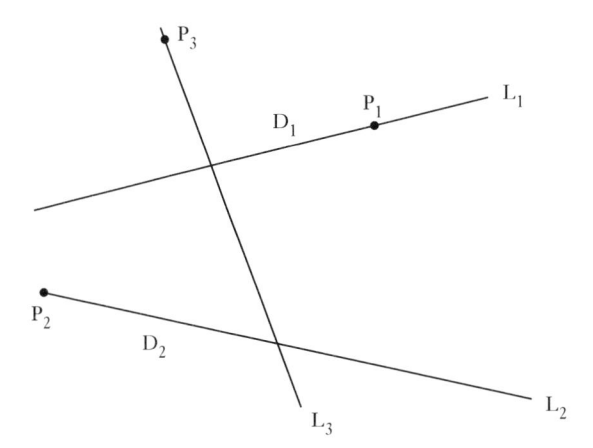

Figure 12.1 Cutting off a section. L_1, L_2, L_3: specified lines; P_1, P_2, P_3: specified points; D_1, D_2: sections cut off.

Halley began with the *De sectione rationis*, or *The cutting-off of a ratio*. It showed how, given two lines, L_1 and L_2, points P_1 and P_2 in each, and a third point P_3 not in the lines, to draw a straight line L_3 through the third point to cut the lines in points such that their distances D_1 and D_2 from the original points in the lines were in a predetermined ratio (Fig. 12.1). The Greek versions have been lost and it survives in an Arabic translation. Bernard had made a copy of the Arabic version in Leyden and had translated about a tenth of it into Latin. Halley says that he acquired his knowledge of Arabic from comparing Bernard's Latin version with the Arabic, although Hearne repeated a story that Halley had got a Mr Jones to do the translation:

> John Keil mentioned that Ld Pembroke had been told that Halley did not translate Apollonius out of Arabic himself but got one Jones to do it, which Mr. Halley cannot but resent as a great indignity.[44]

Others have been doubtful of Halley's Arabic competence (see below), but he did say how he came to understand it, and he had earlier commented on a Latin translation from the Arabic in his study of the work of al-Batānī (Chapter 8). The knowledge he had of Hebrew from school and university could also have helped him. In any event, Halley completed the Latin version, with guidance from the lemmas that Pappus had provided, and the edition is effectively his. No complete version exists of a related work, again in two books, *De sectione spatii* or *The cutting-off of an area*, which deals with a similar problem about the area enclosed by two intercepts (that is, their product) instead of the ratio of the one to the other. Halley attempted to restore the work and he published his version with the *De sectione rationis*. The Latin text with diagrams was published from the Sheldonian Theatre in 1706, as

> Apollonii Pergaei *De sectioni rationis* libri duo ex Arabico MS[to] Latine versi. Accedunt ejusdem *De sectioni spatii* libri duo restituti.

500 copies were printed in octavo to be sold at a price of 2s. 6d.; 122 were remaindered in 1713.

The costs of production are recorded as 'Printing, 22s. 14d., Paper, 10s. 16d.[45]

Apollonius was by no means the first Greek mathematician to write about conic sections.[46] All Greek studies of conics depended on the geometrical methods of Euclid, for the algebraic geometry of Descartes was unknown, although the Greeks knew equivalent forms for many of the important algebraic properties. The *Conics* had eight books, none of which has survived in Apollonius's original form. The first four were esentially a compendium of all that had gone before Apollonius, while the second four contained new work by him. In the sixth century AD Eutocius edited the first four books, with a commentary, and that is the Greek version that survives. Eutocius apparently intended to edit the other four books but never did. The eighth book was lost at some later time and is known only through references to it by Pappus and others. A few diagrams may have been known to the Arabs.[47]

The first satisfactory source of Western knowledge of Apollonius was the Latin version of the first four books by Fredericus Commandinus, published at Bologna in 1566 (earlier imperfect versions had been printed in the fifteenth century in Bologna and Venice). Claudius Richardus published an edition of the first four books with commentary at Antwerp in 1655. Edward Bernard produced a version of the first seven books in 1707 in which he incorporated the work of Commandinus but also used manuscripts in Leiden.[48] Wallis had already published the lemmas of Pappus, and Barrow had written an explanation of the work of Apollonius. Halley used a Greek text that had been copied for Aldrich in 1704 from a late manuscript in the Royal Library in Paris (now in the Bibliothèque Nationale). It reached England in 1706.[49] Eutocius wrote a commentary on the *Conics*; there is a Greek manuscript in the Barocci Collection in the Bodleian Library.[50]

Other and better manuscripts were found after Halley's day, and in the nineteenth century J. L. Heiberg published a Greek text based on all available manuscripts, a Latin translation of the first four books of the *Conics*, and the supplementary lemmas of Pappus and the commentary by Eutocius, all of which Halley had also included.[51] Heiberg based his edition on Greek texts earlier than Halley's single one, but Halley's edition remains of great value and his emendations to his MS were usually accepted by Heiberg.

A Greek manuscript of Books I to VII was acquired by the Banū Mūsā in the ninth century AD. The Banū Mūsā were three brothers who were able mathematicians and served the caliph al-Ma'mūn in various capacities. Their manuscript of books I to VII pre-dated the edition of Eutocius and they could not understand it, even if they knew Greek. Later they obtained a manuscript of Eutocius's edition with its comments which opened the whole to them. They had the first four books translated by Hilāl b. abī Hilāl al-Himsī and Books V–VII by Thābit ibn Qurra (836–901). It may have been state policy at that time to commission translations of Greek works. The lost Book VIII is known only from lemmas and comments by Pappus and from an Arabic epitome.[52]

Only the first four books of the *Conics* were known in the West until Jacobus Golius acquired a codex of the Banū Mūsā version of Books I–VII in 1627. It eventually came to the Bodleian Library and was Halley's principal source for the Arabic text.[53] A second Arabic manuscript in Bodley's Library was brought back from Constantinople by Christianus Ravius about 1640 and there are others in Bodley's Library, in some of which the text of the *Conics* is incomplete.[54] Two further manuscripts of the Banū Mūsā edition have become available in recent years, one now in Istanbul of 1024 AD and one in Teheran of 1290 AD.

Gregory records the inception of the work on the *Conics* in 1706:

> The original Arabick ms. from whence Ravius translates the 5th, 6th and 7th books of Apollonius into latin, & which Mr Baynard has in 8° is now in the Bodleyan Library.

and

> We have two Arabick printed editions of Apollonius's last 3 books, & those with the mss. copies & translations will be sufficient to restore Apollonius's Conicks intirely which work I believe Mr Halley & I shall undertake. The 8th book of Apollonius seemes to have been neglected to be written out, not being elementary, but containing Problemata Conica determinata as Pappus tells us.[55]

Those entries indicate that Gregory and Halley did not, perhaps could not, contemplate the edition of the *Conics* before the Golius manuscript arrived in Bodley's in 1706.

Gregory and Halley began work together on Books I–IV, using the Latin version by Commandinus as the basis for their parallel Latin text. Gregory died in 1708, so that they collaborated for less than two years and the edition as published was essentially the work of Halley. He attached to his edition of the *Conics* one of the *De sectione cylindri* and *De sectione coni* of Serenus (dealing respectively with plane sections of cylinders and cones and how they are related), 'ex Codd. MSS Graeci edidit', for which he used a copy made for Aldrich of a manuscript in Paris.[56]

Halley published his version of the *Conics* from the Sheldonian Theatre in 1710. He gave a brief account of the life of Apollonius in the preface, particulars of the manuscripts that he used, and some account of the history of conic sections in Greek and Arabic mathematics. Each of the first four books takes the same form, first the relevant lemmas of Pappus, then the Greek text and Halley's Latin version in parallel columns, together with some text figures, mostly by Halley. Each proposition of Apollonius is followed by the commentary of Eutocius, also in Greek and Latin parallel columns. The form of Books V–VII is different because there is no Greek text. Halley did not print the Arabic but only his Latin version. However, he had a Greek text of the lemmas of Pappus and he set those for Books V and VI in parallel Greek and Latin columns before the respective books. He printed the lemmas for the two Books VII and VIII before his version of Book VII.

No text survives for Book VIII, neither in Greek nor Arabic, and it is known only through the Arabic epitome by Abd al-Malik al-Shīrāzī and from Pappus's summary. By comparing the *Epitome* with the lemmas of Pappus, Halley was able to identify the lemmas that related to Book VIII but were not needed for Book VII, and on the basis of that material he made a conjectural reconstruction of Book VIII.

Halley had completed his edition and versions by February 1709, the date on which the Vice Chancellor gave his *Imprimatur*. He began with Gregory in 1706, so that the *Conics* took him at the most three years. It was not his only occupation in that time.

400 copies in folio were printed to be sold at a price of 22s. 6d. 95 copies remained unsold in 1712 and a remainder notice in 1713 offered to gentlemen of the University: 'Folio *Apollonius* Conica by Dr *Halley* Gr. Lat. £0.15s.0d'.

Halley also prepared a Latin version of the *Sphaericorum* of Menelaus of Alexandria (AD 100), a treatise on spherical triangles in three books. On 20 April 1711 the Delegates

> Agreed that Mr Professor Halley take ye Care of Mr Grave English tracts & prepare them for ye press That he likewise prepare Menelaos Pharicos.[57]

Grave was probably Greaves, the former Savilian Professor of Astronomy from 1643 to 1648, at which time he was ejected by the Parliamentary Visitors. He was an early Egyptologist. Halley would have been an appropriate editor of his works, which in a number of ways foreshadowed his own activities.[58]

Halley based his edition of Menelaus on Hebrew and Arabic versions that derive from Nāsir al-Dīn al-Tūsī (1201–1274). Two Arabic MSS, two Hebrew MSS, and one Greek MS are in Bodley's Library.[59] Halley made a Latin version and sheets were printed in 1713–1714 as pp. 7–70 of a book that was not published at the time, although there are copies in the British Library and in Bodley's Library. The sheets were put in store in the Sheldonian Theatre and it seems that Halley was intending to publish his edition in 1713. Nicholas Bernoulli, the nephew of Jacques Bernoulli and subsequently professor at Padua, visited London about that time and after his return to Basel he wrote to Monmort in Champagne on 30 December 1712, 'Mr Halley fait imprimer Menelai Sphaericorum Libros Tres' ('Mr Halley is going to print the three books of Menelaus's *Sphaericorum*') and again a year later on 30 December 1713, writing to de Moivre in London,

> Menelai Sphaericorum de M Halley se sont sans doutes imprimés, je vous prie d'assurer ce Monsieur, comme aussi M Newton, de mes tres humbles respects et de leur souhaiter de ma part la nouvelle année avec toutes sortes de prosperité, ce que je souhaite aussi à vous même étant de tout mon coeur et avec un attachement inviolable.[60]
>
> Mr Halley's Menelai Sphaericorum are without doubt printed; I beg you to assure this gentleman, as also Mr Newton, of my very humble respects and to convey my wishes to them that the new year may bring all sorts of prosperity, as I wish for you also with my heart and with unbreakable affection.

But 'Menelai *Sphaericorum*' were not to see the light in Halley's lifetime and it is not known why. Halley was elected a Secretary of the Royal Society in 1713 and it may be that duties in London took up time. Whatever the reason, Halley apparently did no more after the sheets for pages 7–70 had been put in store. In 1758 G. Costard was asked to prepare them for posthumous publication and he provided a preface to fill the pages 1–6. In this he expressed some surprise that Halley should have been able to understand Arabic and Hebrew. The comment rather suggests that Costard did not know Halley very well, for Halley would have had some Hebrew in his last years at school and as an undergraduate, and he had already dealt with the Arabic versions of Apollonius.

Costard's book was published in 1758 in octavo with a print of 250, to be sold at 1s. 2d. 188 copies were unsold in 1799.

No further editions of the Greek mathematicians came from Oxford after Menelaus.

Mathematicians throughout Europe acclaimed Halley's editions. Johann Bernoulli had the *De sectione rationis* of Apollonius from Halley in 1707, as he wrote to P. Varignon:

> J'as enfin reçu le paquet d'Angleterre où il y avoit aussy une Optique latine de Mr Newton un Apollinius Pergaeus de Sectionis Rationes traduit de l'Arabe par M Halley.
>
> I have finally received the parcel from England in which there was also a latin Optick of Mr Newton and a De sectionis rationes of Appolonius of Perga translated from the Arabic by Mr Halley.

In a very complimentary letter of 6 April 1707 he thanked Halley for it and offered a few comments.[61]

Bernoulli had received the *Conics* from William Burnet, a distant relative of Gilbert Burnet, the bishop of Salisbury, who wrote to Bernoulli:

> Mr Halley m'a dit il y'a quelques tems qu'il vouloit vous faire present d'un *Apollonii Conica* imprimé à Oxford sous sa direction; Je vous enverrons doncs ces deux livres.
>
> Mr Halley told me some time ago that he wanted to make you a present of an Appolonius *Conics* printed at Oxford under his direction; therefore I am sending you these two books.

and Bernoulli replied to Burnet:

> J'accept avec beaucoup de joye l'offre de M Halley qui me veut faire present d'un *Apollonii Conica* imprimé à Oxford sous sa direction, je vous prie de l'en bien remercier de ma part, et de luy faire mes complimens; mon Neveu en retournant parroit se charger de ce livre comme aussy de la nouvelle edition du livre de M Newton.[62]
>
> I receive with much joy the offer of Mr Halley who wants to make me a present of an Appolonius *Conics* printed at Oxford under his direction; I beg you to thank him heartily from me and to give him my compliments; my nephew could collect the book on his return, and also the new editon of Mr Newton's book.

The dates of the letters show the leisurely course of Bernoulli's correspondence. The nephew was Nicholas Bernoulli who was about to return from London where he had enjoyed a cordial reception from Newton, Halley, and de Moivre. Halley's editions are mentioned a number of times in letters between de Moivre and J. I. Bernoulli between 1705 and 1712.[63]

Halley's *Conics* has continued to receive high praise, despite the discovery of other manuscripts and more modern approaches to textual criticism and philology. The changes that later editors have made to Books I–IV and V–VII have been essentially minor. Toomer has written of Halley's outstanding achievement in the later books.[64] Halley started with two considerable advantages—he appreciated the subtlety of the mathematics and he wrote good Latin. His emendations of the Arabic, as well as of the Greek, have mostly stood, and have been justified by the witness of manuscripts of which he was unaware.

Halley's reconstruction of Book VIII is a different matter. The evidence for its contents is thin indeed. It may be that the book did contain the matter of Halley's conjectures, but they seem to be less profound than Apollonius's own remarks suggest. There may have been much more to Book VIII than Halley offered. Many years before Halley's time, the noted Arabic mathematician, Al-Hasan ibn al-Hasan ibn al-Haytham (Alhazen) (965–c.1041) had devised a *Completion* of the *Conics* which differs considerably from Halley's. Its basis is apparently less secure than Halley's, but it may be closer in spirit to Apollonius.[65]

Two problems led the Greeks to conic sections. In the construction of a sundial, the cone traced by the shadow of a gnomon is cut off by the plane surface of the dial. The other problem was that of the doubling of a cube—what was the length of the side of a cube of twice the volume of a given cube? The solution was supposed to have been first given by Menaechmus, about 350 BC, and depended on the intersection of an ellipse and an hyperbola.

Greek mathematicians defined the conic sections, as the name indicates, by the sections that a plane cut off a cone and not by algebraic equations. Apollonius gave a completely general definition in terms of any circular cone, whether right or scalene, and was the first to see that the hyperbola has two branches. He derived most of his results by the methods set out by Euclid, arguments dealing with ratios cut off by lines and with areas erected upon them. He showed that all conic sections satisfy the following relation. Let PN be an ordinate drawn from a point P on the curve to a diameter[*] (Fig. 12.2). Let x be the distance from the foot of PN to the intersection of the diameter with the conic. Then

$$PN^2 = x \cdot \xi,$$

where ξ is the *latus rectum* and depends on the form of the conic. It is a constant, p, for a parabola, it is $a(1 - bx)$ for an ellipse, and it is $a(1 + bx)$ for an hyperbola, where a and b are constants.

Much of Apollonius's treatise is a development of the consequences of that relation, derived from the properties of plane sections of cones. Apollonius hardly refers to the definition of the conics by the relation of distances to the focus and latus rectum, nor does he mention the property of the ellipse that the sum of the distances from the two foci is a

---

[*]A diameter is a line that bisects parallel chords and an ordinate is a half chord bisected by the diameter. Diameters and their ordinates are not in general perpendicular.

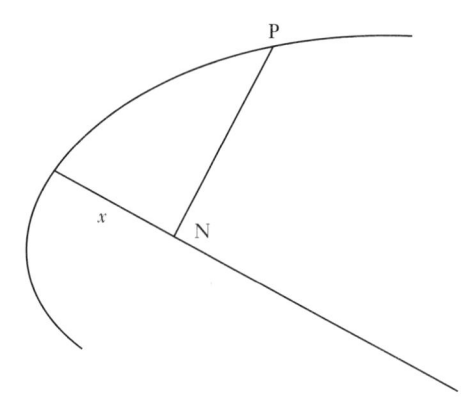

Figure 12.2 General relation for a conic section. PN: semi-diameter.

constant, nor that of the hyperbola, that the difference of the two distances is a constant. Those properties of the ellipse and the hyperbola were known in the mid-seventeenth century, when Halley made use of them in his very first paper. Apollonius does devote considerable space to the properties of tangents and normals, how to draw conic sections touching given lines or through given points, and theorems about limits. Although he does not employ algebraic methods, many of his results, expressed in terms of areas or ratios, are equivalent to algebraic equations referred to oblique axes of tangents and diameters and it is thought that they may have given Descartes his ideas about algebraic equations of curves.

Apollonius's exposition of the Greek treatment of conic sections is an abstract structure that is important in its own right and incorporates many elegant results that are not easily expressed algebraically. In Halley's day it had the very important additional significance that it was one of the bases of Newton's methods in the *Principia*. Greek geometry cannot, however, readily handle changes and for that, since dynamics is about changes and ratios of changes, Newton devised his method of first and final ratios. No one now believes, as Newton came to assert in his controversy with Leibniz, that he first worked out the results of *Principia* by fluxions and then put them into geometrical form. He did something far more profound, he used in effect the idea of limiting processes, through the notion of first and final ratios, that did not enter analysis until Cauchy laid a satisfactory logical basis for the calculus in the form of limiting processes in algebra. Newton's mathematics was geometrical, not algebraical, and he calculated limits by the methods of Euclidean geometry. That may be seen very clearly in the very first result of the *Principia*[66] in which he established that the motion of a body under an attraction to a centre is in a plane and is described with

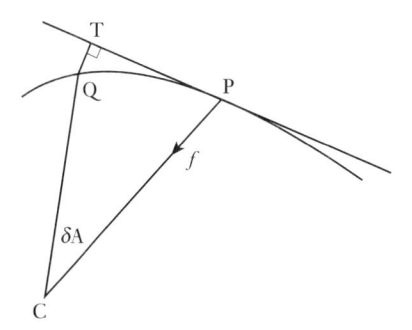

Figure 12.3 Curvature of arc PQ. PT: tangent at P; C: centre of curvature; δA: increment of area; f: force at P.

constant areal velocity, the 'angular momentum theorem' as we would now understand it. The calculations are Euclidean, the approach to the limit is Newton's, and not very explicit.

The areal velocity theorem is one of the two on which Newton based his treatment of orbits. The other is the relation between curvature and force developed in Book I of *Principia*, Prop. 6, Th. 5. Let PT (Fig. 12.3) be the tangent to a curve at the point P. If Q is a successive point of the curve, and if QT is perpendicular to PT, then the limit of QT divided by the arc length PQ is a measure of the curvature of PQ. Let the acceleration to C, a centre of force, be f. Then in time δt the displacement QT will be $\frac{1}{2} f \delta t^2$. But δt is proportional to δA, the increment of area under the curve, and consequently the force on the body at P is the limit of $QT/\delta A^2$. Newton's expression in Prop. 6, Th. 5 relates the force to a geometrical form of that limit and he gives others in the following propositions. His treatment of orbits that are conics involves relations between those geometric forms and the geometrical properties of conics that he takes from Apollonius and from four original lemmas of his own. In that way he developed the theorems in which he proved that the attraction to the centre of an elliptic orbit is directly as the distance from the centre and the attraction to the focus of an elliptic orbit is as the inverse square of the distance. He very rarely refers to Apollonius specifically: one instance is a lemma in a section that deals with the construction of conics when the focus is not given.[67] Geometrical methods were important for the practical derivation of orbits of planets and comets from observations.

Halley himself was familiar with at least the elementary properties of conics from his undergraduate days. Later, writing on the numerical solutions of cubic and quadratic equations, he adopted the common idea that they derived from the intersections of two conics, and he used elementary properties in his studies of cometary orbits in the 1690s. Neither Newton nor Halley required the more advanced and sophisticated results in the later books of Apollonius, and the results that they did use they could have obtained from a book such as the summary by Isaac Barrow that Newton owned.[68]

It is not difficult to see why the treatise of Apollonius would have little influence on later celestial mechanics. Greek conic sections dealt with exact conics; the orbits of celestial bodies are never exact conic sections. The Sun attracts the Moon, planets attract each other, the major planets attract comets, and artificial satellites of the Earth move under an attraction that is not exactly inverse square because the Earth is not a perfect sphere. The methods of celestial mechanics must be able to handle curves of more general forms than exact conic sections, and had to await the application of analytical procedures and the infinitesimal calculus by Clairaut, d'Alembert, and Euler, although Newton went some way towards this with his calculations of the changes of the elements of orbits (see Chapter 13). None the less, Halley's edition of Apollonius remains an achievement in which his linguistic ability and mathematical insight were combined to resurrect and preserve a work of intellectual depth and elegance.

Other of Halley's plans did not come to fruition. Gregory and Hearne had both noted that he planned to bring out an edition of Ptolemy's *Geography* and he had collected material for it. At one time he asked Roger Gale, the son of his old headmaster, for the loan of a manuscript that had belonged to Thomas Gale. On 21 November 1709 Hearne wrote to his friend Dr T. Smith in London, that Halley, 'this great man', had nearly finished Apollonius, thinks of doing Ptolemy's *Geography*, and has been collating Mercator's edition with a MS in the Imperial Library. Ten years later Hearne noted that Halley was laying aside Ptolemy, although he had done much on it, on account of a foreign publication.[69]

The editions of Apollonius and Menelaus were Halley's most obvious achievements during his time at Oxford, but they were far from all in a busy and productive sixteen years.

A calendar of his activities in outline would be:

1705	First publication of *Astronomiae cometicae synopsis*
1706	Publication of Apollonius, *De sectione*
	Begins work on *Conics* of Apollonius with Gregory
1708	Death of Gregory, Halley continues with *Conics* and takes over edition of *Historia coelestis*
1710	Publication of *Conics*
1712	Publication of *Historia coelestis*
	Menelaus in train, may have begun work on lunar tables

1713	Menelaus set in print
	Publication of second edition of *Principia* with final lunar theory
	Elected Sec. R.S.
1713–20	Preparation of lunar tables; revision and extension of cometary calculations
1715	Organisation and discussion of observations of the solar eclipse
1720	Astronomical tables set in print by this year.

The lunar and cometary studies are the subjects of Chapter 13.

Halley lectured on mathematical topics. No complete list of his lectures exists but some lectures were published—they are essentially an exposition of his work on solutions of polynomial equations.[70] In addition, in the four or five years before he went to Greenwich, he published a number of important papers on a wide range of topics, some of them accounts of work that he had done years before, others contemporary.

12.5 *Lights in the heavens*

Halley wrote nothing for the *Philosophical Transactions* in the twelve years from 1702 to 1714, no doubt because he was occupied with the Greek mathematicians, with the Moon (Chapter 13) and, at the end of the period, with the edition of Flamsteed's *Historia coelestis* (Chapter 14). When he was elected Secretary of the Royal Society in 1713 he took over the publication of *Philosophical Transactions*.

Volume 29 of *Philosophical Transactions* has thirteen contributions of his, almost all of which are of abiding interest. He collected observations of a number of meteors and showed that on three occasions they must have been at heights of 40 miles (64 km) or more. Because the atmosphere at that height is very thin, he considered that meteors could not be atmospheric phenomena.[71]

He discussed some French observations of the magnetic variation in the region of South America.[72] He was always keen to obtain data that would confirm or extend his Atlantic and World charts; this paper is additionally of interest for the naming of the Falkland Islands, which is sometimes attributed to him, but it appears that he just reported it:

> Besides I have had in my Custody a very curious Journal of one Capt. Strong, who went into the South Seas in quest of a rich Plate-Wreck and discovered the two Islands he called the Falkland Isles, lying about 120 Leagues to the Eastwards of the Patagonia Coast about the latitude of $51\frac{1}{2}$.

In Number 344 of the *Transactions*, pages 285–94, Flamsteed's observations of the Moon and planets at Greenwich for the year 1713 were summarised, no doubt by Halley by direction of the Visitors, and to Flamsteed's annoyance.

In the same number Halley discussed the saltness of the seas and of lakes without outlets, and showed how a limit may be set to the age of the Earth by relating the size of a lake to the rate at which rivers carried salt into it. He pointed out that if a lake had an original salt content, then its actual age would be less than that so estimated. He wished that the ancients could have determined the saltness of the sea 2000 years ago to see how it changed with time, but that was not possible. He argues that the limit on the age of the Earth obtained from the saltness is a maximum, and so refutes 'the notion of the Eternity of all Things, though perhaps the World may be found much older than some have imagined'. Halley is clearing himself from the old accusation that he denied the finite age of the world.[73]

In an anonymous piece, but clearly an 'editorial' by Halley, he collected the records of new stars that had appeared in the previous 150 years.[74] He distinguished from the rest the two that appeared in 1574 (Tycho's in Cassiopeia) and 1604 (Kepler's); they were exceptionally bright but faded quickly and could not be seen after two years. The three others were much less bright. That in Colli Cygnis, first seen in 1596, appeared and reappeared periodically after the first detection seven times in every six years, while the one first seen by Kirch in 1686 had a very regular period of 404.5 days. Thus Halley distinguished between novae and supernovae, as they are now known. The remains of supernovae are powerful radio sources at the present day.

Halley wrote a further editorial summary about nebulae; that the paper was his seems established by a reference back to the novae paper.[75] He listed nebulae, including the great nebula in Andromeda which was discovered by Huygens in 1656, and two that he had discovered, in Centaur (St Helena, 1677) and Hercules (1714). He concluded from their finite sizes and from the fact that they show no annual parallax that they occupy spaces immensely great, larger than the solar system. He also remarked, in relation to the sequence of creation in the Hexaemeron of *Genesis*, that they show that light was created before the stars—that may have been a theologically significant observation at the time, but it also foreshadows the modern perception of gaseous nebula as the forerunners of stars.

Halley contributed an account of aurorae that he himself saw in March 1716, as he watched from his house (in London?) until 3a.m.[76] A luminous cloud over the horizon with coloured rays, yellow, red, and dusky green, stretched up almost to the zenith. It moved to the north west, and declined after a few hours. He remarked that the rays have been compared to the glory round God in religious paintings and to the

Star of the Garter. The rays were more intense in the north of England and in Scotland. He mentioned previous reports of aurorae, the name of which, *aurora borealis*, he attributed to Gassendi in 1621; he listed those of Camden and Stow in 1574, of others in 1560, 1564, 1575, 1581, of 1707 in Ireland and Rome, and of 1708 in London. His account shows, as do many other of his papers, his interest in the geographical extent of phenomena and in their occurrence in times past, matters that are fundamental to geophysical studies.

Halley considered possible sources of the clouds and rays and argued that, because of their great geographical extent, they could not arise from, for example, volcanic vapours. He showed that the forms of the rays corresponded to the field lines of a uniformly magnetised sphere, as displayed by iron filings, and he argued that the aurorae were the result of circulation of matter in the Earth's field, most intense at the poles where the field lines crowd together (Fig. 12.4). He also pointed out that the magnetic pole of the Earth was displaced from the geographical pole and that the auroral phenomena followed it, and so ruled out every possible cause except that of the magnetic field. Finally he referred back to his idea of the Earth having a shell and mantle, both magnetised, and speculated that the medium in the space between them may exude from the poles into the atmosphere, because at the poles the shell is thinnest on account of the polar flattening of the Earth. That is not so, but his analysis of the structure of aurorae and their correspondence with the Earth's magnetic field remains impressive after nearly three hundred years. He was right, and for more than thirty years the study of how charged particles (of the solar wind) circulate in magnetic fields of the Earth, of Jupiter and other planets, and of stars, has been a major field of space research.

Halley edited the next volume of *Philosophical Transactions*, Volume 30, and prefixed a dedication to Thomas Lord Parker Baron of Macclesfield, Lord High Chancellor: 'as a small Acknowledgement for very Great Favours'. Parker had, as an amateur astronomer, observed the eclipse of 1715 from Crane Court and he maintained an observatory at his seat, Sherbourne Castle near Oxford, that was sufficiently significant for Halley to include its position in the list of sites in his 'Tables'. The dedication may mean that Volume 30 was only published after Halley had become Astronomer Royal, for Parker was one of those who canvassed his succession to Flamsteed (Chapter 14).

Halley published on astronomical topics in both volumes 29 and 30. He explained the use of transits of Venus to determine the solar parallax, as he had first expounded to the Royal Society some years before (Chapter 8).[77] He described occultations of stars by Jupiter and Mars, he called attention to appulses of the Moon by stars of the Hyades, he

Figure 12.4 Halley's sketch of magnetic field lines of the Earth (from *Philosophical Transactions*, **30**, No. 363, 1099–100. (Reproduced by permission of the University Librarian, Cambridge.)

described his detection of a small comet with a 24 ft (7.32 m) telescope, and he collected reports of a notable meteor seen over much of England.[78]

In his most significant paper in Volume 30 Halley demonstrated the proper motions of some stars which he apparently detected while preparing his *Astronomical tables*.[79] He showed that the changes of latitude of three stars, since the determinations by Hipparchos as given by Ptolemy, were inconsistent with the general change of latitude produced by the luni-solar precession of the equinoces since that time. Palilicium

(Aldebaran), Sirius, and Arcturus were all about 30′ more southerly now than they should have been according to Ptolemy, while the bright star in the shoulder of Orion was more northerly. Halley thought he could detect a change of 2′ in the latitude of Sirius since Tycho's observations, and he also noted that Bullialdus recorded an ancient observation at Athens on 11 March AD 509, when Palilicium was nearly eclipsed by Mars. Halley remarked that the bright stars were probably closest to the Earth so that any changes in their positions would be the ones most likely to be seen. His discussion, like that of his suggestion that the Moon was accelerating, depended upon his critical use of historical data and implicitly upon accepting that the cosmos might be changing. It also shows that he had a modern concept of the distribution of stars in space beyond the solar system. The two most important discoveries in astrometry of the time were Bradley's discovery of the aberration of light and Halley's detection of proper motion. Flamsteed prepared his catalogue in ignorance of those two phenomena and they made it, and all previous catalogues, out of date.

Halley was always alert to compare new observations or publications with earlier results of his own and so, after a lunar eclipse had been observed at sea from the ship *Emperor* 180 leagues to the east of the Cape of Good Hope, probably off the present Port Elizabeth, he revisited the question of the longitude of the Cape; he had previously disputed a French determination.[80] The eclipse was not observed in London but, from one observed there 36 years and 20 days (two saronic periods) before on 11 February 1682, Halley concluded that the longitude of the ship was 26° east of London and hence that the Cape was 15° east. He obtained confirmation from observations of the Moon at Table Bay on 4 and 5 August 1694 by a Mr Alexander Brown on his way to reside in India for the East India Company. The Moon was not observed in Europe on those days but Halley found an observation of his own made on 23 July 1676, one saronic interval earlier, from which he deduced that the longitude of the Cape was 16° 30′ east of London. Both values of the longitude of the Cape are too westerly, the actual value being about 18° 30″. The interest of this otherwise minor study is that, because the necessary contemporary observations in London had not been made, Halley used lunar observations from the corresponding times in previous saronic cycles to derive the longitude, on the assumption that the perturbations of the Moon repeated at correponding times in the cycles. He had had that idea since 1682 and when he went to Greenwich, he investigated it more thoroughly (Chapter 13).

Halley resigned as Secretary of the Royal Society when he became Astronomer Royal in 1720. Volume 31 of *Philosophical Transactions* was edited by James Jurin, his successor as Secretary, but he continued to contribute papers on a variety of topics, among them a number of relatively minor points of astronomical technique.[81] He also now published his account of diving, which previously he had only given in a series of talks to the Royal Society (Chapter 9).[82]

Halley's two most original papers of Volume 31 are those on the numbers and light of the fixed stars.[83] He first considered the proposition that the number of the fixed stars is infinite, one argument for that being that if the number were finite they would all eventually coalesce under their mutual gravitational attraction, as Newton had considered. Against that proposition Halley argued that the number of stars cannot actually be infinite for it is certainly definite even if unknown, and so there must be a number which the number of stars, however large, does not exceed. He similarly argued that there must be two stars with the greatest distance between them. Halley's second objection was that people argued that, if the number of stars was infinite, the whole celestial sphere would be luminous, for the number of stars would be greater than the number of seconds on the sphere (here Halley presumably assumed that the diameter of a star is about 1 arcsec as seen from the Earth). That is Olbers's paradox and from the way that Halley stated it, others must have seen it before him.

In the second paper he estimated the number of stars of given magnitude as seen from the Earth. He suggested that all stars have the same intrinsic brightness as the Sun and they appear to us of different magnitudes because they are at different distances, the stars of first magnitude being the closest. He observed that not more than 13 points can be set on a spherical shell equidistant from each other and compared that with the number of stars classed as first magnitude, which was 16.[84] He then estimated the number of stars of other magnitudes and suggested that at 100 times the distance of the Sun there would be 130 000 stars of luminosity 10^{-4} of those of first magnitude. He argued that such an intensity is below the threshold of the eye and so the sky would not be seen as bright.

There are two subtle arguments here. The first is in essence that the universe is isotropic—it looks the same from wherever we are: that is the basis of the calculation of the numbers of stars of different magnitude. The second argument is that the eye as a detector has a threshold and that, therefore, there is a limit to the number of separate stars which would produce any response from the eye. Olbers's paradox is difficult to get round; it does not do to suppose that there is an absorbing medium in interstellar space because the medium would heat up to the radiation temperature of the stars and then itself radiate. Halley's threshold argument also seems to fail: it assumes that we see each star individually, whereas we see diffraction images that overlap and so produce a greater intensity at the eye. Halley speaks of the response of the eye, but of course his argument and the objection to it apply equally to any other detector. Olbers's paradox was discussed by a number of people after Halley, by Kant, and by Olbers who gave his name to it. It was resolved by the demonstration that the galaxies were all moving away from each other so that the frequency of the starlight, and therefore the intensity of radiation, was reduced by the Doppler shift for the more distant objects.

In his years at Oxford Halley was principally occupied with his Greek geometers, with his Astronomical Tables, and with seeing Flamsteed's catalogue through the press. He reverted to some of his earlier interests and published accounts of diving, data on the Earth's magnetic field, and determinations of longitude, but he also published short yet important papers which showed that his insight into the physics of the Earth and the Cosmos was as perceptive as ever. His discussion of aurorae and their relation to the magnetic field of the Earth, his remarks on novae and nebulae and, above all, his discovery of the proper motion of stars and his two short papers on the number and brightness of the stars, show that in his late fifties and early sixties he was as original as anyone and set problems that are still with us today.

12.6 The eclipse of 1715

The total solar eclipse of 1715 was the first to be seen in London since 1104. Halley worked out the path of totality across England and published a map of his predictions (Plate XVI), partly to let people see that there was nothing mysterious or magical about an eclipse, but mainly to help observers to record the event and especially the duration of totality. He also wrote to a number of careful observers and asked them to send reports to the Royal Society.[85] The skies were clear over London and many valuable reports came in.

Never before had an important natural event been recorded by many observers in different places in an organised and coordinated programme. It was very successful. Nowadays we expect that eclipses and other natural geophysical and astronomical phenomena will be observed systematically, but it was long after Halley's day before his example was generally followed.

A number of fellows of the Royal Society gathered at Crane Court, among them the Earl of Abington and Lord Justice Parker. There were foreign guests as well, including de Louville and Montmort from Paris. Montmort met Newton's niece, Kitty Barton, and seems to have been as much taken with her as with the eclipse.[86] Halley himself observed the time of totality, using a quadrant of 30 in (76 cm) radius and a pendulum clock, while his companions used other telescopes. The eclipse began at 8 h 6 min, totality lasted from 9 h 9 min 3 s to 9 h 12 min 2 s, and the eclipse ended at 10 h 20 min. Halley noted bright flashes around the rim of the Moon at totality (Bailey's beads), which he correctly attributed to the irregular surface of the Moon. He also recorded the luminous ring around the Moon with rays of light going out from it, the corona, of which he gave a drawing in the paper.

James Pound FRS, the rector of Wanstead in Essex (where he had an observatory), sent a report to Halley, and so did the rector of Uppingham in Essex, William Derham FRS, who was accompanied by Samuel Molineux, the Secretary to the Prince of Wales. Oxford had cloud. Roger Cotes, the Plumian Professor of Mathematics at Cambridge, missed important features because of 'too great Company'. Greenwich sent little: Flamsteed told Sharp to send nothing to Halley before he, Flamsteed, had seen it, and to refer Halley to Flamsteed for the Greenwich account of the eclipse. The Greenwich observations seem not to have been published.[87]

Halley gave a critical account of the observations he had received from the country, especially from people near the edge of totality. He had estimates of the duration of totality from near the centre of the track, 4 min at Plymouth, Exeter, Weymouth, Lyme Regis, and Daventry, 3 min 53 s at Barton near Kettering, and 3 min 52 m at King's Walden between Luton and Stevenage. He estimated that the centre crossed the coast of Norfolk between Wells and Blakeney. The southern limit was between Newhaven and Brighton on the south coast and between Hern and Reculver on the north coast of Kent. The northern limit lay across the country from Haverford West through Shrewsbury, Congleton, Pontefract, and Doncaster to Flamborough Head. Halley pointed out that the area of totality on the ground was that cut off by the intersection of a cone with the spherical surface of the Earth. If the Sun was high in the sky, it was sufficient to take it to be a plane section of the cone, an ellipse, but when the Sun was low in the sky that was not good enough; Halley calculated the dimensions of the area and the speed with which it moved over the ground.

Halley's organisation of his correspondents, and the new phenomena that he himself observed, would be notable in any case, but they have lately served a purpose of the sort that as an historian he would have appreciated. There are places in his compilation of country-wide observations at which the eclipse was very well recorded as just total but no more. They lie very close to the limits of totality for a Sun of the present diameter and are inconsistent with the limits for a larger Sun. The diameter of the Sun as seen from the Earth has not decreased by more than 0.1 arcsec in 959.63 arcsec, from 1715 to 1988.[88]

12.7 *A mathematician among mathematicians*

Johann Bernoulli considered Halley to be one of the three or four outstanding English mathematicians, along with Newton and de Moivre. Writing to William Burnet, he refers to 'Messrs Leibniz, Newton, Halley et d'autres solides Mathematiciens', and again to Montmort, referring disparagingly to Keill, 'Il me semble que les veritables Mathematiciens d'Angleterre, comme sont Mrs Newton, Halley, Taylor, etc. le devroient chasser de leur Societé' 'It seems to me that the true mathematicians of England, as are Messrs Newton, Halley, Taylor etc., should drive him from their society'. He often expressed his admiration for Halley as a mathematician in letters to de Moivre.[89] Nowadays it may seem difficult to understand why Halley was seen as a great mathematician by his European contemporaries. He made no notable advance in pure mathematics, no developments of series came from him, nothing like fluxions or the calculus. The mathematics that he used was of his time. He made use of series, mainly developed by Newton, for his theory of logarithms and in his schemes for the numerical solutions of polynomial equations. He employed geometrical methods in his calculations of cometary orbits and he used spherical trigonometry in his study of the rhumb line and extensively in navigation. He seems to have made little use of fluxions and none of Cartesian analytical geometry. He followed Newton, but without Newton's originality; Halley was no innovative pure mathematician in the modern sense. In today's characterisation, he was an applied mathematician. He applied the mathematical methods of his day with great effect over a wide range of topics, over a wider range than Newton, and in many he opened up new fields or new ways of seeing old problems.

13

The matter of the Moon

> To behold the wandering Moon,
> Riding now her highest noon,
> Like one that has been led astray
> Through the heav'n's wide pathless way ...
> MILTON: *IL PENSEROSO*, 1, 65

13.1 *The preoccupation of a lifetime*

MOST astronomers in the latter half of the seventeenth century studied the Moon, whether it was its surface appearance, which Hevelius in particular had delineated in great detail, or the libration, which Cassini had defined in three laws, or the orbital motion. Halley never seems to have concerned himself with the surface of the Moon nor with its libration, but he did study its motion in its orbit for many, many years. Theories of the orbit were only descriptive when he began, probably when still at school. The thoughts he had on the orbit at St Helena were purely geometrical (Chapter 3). He observed the Moon from Islington on many occasions in the first two years of his marriage. A reference in the first edition of the *Principia* of 1687 may indicate that he obtained new results on the variation. In Prop. 29 of Book III, on the variation, Newton wrote:

> Halleius autem recentissime deprehendit esse 38′ in Octantibus versus oppositionem Solis, & 32′ in Octantibus solem versus.*
>
> Halley also very recently discovered it to be 38′ in the octants towards solar opposition and 32′ in the octants in the directions of the Sun.

Recentissime ('very recently') seems to suggest a discovery that Halley made at Islington. Three more years passed until he brought the printing of *Principia* to its conclusion and realised its potential for lunar theory and for navigation. Then he urged Newton to

> ... resume those contemplations, wherein you have had such good success, and attempt the perfection of the Lunar Theory, which will be of prodigious use in Navigation, as well as of profound and subtile speculation.[1]

---

*Halley himself changed *quadraturis* to *Octantibus* in the manuscript of *Principia* that went to the printer.

He himself would devote many years of his life to 'prodigious use in navigation' and to 'profound and subtile speculation'.

Halley did not resume lunar studies when he had seen *Principia* through the press, for he was fully occupied with other matters until he went to Oxford in 1704. He was, none the less, well aware of the revisions that Newton was making in his lunar theory from at least 1694. Newton gave some of his new results to David Gregory who incorporated them in the first edition (of 1702) of his *Elements of astronomy*, and Halley praised them to Leibniz when he met him in Hanover.[2] Newton finally published his revised theory in the second edition of *Principia*, omitting that reference to Halley's 'recent' discovery. Halley returned to the study of the Moon before 1710. He constructed tables of the Moon's motion based on Newton's revised theory, and had them set in print about 1720, but did not publish them in his lifetime—that was left for John Bevis. He delayed because he wanted to compare his tables with the long run of observations he started when he went to Greenwich. He also provided tables for the planets, which included the mutual perturbations of Saturn and Jupiter calculated from Newton's treatment of the problem of three bodies.[3]

Halley's extensive studies of the Moon's motion span more than sixty-five years, from his undergraduate days and his year on St Helena, at Islington, to Oxford from 1704 to 1720, and at Greenwich from 1720 to 1739. It seems from the account that he gave in 1731 that he made no observations as Savilian Professor at Oxford, or at least no continuous series of observations, although the material in the archives of the Royal Observatory may include some results of his Oxford years. This then is the place, between his Oxford and Greenwich periods, to set out his contributions to the observation and understanding of the motion of the Moon.

13.2 *Observations of the Moon*

Halley lacked suitable instruments in his first three years at Greenwich and he considered himself, at his age, past all hope of completing a saronic cycle of eighteen years. 'But Thanks to GOD' he observed 'with his own hands and eyes' during one whole period of perigee (nine years) and recorded over 1500 meridian transits of the Moon, not less than the combined total of Tycho, Hevelius, and Flamsteed.[4] He went on for another nine years to complete a full saronic cycle.

Halley almost never used a meridian circle to obtain absolute right ascension and declination until he went to Greenwich. For the most part he derived the Moon's position from those of stars occulted by the Moon or from appulses, close approaches of the Moon to prominent stars of known positions, the utility of which he recommended.[5] He also used lunar eclipses. His papers in the archives of the Royal Observatory preserve three sets of his observations, some from London and Paris from 1679 to early 1682, those from his house at Islington from late 1682 to 1684, and, with gaps, those at Greenwich from 1720 to 1739: he did cover an eighteen-year span at Greenwich. His posthumous *Tables* (1749, 1752) include a summary of his results; the observations themselves were never published but remain in the archives of the Royal Observatory and, for the most part, in a manuscript copy in the Royal Astronomical Society in London.[6] As well as his many observations of the Moon from Greenwich, presumably whenever it was visible, he recorded transits of planets and of a number of stars, typically six or seven each night.

Halley used first a transit telescope at Greenwich and later a mural arc, but he never recorded zenith distances, and it is possible that he was unable to do so, for apparently his instruments had no plumb line or other means of determining the vertical. It is thought that, insofar as he did give declinations, he derived them from declinations of known stars. For the most part, however, he observed only the right ascensions of objects in or close to the ecliptic and, in particular, the right ascension of a limb of the Moon. He also observed some three hundred bright stars in the ecliptic which he intended for use at sea to determine the position of the Moon and its distance from the Sun, and hence the longitude of a ship.[7] He recorded times of transit of a limb of the Sun and hence its right ascension; he also recorded the longitude of the Sun and its mean anomaly. He transferred the observations of the Moon to manuscript books in which he calculated the longitude of the Moon according to the lunar theory of Newton.[8]

13.3 *Principles of lunar theory*

Now, as he is about to begin his Greenwich observations, let us see the problem of the Moon through Halley's eyes, as they were opened by Newton's dynamics. Hellenistic astronomers had identified some of the main features of the Moon's motion and its deviations from a simple path through the heavens. They knew that its path was slightly inclined to that of the Sun, that it took about 27 days to describe that path, that the nodes moved round the ecliptic with a period of 18 years, that its path was eccentric, and that the line of apsides rotated with a period of

9 years. The Greeks knew that eclipses recurred with a period of 18 years (known as the *saronic* period). That is because eclipses can only occur when the Sun and the Moon are in line near the node of the Moon's orbit on the ecliptic. Although the geometry is slightly different at each passage of the Moon through a node, giving rise to eclipses of varying duration or completeness, the circumstances recur when the node returns to its previous position after 18 years. Hipparchos had found that an irregularity known as the *evection* distorted the Moon's motion, as disclosed by vagaries in the times of eclipses. The evection depends both on the mean longitude of the Sun and on the difference of the longitudes of the Sun and Moon, as seen from the Earth. A second major irregularity, the *variation*, that depends only on the difference of the lunar and solar longitudes, was detected by Tycho Brahe. In an elliptic orbit the speed of the Moon would go through one cycle in a revolution, being greatest at perigee and least at apogee. Tycho found that the speed of the Moon went through two cycles in a revolution.[9] Evection and variation could not be explained without a dynamical theory. They are very important for the use of the Moon as a clock in navigation, for each exceeds 1°, equivalent to about 110 km in longitude at the equator.

Horrocks had described the orbit of the Moon as an ellipse with a variable eccentricity and a rotating major axis. Halley thought that was the best way to look at it but allowed the focus to be on a small epicycle.[10] Newton would devote considerable space in *Principia* to the dynamical theory of rotating orbits (below).

Newton's wanted to show in Book III of the *Principia* how all motions of celestial bodies could be accounted for by the universal attraction of gravity. He had already the explanation of the elliptical orbits of the planets to hand when he began the *Principia*, and in the course of composing it he brought comets, the tides, the figure of the Earth, and the precession of the equinoces within the grasp of gravity. He did not change the treatments of those phenomena significantly in the later editions, but he did return to extend lunar theory substantially in the second edition.[11]

Newton set out the principles of the dynamics of three bodies under mutual attractions according to the inverse square law in Prop. 66, Th. 26 of Book I of the *Principia*. In a set of many corollaries he discussed the motion of an inner planet (Jupiter) attracted by an outer one (Saturn), the origin of tides, precession due to an equatorial bulge and how it might be used to investigate the internal structure of a planet and, briefly, the motion of the Moon. All those he developed in Book III. Newton did not give a fully satisfactory theory of the Moon in the first (or any) edition of *Principia*, but did reveal the gravitational attraction of the Sun as the origin of the Moon's complex motion. Thus the effects of the Sun would be the same each time the Moon returned to the same place relative to the Sun, after one complete revolution of the nodes of the orbit on the ecliptic and very nearly two revolutions of perigee, that is, after the saronic interval of 223 lunations, or 18y 11d. There was the justification for the idea that Halley had held since 1682 that numerous systematic observations of the Moon's position, extended over a complete saronic cycle, would enable its position to be found at corresponding times in any other cycle.

Halley intended to check his idea by his programme of observations at Greenwich, and to provide the data needed by seamen to find their longitudes from observations of the position of the Moon relative to the fixed stars. That method, he considered, was then the most practical on a ship at sea: the long telescopes that were required to observe the satellites of Jupiter could not be handled at sea. In support of his belief, he quoted the case of the solar eclipse of 2 July 1684, the circumstances of which he claimed to have predicted very closely from those of the eclipse of 22 June 1666. He also claimed that he had predicted the stages of the lunar eclipse of 19 September 1689 to within two or three minutes from the records of the corresponding eclipse of 8 September 1671.[12]

There are various ways of expressing the effect of the Sun on the Moon. The most direct way is to calculate the position of the Moon in appropriate coordinates by solving the equations of motion of the Moon analytically or numerically. In his analytical theory, Clairaut took the independent variable to be the longitude of the Moon, not the time. Alternatively, equations for changes in the parameters of the orbit are obtained and solved, and the position of the Moon calculated from them. Horrocks's description of the orbit was of that sort and Newton followed him. The nodes, which are where the Moon crosses the ecliptic, are well defined, and so are apogee and perigee, where the speed of the Moon is least and greatest. The difficulty is that the orbit does not lie in a plane, nor is it an exact ellipse, so that other parameters, for instance the inclination to the ecliptic and the eccentricity, are not well defined. Even the well-defined elements are continually changing, so that all the parameters by which orbits are described must be defined in precise and careful ways. Newton in his lunar theory derived the changes in the parameters of the orbit; Euler gave the first algebraic theory of the changes and the method was further developed by Lagrange.[13]

There are two stages in the theory of the motion of the Moon, as disturbed by the attraction of the Sun. First the force of the Sun on the Moon has to be calculated from their positions. Secondly the change in the motion of the Moon has to be calculated from that force. Both steps are complex. The force of the Sun on the Moon is about 1 part in 178 of the attraction of the Earth on the Moon, and simple expressions for the effects are inadequate. The force of the Sun and the effect it has on the Moon vary with their positions, which in turn depend on the force and its effects.

Newton relied on two fundamental results to calculate the effects of the force of the Sun. The first (Prop. 1, Th. 1 in all three editions of the *Principia*) is the areal velocity (or angular momentum) theorem: the areal velocity of a body subject to a force directed to a fixed point is a constant. The area swept out by a radius vector from the centre of attraction is proportional to time, and may replace it in dynamical equations for central orbits. The second result (Prop. 6, Th. 5) relates the curvature of an orbit to the force exerted upon a body, supposing the force to be directed to a centre.[14] Newton also developed a treatment of rotating orbits in a manner that foreshadowed the conservation of energy, although he did not have that concept as such.[15] Strictly speaking, neither of Newton's two theorems applies to the Moon as attracted by the Sun, for the areal velocity of the Moon about the Earth is not constant under the attraction of the Sun, and the force is not directed solely to the centre of the Earth, although there is always a component that is.

Let the vector distance of the Earth from the Sun be $\mathbf{R}_E$ and that of the Moon from the Sun be $\mathbf{R}_M$. The acceleration of the Earth to the Sun is then $\mu_S \mathbf{R}_E / R_E^3$; that of the Moon is $\mu_S \mathbf{R}_M / R_M^3$, where μ_S is the mass of the Sun multiplied by the constant of gravitation. The difference between the two is $\mu_S(\mathbf{R}_M/R_M^3 - \mathbf{R}_E/R_E^3)$, or $\mu_S(\mathbf{R}_E - \mathbf{R}_M)/R_E^3$, since R_E and R_M are very nearly equal. The difference of the force vectors has to be determined from the positions of the Sun and the Moon and resolved into distinct terms depending on them. Nowadays that is done analytically or numerically; Newton did it geometrically.

Newton's diagram for his construction is reproduced in Fig. 13.1. Here T is the centre of the Earth (*Terra*), S is the Sun, and P the Moon. The accelerations of the Earth and the Moon towards the Sun are respectively $\mu_S/(ST)^2$ and $\mu_S/(SP)^2$. The net acceleration of the Moon relative to the Earth is its acceleration to the centre of the Earth under the gravitational attraction of the Earth together with the difference of its acceleration to the Sun from that of the Earth. The point K on the projection of SP is such that SK equals ST. The point L on the projection of SP is chosen so that

$$\frac{SL}{SK} = \left(\frac{SK}{SP}\right)^2,$$

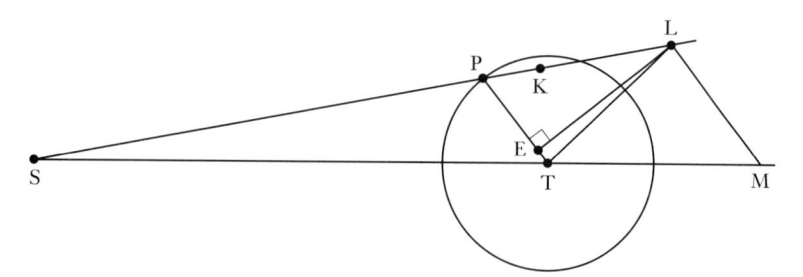

Figure 13.1 The force of the Sun on the Moon. T: Earth; S: Sun; P: Moon. SK = ST.

whence SL is proportional to the gravitational acceleration of the Moon towards the Sun. Newton drew LM parallel to PT, and resolved SL into the components SM in the direction of the Earth from the Sun and LM parallel to the direction of the Moon from the Earth. Since ST is proportional to the acceleration of the Earth to the Sun (as follows when SK and SP are put equal to ST), the perturbing solar accelerations of the Moon's motion are proportional to LM in the direction Earth–Moon and TM (equal to SM–ST) in the direction Earth–Sun. The factors of proportionality depend of course on the masses and distances of the Sun and the Earth from the Moon. An alternative resolution of the forces is into the component LE perpendicular to PT and TE along PT. Then TM and LE will each have two components, one in the plane of the Moon's orbit and the other perpendicular to the plane.

Because the Sun is very distant, Newton took SK to be parallel to ST and therefore TM equal to PL. Now SK is R_E. If PK is put equal to x, defined as the distance of P from the diameter of the lunar orbit perpendicular to SP (Fig. 13.2), then

$$SM - ST = SL - SP = \frac{SK^3}{SP^2} - SP$$

$$= \frac{R_E^3 - (R_E - x)^3}{(R_E - x)^2}$$

$$= 3x$$

since x/R_E is very small.

Newton emphasised that the line ST, and therefore SP, is generally out of the plane of the lunar orbit; LM is, however, always parallel to that plane for PT lies in it. There are in consequence three groups of forces with corresponding effects on the orbit. One is in the plane of the orbit and directed to the Earth, one is directed towards the Sun, and the third is perpendicular to the ecliptic.

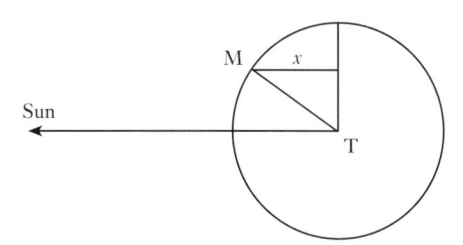

Figure 13.2 The differential force of the Sun on the Moon. T: Earth, M: Moon.

The solar forces acting on the Earth and the Moon depend on the distance from the Sun, which goes through an annual cycle because the orbit of the Earth is eccentric. The solar accelerations therefore vary with the same period. The main orbital motion of the Moon and all its perturbations show annual variations. The amplitude of the annual term in the longitude of the Moon is about 11′ 50″ and the annual variations in the motion of the node and perigee are proportional to it.

According to the relation between curvature and centripetal force in an orbit of constant areal velocity, the less the acceleration, the less the curvature. The *variation* comes about because the acceleration proportional to the component of the vector difference ($\mathbf{R}_M - \mathbf{R}_E$) along the radius vector of the Moon from the Earth varies with the position of the Sun. It is greatest when the Earth, Moon, and Sun are in line at new or full Moon (syzygies) and vanishes when the Moon is in quadrature. It is always negative, so that the average curvature of the orbit is less than it would be in the absence of the Sun and correspondingly the average size is greater than it would be with no Sun (*Principia*, Book III, Props 26, 27, 28). The greatest and least values each occur twice in the orbit, and so the variation is a periodic function of twice the angle between the Sun and the Moon, as seen from the Earth. The radius vector of the Moon has the form

$$a (1 + \alpha m^2 \sin \{2 (n - n') t + \phi\})$$

Here a is the mean radius of the orbit, m is the ratio n/n', where n is the mean motion of the Moon about the Earth and n' is that of the Sun about the Earth. The value of m is about 0.0748. The quantities α and ϕ are constants; α is about $\frac{11}{8}$.

The corresponding amplitude of the variation in true longitude is 2106″ or 35′ 06″; it is a mean value corresponding to the mean distance of the Sun.

Newton's calculation of the variation is, not surprisingly, complex (as are analytical calculations). He gave some numerical values in Book III, Prop. 29 for the magnitude of the variation, ranging from 33′ to 37′ in longitude, according to the distance of the Sun from the Earth. Here as elsewhere the geometrical approach, while making clear the origin of some of the distinct perturbations of the Moon's orbit, is not well adapted to relating them to the parameters of the motion, namely the longitudes of the Sun and the Moon, the longitudes of the nodes of the Sun and the Moon, and the longitudes of their perigees.

Horrocks's model of an elliptical orbit with a rotating line of apsides was a great improvement on any previous representation of the motion of the Moon, and Halley's refinement was even better. In Section IX of Book I of the *Principia* Newton discussed in considerable generality and detail the motion of the line of apsides in the plane of an orbit of a body moving under a central force, not necessarily inverse square.[16] In Prop. 45, Prob. 31 (Cor. 2) he dealt with a body moving in a nearly circular orbit under a central force proportional to the sum of two powers of the distance, A, of apocentre from the centre, namely $(aA^n + bA^m)$. He showed that the rate of motion of the apsides was proportional to

$$\frac{(aA^n + bA^m)}{A^3},$$

and applied that general result to a body moving under an inverse square force ($n = 1$) and perturbed by a 'foreign' force to the centre proportional to $-cA$ ($m = 4$), the case of the solar attraction on the Moon. The rate of rotation of the apsides is then

$$\left(\frac{1-c}{1-4c}\right)^{1/2}$$

per revolution.

Newton estimated that if a foreign force was 1 part in 357.48 of the main inverse square attraction to a centre of force, the line of apsides would revolve by 1° 31′ 28″ in one revolution in the orbit. He then remarked (in the second edition) that the rate for the Moon is twice that; the remark did not appear in the first edition of the *Principia* and was omitted from the third edition. The maximum solar force is in fact 1/178 of the Earth's attraction on the Moon, and Newton's example of half that value corresponds to an average around the orbit. The discrepancy comes about because Newton's theory of the rotation of the apsides includes only the central forces, assumes that the areal velocity is constant, and takes the eccentricity to be negligible. The tangential component of the solar force on the Moon cannot be neglected: the rate of rotation of the apse line depends on both the radial and the tangential forces exerted by the Sun, and the orbit is not an exact ellipse. Newton's theory was thus incomplete.

He had developed a more complete theory in unpublished papers of 1684–1685, but never achieved agreement with the observed motion of apogee. He mentioned those further developments in the first edition of the *Principia* but removed the reference in the second edition.[17] Clairaut embarked upon his analytical lunar theory because he thought that the failure of Newton's theory of the apogee might indicate that gravitational forces were as the inverse cube of the distance; when he carried the algebra to sufficient detail he found that the inverse square law fully accounted for the observations. Newton's result in his unpublished manuscript was equivalent to an hourly mean rate

$$(1 + \lambda \cos 2 \psi) . 8''.9$$

where ψ is the difference of the lunar and solar longitudes, 0° and 180° at syzygies and 90° and 270° at quadrature.

Newton calculated λ to be $\frac{11}{2}$, leading to the low annual rate of 38° 52′, but he then adjusted λ to be 6, corresponding to an annual rate of 40° 43′. Although that is close to the observed value, there is no theoretical justification for it.

Halley based his lunar tables on the observed rate of rotation, not on Newton's theoretical value. He took it to be 40° 23′ 42″ per year, close to Flamsteed's observed rate of 40° 41′ 30″ per year and to the present estimate of 40° 40′ 42″.[18]

The disturbing force on the Moon perpendicular to its orbit corresponds to a couple on the Moon about the line of nodes and, to conserve angular momentum, the nodes move backwards along the ecliptic. That is the modern way of stating things. Newton had no concept of angular momentum and his elegant geometrical argument was complex. As seen above, the force directed to the Sun is proportional to $3x$, where x is the perpendicular distance of the Moon from the diameter of the orbit normal to the direction of the Sun. The greatest value occurs when the nodes are at quadrature and the Moon in syzygy (see Fig. 13.2); it is $3m^2$, about 0.016785. The maximum rate of regression is about 38° per sidereal year and, since the instantaneous rate is proportional to $\sin^2 v$, where v is the longitude of the Moon measured from the ascending node, the mean value over one revolution of the Moon is about 19° per sidereal year. The eccentricity of the orbit makes a small difference and the rate has an annual variation. Newton's final value was 19° 18′ per sidereal year, or a nodal period of 18.65 years.*

The contemporary observed value was 19° 21′ and the period 18.60y. (Props 30 to 33 of Book III). According to modern theory the period of

---

*Subsequently in the third edition of *Principia*, published after Halley had printed (but not published) his Tables, Newton inserted a treatment by Machin and Pemberton which gave a similar value (scholium to Prop. II of Machin).

the secular motion of the node is 18.59y. Newton's theory and contemporary observations were thus close to modern values.

Newton followed Prop. 35 with a scholium in which he set out further gravitational effects of the Sun upon the lunar orbit, but without details of the calculations. The scholium seems to have given rise to a great deal of misunderstanding, for despite Newton's statement in the scholium and elsewhere[19] that he calculated the effects from gravitational theory, people have continued to state that his lunar theory was a failure and that the equations he gave in the scholium were empirical. It will be clear from what has just been explained that his calculations of the motions of node and perigee and of the variation were all entirely theoretical. There is more than just Newton's word to support his assertion that the further equations in the scholium are based on theory. In the first place, they are too small to have been determined empirically at that time, but more significantly, they agree closely with corresponding terms in modern theories.

Newton's final version of his lunar theory is that in the second edition of *Principia*, but he had prepared a summary, *A new and most accurate theory of the Moon's motion*, ten years earlier.[20] It is not truly a theory but a rather brief summary of all the inequalities in the Moon's motion of which Newton was aware. Newton makes clear the status of the various components in the second edition of *Principia*, Book III, where he states that some of them were found from gravitational theory, the notable exception being the motion of the apsides. There are in fact three statements of the results of Newton's developed theory, for in addition to *The theory of the Moon's motion* and the second edition of *Principia*, Halley summarised it in the explanation of his posthumous *Tables*. With one exception, Newton's 7th equation, the versions agree.[21]

Newton's equations do not appear in his form in today's theories because the arguments of the various trigonometric terms differ from those that he used. Currently the arguments are linear combinations of the mean motion of the Moon, M_L, equal to nt, of the Sun, M_S, equal to $n't$, of ξ, the angular distance between Moon and Sun, and of ϕ, the mean longitude of the Moon from the ascending node. Newton used the difference between the longitudes of the Sun and node and perigee and other angles.[22]

The comparisons between Newton's terms and modern ones are shown in Table 13.1.[23] The numbering of the terms is Newton's, the alternative descriptions are Halley's.

The first two terms are the leading terms of the difference between mean and true anomaly in an elliptic orbit:

$$v - M = 2e \sin M - \tfrac{3}{4} e^2 \sin 2M + \cdots$$

Table 13.1 Coefficients of perturbations in lunar longitude in Newton's theory and modern forms

Argument Modern	Argument Newton	Coefficient Modern	Coefficient Newton	Identification
Dependent on lunar eccentricity, e:				
$\sin M_L$	$\sin(v - \omega)$	$2e$		
$\sin 2M_L$	$\sin 2(v - \omega)$	$\tfrac{3}{4}e^2$		Newton's 4th eqn
Dependent on m and e:				
$\sin 2\xi$	$\sin 2(v - v_S)$	39′ 29″	35′ 32″	Variation orbit. Newton's 5th eqn
$\sin(2\xi - M_L)$	$\sin(v + \omega - 2v_S)$	1° 16′ 26″		Evection
$\sin(2\xi - 2M_L)$	$\sin 2(\omega - v_S)$	3′ 32″	3′ 45″	Newton's 2nd eqn: Aequatio semestris prima
Dependent on angular distance of Moon from Sun:				
$\sin \xi$	$\sin(v - v_S)$	−2′ 5″	−2′ 20″	Newton's 7th eqn
Dependent on eccentricity of Earth (solar equation of centre):				
$\sin M_S$	$\sin(v_S - \omega_S)$	−11′ 8″	−11′ 49″	Newton's 1st eqn
Dependent on the direction of the Sun relative to the node of the Moon:				
$\sin 2(\xi - \phi)$	$\sin 2(\Omega - v_S)$	55″	47″	Newton's 3rd eqn: Aequatio semestris altera
Dependent on the direction of the Sun relative to apogee of the Moon:				
$\sin(M_L - M_S)$	$\sin(\xi - \omega + \omega_S)$	2′ 28″	2′ 25″	Newton's 6th eqn: Aequatio quarta lunae

M_L is the mean anomaly of the Moon, M_S that of the Sun, ξ is the angle between the Sun and the Moon, v is the longitude of the Moon and v_S that of the Sun, both measured from their respective perigees, ω is the longitude of perigee of the Moon and ω_S that of the Earth. Newton's 7th equation is proportional to the lunar parallax.

Newton's treatment of the perturbations of the Moon's motion foreshadows in an important respect much later developments. The first analytical theories of Clairaut and d'Alembert, followed by Laplace, derived the time as a function of the lunar longitude. Euler calculated analytically the changes in the parameters of the orbit, node, eccentricity, inclination, and perigee, and Lagrange developed general equations for the rates of change of those variables as functions of the three perpendicular forces acting on a planet (not specifically the Moon). Newton, by his geometrical methods, anticipated them all with his expressions for the motions of the nodes, of the apse line, of the inclination, and of the eccentricity.[24]

People have said how unsatisfactory Newton's theory was, yet for a first excursion into an unknown and complex field and without benefit of modern analytical methods and computers, it both revealed the dynamical basis of the vagaries of the Moon and gave good approximations to the variation and the regression of the nodes and a first attempt on the advance of apogee. Newton's account of the various perturbations, and how they come about, is dense but clear. If the way in which he obtained his numerical values is sometimes veiled, his exposition is much clearer than the involved analysis of Clairaut.

13.4 *Halley's lunar tables*

Sometime after 1704 and before 1720, Halley constructed a large set of astronomical tables, most of which were for calculations of the position of the Moon but which included tables of the motions of the planets, of the satellites of Jupiter, and cometary tables. The lunar tables represent the theory of the motion of the Moon as Newton gave it in the second edition of *Principia*, primarily in Theorem 66 of Book I and Props 25 to 35 and the Scholium following them in Book III. Halley may well have started work on his tables before the second edition was published, for Gregory noted that, in 1708, 'Newton advises Mr. Halley to make tables'; they were probably those that Halley completed by 1720.[25] He had already done his cometary calculations for Newton and later they were in close touch about the publication of Flamsteed's observations (Chapter 14). He must have been well aware of Newton's plans for the second edition of *Principia*, for there are at least three copies of the first edition in which he entered changes that Newton subsequently made in the second edition. Newton completed the lunar theory only in the last months before final sheets of the second edition went to the press. Halley's annotated copy of the first edition in King's College, Cambridge, has many of the changes that Newton made in the second edition, but those sections on lunar theory that Newton revised drastically for the second edition are not annotated (Chapter 6). That suggests that Halley made his annotations before Newton had finished his revisions. It also implies that he probably could not have completed the lunar tables much before 1712. The tables were set up in print before 1720, but Halley did not publish them then, although their existence was known and people expected their appearance.[26] When he went to Greenwich he further put off publication until he should have observations of the Moon's position with which to compare the tabular positions. He himself did not bring them out; John Bevis did so in 1749 as the posthumous *Astronomical tables* of 1749 and 1752. They were not the first published realisations of Newton's lunar theory. Flamsteed had constructed some tables from the theory of the first edition of *Principia* and the amendments of 1702, which were published in 1746 by P.-C. le Monnier, who had received them from Halley.[27]

David Gregory, in his *Astronomiae physicae et geometricae elementa*, summarised the calculations of the advance of the apsides, of the variation and of the regression of the nodes, as they were presented in the first edition of the *Principia*. Then in his Section V, Prop. 21, he listed all the effects of the Sun, incorporating Newton's tract of 1702. In his next Section, VI, he specified the various tables that would be needed for the calculation of the Moon's position according to Newton's theory and the way in which they should be used, much as Newton would prescribe in the second edition of *Principia*. The construction of tables according to that theory was what Newton seems to have urged upon Halley, and what Halley did undertake.[28]

Halley based his tables on the best values of the observed mean motion of the Moon and of her apogee and node that were available to him. He took the annual terms, the variation and the parallactic inequality, together with the additional perturbations introduced by Newton, from Newton's theory.

Halley took the steady and oscillating motions of the apsides and the variation of the eccentricity from Horrocks's model. Newton remarked in the Scholium that Horrocks was the first to suggest the model of the Moon's motion in which the orbit was an ellipse with the apogee rotating about the focus that contained the Earth, and that Halley improved the model by allowing the focus to move round the Earth on an epicycle. Those models, devised when there was no dynamical theory, do have a dynamical basis and Newton went far to showing why apogee should rotate about the Earth. He showed also how to find the lunar eccentricity from Horrocks's model and Halley presented the corresponding tables.

The Tables are the only numerical representation of Newton's developed theory, for although he himself gave numerical values for particular positions of the Moon to illustrate its behaviour, he nowhere gave any for an arbitrary time. Halley provided them. His tables enable the position of the Sun and the Moon to be calculated at every second over a wide span of dates. Newton included in the Scholium of the second edition a paragraph on the order in which the various corrections should be applied in the most precise calculations. Newton omitted the paragraph from the third edition of *Principia*, but Halley followed its rules in the application of his tables.[29]

13.5 *Halley's comparisons with Newton's theory*

The final result from Halley's tables is the true longitude of the centre of the Moon in its orbit at a given time, but Halley observed the right ascension of a point on the illuminated limb. He had to apply corrections to his observations to obtain the true longitude and latidude. First he made a correction for atmospheric refraction which depended on the zenith distance. Then he calculated the position of the centre of the Moon from that of the limb using the apparent diameter of the Moon, which varies with the parallax and which he took from the tabular position of the Moon. Those corrections, along with the latitude of the observatory, gave him the right ascension and declination of the centre of the Moon. Right ascension and declination are not, however, the same as longitude and latitude in and out of the lunar orbit, for the lunar orbit in which longitude and latitude are defined is inclined to the terrestrial equator by which right ascension and declination are defined. Thus a final calculation is necessary to obtain the true orbital longitude to compare with the value found from the tables; the worked example for 5 December 1725 includes the reduction of the observations.

When Halley prepared the *Historia coelestis* of Flamsteed (Chapter 14), he gained access to a great quantity of Flamsteed's results at about the same time as the second edition of *Principia* appeared. With that material he began to investigate the errors in Newton's theory. He found that the differences could reach 5′ in parts of the orbit where Flamsteed had not observed, mainly in the third and fourth quarters of the Moon's age. He remarked that Flamsteed could have observed for two periods of eighteen years, but his lunar observations were sparse and a whole year, 1716, was lacking completely,

> ... so that notwithstanding what he has left us must be acknowledged more than equal to all that was done before him, both as to the Number and Accuracy of his Accounts, they are not sufficient to determine the Moon's position at all parts of her orbit ...[30]

The small book now classified as RGO 2/12 in the University Library, Cambridge, lists the times of Halley's observations of the Moon at Greenwich, the observed and calculated positions, and the error of the calculated position. Four other books (RGO 2/6–8, 10, and 11, originally labelled COMP 1–6) contain the details of the calculations for some of the observations (see note 8). The entries in the table *Lunae Meridianae Longitudines, Grenovici Observatae cum computo Nostro Collate* in Halley's *Astronomical tables* correspond to those in RGO 2/12.

The observations in the printed table run from January 1722 to 27 December 1739 in two series. In his first years at Greenwich,

Halley used a transit telescope instead of a mural arc and he observed the right ascension of the limb of the Moon, but after 5 December 1725, having now his mural arc in use, he obtained absolute positions of the ecliptic longitude of the centre of the Moon; thus up to 4 December 1725 the positions, observed and calculated, are entered as *Ascens Rect Limbi Lunae*, while from 5 December 1725 they are given as *Longitudo Centri Lunae*.

Halley worked out the tabular positions for many observations beside his own. The sheets of computations contain many calculations for observations at 9, 18, 27, or 36 years before his own at Greenwich. Many must be Flamsteed's, for Halley stated in his paper of 1731:

> Comparing likewise many of the most accurate of *Mr Flamsteed* made eighteen or thirty-six Years before (that is one or two *Periods* before mine) with those of mine which tallied with them, I had the satisfaction to find that what I had proposed in 1710 was fully verified; and that the Errors of the *Calculus* in 1690 and 1708, for example differed insensibly from what I found in the like situation of the Sun and apogee, in the Year 1726, the great Agreement of the *Theory* with the *Heavens* compensating for the Differences that might otherwise arise from the Incommensurability and Eccentricity of the Motions of the Sun, moon and Apogee.[31]

That is a clear statement of the source of the earlier observations and of the use to which Halley put them.

The book RGO 2/7 contains calculations for a period, 10 March 1702 to 11 January 1704, when Halley was abroad on the Adriatic; and for the period 9 October 1714 to 19 August 1719 when he could have observed at Oxford. Another comparison in the same book is between November 1718, when he was in Oxford, and November 1700 when he was in London between voyages. In all those instances, as with the observations in the 1690s, it is most likely that the observations are among those that Halley says he took from Flamsteed's records; there is no independent evidence that he made any lunar observations at Oxford although he had an observatory on his house in New College Lane.

It is plain that Halley made the calculations for those earlier observations at the same time as the ones for his own Greenwich observations, for they are often on the same sheets as calculations for Greenwich observations made 9 or 18 or 36 years later. He sometimes notes the difference between the observed and calculated positions for a Greenwich observation with that for an observation 18 years and 11 days or 36 years and 22 days previously, testing his idea that the errors in Newton's theory of the Moon would repeat after a saronic interval of 18 years 11 days. The sheets of rough calculations, with observations made at earlier dates interspersed among current ones, show that Halley assembled the earlier material concurrently with his analysis of his contemporary observations, for until he himself had made an observation, as permitted by the weather, he would not know where to look for a corresponding observation 18 years or another interval before.

Halley occasionally notes on the sheets the comparisons between his own observations and corresponding earlier ones, but his remarks in his paper of 1731 suggest that he found more correspondences at intervals of one or two saronic periods than the few that he has noted. No list of such comparisons exists among his papers, but there are many. They are limited in two ways. Flamsteed had no mural instrument before 1690 and his earlier observations of the Moon, made with his sextant, cannot easily be compared with Halley's observations before 1726. Secondly the calculations in COMP 1–6 do not extend beyond 1731 and were probably made for his 1731 paper on the longitude. Halley clearly made later calculations, for the results are summarised in his *Tables*, but the calculations have disappeared. The most substantial set of comparisons runs from 1725 to 1729, in which period there are 106 observations that can be paired with observations 36 years and 22 days earlier in 1690–1694. On average there are between two and three pairs each month for almost three and a half years.

For each observation Halley calculated the difference, D, equal to $T - O$, the *tabular* value of the Moon's longitude less the *observed* value. Let D_F denote a difference for an observation by Flamsteed in the 1690s and D_H that for an observation by Halley in the 1720s. The scatter diagram of Fig. 13.3 is a plot of values of D_F against values of D_H 36 years and 22 days later. The values of D_H are plainly strongly correlated with the corresponding ones of D_F, as to be expected on Halley's hypothesis.

That conclusion can be taken further. The data afford the following statistics:

	D_F	D_H
Mean value	30″.4	−26″.9
Variance (10^4 sec^2)	1.644	1.839
Standard deviation	2′ 08″	2′ 16″
Covariance (D_F, D_H)		$1.445.10^4 \text{sec}^2$
Correlation coefficient		0.83

The errors, T, of the Tables, are the same for corresponding members of a pair. Accordingly the calculated differences are given by

$$D_F = (T - O_F), \quad D_H = (T - O_H).$$

The errors of observation are clearly independent and therefore

$$\text{var } D_F = \text{var } T + \text{var } O_F, \quad \text{var } D_H = \text{var } T + \text{var } O_H,$$

The matter of the Moon

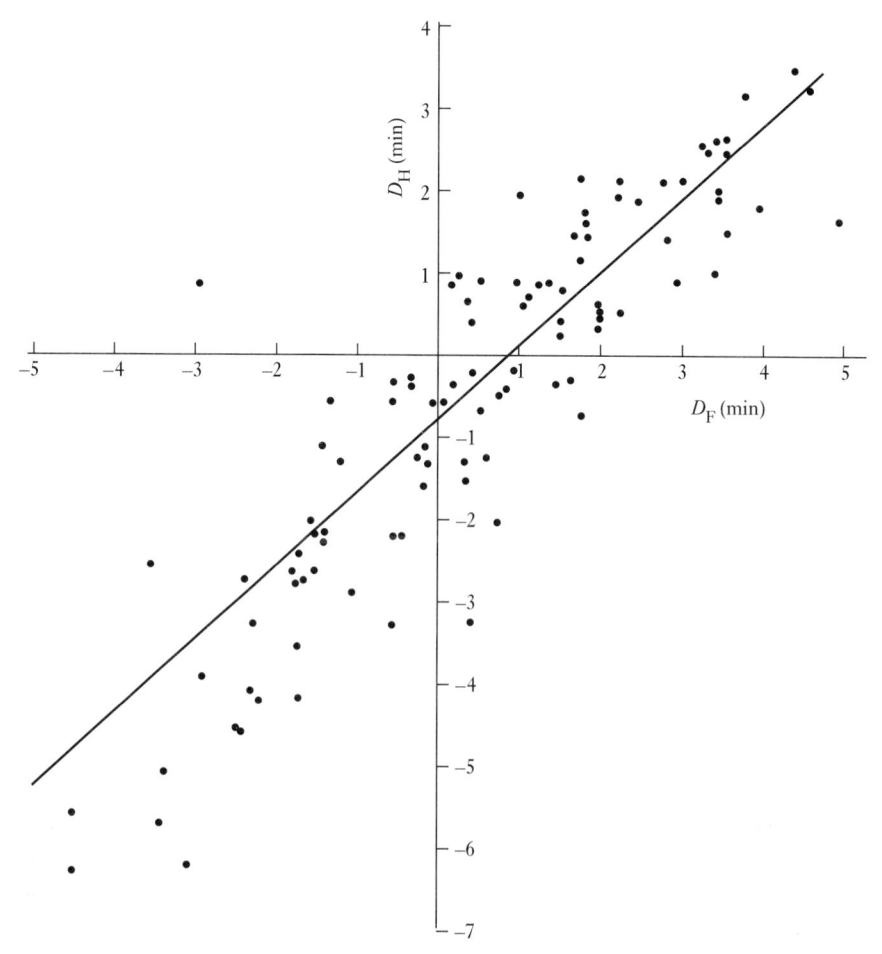

Figure 13.3 Scatter diagram of observations at 36y interval. Abscissa: D_F, differences from tabular values as observed by Flamsteed (in minutes). Ordinates: D_H, differences from tabular values as observed by Halley (in minutes).

while
$$\mathrm{cov}\,(D_F, D_H) = \mathrm{var}\,T.$$

The numerical values are
$$\mathrm{var}\,T = 1.445.10^4 \text{ sec}^2, \quad \text{s.d.} = 2'$$

and
$$\mathrm{var}\,O_F = 0.199.10^4 \text{ sec}^2, \quad \text{s.d.} = 45'',$$
$$\mathrm{var}\,O_H = 0.394.10^4 \text{ sec}^2, \quad \text{s.d.} = 63''.$$

The statistics are well established, for the results for three separate groups into which the observations fall are very similar.

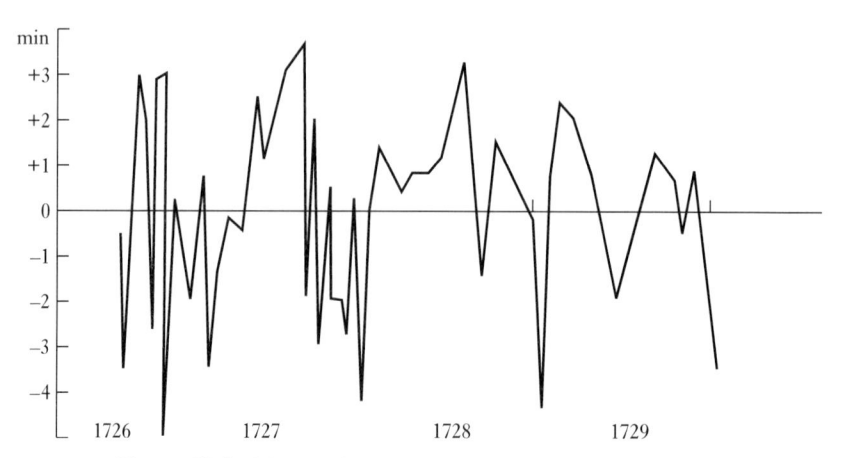

Figure 13.4 Means of D_F and D_H over 3y. 5m. (in minutes).

The first implication of the analysis is that the observations, like the tabular values, do correspond at saronic intervals, for the tabular values are constructed to correspond and the observations follow them as the scatter diagram and the correlation coefficient show.

Secondly there is a systematic difference of about 1′ between the origins of longitude in the 1690s and the 1720s. Thirdly the standard deviation of the later observations (Halley) is significantly greater than that of the earlier ones (Flamsteed).

Finally the tables that Halley constructed according to Newton's theory appear to agree with observation with a standard deviation of 2′. Halley said that Newton's theory was correct to 2′ but how he obtained that estimate without the analysis of variance is a mystery.

According to Halley's saronic hypothesis the mean of corresponding differences, D_F and D_H, should represent the errors of the tables better than either separately; it is shown in Fig. 13.4. Not all 106 values are shown, for where some fall within a few days, a mean value is plotted. The diagram appears to indicate that the dominant deviation has an approximately annual period, perhaps because Newton's 6th equation was omitted.

Halley's result implies that the best predictor for his observed values would be the equation

$$D_H = -54'' + 0.88 D_F.$$

The residuals from that relation have a standard deviation of 1′ 15″.

There are other comparisons to be derived from the manuscript books. Halley himself noted a few, 14 between February and April 1695 and March and May 1731. Some at the 18-year interval may be found and Halley also made calculations for observations approximately 9 and 27 years apart, but none provides such clear cut results as the 106 at the 36y. interval.

Halley's own estimate of the accuracy of Newton's theory and of the errors of his and Flamsteed's observations is close to the above discussion. John Conduitt reported a conversation on 13 May 1730 with Halley who

> ... told me that Flamsteed's observations which he had compared erred about two minutes and a half, his own former ones about 1 min from the truth and Sr I's theory at most 5 minutes from observations which might err at least one minute from the truth.[32]

That conversation took place a little before Halley's more considered estimates in his paper of 1731.

The comparisons over the period up to 1731 are the most favourable for Newton's theory and Halley's saronic hypothesis for, in later years, as the published tables show, some of the deviations between theory and observation become much greater, possibly because some of the parameters that Halley used in constructing his tables were not the best.

Halley has been characterised as the only Astronomer Royal whose observations have never been published. So far as his original records go, that is true.[33] The lunar material that Baily had copied consists of the times of Halley's observations and the observed right ascensions of the Moon, from which its longitude is calculated. They, with the stellar and planetary observations, are the only surviving records of Halley's years of observation as Astronomer Royal. They are not, however, entirely unpublished, for the dates and times of observation and the observed right ascensions for all his lunar observations are given in the table in which the observed and tabular longitudes are compared. What we do not have are the various corrections such as the clock errors which had to be applied to obtain the true right ascension from the observed time of transit according to the clock Halley was using. Halley's observations, it seems, were slightly but significantly less accurate than Flamsteed's.

13.6 Jupiter and Saturn

Halley wrote (Chapter 3) that it was the large discrepancies between the observed positions of Jupiter and Saturn in his youth, and the values in the existing tables, that first turned his mind to astronomy. When he came to calculate the orbits of comets he, with Newton, appreciated that comets would not follow exactly elliptical paths on account of the attractions of Jupiter and Saturn. Now in his *Tables* he returned to the mutual perturbations of the two major planets. In *Principia* Book I, Prop. 66, Th. 26, Newton discussed in qualitative terms how lesser bodies in orbit around a greater, such as Jupiter and Saturn around the Sun, would influence each other. The perturbations of an inner satellite by an outer are set out in general terms in Corollaries 6 and 7 to the Proposition. The main point that Newton makes is that the attraction of the outer planet reduces the force to the Sun when the two planets are in conjunction or opposition and has no effect when they are in quadrature. Consequently the areal velocity varies round the orbit of the inner planet and its line of apsides rotates, as with the Moon. Similar effects occur for the outer planet, except that the attraction of the inner planet increases the force on the outer at conjunction and opposition. In Book III, Prop. 13, Newton estimated the gravitational attraction of Jupiter upon Saturn at conjunction to be 1/211 of the attraction of the Sun on Saturn, and likewise, the attraction of Saturn on Jupiter as 1/2409 of the attraction of the Sun on Jupiter at conjunction. He made no calculations of the effects of those forces, but thought that the greatest error would not exceed 2' yearly in the mean motion. The periods of Jupiter and Saturn are of course many years, so that the actual errors might be much larger.

Halley did calculate the mutual effects and tabulated them. He did not describe how he did so, nor is there much in *Principia* to enlighten us, although Newton's discussion of the perturbation of the line of apsides by a foreign force in Book I, Section IX, provides the basis for a calculation. Similarly, Corollary 11 to Prop. 66, taken with Newton's theory of the motion of the nodes of the Moon, would enable Halley to calculate the regressions of the nodes of Jupiter and Saturn.

The geometry is shown in Fig. 13.5. The attraction of Saturn upon Jupiter may be resolved into a force in the direction of Jupiter to the

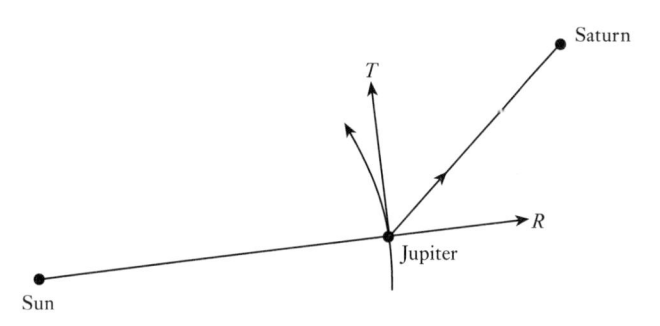

Figure 13.5 The mutual attraction of Jupiter and Saturn. R, radial component, T, tangential component of attraction of Saturn on Jupiter.

Sun and one tangential to the orbit of Jupiter. Halley tabulated the mean motions of Jupiter and Saturn and the logarithms of their distances from the Sun, and of their aphelia and nodes from 1661 to 1800, by day of the month throughout the year and by centuries. From those he could have calculated the forces upon Jupiter, and hence the changes in velocity and longitude of the planet. He could similarly have calculated the effects upon Saturn. He tabulated the equations of Jupiter's mean motion against its mean anomaly and gave comparisons with observations of conjunctions of the Sun and Jupiter from 1657 to 1719. He found the deviations to be mostly between 1′ and 2′, the greatest being −5′ 28″ in July 1666 and +7′ 14″ in July 1700. He constructed similar tables for Saturn and found generally larger deviations, the greatest being +4′ 43″ in August 1669 and −9′ 44″ in July 1696.

13.7 Conspectus

Halley spent more than thirty years of his life upon the matter of the Moon. His labours demonstrate that Newton's lunar theory, as set out in Book III of *Principia*, agreed with observation to within about 2′ over two periods of some five years each. That was a very satisfactory result from years of sustained computation and observation, much of it done at a time of life when few people would expect to survive. Halley never summarised his results, they were never publicised in the way that his cometary predictions were, yet as a demonstration of Newtonian celestial mechanics, they are at least as compelling as the cometary studies. Newton's geometry was almost sufficient for the theory of the Moon, whereas comets demanded the more sophisticated analysis of Clairaut, and even then the attraction of the major planets did not allow a very exact prediction of the return of the comet. The observations and predictions of the places of the Moon to better than 2′ were of a different order. Newton's own summary in *The theory of the Moon's motion* did attract a great deal of attention over many years,[34] but his disappointment with his theory, the later successes of Clairaut, d'Alambert, and Euler, and the lunar tables of Mayer, have dominated assessments of the state of lunar theory. Halley's achievement, effectively unpublished, remained hidden. He took Newton's geometrical method as far, probably, as it could be taken. His results, being concealed, did not compel belief in dynamics as the return of the comet did, and so had no effect on the general view of nature.

The return of Halley's comet rightly made a great stir and helped to convince people that Newton's dynamics was valid, but Halley's demonstration of the concurrence of lunar theory with observation would have been beyond laymen. Halley, by his observations and numerical comparisons did as much, if not more, to establish Newtonian dynamics here as with comets, even if he died with his labours incomplete, unpublished, and unknown.

Flamsteed's lunar tables (RG01/50I) follow Newton (1702) and are similar to but not identical with Halley's of 1749 and 1752. Flamsteed (RG0.1/50K) made a copy of tables of Saturn made by Halley for Jonas Moore, hence constructed before 1679.

14
Astronomer Royal

> Tanto ch'i vidi delle cose belle
> che parta'l ciel, per un pertugio tondo;
> e quindi uscimmo a riveder le stelle...
> [So I saw as through a round window beautiful heavenly things
> and we came forth to see the stars again...]
>
> DANTE: *INFERNO*, XXXIV, 137–9

14.1 *The Royal Observatory*

LOUISE DE LA KÉROUALLE, the Duchess of Portsmouth, had among her clients at the Court of Charles II a French gentleman, le Sieur de St Pierre. In the last days of 1674 he told the King that he had 'found out the true Knowledge of the longitude, and desires to be put on Tryall thereof'. Thereupon the Secretary of State, Sir Joseph Williamson, by command of the King, appointed Lord Brouncker, Seth Ward, the Bishop of Salisbury, Sir Samuel Morland, Sir Christopher Wren, Colonel Silius Titus, Dr John Pell, and Robert Hooke, to provide St Pierre with astronomical observations so that he might try to determine where they had been made. They were to report the results to the King and whether the method would be practical and useful to the public.[1] Pell, like Brouncker and Wren, was a founding fellow of the Royal Society. Titus (1623–1704) was a Commonwealth soldier who must have been reconciled to Charles II, for in 1671 he was 'of the Bed chamber' and, with Evelyn, a member of the Board of Plantations which oversaw the affairs of the colonies. In later years he was prominent in the parliamentary struggle against Charles in 1679–1682.[2]

Morland, Pell, Titus, and Hooke met on 12 February 1675 in Titus's house off Chancery Lane. They asked John Flamsteed to assist them. He was already known to be a skilful astronomer, and had recently come to London as the guest of Sir Jonas Moore in the Tower. Flamsteed criticised the proposal but agreed to produce the observations that St Pierre wanted. When he did so, St Pierre was not satisfied. That brought the King to see that astronomical quantities required for navigation were very uncertain. Charles knew something of Flamsteed's abilities from Moore and he immediately appointed him his astronomical observer:

> ... forthwith to apply himself with the most exact care and diligence to the rectifying of the tables of the motions of the heavens, and the places of the fixed stars so as to find out the so much-desired longitude of places for perfecting the art of navigation.[3]

Flamsteed demolished St Pierre's scheme, showing that existing constants of astronomy were unreliable, and that they had to be thoroughly revised. The King thereupon ordered that an observatory be built at Greenwich on a site recommended by Sir Christopher Wren. The Royal Warrant of 22 June 1675, addressed to the Master General of the Ordnance, begins:

> Whereas, in order to the finding out of the longitude of places for perfecting navigation and astronomy, we have resolved to build a small observatory within our park at Greenwich, upon the highest ground, at or near the place where the castle stood, with lodging rooms for our astronomical observer and assistant, Our Will and Pleasure is that according to such plot and design as shall be given you by our trusty and well-beloved Sir Christopher Wren, Knight, our surveyor-general of the place and scite of the said observatory, you cause the same to be built and finished with all convenient speed ...[4]

Halley, still an undergraduate, now enters the history of the Royal Observatory. The very next day he and Flamsteed consulted Hooke at the Tower about the design of a mural quadrant, probably for the new observatory, and then on 2 July Halley went with Wren and Hooke to view the site in Greenwich Park.[5]

Charles II laid the foundation stone in August 1675. Flamsteed moved out of the Tower to lodgings in Greenwich to supervise (Fig. 14.1). He continued his observations from the Queen's House, now part of the National Maritime Museum. His apartment in the Observatory was ready on 10 July 1676, about the time that Halley was back there with Hooke and Sir John and Lady Hoskins and the King's plumber Lingren.

Flamsteed began to observe from the Great Room on 19 September. Sir Jonas Moore provided his first instruments, his micrometer, his pendulum clock, his first mural arc, and his sextant. The mural arc, to a design of Hooke, was unsatisfactory (Chapter 3) and Flamsteed could not use it. He thought that Hooke had made it deliberately so to obstruct him. They were never on the best of terms. Flamsteed then had a wooden arc made locally but it was too weak, although he did take meridional altitudes with it from 1683 to 1686. He had to depend on the sextant until 1689. When he received a benefice and his financial

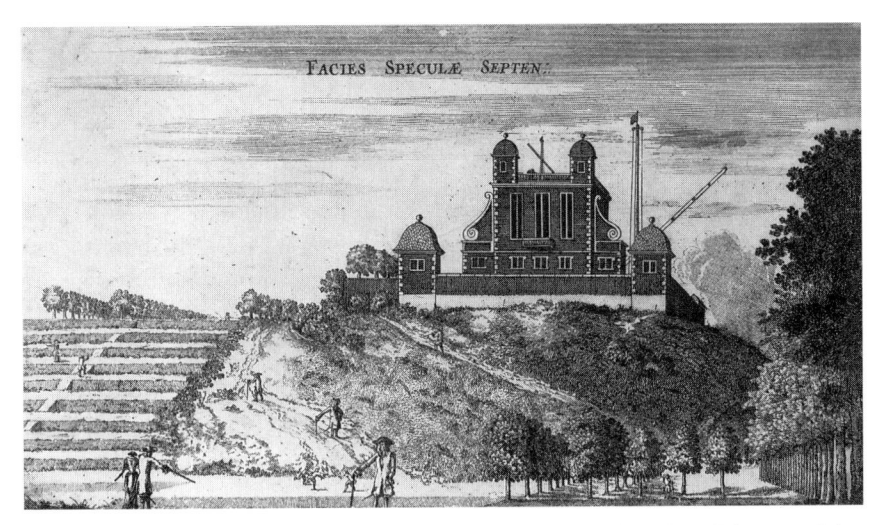

Figure 14.1 The Royal Observatory (reproduced by permission of the Director of the National Maritime Museum).

position improved, he paid for a better brass mural arc of 140° amplitude, made by his assistant, Abraham Sharp. It was mounted in the summer of 1689. He used it for fundamemtal positions from the beginning of 1690. When he died in 1720 Mrs Flamsteed removed it. Molyneux thought to purchase it in 1721 and it has not been heard of since. The radius was 6 ft $7\frac{1}{2}$ in (2 m) and it had a telescope 7 ft (2.1 m) long. There is a contemporary picture of the arc with Flamsteed and his assistant, Thomas Weston, in a painting by Sir James Thornhill on the ceiling of the Painted Hall in what is now the Royal Naval College.

Tower smiths made the 7 ft sextant that Flamsteed used from 1676 to 1689, and Thomas Tompion, the clockmaker, graduated the arc. The sextant that Halley took to St Helena was probably a twin. Flamsteed mounted his sextant equatorially and had telescopic sights as Auzout and Hooke had urged. That led him into controversy with Hevelius (Chapter 4). He first used the sextant in 1676 and abandoned it after 1714 when it was too worn and inaccurate.

Flamsteed had other instruments, including telescopes of 8, 16, 27, and 60 ft (2.4, 4.9, 8.2, and 18.3 m) focal lengths and two small quadrants, and a third borrowed for two or three years from the Royal Society. He had seven clocks, two of them of a special design due to Hooke and Towneley intended to reduce the angle of swing of the pendulum and therefore the circular error; Mrs Flamsteed removed the two special clocks in 1720. One is now in the British Museum and the other at Holkham Hall. One of the other clocks was adjusted to sidereal time, displayed in degrees, not hours, and is in the restored Sextant House at Greenwich.

Flamsteed undoubtedly had instruments of the most advanced design of his day, but Halley would probably have had to replace them even had Mrs Flamsteed not removed them. The great mural arc became defective and the brick wall on which it was mounted was unstable. Clocks were much improved by the time that Halley came to the Observatory: two that George Graham made for him were still in use at the beginning of the twentieth century and remain in the Royal Observatory.

14.2 Halley and Flamsteed

There are few periods after 1675 in which Halley and Flamsteed did not interact in some way. The history of their relations is as much part of the history of the first years of the Royal Observatory as it is of their separate biographies. They collaborated in observations for almost ten years, until Halley became involved with Newton in the *Principia*. Then Flamsteed seems to have disapproved of Halley more and more, until in 1710 he was infuriated by Halley's bringing out the first version of the *Historia coelestis*. Relations remained bad to the end of Flamsteed's life.

For most of the time we know only one side of the story. After Halley first wrote to Flamsteed while still an undergraduate, his observations with Flamsteed and the results he sent him from Oxford are to be found in Flamsteed's records and in his correspondence with William Molyneux, while it was Hooke who related that Halley was present when the site of the Observatory was chosen. Flamsteed wrote to a number of his correspondents in those first years about Halley's abilities and referred to him as a good friend; there is almost no surviving first-hand evidence of Halley's opinion of Flamsteed after his first respectful and admiring letter. Even at first there was some friction in their relation, for Hooke noted in his *Diary* for 7 January 1676, 'Flamstead and Halley fallen out'.[6] To the end of his life Halley wrote appreciatively of Flamsteed's achievements, while critical of his neglect of the Moon. Flamsteed certainly set down from time to time what he thought Halley and his associates thought of him, but that is scarcely evidence of what Halley did think.

Flamsteed had kept in touch with Halley's progress, or lack of it, in St Helena, and on 28 October 1677 he noted that the transit of Mercury was obscured by clouds at Greenwich but he recorded the time of egress as seen by Towneley at Towneley in Lancashire and by

Gallet at Avignon, as well as by Halley on St Helena, so that he was able to calculate the differences of longitude between Avignon and St Helena and Towneley.[7] Flamsteed's assessment of Halley's campaign in St Helena, quoted in Chapter 3, was mixed. When Halley was away on St Helena, Flamsteed defended his priority, against Cassini, in the matter of the rotation of the Sun (Chapter 2).

Flamsteed fell sick in July 1678 after Halley had returned from St Helena, and retired to the country until the end of October. Halley stood in for him and made a number of observations at Greenwich, not regularly, but apparently for special events.[8] Flamsteed was back at the end of October in time for a lunar eclipse which he observed at Greenwich along with Halley, Perkins, and 'Faber' his servant (*anglice* Smith); he recorded reports from Haynes in London at Aula Basingensi (Basing Lane near St Paul's), Colson at the Hermitage at Wapping, and Cassini, Römer, and Picard at Paris, and they determined the difference of the longitudes of Greenwich and Paris.[9]

Flamsteed, like Cassini, thought it should be possible to find the distance of the Sun from observations of Mars at apposition close to the Earth and on 14 January 1679 he took advantage of clear skies at Greenwich to observe Mars in the morning, about 7 h 45 min, and again in the evening, about 18 h 15 min. Halley joined him and observed frequently and accurately with him.[10] The observations seemed to show that the parallax of the Sun could not exceed 10″, but Halley thought that neither they nor the observations of Cassini were sufficiently certain to give valid estimates and no doubt he called on his practical experience when he discussed the method later.[11]

Halley was in Danzig from 26 May to early July of 1679 and he could not have been still in England when Römer visited the observatory on 14 May. He wrote to Flamsteed on 7 June with news of his activities at Danzig, with an account of Hevelius's instruments, and a report of their observation of an occultation by the Moon.[12] He sent Flamsteed a report from Oxford of an occultation of the Moon on 28 March 1680. In Paris he obtained a large map of the Moon for Flamsteed, he sent him observations of the comet of 1680/1[13], and later of the eclipse that he and Gallet had observed at Avignon (Chapter 3). Flamsteed in turn wrote to him about observations of the comet of 1680/1 at Greenwich and asked for information about the instruments in the Observatoire in Paris.[14]

Shortly after Halley returned from Italy he joined Flamsteed and 'Faber' (Smith) at Greenwich in observing the lunar eclipse of 11 February 1682 which was also observed by Edward Haynes in Basing Lane in London.[15] When Flamsteed, in one of his first letters to William Molyneux in Dublin, reported his observations of that eclipse, he wrote

'My freind Mr Halley, a very ingenious person whom I believe you have heard of, was with me'.[16] After his marriage Halley sent Flamsteed reports of observations he made at Islington, the last sighting of the comet of 1682, obscured by clouds, on 11 September, the second satellite of Jupiter on 24 October, and the third satellite on 19 December. Flamsteed had met Mary Halley by the end of the year for in a letter of 21 November he sends her his respects.[17] Halley continued to communicate results, that Flamsteed noted in *Historia coelestis Britannica*, of a lunar observation on 22 March 1683 and of the last sighting of the comet of 1683 on 26 August.[18] The last sighting of the comet is also the last communication between Flamsteed and Halley for some time.

Halley and Flamsteed had collaborated regularly if not closely over almost eight years and without overt friction, apart from Hooke's early note that they had fallen out.[19] Now their correspondence ceased. Halley became involved in his family disputes and the promotion of *Principia*, with his duties at the Royal Society, and with going to sea. Positional astronomy, although it retained his interest, he laid aside; he saw much less of Flamsteed.

Sometime between 1684 and 1686 Flamsteed became very hostile to Halley as an astronomer and socially. Some of the correspondence in which Flamsteed expressed his very strong disapproval and resentment has been summarised in Chapter 6. He thought Halley had stolen observations and ideas from Peter Perkins; he resented Halley's disagreement with his explanations of the tides at Dublin; an acquaintance of them both had told him he should resign in Halley's favour; and he considered Halley had entertained him meanly. He attributed what he saw as graceless behaviour by Halley to Halley's being close to Hooke.[20]

In 1695 Flamsteed wrote of Halley on a letter he had received from Newton

> Whatever he may say to you to the contrary, his behaviour towards me has been the most impudently and ungratefully base. I know him and you do not, therefore am resolved to have no further concern with him.[21]

Flamsteed disapproved of others beside Halley. He showed deep animus to Hooke, he came to regard David Gregory as an enemy, and he thought poorly of Pound. In his later years he was suspicious of anyone he thought was a dependent of Newton, and Newton himself aroused his particular rage and antipathy. Even so he had written to Pound in 1704 with a balanced criticism of Halley's southern catalogue (Chapter 3).

The bitter dispute between Flamsteed and Halley and Newton, in the years that Halley was at Oxford, has, for many people since, held the stage. The seeds were sewn years before, when Newton was writing

Principia. Newton was keen to have positions of the Moon as a basis for his theory of the Moon's motion, but had great difficulty in getting them from Flamsteed. Flamsteed offered him reduced observations whereas Newton wanted them without any adjustment. Relations were exacerbated when Newton learnt that Wallis intended to publish a reference to Newton's having received 150 places of the Moon from Flamsteed.[22] Newton and Flamsteed also differed over the programme of observations that Flamsteed should be pursuing with the overall aim of improving navigation and especially the determination of longitude at sea: should observations of the Moon take first place, as Newton wished, or should they await a firm catalogue of places of stars, which was Flamsteed's more systematic approach? Under those differences of attitude and purpose there lay the greater difference between Flamsteed, who regarded the observations as his personal property, especially as he had provided some of the instruments at Greenwich himself, and Newton, who considered them public property since the King paid Flamsteed's salary. Newton was and remained deeply resentful of Flamsteed's reluctance to hand over results; so Stukeley remembered his conversation of 23 February 1721, not long after Flamsteed's death, when he and Newton breakfasted together:

> Sir Isaac among other discourse mentioned the poverty of materials he had for making his theory of the moon's motion. He said Mr Flamsted would not communicate his observations to him so that what he did, was from 3 or 4 observations only of Mr Flamsteds; for which he owed him no thanks as not designed for him. But he said that now he could finish that theory if he would set about it; but he rather chose to leave it to others.[23]

The statement is less an objective historical account, more a revelation of Newton's bitter attitude. Newton, however, was not the only one to be critical of Flamsteed's delays, for on 29 May 1699, James Gregory (David's brother) wrote to the Revd Colin Campbell in Edinburgh, in a letter already quoted in Chapter 10:

> Mr Flamsteed has rectifyed above 3,000 fixed stars: but is so perversly wicked that he will neither publish nor communicat his observations.[24]

Almost twenty years earlier, on 16 January 1679, when Moore reported Halley's progress at St Helena to the Royal Society, the President asked when they would see Flamsteed's observations. The matter came up again at the next meeting on 23 January, when Moore said that some were ready for the press, but printers were unwilling to handle them. Moore took the opportunity to report the result of the observation of the parallax of Mars by Flamsteed and Halley, which he thought to be the best value for the solar parallax so far.[25]

Matters took a most unfavourable turn for Flamsteed at the very end of 1703. At the Anniversary Meeting of the Royal Society on 30 November Newton, already influential as Master of the Mint, was elected President. Flamsteed had previously had some indication that Queen Anne's consort, Prince George of Denmark, might pay for the publication of his catalogue, the fruit of dedicated work and thought over twenty years. Now Newton brought a firm promise of that support. He dined with Flamsteed at Greenwich on 11 April 1704, evidently to discover the state of Flamsteed's star catalogue which Flamsteed had told Dr Smith of Oxford was ready for the press, if he could have assistants to copy the entries and complete the calculations. If not, people would have to wait until he could do all that himself.[26] Flamsteed was asked to give an estimate of the size of his catalogue and provide a plan for publication; he did so and they were read to the Royal Society on 15 November 1704 and were unanimously commended. A short while later, after Prince George had been elected a Fellow of the Society, he did offer to pay for the publication. Accordingly a committee of referees was appointed (or came into being), Newton, Wren, David Gregory, Francis Roberts, and John Arbuthnott, to inspect Flamsteed's manuscripts and assess the prospective publication. Later Francis Aston joined the committee of referees.

Arbuthnott (Plate XI) was the son of a Scottish clergyman who, like David Gregory, had been ejected as an episcopalian. Arbuthnott and Gregory were both undergraduates at Aberdeen and were close friends at Oxford. Arbuthnott was a fellow commoner of University College, and kept up a correspondence with Charlett, the Master. He had a medical practice in London, as did Gregory. Gregory, as tutor to Queen Anne's son, the Duke of Gloucester, may have introduced Arbuthnott to the Court, where he was appointed first a physician extraordinary, and later, in 1709, physician ordinary. Halley knew him before that, and when at sea in 1699 had asked him to act as Clerk to the Royal Society (Chapter 10). Arbuthnott was in Queen Anne's train when she visited Cambridge in 1705 and dubbed Newton knight. He was one of those on whom the University conferred the degree of Doctor of Physic. He was a competent mathematician and had written on the history of the Earth. He was elected a fellow of the Royal Society in 1704. He seems to have been the link between Prince George and the Royal Society in the Flamsteed business.

Arbuthnott shows how wrong it may be in his day to isolate particular spheres of activity, such as natural philosophy, or religion, or literature, or music, and discuss them and their practitioners in isolation. He was influential at the Court of Queen Anne, he was a friend of Swift, and he wrote some or all of the *John Bull* satirical pamphlets. He surely knew Kitty Barton, Newton's niece and close friend of Swift. He was a musician, a friend of Handel, and in the Cannons circle of the Duke of Chandos. He probably joined Handel and Pepusch in writing music for that circle and he may have written some of Handel's early oratorio libretti. In 1719 and 1720 he was a director of the Royal Academy of Music that put on Handel's operas. Halley was closely associated with Arbuthnott in the Flamsteed affair and also over the Leibniz controversy, and through him could have known Swift and Kitty Barton and Handel. There is no evidence, but given the small compact society that London was, Halley, genial and sociable, was more likely than not of the society of such lively people.[27]

The referees recommended publication of Flamsteed's work and he was asked to prepare his material for the press. He handed over a copy of his observations and the catalogue, incomplete, that he had derived from them. He intended that only as a guarantee that he would complete the work, for he demanded that nothing should be done to print it until he provided the final copy. A letter from Flamsteed to his assistant Abraham Sharp shows that David Gregory had the main responsibility for preparing Flamsteed's material for the press; Gregory was one of the referees, as Halley was not. The printing of the observations began in May 1706 but went very slowly and finally came to a stop with the death of both the Prince and Gregory in 1708.[28] Halley was involved in some ways before Gregory died. He was told of a request that had been made to Römer for help in obtaining the original material from Denmark for Tycho Brahe's catalogue, which Flamsteed proposed to include with his own. Arbuthnott added a postscript to his letter to Newton about the approach to Römer, 'I know you will not lett Flamsted know that you consulted Mr Hally'; the referees were even so early proceeding in the underhand manner of their later behaviour.[29]

Flamsteed did think that Halley was manipulating matters as an *eminence grise*, and besides Arbuthnott's letter, other evidence suggests that Flamsteed had grounds for his suspicions. In 1713, Keill, Newton, Halley, and Pound recommended Machin for the post of Gresham's Professor of Astronomy and referred to his having been deputed by the managers to examine Flamsteed's manuscript before it went to press. Halley may already in 1706 have had access to Flamsteed's lunar observations of 1705, for on 23 July 1706 Machin wrote to someone, Halley according to Rigaud, to explain why some lunar observations for August and September of 1705 were missing from Flamsteed's records (they should have appeared on p. 181 of *Historia coelestis* of 1712).[30] It is not certain that Halley was involved before Gregory died; the main work was undertaken by Gregory and Machin, but he may have been consulted about it and may have prepared some of the material, especially the lunar observations.

In 1710 the Queen appointed the President of the Royal Society and other members of the Council chosen by him to be a Board of Visitors to the Royal Observatory. Visitors to corporate institutions such as the universities and their colleges were quite familiar in the seventeenth and eighteenth centuries in England, and did not confine their activies, as nowadays, to resolving, when called upon, the rare internecine dispute and to dining on festive occasions. They intervened forcefully to change people and policies, especially at the Restoration and during the reign of James II. Flamsteed deeply resented having Visitors imposed upon him to oversee work he had so far carried out on his own. His great predecessors, Tycho and Hevelius, had no Visitors and had done very well without them. It was a clear judgement on his labours over thirty years, apart from whom the Visitors were. Flamsteed can have thought that no worse choice was possible than a Board headed by Newton, who had been hounding him for his results for more than twenty years and who had broken off correspondence with him. Arbuthnott was a Visitor and so was another fashionable London physician, Dr Mead, Newton's physician, who also had influence at Court and was a correspondent of Hearne. Francis Roberts, Abraham Hill, Sir Christopher Wren and Mr Wren, and Halley were the others. With what feelings must Flamsteed have learnt that his Visitors were to include the man who to him seemed to dog and denigrate him at every turn.[31] No wonder that Flamsteed glowered in Greenwich, the while Newton fumed in arrogant frustration at the Royal Society.

Printing resumed in December 1710. The Visitors, Newton in effect, asked Halley to see Flamsteed's work through the press and to correct and edit it. That was certainly grossly high-handed. It was more than high-handed, it was deceitful, for Arbuthnott assured Flamsteed that the catalogue was not in the press when in fact it was, and with Halley's changes. While Halley corrected copying errors, it is unlikely that he recalculated any of Flamsteed's reductions, but he changed the descriptions of many stars. To Flamsteed's fury, the catalogue was published in 1712 in the manner that Newton had always intended, before instead of after the sextant observations, and with far fewer original observations than Flamsteed wanted.

When Queen Anne died in 1714, Newton's influence at Court declined and Flamsteed was able to get all the undistributed copies of the Halley version and burn them as 'A Sacrifice to Heavenly Truth: If Sir I Newton would be sensible of it, I have done both him and

Dr Halley a very great kindness'. Flamsteed sent Sharp a copy of 'my Catalogue of the fixed stars, as it is corrupted and spoiled by Dr Halley', on which he marked all the faults.[32] Halley seems not to have been involved thereafter but Newton and Flamsteed remained on the worst of terms until Flamsteed's death. Flamsteed had retained a copy of all his material and he now printed at his own expense the first two volumes of the *Historia coelestis Britannica* as he wished it.[33]

There had been a monumental scene on 26 October 1711. Flamsteed was bidden to meet the Council of the Royal Society to discuss repairs and additions to his instruments. Only Newton, Sloane, and Mead met him but Halley was present in Crane Court, and either before or after the meeting invited Flamsteed to take coffee with him. The invitation was declined with deep disdain. The meeting was very heated and culminated in Newton accusing Flamsteed of calling him an atheist, a term to which, for reasons we now know but Flamsteed did not, he would be very sensitive.[34] Halley seems to have tried to be conciliatory, and his call on Flamsteed on 18 June 1712 may have been with the same purpose, but to no avail. Flamsteed recalled,

> ... the impudent editor, with wife son and daughters attending him, and a neighbouring clergyman in his company, came hither, I said little to him. He offered to burn his catalogue (so he called his corrupted and spoiled copy of mine, of which I had now a correct and enlarged edition in the press, and the second sheet printing off) if I would print mine. I am apt to think he knew it was so, and was endeavouring to prevent it. But to render his design ineffectual, I said little to him of it: so he went away not much wiser than he came.[35]

Oh to have been a fly on the wall! In 1712 Halley's son and daughters were in their late twenties. In earlier days, not long after Halley had married, Flamsteed visited his house: perhaps Halley hoped the family party might lighten the mood. It did not.

Flamsteed's final grievance was that Halley received £150 for his work on the catalogue, whereas he himself had spent large sums of his own over many years and got less in return.[36]

It is pointless at this stretch of time to apportion blame for the unhappy history, although Newton's biographer considers that the main cause was that Newton was determined to get his way in the publication of Flamsteed's material.[37] Flamsteed extended his invective to all whom he considered to be associates or clients of Newton and his antipathy to Halley went back nearly forty years; the events are the culmination of long-felt grievances of Flamsteed against Newton and Halley, and of the frustration of Newton.

14.3 Two Histories

The *Historia coelestis* of 1712[38] has two books in a single volume. Book I comprises a Preface, a Catalogue of fixed stars to the year 1690, and Observations with the sextant from 1676 to 1689. Book II has some of the Observations with the mural arc to about 1705. Flamsteed's own *Historia coelestis Britannica* of 1725[39] is in three volumes: the first has an extensive *Prolegomena* followed by the sextant observations from 1676 to 1689, the second contains the observations with the mural arc from 1690 to the end of Flamsteed's career, and the third is made up of a number of earlier catalogues and Flamsteed's own.

Flamsteed died at the end of 1719 before his third volume was complete. His former assistants, Joseph Crosthwait and Abraham Sharp, and especially Crosthwait, undertook to bring it out and to get all the charts of the constellations engraved. The story can be followed in the letters from Crosthwait to Sharp after Flamsteed's death. A hard time Crosthwait had of it, not least in difficulties with Mrs Flamsteed, from whom in the end he received no recompense in her will for all his time and labours. She may have lost money in South Sea stock.[40]

When Halley took over the *Historia coelestis* he had the sheets of the sextant observations which Gregory had printed. He incorporated them into his book. Flamsteed kept the sextant observations when he got hold of his 300 copies of the *Historia coelestis* and they made up his Volume 1; he probably received unbound copies and could easily have separated the sheets he wanted. The sheets of the sextant observations in copies of the 1712 and 1725 books are plainly from the same setting and on the same paper. The treatment of the mural arc material is quite different. In his last years Flamsteed seems to have assembled all his observations in the form in which they now appear in his Volume 2 and they were printed under his direction in his very last years. Halley in 1710 had no such compilation but he did have access to some of Flamsteed's observations. He selected planetary and lunar material up to 1705 and printed it as Book II, bound in with Book I.

The treatment of the observations in the two *Histories* is fairly straightforward, that of the catalogues is anything but transparent. Stars in zodiacal constellations appear first in both forms of catalogue and constellations to the north and to the south of the zodiac then follow them, all being restricted to stars that can be seen from Greenwich. The Crosthwait version puts the southern constellations first, Halley the northern ones. There are other important differences. Flamsteed complained bitterly that Halley had altered the descriptions of stars from the traditional ones of Ptolemy and so muddled their identification, and he also complained that Halley had changed the order in which stars were listed within constellations. The second complaint is not entirely justified: the arrangements in Aries and Taurus and indeed many constellations seem to be the same in the two versions. However, there are substantial differences in the assignment of certain stars to constellations, for instance, some that were in Ursa major in 1712 were moved in 1725 to new constellations that Hevelius had erected, but which Halley did not adopt.

The matter of descriptions is more serious. In today's catalogues stars are identified by their coordinates, but in earlier days the only way of identifying a star was by its relation to some well known bright stars such as Sirius or the Pleiades or Aldebaran or the Pole star. Differences between the coordinates assigned to some stars in the earlier catalogues in Volume III of *Historia coelestis Britannica* were so large that any attempt to identify stars from coordinates would have been quite futile. Thus the verbal description had to be clear and unambiguous and it would be helpful for stars in a constellation to appear always in the same order in a list. Halley, supported by Arbuthnott who was no negligible classical scholar, made many changes in the verbal descriptions. Flamsteed, for instance, identified a star as being to the left or right of some other one, whereas Halley described it as preceding or following the reference star, which is less ambiguous. There was a passionate exchange of letters about fourteen instances.[41]

The matter of the descriptions of the constellations is not negligible and Flamsteed had long been concerned about it. Ptolemy originally delineated the present constellations but his successors often garbled his descriptions because the old translators misunderstood him. Flamsteed made a new translation from the Greek of Ptolemy and drew the figures anew according to Ptolemy's description. He found that the earlier catalogues mostly agreed with his new figures, with one major difference, the way in which human forms are represented. Ptolemy, he says, always shows them with their faces towards the viewer but some, and he particularly criticises Bayer, drew them with their backs to the viewer. Flamsteed thought the difference came about because while everyone supposed the constellations to be drawn on a globe, some thought them drawn on a solid globe and viewed from outside, others thought of them as drawn on a spherical framework or astrolabe and viewed from inside, from the centre. Flamsteed considered that Ptolemy had followed the second convention. He set out the argument quite fully in a letter to Pound in which he described the numerous and confusing systems of his predecessors.[42] Halley, according to Flamsteed, drew the constellations the wrong way round in his southern planisphere and when Flamsteed taxed him with it, passed it off rather flippantly as a young man's fault. Flamsteed did not appreciate joking about serious matters.

There is another major difference between the Crosthwait and Halley catalogues. Halley included many stars that Flamsteed had observed with the mural arc but not with the sextant, 500 in all according to Newton's accounts for the 1712 catalogue. Thirteen stars at the beginning of Andromedae show that clearly. In Halley's catalogue they are listed as 1–13. They are not in the Crosthwait catalogue and they were not among Flamsteed's sextant observations of Andromeda as printed in Volume I of *Historia coelestis Britannica*, but they were observed with the mural arc in the autumn of 1691 as listed in Volume II.

What do the differences between the two versions of the catalogue imply? The development of the two printed catalogues can be followed, at least in broad terms, in documents in the Flamsteed papers in the archives of the Royal Observatory, together with the copy of the 1712 catalogue that Flamsteed annotated and sent to Sharp (See Note 32). RGO 1/25 is a fair copy of Flamsteed's catalogue that he compiled over some years in the first decade of the eighteenth century, as shown by sheets dated in his own hand. Documents in pieces RGO 1/26, 27, and 28 appear to be preliminary material. RGO 1/25 may be the master copy from which the printer's copy for the 1725 catalogue was prepared. The University Library at Cambridge also holds the manuscript catalogue which had been presented to the Royal Greenwich Observatory in 1935 by the Radcliffe Observatory in Oxford. It is not known how it reached the Radcliffe.[43]

The Radcliffe manuscript consists of folio sheets originally stitched together and bound between board covers. The stitching and binding have given way in many places. The original manuscript on the recto side of each folio is covered with corrections, additions, and other changes, and many notes and calculations are written on the verso sides. The annotations and changes seem to be mostly in Halley's hand. Some of the sheets are in poor condition, torn, faded, blotched, and dirty, but the writing is generally quite legible. The manuscript in fact comprises two documents, an original catalogue and a revised catalogue produced by overwriting the original and inserting additional sheets. They will be referred to as M_1 and M_2 respectively. In a similar way the published versions will be called P_1, for the 1712 version, and P_2 for the 1725 version. The relations between the various versions are analysed in Appendix 9 and lead to the following account of their production.

Flamsteed drew up a catalogue of the stars he observed with his sextant up to 1689 which may be denoted by M_0, and which is probably to be identified, in essence if not as an actual manuscript, with the sealed version deposited with the referees. Flamsteed continued to develop it in parallel with his mural arc observations, as recorded by the dates in the fair copy of RGO 1/25. It eventually became the catalogue of 1725, no doubt in the form it had at Flamsteed's death. The arrangement of stars and the descriptions were Flamsteed's. Meanwhile Halley was working on the incomplete catalogue M_0 which someone had copied as M_1, with the constellations in the same order as they appear in P_2. The catalogue M_1 is unfinished, in particular there are substantial omissions of latitudes and longitudes of whole blocks of stars in the constellations to the north of the zodiac. That and dates on two folios point to Gregory as the compiler: M_1 was unfinished when Gregory died in 1708. Subsequently a redactor, Halley almost certainly, produced M_2. He changed the order of the northern and southern stars to that in P_1, he altered the order of constellations especially among the northern stars, he changed the titles of the groups and constellations, he added a great many stars that Flamsteed had observed with the mural arc, he completed the catalogue entries missing in M_1, and he made many alterations to the descriptions of stars and corrected a few values of coordinates in M_1. The result was that M_2 is effectively identical with P_1. It cannot, however, have been the printer's copy for P_1 for it is in too much of a mess; the printer probably had a fair copy from it. When Flamsteed got hold of the 1712 printed catalogue sheets he sacrificed all but a few, binding up one set and sending it to Sharp with all Halley's 'corruptions' marked. Flamsteed's changes correspond to the form in which Crosthwait and Sharp would publish P_2, in particular, the descriptions revert to Flamsteed's original forms and many stars are moved from Halley's constellations to others.[*]

Crosthwait and Sharp had not finished with Halley when Flamsteed died. Flamsteed included previous catalogues in his *Historia* to show where he stood in the history of astronomy. Halley's catalogue of the southern stars was one he intended to include for he listed it in

> An Estimate of the Number of Folio Pages, that the *Historia Coelestis Britannica*, may contain when Printed

that he drew up for the referees on 8 November 1704, as

> Mr *Halley's* Catalogue of the Southern Fix'd Stars, to the Year 1677; with the *French* small Catalogues of them, in P.7.[44]

Halley's observed distances were retained in the 1725 version but the places were recalculated by Sharp, and based on Flamsteed's catalogue

---

[*]By a fortunate chance, Professor Owen Gingerich was able to purchase the copy that Flamsteed sent to Sharp, and I am grateful to him for letting me see it and for discussing it and other documents with me (see note 32).

instead of Tycho's. The relation of Sharp's revision to Halley's original catalogue was analysed in Chapter 3 and Appendix 5. Halley was not pleased when he heard about Sharp's new calculations, but he was hardly in any position to complain. Indeed, when he had published his southern catalogue, he had printed the observations precisely so that others might recalculate his places when Tycho's were superseded. It is true that he was thinking mainly of known errors in Tycho's elements of the ecliptic, which were indeed the major source of the changes, but it was reasonable to recalculate his places when Flamsteed's catalogue became available.[45]

Flamsteed intended that his *Prolegomena* should be both an history of astronomy and an account of his life's work. He never saw it printed. He had written an English version when he died and Crosthwait and Sharp got it put into Latin, with considerable difficulty.[46] Flamsteed's *Prolegomena* has a substantial account of astronomy up to the time of Hevelius, with much emphasis given to Tycho. He followed it with a description of the instruments he himself used, and of his early observations of the Earth's motion in its orbit round the Sun. He had shown that the Earth spun at a steady rate and that variations in solar time arose from the ellipticity of the orbit. He saw that as confirming the Copernican heliocentric system. He described how he adjusted his instruments and how he employed his calculators, Luke Leigh, a distant relative of Halley's, and James Hodgson. He also went back to an old controversy with Newton about the comet of 1680 which Newton had at first thought was two comets. Flamsteed always thought there was just one and although Newton soon changed his mind, Flamsteed could never forget that he had been right and Newton wrong.

Halley's *Prefatio* to his 1712 version was much shorter. There was a very condensed history and an account of Flamsteed's observations, which had extended over thirty years at Greenwich

> ... but nothing has emerged from the Observatory to justify all the equipment and expense so that he seemed, so far, only to have worked for himself, or at any rate for a few of his friends.

That graceless and not wholly accurate passage is so close to the views that Newton expressed on more than one occasion that, even if Halley wrote it, the opinion is Newton's. The *Prefatio* also says that there were imperfections in the catalogue that Flamsteed handed over and that it was given to Halley to revise, who had to do all the calculations again, costing him a great labour.

14.4 *Halley returns to the Observatory he saw set out*

John Flamsteed died on 31 December 1719 and on 10 January 1720 Hearne learnt that Halley was to succeed him. Halley's friends had evidently foreseen Flamsteed's death, for on 11 December 1719, Hearne wrote to Mead, 'I must heartily congraulate you upon the success you have had in behalf of Dr Halley. This great Man had been neglected too long', and noted on 10 January 1720:

> Mr Flamsteed is dead and Dr Halley hath got his place at Greenwich for wch he is obliged to Dr Mead. For so Dr Mead writes me in a letter of the 9th of this Month (viz) I have bin so happy to gett Flamsteed's place for Dr Halley by means of my Lord Sunderland.[47]

Mead had the ear of those who could influence royal appointments. Sunderland was the son of the second earl who had been responsible for carrying out many of the arbitrary acts of James II. According to the *Memoir* of Halley the Earl of Macclesfield also recommended Halley. Macclesfield's son, also Earl of Macclesfield, who supported Bradley as Halley's successor, confirmed his father's part when he wrote of the pleasure his father had in getting Halley appointed as Astronomer Royal.[48]

Halley lost no time in getting Mrs Flamsteed to leave the Observatory. He found it bare. Mrs Flamsteed had removed all the instruments on the grounds that they were Flamsteed's personal property.[49] The Government accepted that was so. Halley had to re-equip the Observatory, for which he obtained a grant from the Government of £500 to replace the mural arc, although he did not receive it until 1724.

Halley's new instruments were far better than Flamsteed's. While waiting for the new mural arc he set up a transit telescope, the first to be made in England. The transit telescope had been invented by Römer. A telescope was mounted on a horizontal east–west axis and in principle should serve the same purpose as a mural arc. However, the axis could not be exactly adjusted and it went out of true with changes of temperature. Halley is supposed to have used a telescope that was made by Hooke some twenty years earlier; it was five-and-a-half feet (1.7 m) long with an aperture of $1\frac{3}{4}$ inches (4.4 cm). The axis was three-and-a-half feet (1.1 m) long and the telescope was mounted off-centre as in the original, somewhat unsatisfactory, design of Römer. The telescope was mounted in 1721 and remained in use until 1774; Halley's arrangements were described in the Visitors' report of 16 May 1726:

That he accordingly first provided a meridian instrument, being an axis fitted to an excellent 5ft telescope, made with very great care substantially supported with free stones, and placed in a little room built on purpose, adjoining to the west side of the Observatory and fitted up with proper openings both at the top and each end, for directing the telescope to all parts of the meridian.

The expenses were:

The instrument itself and telescope:	£30 0 0
The stone and mason's work:	£1 10 0
A plain week clock to stand by it:	£5 0 0
Building and fitting up the little room:	£25 0 0,

in all £61 10s. 0d.[50]

Halley observed with the transit instrument until he had his new mural arc, after which he did not use it regularly. Bradley refurbished and readjusted it and made regular transit observations from 1743 until 1750; it is now in the Old Observatory.[51]

When Halley got his government grant in 1724 he had a new mural arc made by Jonathan Sisson to the design of George Graham (1673–1757), the renowned clockmaker, who adjusted it and divided the limb. The arc was iron, very rigid, and mounted on a new wall of just nine stones (Fig. 14.2). Flamsteed's brick wall was too close to the edge of the cliff on which the Observatory stands and it gave continual trouble as it settled, something that Flamsteed seems to have tried to conceal from Newton. The third important feature of the new arc was that Graham himself divided the scale on the limb far more accurately than anything before. The design and the care in the construction of Halley's arc were the basis of the high repute and great success of English instrument makers in the eighteenth century.

The costs were:

Mason's work:	£42 17 0
The Quadrant:	£204 8 6
The house:	£72 1 10
A month clock:	£12 0 0
Telescope lenses:	£3 0 0.[52]

Halley used the arc continuously up to his last years, after which Bradley and others made various improvements and modifications so that the instrument remained in use until 1812; it is now preserved in the Observatory as shown in Fig. 14.2.

The accounts for the transit telescope and the mural arc include clocks. Halley had three made by Graham, of which one, the week clock, has disappeared. Graham 1 and Graham 2, the month clocks, were in use up to the beginning of this century and are still keeping time in the Royal

Figure 14.2 Halley's mural arc (reproduced by permission of the Director of the National Maritime Museum).

Observatory to within a few seconds a week. Graham 1 was borrowed by the Revd W. Hirst when he observed the Transit of Venus from the Inner Temple on 12 June 1769, a nice connection with Halley.[53]

14.5 *Knowledge of the longitude*

Halley had been concerned with the longitude for the whole of his long scientific life, and over many years he collected observations from travellers and sea captains from which he could derive longitudes. Most of those results are in Table 14.1, which shows both the range of the methods that were employed and how reliable the results were.

Table 14.1 Determinations of longitude by various methods

Date	Place	Methods	Value			Modern value		Reference
			deg	min		deg	min	
1677	Avignon	L.E.				4	45	
	St Helena	L.E.	6	30	W	5	42	
1678	Paris				E	2	20	
1708	Buenos Aires	L.O.	58		W	58	21	a
1722	Port Royal	L.E.	76	37.5	W	76	40	b
1722	Carthagena	J.S.	75	38	W	75	34	c
1727	Vera Cruz	S.E.	97	30	W	96	05	d
1726	New York	J.S.	74	04	W	74	00	e
1726	Lisbon	J.S.	09	07.5	W	09	11	e

Methods: L.O., lunar occultation; L.E., lunar eclipse; J.S., Jupiter's satellites; S.E., solar eclipse.
References: a: Halley 1722a; b: Halley 1723a; c: Halley 1723b; d: Halley 1728; e: Bradley 1726.

Halley derived the value for Buenos Aires from the immersion of a star in the Moon reported in the *Memoirs* of the Royal Academy of Paris for 1711. He found that the day had been incorrectly recorded and the star misidentified.

He found the longitude of Port Royal from an eclipse that was obscured at London, so he used an observation by Christfried Kirck in Berlin. He said that the value agreed with one that he had obtained many years before from observations of Charles Boucher.

Although Halley was so deeply concerned with the longitude throughout his life, he was in that but one of many. In 1713 William Whiston (1667–1752), Newton's successor in the Lucasian Chair at Cambridge, along with Humphrey Ditton, a mathematics teacher at Christ's Hospital, said they had a scheme for ships to discover longitude at sea, although they did not disclose what it was.[54] In April 1714, they petitioned Parliament to award a prize for it, as part of a project to get Parliament to encourage the discovery of methods for finding longitude at sea. A group of Royal Navy and Merchant Marine seamen, along with City merchants, also petitioned for a prize in May of that year. The House of Commons set up a committee, including Joseph Addison, the essayist and editor of *The Guardian*, and Lord James Stanhope, a friend of Whiston, to examine the proposal, and the committee called eminent scientists before it for advice: Newton, Halley, Cotes, and Samuel Clarke.

Whiston and Ditton, as they had revealed their scheme to the House Committee and their advisers, proposed to have ships stationed at anchor at regular intervals along trade routes. The ships were to fire star shells at midnight to burst at a predetermined height of 6 440 ft (1963 m). Mariners would observe the bearing of the star shell and would calculate their distance east–west from a signal ship, and hence their longitude, either from the elevation of the burst or from the time between seeing the light and hearing the sound of the burst. Cotes spoke favourably of the scheme, and so, after some persuasion, did Newton. Halley said that the various items in the proposal should be examined experimentally before he would give his opinion and he thought that would be expensive, but could not say how expensive.[55] The Committee unanimously approved the plan for a prize. A bill 'for Providing a Publick Reward for such Person or Persons as shall Discover the Longitude at Sea' passed the Commons on 3 July 1714 and the Lords on the 8th, and Queen Anne gave her Assent on 20 July (12 Anne Cap 15). Clearly the proposal of Whiston and Ditton for a prize was timely, however impractical their own solution.

The Prize was to be £10 000 for a method that gave an error of less than 60 nautical miles on a voyage to the West Indies and back, £15 000 for an error of less than 40 miles, and £20 000 for an error of less than 30 miles. 30 miles corresponds to an error in time of about 12 s. The Act established Commissioners to consider projects and they formed the Board of Longitude. They did not award the prize to Whiston, neither for his signal ships, nor for a scheme using the magnetic dip, nor for one using observation of eclipses of the satellites of Jupiter. Halley was an original *ex officio* member of the Board, first as Savilian Professor and then as Astronomer Royal. The Board did not meet for many years.

Some while after the passage of the Act, John and James Harrison, two carpenters of Barrow-on-Humber who made clocks with wooden movements, heard of the Prize and John came to London about it, probably in 1730. He got in touch with Halley who advised him to consult George Graham, the leading clockmaker of the day and a fellow of the Royal Society. Graham urged Harrison to construct a clock according to his plans. The Harrisons then made the timekeeper H-1 with a wooden movement. It was completed in 1735 and tested in ships of the Royal Navy on voyages to Lisbon and back. The performance was spectacular and, following the return of the timekeeper to England, the Royal Society gave a certificate, signed by Halley, Smith, Bradley, Machin, and Graham, that the Harrisons deserved public encouragement. Harrison then constructed H-2 with a brass movement. In 1741 he began H-3 for which he made many changes. The later history of the Harrison timekeepers, the long time it took to build H-3, the construction of H-4, and the test with the Kendal copy on Cook's second voyage from 1772, extends thirty years beyond Halley's lifetime.

Harrison was, however, to recall with gratitude the early support he had had from Halley and Graham.[56]

The alternative method of finding longitude at sea from the position of the Moon also became practical about 1770. Euler at last developed the theory of the Moon's motion to an adequate state and Mayer produced tables of Euler's theory. At the same time, Hadley devised the modern sextant, an improvement over Halley's instrument, and it became possible to make reasonably accurate astronomical observations from a ship at sea. Maskelyne, then Astronomer Royal, was a strong advocate of the lunar method and Cook had placed great trust in it on his first voyage, but on his second voyage, he found the Kendal watch his secure guide.[57]

Cook's voyages brought together three dominant concerns of Halley. The first voyage was the direct outcome of his prediction of the transit of Venus. The lunar method for longitude that Cook employed on that voyage had preoccupied Halley for almost the whole of his life. The Kendal watch that Cook took on his second and third voyages was the descendant of the first Harrison chronometer that Halley and Graham had encouraged. In all those three strands, as in much else, Halley had a vision of the future, of what might be and what should be, that matched his appreciation of the past.

14.6 Tidying up

Halley continued to publish for seventeen years after he became Astronomer Royal, for all that he was then sixty-five. His papers in *Philosophical Transactions* were for the most part rather minor. Among them were comments on magnetic and astronomical observations that had been sent by sea captains or others, and papers of 1721 on diving, on finding the positions of planets from appulses to fixed stars, and on a parhelion. Almost his last paper was a discussion of observations of latitude and magnetic variation taken on board *Hartford* on passage from Java Head to St Helena.[58]

Two papers were concerned more particularly with his duties as Astronomer Royal. In 1721 he discussed atmospheric refraction as it affected astronomical observations and published a table constructed by Newton.[59] He pointed out that refraction makes all objects appear higher in the sky than they really are and showed how to arrange observations to reduce the consequent errors in navigation. In 1725 he described a transit of Mercury from which he derived the mean motion and the longitudes of the nodes of the orbit, returning to one of his earliest studies on St Helena.[60] In addition to the observations of the lunar clipse that he used to find the longitude of Port Royal, he reported his own observations of the solar eclipse of 1722 and of the lunar eclipse of 1736.[61]

In 1694 Halley had read two papers to the Royal Society on possible causes of the Universal Deluge—Noah's Flood—as recounted in Genesis. He did not publish them at the time because he thought them too speculative, 'and apprehensive least by some unguarded Expression he might incur the Censure of the Sacred Order', but deposited the manuscripts sealed up with the Society. Now after thirty years a committee of the Society desired him to publish. He considered that the scriptural account demonstrated that there had been a great catastrophe and that it could have been the cause of such things as fossils high up in cliffs, as he himself had seen at Harwich, but he thought that the suggestions for its cause that had been put forward by others were inadequate. He proposed that the shock of a comet colliding with the Earth might have so changed the face of the Earth as to produce the Deluge and to leave great craters such as the Caspian Sea. In his second paper, 'Some further Thoughts', he said that someone had suggested to him that his speculations might really apply to a previous world destroyed before the Biblical Creation. Halley thought there might be some advantage in the destruction of a previous world to enable ours to be recreated in a better form, perhaps more suitable for agriculture and other pursuits. He went on

> This may, perhaps, be thought hard, to destroy the whole Race for the Benefit of those that are to succeed. But if we consider Death simply, and how that the Life of each Individual is but of a very small Duration, it will be found that as to those that die, it is indifferent whether they die in a pestilence out of 100000 per Ann or ordinarily out of 25000 in this great City, the Pestilence only appearing terrible to those that survive to contemplate the Danger they have escaped. Besides, as Seneca has it
> Vitae est avidus quisquis non vult
> Mundo secum peruente mori.[62]

Halley's speculations about the interior of the Earth, the nature of fossils, and the changes that might have happened on the surface were not exceptional in his day. Hooke, Thomas Burnet, and Arbuthnott are just a few who had written on them. Notions of other and earlier worlds also were far from novel; Queen Christina, almost a century previously, was but one who had entertained them. It may readily be understood that Halley was hesitant to publish such speculations in the 1690s when they would have attracted the ire of the High Church, but in the context of millenial ideas that were prevalent in the seventeenth and early eighteenth centuries they do not seem so remarkable.[63] The notion that a comet may have been the cause of great disturbances is still with us, reinforced in the present time by the impact of a comet on Jupiter. Hudson's Bay is one of the most prominent crater features that have been thought to be the results of meteor or comet impacts, and the extinction of dinosaurs and other families, and the iridium layer at the base of Cretaceous strata are also ascribed to such catastrophic causes. Halley shows a sceptical and stoic turn of mind in his observation that it is those who survive to contemplate them who are terrified by great epidemics.

In 1727 Halley discussed an article published in Paris by Father Étienne Souciet who argued against Newton's biblical chronology. Newton's chronology appeared posthumously in 1728 but earlier, towards the end of his life, Newton had drawn up a summary as desired by Caroline of Anspach, then Princess of Wales, for her own use. By some means it came into the hands of a French printer who published it with adverse comments, to which Newton replied in protest. Shortly after Newton had died, Souciet published five 'Dissertations' fiercely criticising Newton. He thought that the dates that Newton had assigned to the Siege of Troy and the Voyage of the Argonauts were too recent and used astronomical arguments to set them earlier. Halley says that he was induced to take up the matter in part because of his position as Astronomer Royal, which justified his discussion of ancient astronomy, and 'also from the long Acquaintance and Friendship that has subsisted between the Deceased and myself'. The astronomical argument turns on the identification of stars that Hipparchos and Eudoxus had placed at the origin of the constellation Aries. Souciet and Newton both started from the proposition:

> ... what seems the most exceptional part of the whole System, viz. that Chiron the Centaur fixt the Colures[*] in the ancient Sphere of the fixt Stars, in the same place as Hipparchos tells us they had been supposed by Eudoxus many Centuries of Years after Chiron.

Halley writes 'I find the Dispute to be chiefly over what part of the Back of Aries the Colure past', that is, over the description of the constellation, and also the identification of stars mentioned by Hipparchos, just the same issues over which Flamsteed was so concerned in his publication of the *Historia coelestis Britannica*. Halley wrote his second paper because he had obtained a copy of a Florentine edition of Hipparchos and, from a close reading of the Greek, he derived the right

---

[*] A *colure* is a great circle through the poles; the colures of which Halley is writing are the equinoctial and solstitial colures, the former passing through the equinoces and the latter through the solstices.

ascension and declination of the first star in Aries. He disagreed with Souciet and supported Newton's more condensed chronology.[64]

Halley used astronomical and other data in the assessment of the implications of three groups of literary texts: the accounts of the landing of Julius Caesar by Caesar and Dio Cassius, the Gospel accounts of the Crucifixion, and the Greek tales of Chiron and the Trojan Wars. He treated them all similarly—did evidence of the natural world—tides in the Channel, a seamless robe brought from India, the precession of the equinoces—help in elucidating the meanings of those texts? We today have different perceptions of the three groups of texts. We accept that Caesar's account is that of a participant, that the Gospel story may be secondhand and may or may not relate to an historical event, while the stories of Chiron and the Trojan War are mythical and can bear no relation to actual celestial phenomena.

His last papers reveal how distant Halley's preconceptions of the world were from ours, even to his life's end. True he was using astronomical observations to fix past events, just as people do today for biblical and other events,[65] but as a result of literary and historical criticism such as Newton and others were starting to apply to the Bible and to the ancient writers, the status of many recorded events has greatly changed, and many that were taken to be real by Halley and his contemporaries are now seen as legendary or mythical. The Flood and the Hexaemeron have joined Chiron, the Argonauts, and the Siege of Troy as expressions of beliefs about the world, not records of history that match astronomical occurrences. Halley handled historical astronomy and geophysics much as we do today, but his attitude to human history and myth, like that of most of his contemporaries, was very different from ours.

Halley's life comes full circle as he returns in almost his last paper to the longitude, the subject which prompted the founding of the Royal Observatory while he was still an undergraduate, which sent him off to St Helena, and on which he was working to almost his last days with his observations of the Moon (Chapter 13).[66] Not often do we find such a consistent endeavour in a scientist who at the same time explored and opened up many of the most significant fields of physical science of his day, not forgetting other major branches of knowledge that occupied him.

14.7 Last days

In his last years at Greenwich Halley remained active in the Royal Society and was Vice-President in 1731. He went regularly to London and began the custom of fellows of the Royal Society dining together, from which sprang the present Royal Society Dining Club.[67] His correspondence shows that he kept in touch with scientific men, for example, he received through John Bevis a packet from de la Condamine in Peru who was measuring the degree of latitude there which, together with the other measurements in Lapland, would demonstrate once for all that Newton was right about the oblate (Dutch cheese-shaped) figure of the Earth. It was half a century since Halley on St Helena and Richer in Cayenne had found the changes in the lengths of the pendulums of their clocks that Newton had used in the *Principia*. De la Condamine's packet may have contained medicinal Peruvian bark and perhaps also the letter from Bouguer and la Condamine at Quito, dated May 1738, in which they reported their measurement of the obliquity of the ecliptic and the problems that they were having:

> M de Maupertuis qui doit être de retour de son voyage du nord, où lui et les autres mathématiciens nos confrères, n'ont eu à vaincre que les difficultés du froid du climat, qui nous ont été commune avec eux contre tout apparence, mais òu ils n'ont pas sans doute trouvé les obstacles que l'ignorance, les préjugés, la barbarie et l'impossibilité de faire éxécuter les ordres du souverain nous sucitent ici à chaque pas.
>
> [Mr de Maupertuis, who must be about to return from his voyage in the north, where he and our colleagues the other mathematicians had nothing to conquer but the difficulties of the cold of the climate, a thing which we, against all odds, had in common with them, but where they undoubtedly did not find obstacles such as ignorance, prejudice, barbarity, and the impossibility of carrying out the orders of our sovereign which have dogged us at every turn].

Halley was still in correspondence with Newton about cometary calculations in 1724/5.[68]

The Visitors to the Royal Observatory came there from time to time but there are only few notices of their visits. Stukeley records an occasion on 13 May 1726 when Martin Folkes, President of the Royal Society after Newton's death, Taylor, Machin, Graham, and Stukeley himself, came to examine the instruments provided from the grant of £500 from the Board of Ordnance.[69] Halley showed them the meridian transit of Sirius, presumably in his transit telescope. The most notable visitor was Queen Caroline (the *Minerva of the Age*) in 1729. When she learnt that his salary was £100 a year, she said, so the story goes, that she would have it increased, but he dissuaded her, saying that others would seek the position for the money and not from the love of knowledge. However, hearing that he was a post-captain in the Royal Navy, she arranged that he should have the captain's half pay to which he was entitled and had not so far received.[70]

Halley became slightly paralysed in his last years, but that did not prevent his observing, nor did it prevent his visiting London, attending the Royal Society, and dining with other fellows.

Mary died on 30 January 1736 and was buried in the church of St Margaret, Lee, close by the Observatory, on 14 February 1736.[71] The *Memoir* of Halley says that Mary was 'an agreeable young Gentlewoman and a Person of real merit' but almost nothing is known of their married life or domestic circumstances between the record of their marriage and that of Mary's death, and that little comes, ironically, from Flamsteed when, shortly after their marriage, he wrote to thank Halley for a visit he paid them and sent his respects to Mary.[72] Much later, in the midst of the quarrel over the printing of the *Historia coelestis* he notes, as recounted above, that Mary accompanied Halley with their children on a visit to Greenwich. For the rest, nothing remains, not even a will from Mary. Halley, in his will, asked to be buried beside Mary at St Margaret's, Lee.

Halley died peacefully after a glass of wine, on 14 January 1742, at the age of eighty-five. His daughters set up the inscription on his tomb:

>   Sub hoc marmore
> Placide requiescit, cum uxore carissima
>   EDMUNDUS HALLEIUS, LL.D.
>   Astronomorum sui seculi facile princeps
>   Ut vero scias lector
>   Qualis quantosque vir ille fuit,
>   Scripta ejus multifaria lege;
>   Quibus omnes fere artes & scientias
>   Illustravit ornavit amplificavit
>   Aequum est igitur
>   Ut quem cives sui vivum
>   Tantopere coluere
>   Memoriam ejus posteritas
>   Grata veneretur
>   Natus   MDCLVI
>   } Est A.C {
>   Mortuus   MDCCXLI
>   Hoc saxum optimis parentibus
>   Sacrarunt duae filiae pientissimae
>   Anno MDCCXLII

Quietly under this marble, Edmund Halley, LL.D., easily the prince of astronomers of his age, rests with his dearest wife. Reader, to know how truly great that man was, read his many writings; wherein he illustrated, improved, and enlarged almost all the arts and sciences. It is just therefore that as in life he was highly honoured by his fellow citizens, so his memory should be gratefully respected by posterity. Born 1656, died 1741. His two daughters most dutifully consecrated this stone to the best of parents.

His daughters, Margaret Halley and Catharine Price, with Henry Price his son-in-law, and John Pond (1767–1836), a later Astronomer Royal, are buried in the same tomb. St. Margaret's, Lee is in Blackheath about 1 mile south of the Old Royal Observatory. The church Halley knew was torn down in the nineteenth century; only the base of the tower is now standing. The present church was built in 1840.

The tomb deteriorated and in 1854 the Lords Commissioners of the Admiralty restored it and removed the memorial slab to the Observatory and built it into the wall there, where it is still to be seen. The Admiralty further repaired the tomb in 1909.[73]

For almost 250 years Halley had no other memorial. Newton was buried with great pomp in Westminster Abbey, Wren has St Paul's as his memorial, Halley was buried, as he wished, beside Mary at Lee. There is no hint in Halley's life that he sought honours: he may, like Evelyn, have declined them. In death as in life he remains a modest gentleman. In 1986 the Dean and Chapter of the Abbey agreed to a memorial plaque being placed in the cloisters, where it was dedicated at a special service on 13 November 1986. Halley's career as Savilian Professor is commemorated in Oxford by a plaque on No. 8 New College Lane and his visit to Avignon by the rue Halley in that city.[74]

15

The improvement of natural knowledge

ἀνδρων γαρ ἐπιφανων πασα γη ταφος
[For famous men have the whole Earth as their memorial...]
PERICLES: THUCYDIDES, *HISTORIES*, II 43,3

15.1 *Halley in his own day*

HALLEY'S reputation was high in his own day. Hooke and Hearne, in their various ways, testify to his distinction and to the esteem in which he was held. Newton thought highly of him, so did Pepys and others in the service of the Crown. He was admired in Europe as a mathematician. There is not much on which to base an assessment of his character but it seems that he was genial and sociable, drinking tea with Hooke, sharing a bottle of wine with a friend, dining in Oxford, taking coffee after meetings of the Royal Society, and going to London for dinners when he was at Greenwich.[1]

His family life leaves little record. Mary lived to a great age and Halley wished to be buried beside her—that is almost all that is known about their married life. He had at least one friend, Robert Nelson, from schooldays to late in life. There are occasional indications of domestic interests, such as his observation that he had preserved rosemary in his garden throughout the severe winter of 1688/9 by watering it with soapy water.[2] The sale catalogue of his books, though ambiguous (Appendix 8), suggests that he was a great collector with very wide interests. For the rest we have little impression of the man, and the reason is that he left a meagre correspondence and no diary. What did Mary do when he was at sea? Did she accompany him on some of his travels, as Elizabeth Pepys and Mary Evelyn sometimes joined their husbands? There is no answer.

Halley was out of London a great deal in the decade from 1695 until he became Savilian Professor. He then had a house in London and spent much time there—the houses in New College Lane in Oxford were meant to encourage him and Gregory to stay more in Oxford. He was often at the Royal Society and on the Council from time to time.

Conduitt, Stukeley, and others mention him as having dined or breakfasted with Newton and other company. He would surely have known Newton's *châtelaine*, the 'pretty witty Kitty' Barton.

Catherine Barton, the elegant daughter of Newton's half-sister, was a close friend of Jonathan Swift, a toast of the Kit-Kat Club* and a very intimate companion of Newton's patron, the Earl of Halifax, from whom she received a considerable inheritance. Catherine was also a close friend of Jonathan Swift who was a friend and literary colleague of John Arbuthnott, and Catherine and Arbuthnott were most likely acquainted. Halley, close to Newton and knowing Arbuthnott well, was probably in Catherine's circle. Catherine would entertain Newton's visitors from abroad, one of whom, Voltaire, started a scurrilous but chronologically impossible canard about her influence with Halifax on Newton's behalf. She made a great impression on the foreign guests of the Royal Society who came to see the solar eclipse of 1715 in London (Chapter 13). Rémond de Montmort sent Newton fifty bottles of champagne afterwards and received back the *Opticks* from him and gloves and other items for his wife from Catherine. Halley was doubtless among the social company, even as he organised the astronomical activities.

One thing we do know about Halley, he was firm in maintaining his own interests and those for whom he was responsible. He pursued his own Chancery actions and those of Mary. He ensured that the flag he wore as a naval captain in *Paramore* was not insulted. He insisted on the Imperial Commissioners on the Adriatic carrying out their duties in the interests of his Queen. He was alive to the shortcomings of others, for instance when he remarked on the failure of his people to recognise the Isle of Wight from *Paramore* and when he dismissed as a publicity stunt the French claims to a great deed in burning Aquilea. He was not always infallible. In later life especially he seems to have slipped up sometimes, as when he admitted to Sloane in 1722 that he had mislaid a note that he had made of Michael Maestlin's observation of the comet of 1580, or when he took the motion of the Earth in the wrong sense in computing the orbit of a comet.[3]

---

*One Kit-Kat Club verse about her was:

At Barton's feet the God of Love
His arrows and his Quiver lays
Forgets he has a Throne above
And with this lovely Creature stays

Not Venus' Beauties are more bright
But each appear so like the other
That Cupid has mistook the Right
And takes the Nymph to be his Mother

(John Dryden, *Miscellany poems*, 5th edn (London 1727), vol. 5, 61—probably not by Dryden.)

Halley was born three years before Purcell and could well have attended some of the great State occasions for which Purcell provided the music. Acquaintances were actively musical, Pepys, Williamson, Sir Gabriel Roberts, Arbuthnott. Unless Mary and he were strongly puritanical, they probably saw some of Purcell's semi-operas and later they could have seen Handel's Italian operas and heard his earlier oratorios, to which Arbuthnott possibly contributed. The catalogue of his book sale contains a large group of operas, but whether they were his or just sold with his own books cannot be known.

Halley evidently had good health for most of his life. Apart from the disabilities of his last years when he is supposed to have been slightly paralysed and when he lost all his teeth so that he had to eat fish, his only recorded illness was the fever he contracted in the West Indies.

15.2 *Religion and politics*

In his own day Halley was supposed to be irreligious, and his alleged scepticism or indifference to religion is still accepted, praised indeed on occasion as that of a rational man against the superstition and intolerance of religion. Rigaud on the other hand defended Halley against the charge of infidelity.[4] Accusation and defence alike are insecurely based on circumstantial evidence. The event on which they seem first to have turned was the election of Gregory instead of Halley as Savilian Professor of Geometry in 1683 (Chapter 9). Newton supported Gregory while the Royal Society gave Halley a recommendation and Caswell was not in serious consideration. Possible reasons for the electors' choice, other than religious considerations, were suggested in Chapter 9, but there is no doubt that some thought at the time that religion came into it. Halley himself said that a caveat had been entered against him on the grounds that he had asserted the infinite existence of the world, while Fatio de Duillier wrote to Huygens that Halley was thought to have lost the election on account of his reputed irreligion. There can be little doubt that Flamsteed was responsible for the caveat, for he effectively claimed it himself.[5]

Later, when Halley was at Oxford as Savilian Professor of Geometry, Hearne recorded a number of tales about his irreligion or atheism (Chapter 12). Flamsteed may well have been the source of Hearne's anecdotes, which only appeared after Halley had published his version of the *Historia coelestis*.

There was also a story, said to come from Catherine Barton, that Newton had upbraided Halley for his irreligion. That is rather simplistic. Conduitt's words are

He [Newton] would not bear to hear anyone talk ludicrously of religion, often angry with Dr Halley on that score and lessened his affection for Bentley.

Bentley, who was supposed to have denounced Halley to Stillingfleet for irreligion (Chapter 9), had been chosen to give the first Boyle lectures on natural philosophy and religion. It might be thought odd that he is bracketed here with Halley as attracting Newton's displeasure. Newton, who concealed his Unitarian opinions, was very sensitive to anything that might cast doubt on his own orthodoxy; he could well have been anxious if he thought that Halley was an associate with heretical beliefs. A less probable version of the tale has it that Newton reasoned 'gently' with Halley.[6]

In the first half of 1734 Bishop Berkeley published *The analyst or a discourse addressed to an infidel mathematician*.[7] He argued that the foundations of the fluxional and differential calculus depended on illogical and unclear notions of infinitesimals, that they were therefore obscure and unsound, and that mathematicians who accepted them should not presume to criticise matters of religious belief on the same grounds of illogicality and obscurity. His criticism of Newtonian dynamics, in other tracts as well as in *The analyst*, went much further than the formal basis of the infinitesimal calculus. He argued from his principles of immaterialism and idealism that bodies were inert and therefore that their motions were the results of influences external to them, in particular, of a Prime Mover—a position akin to that of Leibniz before him. He accepted Newtonian dynamics and universal gravitation as an account of how things moved but not as one of why they moved as they did. Berkeley's mathematical criticisms were valid and continued to exercise mathematicians until Cauchy produced a consistent basis for the calculus through limiting processes. Berkeley's other objection to Newtonian dynamics, that it is description but not explanation, still comes up in various guises in current discussions of the nature of gravitation. His criticisms were metaphysical but they were set in the context of the issues that set latitudinarians against High Church men in those decades.

The introduction to *The analyst* was thought to be addressed to some one in particular:

> Though I am a stranger to your person, yet I am not, Sir, a stranger to the reputation you have acquired in that branch of learning that hath been your particular study; nor to the authority that you therefore assume in things foreign to your profession, nor to the abuse that you, and too many more of the like character, are known to make of such undue authority, to the misleadng of unwary persons in matters of the highest concernment, and

whereof your mathematical knowledge can by no means qualify you to be a competent judge.

Who was that mathematician? In a sequel, *A defence of free-thinking in mathematics*, Berkeley wrote[8]

> I can truly say the late celebrated Mr Addison is one of the persons whom you are pleased to characterise in those modest and mannerly terms. He assured me that the infidelity of a certain mathematician, still living, was one principle reason assigned by a witty man of those times for his being an infidel.

The identities of the witty man and the infidel mathematician were disclosed by Berkeley's first biographer:

> The occasion [of *The analyst*] was this: Mr Addison had given the bishop [Berkeley] an account of the behaviour of their common friend Dr Garth's behaviour in his last illness, which was equally unpleasing to both those excellent advocates for revealed religion. For when Mr Addison went to see the doctor, and began to discourse with him seriously about preparing for his approaching dissolution, the other made answer, 'Surely, Addison, I have good reason not to believe those trifles, since my friend Dr Halley who has dealt so much in demonstration has assured me that the doctrines of Chritianity are incomprehensible and the religion itself an imposture'.[9]

Garth died in January 1719 and Addison six months later, while Stock wrote his *Life* many years after *The analyst* appeared, and it is not clear that Stock had the story directly from Berkeley.

Who was Garth? Samuel Garth was a distinguished physician, of much the same age as Halley. He was a firm supporter of the constitutional settlement of 1689 and of the Hanoverian succession. George I appointed him physcan-general to the Army and knighted him. He was a member of the Kit-Kat Club and a competent poet in Latin and English who wrote verses for the toasting glasses of the Club to Lady Essex, Lady Hyde, and Lady Carlisle. Pope wrote that Garth, Vanbrugh, and Congreve were the 'three ... most real good men of the poetical members of the Kit-Kat Club' and Garth, with Addison and Congreve, first heard Pope read his *Illiad* at the house of Lord Halifax.[10] Garth's verses were attacked by a Tory journal but defended in a spirited reply by Addison.[11] He was responsible for arranging for Dryden to be buried in Westminster Abbey. As to his last days, Pope wrote:

> The best natured of men, Sir Samuel Garth has left me in the truest concern for his loss. His death was very heroical, and yet unaffected enough to make a Senator or Philosopher famous. But ill tongues and worse hearts, have branded even his last moments as wrongfully as they did his life, with

irreligion. You must have heard many tales on this subject; but if ever there was a good Christian without knowing himself to be so, it was Dr Garth.[12]

Pope, who apparently had some particulars of Garth's last illness from Towneley, went further, according to the anecdotes of Pope recorded by Joseph Spence. He told Spence that Garth was less libertine[*] in his last years and died a papist. Halley does not come into Pope's account of Garth's death and appears only once in Spence's anecdotes in the statement of the Chevalier Ramsay that he aimed to find the parallax of the fixed stars from observations of the Moon, that he had observed for thirteen years and would require seven more.[13]

Garth spent a cheerful life as a distinguished physician and a sociable and tuneful poet with strong Whig connections. He and Halley could well have been friends, although there is no evidence for it except Stock's late story. They did know some of the same people and Garth's doctor was Mead who also attended Newton in his last illness. Their rather insouciant and cheerful attitudes to life were similar. In his last years Garth found life more and more wearisome and looked forward to death, but whether with stoic resignation or Christian acceptance, it is scarcely possible to say now.

If Garth's irreligion might have been the construction of malicious opponents, Halley's irreligion, so far as it was based at late second hand upon Garth's, is yet more insecure. Stock's identification of Halley as Berkeley's infidel mathematician seems pure speculation. There are other reasons for doubting that Berkeley aimed at Halley. Halley never promoted irreligion or heresy in print and there is no record that he ever wrote against Addison. The evidence for his sceptical or agnostic views is entirely oral, unless his misinterpreted ideas on the age of the Earth were so taken. He never did any original work on fluxions, certainly not on the foundations, nor did he make very much overt use of them in his publications. There are only two associations with Berkeley's attack, Halley's paper on orders of infinity which Berkeley had studied and disagreed with, and Halley's adoption of universal gravitation. Those however were not prominent in *The analyst*. Whiston is a more credible target than Halley. Still alive when Berkeley wrote, he had written on fluxions, he held millenian views, and he was ejected from his Cambridge chair for heretical Arian views that he never concealed.

If contemporary and later opinions are little more than malicious gossip, and if Rigaud's pleading is rather special, can more definite evidence be found in Halley's own statements? There were two issues, one of natural theology and the other of Christology. The question in

---

[*]'Libertine' often meant free-thinking rather than dissolute.

natural theology was how to read the holy scriptures in the light of natural knowledge; the issue in Christology was whether Jesus was God as well as man, the orthodox Trinitarian position, or whether He, with the Holy Spirit, was in some way subordinate to God the Father.

The place of natural knowledge had occupied theologians from at least St Augustine onwards and was a lively subject in Halley's day. Newton paid much attention to it in the later editions of *Principia* and he was far from alone. Halley seems to have had no difficulty in harmonising scriptural accounts with natural knowledge. He attended for the most part, as did Newton, to chronology. He estimated the age of the world, which he emphasised was finite, from physical arguments, and he supported Newton's chronology against a French theologian by astronomical evidence. Few in his day would have been troubled by that or thought it exceptional. His remarks upon the seamless shirt brought from India and its relevance to the Crucifixion narrative show a similar attitude, a critical attitude to details of the scriptures, readiness to be persuaded by relevant evidence and implicit acceptance of the essential elements of revelation. Then, as now, those who believed in the literal truth of the scriptures might have cavilled at Halley's approach, but it would have been quite acceptable to many in the Church of England and in particular to Tillotson and his latitudinarian followers. It was close in spirit to that of contemporary literary scholars who were applying the methods of literary criticism to the sacred scriptures.

There is less evidence for Halley's Christology, or even that he had any view distinct from acquiesence in the Trinitarian Articles of the Church of England. All that can be said is that when he refers to matters of religion, which he does very rarely, it is in conventional terms. Thus writing of the seamless shirt, he refers to the coat of 'our Saviour', as any Christian would. He is said to have protested, when he was a candidate for the Savilian Chair in 1693, that he declared himself a Christian and asked to be considered as such, but that, like so much else, is a second-hand anecdote. He used conventional Christian expressions in letters to Hevelius and others, and drew up his will in the common Christian form. He offers occasional thanks to God, as when he escaped from danger in southern ice or storm ('Fortunately the water ran clear through the gunwales and *it pleased God she wrighted again*'.—Chapter 10) or when he lived long enough to make good progress with his lunar programme ('But Thanks to GOD' he observed 'with his own hands and eyes' during one whole period of perigee (nine years)—Chapter 13), and he hoped that God might help him to demonstrate the acceleration of the Moon ('and shall (God willing) make it one day appear to the Publick'—Chapter 8). Those are the conventional phrases that most of his contemporaries might have used, and as Flamsteed did in much the same way. It would be difficult to make a case for strong Christian or for strong anti-Christian views from such meagre and anodyne material.

Halley, like many adherents of the Church of England then and now, was not, it seems, greatly exercised by Christology. Newton's Arian views were related to his alchemical practices and his critical studies of the Holy Scriptures, neither of which occupied Halley. Newton, Locke, and Boyle, all older than Halley, were all three adepts and Locke became, like Newton, an Arian. Halley was never an alchemist and almost certainly never entertained Newton's idea that the Holy Spirit acted in alchemical transformations. Alchemy was going out of fashion in Halley's day and with it one of the rather mystical elements in Arianism. Biblical criticism, on the other hand, was receiving increasing attention, but Halley did not engage with it as he did with Greek mathematics. Whiston, however, came to his Arian views as a result of patristic studies. Halley pursued none of the studies that led Locke, Newton, and Whiston to their Arian opinions.

Why should Halley have been castigated as irreligious if he adhered to the Church of England without undue difficuly or questioning, but also without great passion? Part of the reason must be malicious stories put about by Flamsteed but, as with Garth, they may have been given some credence by Halley's own character. If he held serious views about the world and religion and God, he did not express them ponderously and continuously as Flamsteed did, nor with the ferocity of some extreme High Church men. He was rather more conscious of the foibles and absurdities of his fellows, even the pious, as some of his correspondence shows; and the seriously pious were, and are, apt to be very offended by any idea that in some inessential matters they might be mistaken or absurd. 'Atheist' was a term of abuse used especially by Jacobites and men such as Hearne and Proast for their opponents. It cannot be taken as a serious assessment of theological views. Flamsteed need not have been the only one in the passionate society of post-Revolution London who took Halley's possibly sardonic wit for flippant paganism. 'Scoffing' in coffee houses was generally seen as a prime mark of the atheistical wit. At the same time Halley maintained lifelong friendships with committed churchmen such as Nelson, and in his earlier years at least was highly regarded by Tillotson who, if a latitudinarian, strongly advocated the need for apologetic writing in 'this degenerate age which has been so miserably overrun with scepticism and infidelity'. Tillotson wrote that five years before Halley and Nelson stayed with him on their way to France.[14]

Hearne characterised Halley as a trimmer in politics, a stance that he shared with the great majority of his contemporaries in those turbulent times. Our evidence for his real views is even less definite than for his religion. All we know is that he served the regime in power effectively from time to time.

15.3 Two revolutions

The two years from 1687 to 1689 saw two profound revolutions, the 'Newtonian Revolution' with the publication of *Principia* and the 'Glorious Revolution' when William and Mary came in. There may be argument about whether they were really revolutions, but Newton's contemporaries and immediate successors, Clairaut for instance, saw the 'Newtonian Revolution' as a revolution. In the same way, the term 'Glorious Revolution' became common, among whigs at least, within months.[15]

Particular circumstances may trigger revolutions but growing dissatisfaction over years before prepares them. Revolutions do not suddenly cease, their consequences work out over years, decades, even centuries. Revolutions may change ways of doing science, or the constitution and politics of a country, but often the same people pursue and operate the new ways as did the old. Beneath the turbulence of the French Revolution administrative life continued with much the same administrators.[16] The policies of Charles II and James II, and fears of Louis XIV, had over decades prepared the Glorious Revolution. Domestic and foreign events of 1688 triggered it, and it developed in the years up to the Hanoverian settlement. It made little change in the personalities of State. The few extreme left office, the many moderate continued in and out. There are parallels with the Newtonian Revolution. New thoughts at home and abroad, and new observations with more effective instruments, had prepared the ground. A quite specific train of events in 1684 triggered it, and the developments and applications have never since then ceased to extend. No one stopped doing natural philosophy after 1687 because of *Principia* but the old hands tackled it in new ways.

Halley's part in the Newtonian Revolution was fundamental. Without him it would not have happened when it did and perhaps not in the form it took. That was not all. His year in St Helena was a significant part of the preparation. He made the first astrometric catalogue with the new telescopic instruments, he extended Horrock's theory of the lunar orbit as a rotating ellipse, and he tried to find the distance of the Sun from the transit of Mercury, all phenomenological novelties that extended men's ideas of the cosmos and its kinematics. Later in 1680–1682 he observed comets and discussed them with Flamsteed and

Newton. After 1688 Halley worked on direct applications of Newton's celestial mechanics, in the public eye with his cometary calculations, in obscurity in his lunar studies. His interests were much wider than the celestial mechanics of *Principia*: his studies of geomagnetism and of meteorology display the principles of the Newtonian Revolution at work in the formulation of an abstract model to simulate physical phenomena, and its comparison with observation. Halley rarely made explicit his methodology or metaphysics, he was not philosophically inclined as were some of his fellow Newtonians in Europe, but there was scarcely anyone more practically devoted to Newtonian methods and more effective in applying them over a wide range of topics.

It may be thought artificial to link Halley with the Glorious Revolution. He was abroad in critical years of the reign of Charles II, from 1679 to 1682, but the family problems that followed from his father's chance connection with the death of the Earl of Essex in 1684 may have in part determined the time of his visit to Newton that August. Links, whether causal or casual, between the two revolutions cannot be avoided, for even as *Principia* was published Newton became involved in the resistance of Cambridge to the intrusions of James II, and after the invasion of William, Newton sat in the Revolution Convention of 1689. Halley had no such political part but his survey of the Thames approaches seems to have been connected in some obscure way with William's invasion. Newton eventually received his reward as Master of the Mint for his support of the new regime, while Halley went on to display his many talents in the service of William and more especially Anne. Two revolutions, two great men involved in different ways in both, the strands were all twisted together in August 1684.

15.4 *The natural philosopher*

When Halley was at Oxford he was seen by Bernoulli and other continental mathematicians as one of the leading English mathematicians. His credentials were considered in Chapter 12. His contemporaries regarded him highly, his editions of the Greek mathematicians remain impressive works of scholarship, and he was clearly a very skilled computer, but he has left little of his own in the structure of mathematics today. By contrast his stature as natural philosopher stands, if anything, higher than ever as the originality of his ideas is more and more appreciated in the light of current developments in the natural sciences.

Halley was not original in the topics that he studied, he was very much a man of his times. Astrometry, comets, the size of the Earth, its magnetic field, were all of great contemporary interest. Flamsteed made some geodetic observations in the neighbourhood of Greenwich and he recorded observations of the magnetic variation made by seamen. Many sailors observed the variation and many of the results were unknown to Halley (Chapter 10). It was in his approach to topics of his day that Halley was original. Flamsteed collected magnetic observations from seamen, but did little about them; Halley made his own observations and constructed a map. All seamen knew about tides; Halley observed them and produced a chart. Sir Edward Bullard thought that his great contribution lay in putting large quantities of data into an organised form, as he did in his geophysical studies and in his demographic papers.[17] That is certainly a first major step in geophysics, but Halley went beyond that, especially when he interpreted historical material. Many may have pursued the same topics, but apart from his great contemporary, we are indebted to him for more, and more original, contributions to knowledge than to most of his contemporaries.

There are professorships of natural philosophy in the older Scottish universities and in Cambridge and Oxford. To many in our technocratic age the title may seem old-fashioned. In Scotland it has long stood for physics, while in Cambridge the Jacksonian Professorship is now assigned to the Department of Physics. It has not always been so and the founder of the Jacksonian chair envisaged much wider possibilities. Physics is perhaps the outstanding field in which observations of natural phenomena are organised in a rigorous abstract logical structure, but it is not the only one; Rutherford's remark that there was physics and there was stamp collecting seems arrogant, yet it reflects the fact that as the sciences develop they become closer to the patterns of physics. That may be seen plainly in Halley's corpus as he applied the methods established by Newton in the *Principia* to other natural phenomena than celestial mechanics.

Halley's early work before *Principia* was naturally essentially phenomenological, for there was then no dynamical theory into which the motions of the Moon or of comets could be set, but his observations and studies of transits of Venus and Mercury had well known and far-reaching consequences in geography as well as in astronomy. Lunar studies did not achieve his aims, whereas the months he spent on cometary calculations bore unexpected fruit. It was of course satisfying to identify previous apparitions of certain comets, especially of 'his', but the most far-reaching consequences followed after his death when 'his' comet reappeared within days of Clairaut's prediction. Halley had the confidence in the regularity of Nature to predict events years after his own death. The return of the comet made a far greater stir than the transit of Venus but both showed clearly, the comet the more spectacularly, the self-contained reproducible mechanical structure of the solar system moving under the sole attraction of mutual gravitation. God was pushed out.

God was pushed further out as people recognised the extent and mutability of the cosmos; Halley again took some crucial steps. His recognition of the acceleration of the Moon in its orbit, of what became known as Olbers's paradox and the difficulties to which it gave rise, his identification of nebulae and variable stars (though others had made the very first observations), and lastly his discovery of proper motion, all contributed to overturn the picture of a cosmos maintained in a regular condition by God.

Halley investigated the Earth and its immediate surroundings as a physical object. It is a very extensive subject. It requires its students to work and to cooperate world-wide and Halley was one of the first to do so.

The first step is to measure the size and shape of the Earth. Such measurements were in a rather rudimentary state in Halley's day and his own contributions were not great. He observed the changed rate of his pendulum clock on St Helena which Newton later used in his theory of the figure of the Earth in *Principia*, and he looked into the possibility of a triangulation north of London but did not pursue it. He took every opportunity to observe or collect observations of longitude and so improve the mapping of the globe as a whole. One of his most interesting studies was historical, the critical investigation of the Antonine Itineraries and the determination of Greek and Roman measures in Rome. The major works of eighteenth century geodesy were elsewhere. France, Lapland, and Peru were where the size of the Earth was established with good precision and where the Earth was shown to be flattened at the poles as Newton had calculated, and not pointed as Cassini argued. Halley's contributions by comparison were slight.

Hooke was the outstanding student of earthquakes in Halley's day, but Halley's theories of the magnetic field of the Earth and of aurorae led him to speculations about the interior of the Earth. He lived long before anyone had measured the time that the shock of an earthquake took to travel through the Earth and so one essential basis for investigating its interior was denied him.

Halley's contributions to meteorology and hydrology were substantial and set courses for his successors. He was for many years regarded as the pre-eminent authority on the barometer. His chart of the trade winds was the first serious attempt to establish global patterns of the circulation of the atmosphere. His experimental studies of evaporation, coupled with estimates of the global flow of rivers and of evaporation from the sea, showed that evaporation from the seas could produce the flow of rivers, and established the hydrological cycle on a quantitative basis. He underestimated the contribution of rain to rivers and overestimated that of dew on high hills, something that had greatly impressed him on St Helena. He realised that the Sun's heat drove the general circulation of the atmosphere but without the concept of the Coriolis force his theory of how it worked was wrong.

As a result of his measurements of longitude in the course of his Atlantic cruises, Halley gained some idea of currents in the deep oceans, especially the equatorial currents, but the earlier studies of Athanasius Kircher were more penetrating. Halley's great work was the study of tides. It started with his paper on the tides at Tonquin and continued with his exposition of the theory of tides for James II. He must have made some tidal observations in the course of his hydrographic survey of the approaches to the Thames in 1688 or 1689, for otherwise he could not have had a reliable datum for the depths he measured. He observed some tides around Atlantic islands and then followed his Atlantic cruises with his definitive survey of the tides in the English Channel. It may seem to be going too far to call 'definitive' that first survey of all, but since, as Proudman showed, it agreed well with much later surveys and would hold the field for more than a century (Chapter 9), the description seems not inapt. Newton had shown how tides are raised by the attraction of the Moon and the Sun, and Halley had made a detailed survey of tides in one shallow sea but there was something missing, the theory of the response of the oceans as a whole to the tide-raising force. The gap could not be filled until the principles of hydrodynamics and the partial differential equations by which they were expressed had been set out and until the effects of Coriolis forces were understood in oceanography as in meteorology. Only then, with the tidal theory of Laplace and its development by G. Darwin, did the study of the tides advance much beyond the state in which Halley had left it.[18]

The tidal survey and the geomagnetic surveys of the Atlantic together rank Halley as an outstanding geophysicist. But he did more in geomagnetism than make surveys, he established for centuries to come the essential nature of the magnetic field of the Earth and the conditions that a realistic theory had to satisfy. Two things stand out in his magnetic studies, his grasp of the global patterns of the field shown, as Celsius put it, by 'crooked lines', and his insight into the implications of those patterns, that the origin of the field lay deep and that the motions which the westerly drift implied should not affect the external dynamics of the Earth. A realistic theory that conformed to Halley's had to wait more than two and a half centuries for Sir Edward Bullard's account of the westerly drift.[19] Many more observations had intervened but theoretical understanding had hardly advanced, and was not possible before the development of hydrodynamics and magnetohydrodynamic theory, along with understanding of the core–mantle structure of the Earth. None of that was available before the early years of this century. The opinion of Halley's studies that Celsius wrote to Swedenborg is as true now as when he wrote in the middle of the eighteenth century and if anything strengthened by historical perspective; it bears repetition:

> I believe that Dr Halley has more than any other applied the deepest thought in establishing a theory of the declination of the magnet which the learned have been seeking to improve ever since; yet he does not venture to determine by geometry the situation of the magnetic poles upon the earth and to establish rules for computing the declination. Meanwhile however, he has empirically constructed crooked lines representing the declinations of the magnet on the largest ocean of the world, and he has had the good fortune to see those lines, which were constructed mostly on the basis of the observations taken during his voyages confirmed more and more by later experiments.[20]

Organising large amounts of data and seeing patterns in observations is indeed part of science, but not the end. One must look through and beyond the patterns and conceive the abstract relations that lead to them. Halley's study of aurorae was as perceptive as that of the westerly drift. He identified the structure and saw the physics under it, that the aurorae were linked to magnetic field lines. At the same time he was limited by lack of a wider physics. Daniel Bernoulli put his finger on what was missing—in a correspondence with Jallabert he pointed out that Halley had not accounted for the colours of aurorae.[21] No one then knew about the ionisation of atoms and molecules and the streams of radiation and particles from the Sun and the way in which they produce electrified particles that circulate around field lines and radiate when they recombine. All that has become clear only in the last half-century.

Halley did not express his philosophy of science explicity in his writings, but clearly accepted Boyle's principle, that for an hypothesis to be useful it must not only account for what is already known, but must also predict what has yet to be discovered. That lies behind drawing 'crooked lines' where there are no observations, behind predictions of the return of his comet and the future observation of the transit of Venus, and the course of his argument from the rate of flow of water out of a vessel to the force on a man in a tidal stream, to the force on a sail and then to the speed with which a bird must flap its wings.

In the *Principia* Newton demonstrated how Boyle's principle should be applied, by observing nature, by forming an hypothesis in a mathematical form, by rigorous formal deduction of the consequences of that hypothesis, and finally by testing those consequences in numerical comparisons with observation. So Newton established the action of the inverse square law of gravitation throughout the solar system. He then went on to prediction, to the figure of the Earth, to the luni-solar precession of the Earth and to the behaviour of the tides. Impressive as are the power and range of Newton's works, they do not encompass the whole of classical dynamics. In particular Newton had no explicit concept of energy, nor did he develop the dynamics of rigid bodies. Neither he nor Halley nor any of their contemporaries seem to have asked why the Moon always turns the same face to the Earth. Either Newton or Halley could probably have tackled the question by an extension of the method that Newton used for the luni-solar precession, had they realised that the Moon need not be axially symmetrical. Further, since Halley had suggested that the Earth was slowing down and the Moon speeding up, he might have asked, but did not, why the Moon's spin period now today exactly matched its orbital period. Such comments detract nothing from Newton's great achievement but they show that in such matters as rigid dynamics, as well as in the development of algebraic methods and the differential calculus, there was much to do in extending and deepening his work—and that is far from finished as even the local solar system continues to reveal unexpected mysteries.

Debts to the *Principia*, direct or indirect, are to be found in a very great deal of Halley's work after its publication, and Halley's contribution to natural knowledge as a natural philosopher would surely have been far feebler had the *Principia* not appeared. It determined the direction of his astronomy, it gave him the basis for his hydrodynamical studies, and it undoubtedly informed the style of his magnetic investigations. That is just to recognise the great effect that the *Principia* had on him as it has had on all natural philosophers from his day onwards. His own contribution to natural knowledge lives on to this day by reason of his part in the production of the *Principia*, even more than by his geophysics or astronomy or any other of his multifarious studies.

15.5 *The navigator*

Samuel Pepys, who had good reason to judge, wrote of Halley:

> Mr Hawley—May he not be said to have the most, if not to be the first Englishman (and possibly any other) that had so much, or (it might be said) any competent degree (meeting in them) of the science and practice (both) of navigation. And the inferences to be raised therefrom.[22]

In the light of all the developments of navigation since Pepys wrote, his judgement stands. Halley certainly knew the theory, especially of the problem of the longitude, to which he returned so often. He understood the variation of the magnetic field, how it differed over the Earth and changed in time. He knew how errors in longitude, combined with a neglect of the changes in the magnetic variation, could lead to seriously wrong estimates of position and disaster in sailing into the English Channel, and he set out his conclusions in advice to seamen. He had done some interesting mathematics on the theory of rhumb lines and how they were related to navigation. He had introduced isolines. He had published tables for the occultations of the satellites of Jupiter. Finally at the end of his days he produced tables to calculate the position of the Moon, so essential in the determination of the longitude by lunar distances.

The theoretical navigator may be able to work from his armchair; Halley was a sailor. It is one thing to know the theory, another to apply it when the skies may be clouded over for days at a time, when the waves are too high to allow noonday sights of the Sun. The moons of Jupiter are an admirable clock but the long telescopes that had to be used to observe them were impractical at sea. Above all there was no mechanical clock that could be used at sea. Maps and charts were unreliable and longitudes might be seriously wrong. In large tracts of the oceans, even after Halley's own cruises and the charts he drew from them, the magnetic variation was very poorly known so that courses set by compass could be wrong by many degrees. A ship's speed through the water was found from the length of log line let out in an interval timed by a sand glass. That was an erratic method at best; the speed over the sea bed was often seriously different from the speed through the water on account of unknown currents. Halley also recognised from practical experience that the speed through the water shown by the log line could be wrong in high waves. The practical navigator had to supplement his often imperfect, frequently erroneous, astronomical observations by other signs and symptoms, such as the presence of land birds or plants, the colour of the sea, the scent of flowers borne on an offshore breeze.

As commander of *Paramore* in the Atlantic, Halley was not meant to navigate the ship, that was the duty of his lieutenant or master, yet when his lieutenant was difficult, Halley did put his knowledge into practice and brought the ship from Newfoundland to England. In the Channel survey he had the services of a pilot in some places, but apparently not always. His logs of sailing in the Channel imply a good knowledge of those waters and of the ports upon them, perhaps acquired from his work on the frigate *Guiney*. No desk-bound navigator, Halley knew the waters and their dangers, how ships behaved and how men could misbehave: he was a practical seaman.[23]

15.6 The practical man

Throughout his life Halley stands out as someone who got things done. Science is not just thinking about the world, it is putting thoughts into practice. Every one of his major endeavours shows that Halley was an effective scientist. When he set off for St Helena he knew what Flamsteed was preparing at Greenwich, but with just one friend and no doubt local labourers he constructed an observatory and observed a considerable number of stars, doing more in that year than Flamsteed with greater deliberation did in England. As Newton's editor and publisher, he was continually at the author and with the printer, handling technical matters of printing, urgent to produce the great work.

St Helena and *Principia* might be thought to show the ardour of a young man that would cool off with time. Halley remained a young man all his life, in his drive, practicality, and enthusiasm. On St Helena he had to get on with and manage people to get his observatory built, with *Principia* he had difficult men to handle, and his work on *Guiney* involved getting seamen to engage in hard and dangerous activities. He did not just direct, he tried out equipment and clothes himself, he led by example. He showed his practical qualities in full measure on the Adriatic: the energy with which he made his surveys, his critical assessment of colleagues, and his drive as a site engineer. The end of his life shows no less his determination to get things done when, late in a normal life, he re-equipped the Observatory at Greenwich with up-to-date instruments and embarked upon and concluded his programme of observations over almost twenty years.

Some have planned their life and careers and have achieved greatness, others have achieved greatness from unforeseen opportunities. Halley began with plans, plans for St Helena, for travel in Europe, for observations at Islington. Chance intervened, and led to the *Principia*. We do not know if it was of his own volition or from outside demands that he turned to the survey of the Thames approaches, but he became involved in diving because of the chance foundering of *Guiney*. At the same time he was planning world-wide cruises for the magnetic variation. Circumstances, war-time perhaps delayed him, but he eventually achieved part of his plan. The prospect of renewed war may have led to his Channel cruise and certainly took him to the Adriatic. Still thinking of the Pacific as he returned to England, he found Wallis dead, and so went to Oxford. His editions of the Greek mathematicians seem more like an obligation that he accepted, not something he planned, but Oxford gave him the opportunity to return to the study of the Moon, and there and at Greenwich he seized his chances. Chance and circumstance were evidently important in deciding Halley's pursuits, but he did have persistent aims and perhaps the only chances we know of are those he seized because they enabled him to do something he wanted to do.

15.7 Halley in his time

Halley is distinguished from his notable contemporaries by his historical sense, by his awareness of change. The magnetic variation, the saltness of the seas, the motion of the Moon, new stars, proper motion of stars, all particular concerns of his, all witness to a natural world with an history as much as the affairs of mankind. Halley elucidated those historical matters by combining written and archaeological records with physical principles. He also applied physical principles to investigate human history, in the descent of Julius Caesar, the Antonine Itineraries, and the Gospel account of the Crucifixion. While his studies themselves were exceptional, the historical sense that he showed was not; his age, the late Renaissance and the beginnings of Early Modern times, was very conscious of history. In that he was a man of his day.

Halley lived at a time when the intellectual issues between the older literary scholarship, with its historical foundations, and the modern empirical natural philosophy were fiercely debated, as in the 'Battle of the Books'. The 'Two Cultures' were very real. Halley does not fit easily into those categories. He was an outstanding empirical natural philosopher, a modern. He was also a very learned classical scholar of wide interests, able to take his place with the ancients. His wide interests and abilities should not be taken as evidence that the issues between ancients and moderns were just matters of polemics. They were substantial, just as the similar issues raised by the concept of the Two Cultures are substantial. They are indeed much the same as they were then: to which should we give greater weight in intellectual matters, to the empirical study of the natural world, or to the humane study of men and women? To some extent perhaps, and unconsciously, Halley affords an answer, that the study of the natural world is a part of the study of mankind, for we live our lives in a world that both sets constraints on human potentialities, and affords them opportunities.

15.8 Servant of knowledge, servant of the realm

Of all Halley's contributions to science and to our understanding of the world around us, his part in the evolution and publication of the *Principia* stands out. Few other works of natural philosophy have, directly and indirectly, so determined the course of the physical sciences or the appreciation of the autonomous mechanical nature of the physical world. Halley well understood what he had done when he compared himself to Ulysses, and de Morgan stated the position once for all in his summary:

> But for him [Halley], in all human probability, that work would not have been thought of, nor when thought of written, nor when written printed.[24]

Halley did not himself add much to the content of the *Principia*, he was the servant without whom it would not have seen the light of day. It seems that the importance of Newton's work so struck him as a revelation that he forwent all else to see it perfected, his other interests and commitments notwithstanding. A devoted disinterested voluntary servant, he saw clearly how he could advance knowledge, and did so.

He served knowledge then and later as the Clerk of the Royal Society. Subordinate though that position was, the administrative and scientific support that he provided to the older and sometimes more contentious and less conscientious colleagues can be seen to be invaluable at that stage of the Society's evolution; without it the Society might well have faded as others in other lands did. In later years, as Secretary himself, he again served the Society by his administration in times marked by fierce personal quarrels. The lasting memorial of his years as Clerk and Secretary is the sequence of *Philosophical Transactions* that he brought out as editor and to which he made many contributions.

In supporting Newton and in serving the Royal Society, Halley showed a readiness to put the interests of others first and a lack of that fierce possessiveness and secretiveness over their own results that so unpleasantly characterise some of his contemporaries. We readily identify the lasting contributions to knowledge that he himself made, but there were others also who could contribute because of his self-effacing and cooperative concern for the advancement of learning and knowledge.

Servant of knowledge, and servant also of the realm: if evidence is needed that Halley served the realm, Nottingham provided it when he put Halley's case for election to the Savilian Chair (Chapter 12):

... for I have seen his zeal in the publick service (on which he is now abroad) the which added to his extraordinary skill above all his competitors obligd me to be very forward in promoting him to this place.

So likewise did Ormonde in recommending him for the degree of DCL (Chapter 12):

Mr Edmund Hally having been ... often employed by her Majesty and her Predecessors in the Service at Sea in the remotest parts of the World to the Great Satisfaction of the Lords of ye Admiralty and others of the first Quality in the nation; ... I have thought fitt in consideration of his great merit and the Service he hath done to the publick both at home and abroad to recommend it to You that you would conferr the Degree of D^r of Civil Laws on him without fees ...

In Halley's day the government had no permanent officials who could advise it on technical matters and it had to appeal to such as Halley for information and to carry out special investigations. Halley was far from alone in being so asked, but few of his contemporaries were singled out as Halley was by Nottingham and Ormonde, or rewarded as he was by William III, Anne, and the Emperor. As in his capacity as Clerk to the Royal Society he acted as the Executive Secretary is now expected to, so in the public service he filled unofficially the roles of the Hydrographer and Chief Scientist to the Royal Navy, and so he was recognised by Pepys. Halley, however, did not confine himself to technical matters, his Adriatic year reveals him also as a natural diplomat who gained the confidence of his Queen and her Secretaries, of the ambassador in Vienna, of the Emperor, and of Prince Eugen.

Hearne, like many of his contemporaries, wrote of Halley as a 'great man'. His Atlantic voyage into the southern ice and his edition of Apollonius had given him that reputation, not the return of the comet that attracts attention today. Whatever our assessment of his achievements, we compare him with his peers; conversely, the study of his life enlarges our appreciation of those others. The standing of Christopher Wren as mathematician and natural philosopher seems more substantial and Robert Boyle appears more clearly as the philosopher by whom others were guided. The characters of John Flamsteed and Robert Hooke, their abilities and their defects, come into sharper focus. Above all, how does Newton stand in company with Halley? They pursued natural philosophy in different but complementary ways. Newton thought deeply about dynamics and celestial mechanics and optics, and established the rational mathematical form of enquiry. Halley turned his gaze not only upon the whole Earth but upon the entire cosmos, not only upon the here and now, but upon the past and into the future, applying to all the principles that Newton had set out. Three centuries of development of Newton's dynamics have given us a better appreciation of his mathematical approach; only in the present century have Halley's examples been followed in world-wide geophysical surveys at the public charge and in carefully organised cooperative observations of special events. Only now do we appreciate the depth of his insight into such matters as the origin of the Earth's magnetic field or the nature of aurorae. Newton thought deeply, Halley looked widely, but not less acutely. Natural philosophy depends on both.

Edmond Halley was ever an adventurer. His voyage to St Helena was a great adventure. His plan to observe the Moon over eighteen years at an age when most people were long dead, was no less an adventure. Adventures occupied his middle years—the financial risk of the *Principia*, diving on *Guiney* in danger of death, sailing round the Atlantic into the southern ice in a small ship alone, editing *Apollonius* as a tyro in critical studies. He was perhaps an optimist, possibly an unrealistic optimist. He surely never saw a risk of failure as a reason for not venturing.

Appendix 1
The ellipse and the parabola

The elliptic orbit

It is usual in celestial mechanics to define the position of a body moving in an elliptic orbit under the attraction of a larger body by polar coordinates about the focus occupied by the larger body or, more strictly, at the centre of mass of the two bodies. Coordinates of that form are most closely related to observations of bodies in elliptical orbits.

Let F (Fig. A1.1) be the focus, r the distance of a point of the orbit from it, and θ the angle measured from a convenient direction, now usually pericentre, as in this Appendix, but often apocentre in Halley's day. The angle θ is the true anomaly. The equation of an ellipse in those coordinates is

$$r = \frac{a(1 - e^2)}{1 + e\cos\theta},$$

where e is the eccentricity.

An alternative description is by means of the eccentric angle, E, defined as follows. Let the semi-major and semi-minor axes of the

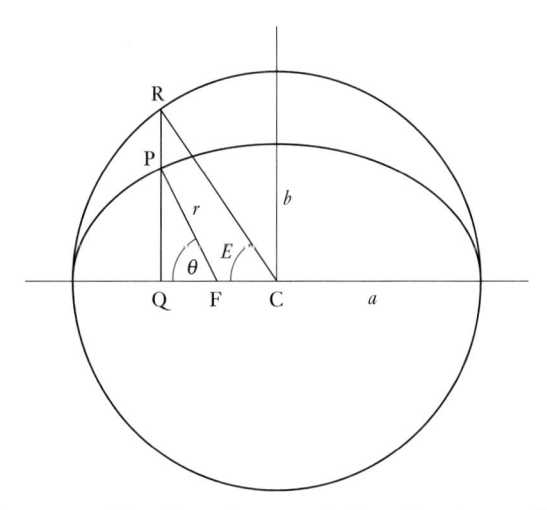

Figure A1.1 Geometry of the ellipse. C: centre; F: focus; P: point on ellipse; R: point on eccentric circle; (r,θ): coordinates of P; E: eccentric angle; a, b: major and minor semi-axes.

ellipse be a and b. Upon the major axis as diameter draw a circle of radius a and centre at the centre of the ellipse. Erect a perpendicular upon the major axis to intersect the ellipse in P and the circle in R. Let the foot of the perpendicular be Q. Then E is the angle QCR. On taking CQ and PQ to be the rectangular coordinates of the point P, the equation for the ellipse is

$$\frac{CQ^2}{a^2} + \frac{PQ^2}{b^2} = 1.$$

Thus $PQ^2 = \left(\frac{b^2}{a^2}\right)(a^2 - CQ^2) = \left(\frac{b^2}{a^2}\right)(a^2 - a^2\cos^2 E),$

so that $PQ = b \sin E$. But $CQ = a \sin E$ and so $PQ/CQ = b/a$.

Now $r \cos \theta = a \cos E - ae$ and $r \sin \theta = b \sin E$, and so, on squaring and adding,

$$r = a(1 - e \cos E).$$

Two expressions for r have now been obtained, one in terms of θ and the other in terms of E. The rates of change of r according to the two must be the same, that is

$$ae \cdot \sin E \frac{dE}{dt} = ae(1 - e^2)\sin\theta \; r^2 \frac{d\theta}{dt}.$$

But $r^2 \dfrac{d\theta}{dt}$ is the constant aereal velocity, so that

$$r^2 \frac{d\theta}{dt} = na$$

where n is the mean angular velocity of the particle in the orbit. Thus

$$(1 - e \cos E)\frac{dE}{dt} = n,$$

and so

$$E - e \sin E = nt + \varepsilon = M,$$

where ε is a constant and M is called the *mean anomaly*.

The relation between M and E is known as *Kepler's equation*.

The difference between M and θ cannot be expressed so simply, but to the order of e^2 it may be written as

$$\theta - M = 2e\sin M + \frac{3}{4}e^2 \sin 2M.$$

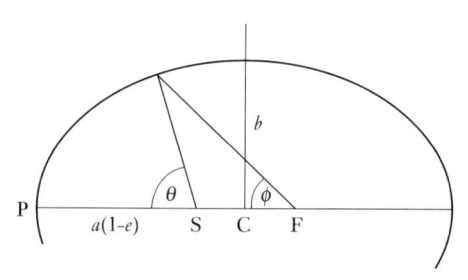

Figure A1.2 Motion about the empty focus. C: centre; S: Sun; F: empty focus; a, b: major and minor semi-axes; e: eccentricity.

In that form it is the *Equation of the Centre*, and it relates the true anomaly, which does not increase uniformly with time, to the mean anomaly, which does increase uniformly with time. In Halley's *Tables* (Chapter 13) the mean anomaly of the Moon is first obtained from the time and then corrected for the various perturbations by the Sun, after which the equation of the centre is applied to obtain the true anomaly or longitude. The *Equation of the Centre*, when applied to the orbit of the Earth about the Sun, is also known as the *Equation of Time*, for it relates time of the year, by the revolution of the Earth about the Sun, to the length of the day or other independent measure of time.

The angular velocity about the focus of attraction is not constant, for it is the areal velocity that is strictly constant about it, but the angular velocity about the empty focus is nearly constant in an orbit of small eccentricity. Since the areal velocity is $r^2 d\theta/dt$, the angular velocity varies as r^{-2}, or as $(1 + e \cos \theta)^2$, or $(1 + 2e \cos \theta)$ when e is small.

If ϕ is the angle between the particle and pericentre at the empty focus (Fig. A1.2), then

$$\sin\phi = -\frac{(1-e^2)\sin\theta}{1+2e\cos\theta + e^2},$$

from which it will be found that when the eccentricity is small,

$$\frac{d\phi}{dt} = -\frac{k}{a^2}(1+3e)$$

to first order in e, where k is the constant areal velocity about the centre of attraction.

The orbit in space is determined by the disposition of the plane of the orbit and the orientation of the orbit in it (Fig. A1.3). The plane is defined by the angle it makes with a reference plane, usually either the

The ellipse and the parabola

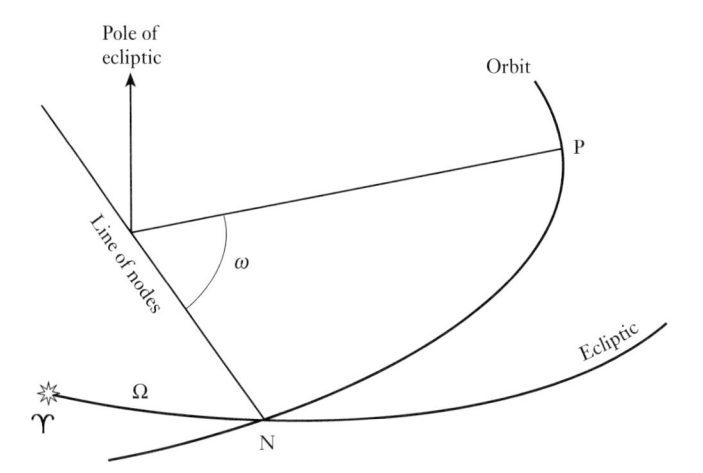

Figure A1.3 The orbit in space. P: perihelion; N: node; Ω: longitude of node; ω: longitude of perihelion, ♈: first point of Aries.

ecliptic or the equatorial plane of the Earth, together with the orientation of the line in which it intersects the reference plane. The angle between the two planes is the *inclination* (sometimes *obliquity*) and the line of intersection is the *line of nodes*. The angle of the line of nodes is usually measured from the vernal equinox, or *First Point of Aries*, the node of the equator on the ecliptic at a particular epoch.

The orientation of the orbit in its plane is defined by the angle between pericentre (perihelion, perigee, in the most common cases) and the ascending node.

The orbits of two spherical bodies around each other are ellipses of constant eccentricity, constant inclination, constant longitudes of the node and pericentre, but as soon as there is any departure from that simple state, for example, by the attraction of the Sun upon the Moon in its orbit about the Earth, all those parameters undergo changes, as do the eccentricity and semi-major axis. In particular, the attraction of the Sun on the Moon produces steady, secular, motions of perigee and node with periods of 9 and 18 years respectively.

Parabolas

The geometrical equation for a parabola in polar coordinates (r, θ) about the focus (the Sun) as origin is

$$r = \frac{2p}{1+\cos\theta}$$

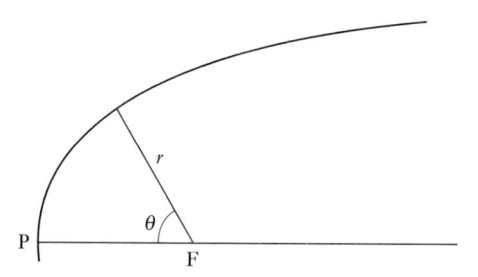

Figure A1.4 Position on a parabola. P: perihelion; F: focus (Sun); (r,θ): polar coordinates.

where p, the semi-latus rectum, is the distance of perihelion, P, from the focus (Fig. A1.4); thus the ratio (r/p) is the same function of θ for all orbits.

Let M be the mass of the Sun and G the constant of gravitation. If v is the velocity at perigee, $v^2 p = 2GM$, corresponding to the relation $a^3 \omega^2 = GM$ for an elliptical orbit.

The areal velocity at any point (r,θ) is $\frac{1}{2} r^2 \frac{d\theta}{dt}$ and at perihelion is $\frac{1}{2} pv$, or $\frac{1}{2} k p^{1/2}$, where k^2 is $2GM$. Since the areal velocity is constant everywhere round the orbit,

$$\frac{d\theta}{dt} = \frac{1}{4} \frac{k}{p^{3/2}} (1 + \cos\theta)^2 = \frac{k}{p^{3/2}} \cos^4 \frac{\theta}{2}.$$

Then

$$\tan\frac{\theta}{2} + \frac{1}{3} \tan^3 \frac{\theta}{2} = \frac{k}{2 p^{3/2}} (t - T_0),$$

where t is the time at the position (r,θ) and T_0 is the time of passage through perigee.

That equation is known as Barker's equation and Halley's general table is a numerical solution. He did not know it in Barker's form and could well have constructed his table by extrapolation using the expression for $d\theta/dt$ and one of Newton's finite difference procedures.

Appendix 2

Genealogies

A: Halley

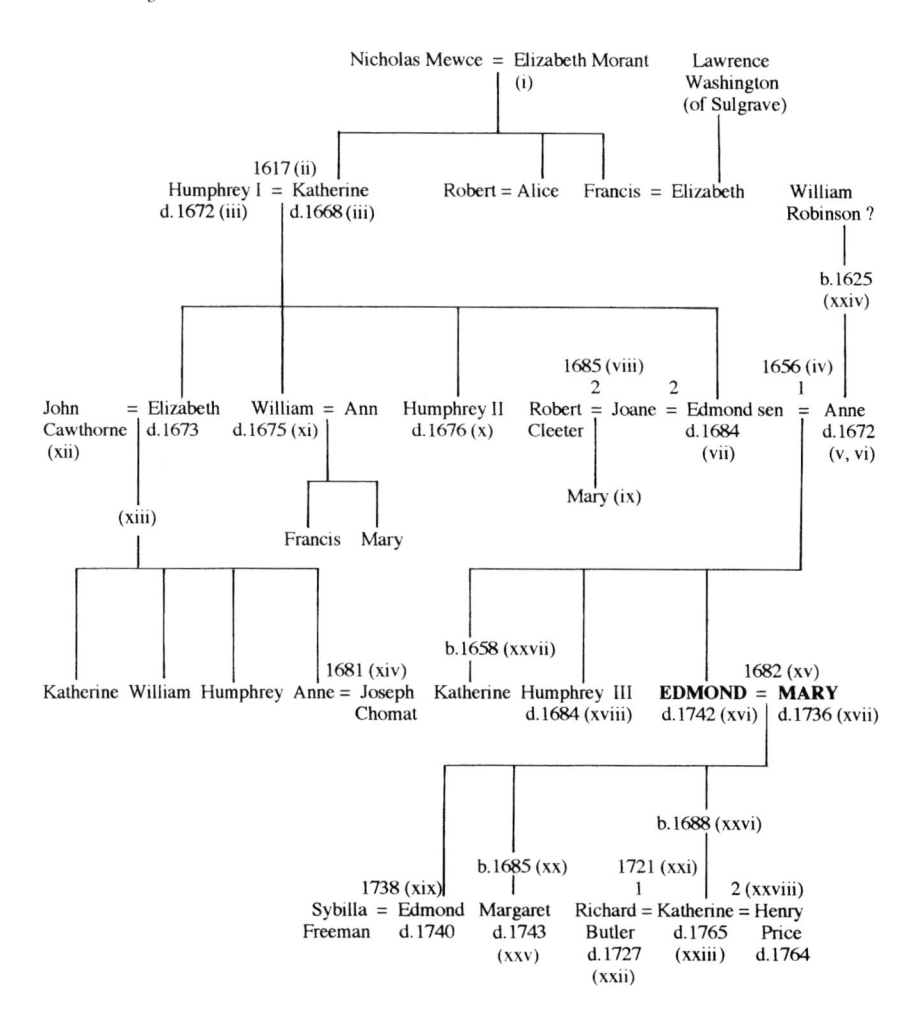

```
                    Nicholas Mewce = Elizabeth Morant      Lawrence
                                (i)                        Washington
                                                           (of Sulgrave)
         ┌──────────────────────┬──────────────────┐            │
     1617 (ii)                                                  │
Humphrey I = Katherine     Robert = Alice   Francis = Elizabeth   William
d. 1672 (iii)  d.1668 (iii)                                      Robinson ?
                                                                   │
                                                                 b.1625
                                                                 (xxiv)
                              1685 (viii)        1656 (iv)
                                  2           2      1
John   = Elizabeth  William = Ann  Humphrey II  Robert = Joane = Edmond sen = Anne
Cawthorne  d.1673   d.1675 (xi)    d.1676 (x)   Cleeter          d.1684       d.1672
(xii)                                                            (vii)       (v, vi)
 │
(xiii)                        Francis  Mary                    Mary (ix)

                                          b.1658 (xxvii)
                              1681 (xiv)                               1682 (xv)
Katherine William Humphrey Anne = Joseph  Katherine Humphrey III   EDMOND = MARY
                                Chomat              d.1684 (xviii) d.1742 (xvi) d.1736 (xvii)

                                                                b.1688 (xxvi)
                                          1738 (xix)    b.1685 (xx)   1721 (xxi)
                                               1                          1        2 (xxviii)
                                     Sybilla = Edmond  Margaret   Richard = Katherine = Henry
                                     Freeman   d.1740  d.1743     Butler    d.1765    Price
                                                        (xxv)     d.1727    (xxiii)   d.1764
                                                                  (xxii)
```

(i) For Mewce, see visitation of Northants, Metcalfe (1887).

(ii) 24 November 1617, marriage register of St Margaret's, Barking, Essex.

(iii) Katherine buried at Alconbury, Hunts, 12 September 1668; burial of Humphrey Halley and wife Katherine (re-burial), 16 October 1672, register of St Margaret's, Barking, Essex.

(iv) 9 September 1656, register of St Margaret's, Westminster.

(v) Anne, wife of Edmund Halley, gent, burial 24 October 1672 to Barking, register of St Giles's, Cripplegate.

(vi) 24 October 1672, burial of Anne, wife of Edmond Halley, register of St Margaret's, Barking, Essex.

(vii) 22 April 1684: 'Mr Edmund Halley of London Merchant murthered buryed in linen, 46/20 pd to this parish for ye use of ye poor', register of St Margaret's, Barking, Essex.

(viii) 2 June 1685, register of St George, Southwark.

(ix) Mary Cleeter was living with her parents in the parish of All Saints, London Wall in 1695 (Glass 1966).

(x) Will of Humphrey Halley II, PCC reg Bench f. 66.

(xi) Will of William Halley, PCC reg Dycer f. 146.

(xii) Possibly John Cawthorne of Huntingdon who matriculated at Trinity College, Cambridge, 1646/7.

(xiii) The four children of John and Elizabeth Cawthorne received legacies by name in the will of Humphrey Halley II—see note (x).

(xiv) Marriage of Joseph Chomatt, merchant, and Anne Cawthorne, spinster, 9 December 1681, register of St James's, Duke's Place.

(xv) Marriage of Edmond Halley and Mary Tooke, 20 April 1682, register of St James's, Duke's Place.

(xvi) Halley died 16 January 1742—*Daily Advertiser*.

(xvii) Mary buried at St Margaret's, Lee, 14 February 1736—register.

(xviii) Administration of estate of Humphrey Halley III, died beyond the seas, by Edmond Halley, PCC Admon Book, 1684, 31 October 1684, 30 April 1685.

(xix) Edmond Halley married to Sybilla Freeman, widow (aged 40), *Notes and Queries*, **164** (1933), 246. (Morden Coll., Charlton, 4 May 1738.)

(xx) Margaret, daughter of Edmund and Mary Halley, baptised 1 May 1685, register of St Benet's, Paul's Wharf.

(xxi) Marriage of Katherine and Richard Butler, widower, 2 October 1722, register of St Margaret's, Lee.

(xxii) Richard Butler buried, 28 October 1727, register of St Margaret's, Lee.

(xxiii) Katherine Price died 1765—will, *Notes and Queries*, 10 Ser., 3, 6.

(xxiv) Anne Robinson baptised 13 July 1625, register of St Margaret's, Westminster.

(xxv) Margaret Halley died 13 October 1743—inscription on monument to Halley. The inscription states that she was aged 55, which is inconsistent with her birth in 1685—see note (xx). The year of her birth may have been confused with that of her sister Katherine by the author of the inscription.

Genealogies

(xxvi) Baptism of Katherine.

(xxvii) 17 February 1658, baptism of Katherine, register of St Giles', Cripplegate. Died before 1684 when Halley was the sole surviving child of his parents—see the accounts of Chancery actions in Chapter 5.

(xxviii) Marriage of Katherine and Henry Price.

Note: An Elizabeth Cawthorne was married in Grays Inn Chapel on 7 September 1700 (Foster 1889).

B: Kinder–Tooke

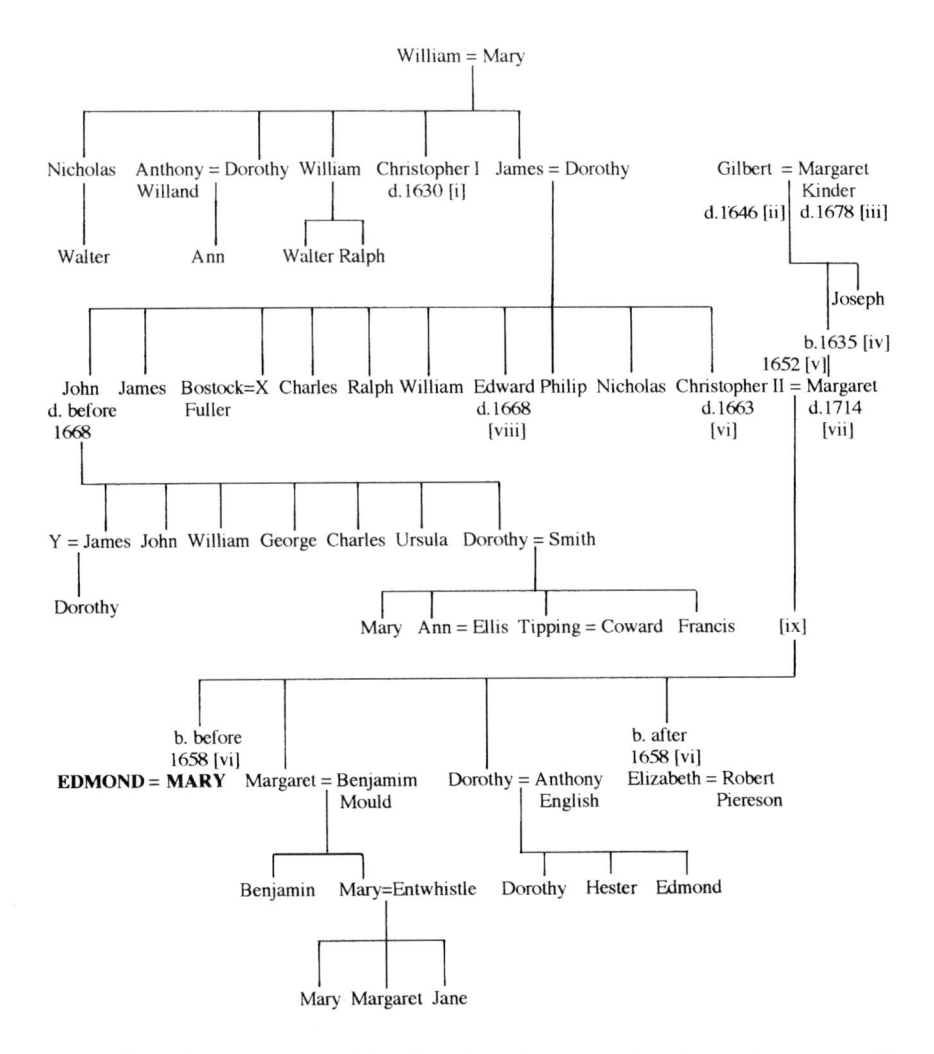

[genealogy chart]

[i] For descendants of his brother James and other relatives: will of Christopher I, PCC PROB 11/158 ff. 71,72, proved 6 August 1630.

[ii] 29 June 1646, burial of Gilbert Kinder in the church—register of St Helen's, Bishopsgate.

[iii] 20 February 1678, 'Margaret Kinder in the chancel close to lower stepp'—register of St Helen's, Bishopsgate.

[iv] 6 January 1635, baptism of Margaret Kinder—register of St Helen's, Bishopsgate.

[v] 10 June 1652, marriage of Christopher Tooke II and Margaret Kinder, St Vedast, Foster Lane.

[vi] Will of Christopher Tooke II, 9 February 1658, PCC reg Juxon f. 42, proved 3 March 1662/3.

[vii] The Margaret Tooke buried in the vault 8 April 1688 (register of St Peter-le-Poer) was not the mother of Mary Halley. Mary's mother made her will in 1710 (*Notes and Queries*, 8 (1907), 221, 222, 373 and **154**, (1928), 208) and died 9 October 1714, register of St Helen's, Bishopsgate.

[viii] Will of Edward Tooke, PCC reg Hone f.80. The will of Edward Tooke refers to his four brothers John (deceased), Charles, Ralph, and Philip, to the children of John who had £200 each, and to the four daughters of Christopher II who also had £200 each (but see Chapters 2 and 5).

[ix] For Tooke descendants see wills of Christopher II (note [vi]), Edward (note [viii]), and Katherine Price, and the Chancery action PRO C.5, Bridges 367/19.

The following Tookes were members of the Inner Temple:

Ralph, admitted 1629, called to the bar 1637, bencher 1652; second son of William.

John, admitted 1635, called to the bar 1644; son of James.

Edward, admitted 1638, called to the bar 1648; Auditor and Librarian to the King, third son of James.

Charles, admitted 1653, called to the bar 1663; second son of James, Auditor of the Court of Wards.

(Inner Temple, 1877)

Appendix 3

The personal estate of Halley's father

The income from the personal estate of Halley's father at his death in April 1694 is given in the schedules to two actions in Chancery, that of Cleeter and Cleeter against Buckworth, Young, and Halley of 1686 and that of Young against Halley, Cleeter, and Cleeter of 1693. The first was made about two years, and the second more than nine years after the father's death, so that they are slightly discrepant, but for the most part they agree well. The net value of the personal estate is given as about £4000.

The schedule to the first action contains the accounts of the moneys received and disbursed by Halley under the authority of the trustees, Buckworth and Young, in the nine quarters up to the date of the action.[1]

Halley received the rents of twelve houses in Winchester Street, in all £320 per annum gross, of which £96 6s. 8d. was Sir John Buckworth's ground rent, leaving a net annual income of £223 13s. 4d.

The annual rents of the twelve houses in Winchester Street ran from £15 to £100 per annum.

He also received £26 per annum interest from the Cole Office but had to pay £6 interest on a debt of £100 owed to Mr Tooke, leaving a net income of £20.

In the first years after the father's death Halley had to pay sundry debts; he also made a net loss on the Dog Tavern in Billingsgate which was let for £100 per annum but had been leased by the father at £150 per annum.

Allowing for funeral expenses, Halley's annual income in 1685 and 1686 would probably have been less than £200 but more than £150.

In the same period, Joane received annually £89 for the net rents of two houses in Canning Street, £68 from the general lottery, and £24 for the rent of the house in Winchester Street in which she continued to live after the death of the father, in all £181 per annum. The division of the estate made by the trustees was evidently intended to give Halley and Joane approximately equal incomes, with Halley managing the property.

Halley later sold the lease of the Dog Tavern and so was free of its burden before the actions of December 1693.[2] The schedules to the Complaint of that action give the sums received and paid out by Joan and Halley in the nine-and-a-half years from June 1684 to December 1693. Joane's receipts were £1514 and payments £424, net £1090, and Halley's were £4030 13s. 6d. gross, payments £2689 12s. 2d., net £1341 1s. 4d. The payments include ground rents, while Halley's accounts include the receipts and payments for the discharge of the lease of the Dog Tavern which seem to have been covered in part by the sale of two houses. Joane's accounts do not appear to include the rent of the house in which she and Cleeter lived in Winchester Street. She paid the funeral expenses and the reward of £100 for the father's discovery. After the Dog Tavern had been disposed of and after the sale of two houses, Halley's net rental income would probably have been about £180 per annum, to which has to be added the net interest from the Cole Office of £20.

In summary, Halley would have had an income of about £200 from the father's personal estate after all the debts and expenses of the settlement had been met, but in the first five or six years after the father's death it would have been less, though probably not much less than £150 per annum, and would have fluctuated as the various obligations were met and discharged.

(1) PRO C 10 (Whittington) Bundle 222, No. 27, Cleeter, Robert and Joane his wife *vs.* Buckworth, Sir John Kt., Young, Richard, and Halley, Edmund, April 1686.
(2) PRO C 9 (Reynardson) Bundle 142, No. 57, Richard Young *vs.* Robert and Joane Cleeter and Edmund Halley.
PRO C.6 (Collins) 301/45, Edmond Halley *vs.* Cleeter, Cleeter, and Young.
PRO C.7 (Hamilton) 181/90, Edmond Halley and Richard Young *vs.* Cleeter and Cleeter. (Answer only)
PRO C.7 (Hamilton) 60/53, Cleeter and Cleeter *vs.* Halley and Young.

Appendix 4

Chronology of Edmond Halley

Date		Halley's life	Nation and world
1656	Sept	parents married born 26 October	
1660			Restoration
1665		enters St Paul's school?	2nd Dutch War; Plague
1666		St Paul's school burnt	Great Fire
1667			Dutch in Medway
1668		grandmother dies	
1671		St Paul's reopens	
1672		mother, grandfather die	
1673		enters Queen's College, Oxford	
1677		to St Helena	
1678		from St Helena	
1679		visit to Hevelius	
1680		to Paris	
1681		France and Italy	
1682		meeting with Newton	
	April	marriage to Mary Tooke	
1683			Rye House Plot; relief of Vienna
	July		death of Earl of Essex
1684	Jan	discussion with Wren and Hooke	
	Mar	father's death	
	Aug	visit to Newton	
1685			death of Charles II; accession of James II
1686		Clerk to Royal Society	
1687		publication of *Principia*; presentation to James II	
1688	Nov		William III lands
1689	Jan	'concerned at seazing persons about Essex'	William and Mary accept crown; Nine Years' War
1689	June	Chart of Thames Approaches	

1691	April	loss of *Guiney* frigate; election of Gregory to Savilian Chair	
1693	June		Lagos disaster
1695		first cometary calculations	death of Queen Mary death of Purcell
1696		Mint at Chester	
1697			Treaty of Ryswick
1698/9		first Atlantic cruise	
1699/1700		second Atlantic cruise	
1700		re-elected FRS	
1701		Channel cruise	
1702	March		death of William III; War of Spanish Succession
	Dec	to the Hague	
1703		Adriatic surveys	deaths of Hooke, Pepys, Wallis
1704	Jan	election to the Savilian Chair	Newton PRS
1707			Union of England and Scotland
1708		death of Gregory; *Apollonius* published	
1710		Oxford LL.D.	Handel in London
1712		*Historia coelestis* published	
1713		Secretary, Royal Society	Treaty of Utrecht
1715		solar eclipse; moves Arundel marbles	death of Anne, accession of George I; Jacobite rebellion
1720		Astronomer Royal; FSA	
1721			Walpole Prime Minister
1725		*Historia coelestis Britannica* published	
1727			death of Newton; death of George I, accession of George II
1729		visit of Queen Caroline	
1736		death of Mary	
1739		death of son	
1742		Halley dies	Walpole resigns

Appendix 5
The southern stars

Determination of positions from sextant observations

The calculation of positions from sextant measurements involves some straightforward solutions of spherical triangles. Let EE be the ecliptic and N its north and S its south pole. Let A and B be the positions of two known stars and let their north polar distances be θ_A and θ_B and their ecliptic longitudes λ_A and λ_B. Let the unknown star be X and let its angular distances from A and B be ϕ_A and ϕ_B. Put $\lambda_A - \lambda_B$ equal to λ and the arc AB equal to ψ. Then ψ is found from the cosine rule in the spherical triangle NAB.

The sine rule in the same triangle gives the angles NAB and NBA. Next the angle AXB is found from the cosine rule in the form:

$$\cos AXB = \frac{(\cos\psi - \cos\phi_A \cos\phi_B)}{\sin\phi_A \sin\phi_B},$$

while the angles BAX and ABX are obtained from the sine rule in the same triangle AXB.

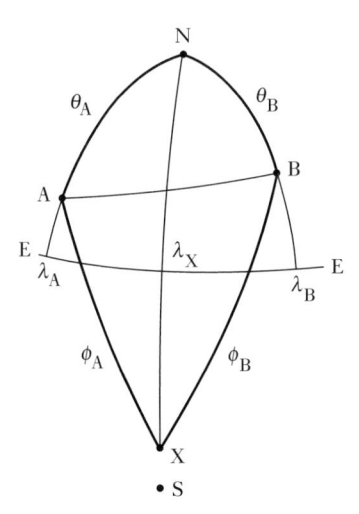

Figure A5.1 Calculation of coordinates of a star from distances from two reference stars. N, S: north and south poles of ecliptic; EE: ecliptic; A, B: reference stars; X: object star; θ_A, θ_B: co-latitudes of stars A and B; λ_A, λ_B: longitudes of stars A and B; ϕ_A, ϕ_B: observed distances of star X from A and B; λ_X: longitude of star X.

In the triangle NAX, the angle NAX is NAB + BAX and the north polar distance, NX or θ_X, is obtained from the cosine rule, and the angle ANX again from the sine rule.

Finally $\lambda_X = \lambda_A + ANX$.

θ_X and λ_X may also be found from the triangle NBX, and should agree with the values found from the triangle NAX, for the two sets are not independent—there are just two observations, the two angles ϕ_A and ϕ_B, from which two unknowns, θ_X and λ_X, are to be calculated.

Comparisons between original and recalculated positions of southern stars from Halley's observations on St Helena

Abraham Sharp recalculated the ecliptic and equatorial coordinates of most of the stars observed by Halley and published them in the *Historia coelestis Britannica*, Vol. III, as:

> Abraham Sharpius, de villa vulgo vocata Little Horton, prope urbana de Bradford in Agro Eborascensi. Matheseos Vir perquam periti; postquam antecedentem Fixarum Catalogum Flamsteedianum ad Annum 1726 provectum habuisset; Exinde atque a Fixarum Distantis quibusdam in Insula Sancte Helenae observatis, sequentem Fixarum quarundam Australiam in Nostro Hemispherion adspectabilium Catalogum Nobis exaravit ad Annum 1726 adaptavit.[*]

He apparently used Halley's observations of distances and recalculated the coordinates from Flamsteed's positions of the Tycho stars on which Halley's catalogue was based. The differences between Sharp's positions and those originally published by Halley will therefore arise from a number of causes, changes in the positions of the reference stars, errors in the calculations of Halley or Sharp, and misidentifications of stars.

Halley published equatorial coordinates for only a few of his stars and so the only relevant comparisons are between the ecliptic positions in the two versions of the southern catalogue. They are summarised below for the constellations having the largest numbers of common stars. Sharp did not include every one of Halley's constellations, for example, he omitted Canis Major and Hydra, and some constellations have only very few stars in Sharp's list.

---

[*]Abraham Sharp, of the town known in the vernacular as Little Horton, near the city of Bradford in the land of York. A man skilled in mathematics; after he had brought the preceeding catalogue of fixed stars of Flamsteed up to 1726, next produced from the distances between fixed stars observed by someone in the Island of St. Helena the following catalogue from southern stars visible in our hemisphere adjusted to 1726.

The southern stars

Table A5.1 Mean differences by constellations between coordinates of southern stars as calculated by Halley (1678) and revised by Sharp (1725); Δ: (Sharp–Halley)

Constellation	Number of stars	Longitude Δ min	Longitude Δ sec	Longitude s.d. sec	Latitude Δ min	Latitude Δ sec	Latitude s.d. sec
Scorpius	8	43	00	103	03	07	244
Eridanus	11	43	11	360	00	27	180
Columba Noachi	6	34	40	31	00	40	49
Argo Novis	36	42	17	126	–00	08	102
Robor Carolina	13	41	10	41	–01	46	27
Centaurus	30	40	00	120	03	24	29
Lupus	17	40	56	30	03	42	28
Ara	8	42	13	27	04	47	27
Corona	10	43	42	90	01	36	70
Grus	13	42	37	72	03	20	29
Phoenix	10	44	48	25	01	06	44

The mean values are:

	Longitude	Latitude
Mean difference	41′ 42″	1′ 54″
Standard deviation	2′ 48″	1′ 56″

Table A5.1 lists the constellations, the numbers of stars for which there are comparisons, the differences, Δ, between the longitudes and latitudes as given by Halley and Sharp, and the standard deviations, s.d., of those differences.

Most of the standard deviations of individual positions within constellations (s.d.) are less than 1′, and can be accounted for by the fact that Halley gave his positions to 1′. The much greater variance between constellations clearly follows from changes in the positions of the reference stars from Tycho to Flamsteed. That is probably the source of the much greater variance within those constellations for which Halley used more than one reference.

Halley was evidently observing to better than one minute of arc and the gross difference between his positions and those of Sharp, 41′ 42″ in longitude, arises from the different values of the longitude of the node of the equator on the ecliptic as used by Tycho and Flamsteed. The overall difference of latitude is not significant.

Appendix 6

Halley's Ode to Newton

In
viri praestantissimi
D. ISAACI NEWTONI
opus hocce
mathematico-physicum
saeculi gentisque nostrae decus egregium

En tibi norma Poli, et divae libramina Molis,
Computius acque Jovis; quas, dum primordia rerum
Pangeret, omniparens Leges violare Creator
Noluit, aeternique fundamina fixit.
Intima panduntur victi penetralia caeli,
Nec latet extremos quae Vis circumrotat Orbes.
Sol solio residens ad se jubet omnia prono
Tendere descensu, nec recto tramite currus
Sidereos patitur vastum per inane moveri;
Sed rapit immotis, se centro, singula Gyris.
Jam patet horrificis quae sit via flexa Cometis;
Jam non miramur barbati Phaenomena Astri.
Discimus hinc tandemqua causa argentea Phoebe
Passibus haud aequis graditur; cur subdita nulli
Hactenus Astronomo numerorum fraena recuset
Cur remeant Nodi, curque Auges progrediuntur,
Discimus et quantis refluum vaga Cynthia Pontum
Viribus impellit, dum fractis fluctibus Ulvam
Deserit ac nautis suspecta nudat arenas;
Alternis vicibus suprema ad litora pulsans.
  Quae toties animos veterum torsere Sophorum,
Quaque Scholas frustra rauco certamine vexant
Obvia conspicimus nubem pellente Mathesi.
Jam dubios nulla caligine praegravat error,
Queis Superum penetrare domos atque ardua Caeli
Scandere sublimis Genii concessit acumen.
  Surgite Mortales, terrenas mittite curas;
Atque hinc caelingenae vires dignoscite Mentis,

A pecudum vita longe lateque remotae.
Qui scriptis jussit tabulis compescere Caedes,
Furta et Adulteria, et perjurae crimina Fraudis;
Quive vagis populis circundare moenibus Urbes
Auctor erat; Cererisve beavit munere gentes;
Vel qui curarum lenimen pressit ab Uva;
Vel qui Niliaca monstravit arundine pictos
Conosciare sonos, oculisque exponere Voces;
Humanam sortem minus extulit; utpote pauca
Respiciens miserae solummodo commoda vitae.
Jam vero Superis convivae admittimur, alti
Jura poli tractare licet, jamque abdita caecae
Claustra patent Terrae, rerumque immobilis ordo,
Et quae praeteriti latuerunt saecula mundi.
   Talia monstrantem mecum celebrate Camaenis,
Vos qui coelesti gaudetis nectare vesci,
Newtonum clausi reserantem scrinia Veri
Newtonum Musis charum, cui pectore puro
Phoebus adest, totoque incessit Numine mentem:
Nec fas est propius Mortali attingere Divos.

EDM, HALLEY

Appendix 7

Correspondence of Halley not listed by MacPike (1932)

A: Letters from Halley

Date	Correspondent	Subject	Reference
1675 April	Flamsteed	refraction	*Flam. Corr.* **1**, 332
April	Flamsteed	refraction	*Flam. Corr.* **1**, 348
1676 March 24	Flamsteed	Moon	*Flam. Corr.* **1**, 438–9
April 29	Flamsteed	observations	*Flam. Corr.* **1**, 439–41
June 26	Flamsteed	Moon	*Flam. Corr.* **1**, 473–4
July 25	Flamsteed	occultation	*Flam. Corr.* **1**, 496–8
1677 Oct	Flamsteed	St Helena	*Flam. Corr.* **1**, 568
Nov	Flamsteed	St Helena	*Flam. Corr.* **1**, 580
1681 June 22	Flamsteed	comets	*Flam. Corr.* **1**, 751
Feb	Flamsteed	comets	*Flam. Corr.* **1**, 771
1686? June 8	Cassini	correspondence	OdP MS B4-10[*]
1696 Nov 28	Thos. Molyneux	Mint	RSL MM. 5. 40.
1697 Aug 2	Thos. Molyneux	Mint	RSL MM. 5. 42.
1702 Dec 4/15	Nottingham	Adriatic	PRO SP 84/224, f. 232
Dec 30	Nottingham	Adriatic	PRO SP 80/19, f. 434
1703 Jan 13/24	Nottingham	Adriatic	PRO SP 80/20, f. 50
Feb 4	Stepney	Adriatic	PRO SP 80/20, f. 109 and PRO SP 105/67, f. 121
Feb 6/17	Stepney	Adriatic	PRO SP 80/20, f. 159 and PRO SP 105/67, f. 294
Feb 14/25	Nottingham	Adriatic	PRO SP 80/20, f. 145 and PRO SP 105/67, f. 300

[*]There is also a copy in the Biblioteca dell' Archiginnasio, Bologna at MS A.431 cc67r–68v.

Feb 27	Nottingham	Adriatic	PRO SP 80/20, f. 147
July 14/25	Nottingham	Adriatic	PRO SP 80/21, f. 168
Aug 0/5	Stepney	Adriatic	PRO SP 105/69*
Aug 0/11	Nottingham	Adriatic	PRO SP 105/69
Aug 11	Nottingham	Adriatic	PRO SP 105/69
Aug 18	Stepney	Adriatic	PRO SP 80/21, f. 276
Sept 0/1	Nottingham	Adriatic	PRO SP 80/21, f. 302
Sept 8	Stepney	Adriatic	PRO SP 105/69
Sept 0/15	Stepney	Adriatic	PRO SP 105/70, f. 29
Sept 15	Stepney	Adriatic	PRO SP 105/69
Sept 11/22	Nottingham	Adriatic	PRO SP 80/21, f. 345
Oct 2/13	Nottingham	Adriatic	PRO SP 80/21, f. 416

B: Letters to Halley

Date	Correspondent	Subject	Reference
1675 March	Flamsteed	refraction	*Flam. Corr.* **1**, 331
April 22	Flamsteed	refraction	*Flam. Corr.* **1**, 332
1676 Aug 10	Oldenburg	Halley's paper	*Oldbg. Corr.* **13**, 30
1678 Aug 23	Hevelius		
1681 Jan (late)	Flamsteed	comets	*Flam. Corr.* **1**, 751–2
Feb 17	Flamsteed	comets	*Flam. Corr.* **1**, 760–4
spring	Flamsteed	comets	*Flam. Corr.* **1**, 789
1695	de Moivre		Schneider 1968
1702 Nov 4	Q. Anne	instructions	NAN FH 275, p. 204
1705	de Moivre		Schneider 1968
1706 July 7	Machin	to Halley?	*Rigaud*, **1**, 255
1707 April 6	J. I. Bernoulli	Greek maths	BAB MS L Ia 674.66
1715 May 16	John Senex		ULC RGO 2/16
1736 Feb 4	Royal Society		ULC RGO 2/17
1736 Sept 10	John Bevis	lunar eclipse	ULC RGO 2/18
1739 Dec 21	Philip Miller	a gardner	ULC RGO 2/17
Dec 25	Gael Morris	calculations	ULC RGO 2/14

Note

The following editions have appeared since MacPike's book:
  The Correspondence of Isaac Newton
  The Correspondence of Henry Oldenburg
  The Correspondence of John Flamsteed, Vols 1 and 2.

---

*The papers in SP 105/69 are unfoliated.

The full references are given in the Bibliography.

The later editions are more complete and reliable than MacPike (1932) and should be consulted where they overlap with him and for letters unknown to him.

There is a further letter of Halley to G. D. Cassini, of 9 July 1681, in the Biblioteca Universitaria, Pisa, MS423.20/3.

Appendix 8

The sale of Halley's books

Halley's books were put up for sale at T. Osborne's shop in Gray's Inn from Thursday 20 May 1742. The catalogue of the sale has been printed in Vol. 11 of *Sale catalogues of eminent persons* (Munby 1975). Halley's books were sold along with those of a 'late eminent Sergeant at Law' and the catalogue does not distinguish them. Altogether there were 5825 items in the catalogue, as follows:

Folio	English	418
	Latin	517
	French	117
	Italian & c.	71
Quarto	English	301
	Latin	659
	French	151
	Italian & c.	115
8vo & 12mo	English	1855
	Latin	954
	French	667

('Latin' includes some Greek texts and 'Italian &c.' includes a few Spanish works.)

Some of the books are clearly the Sergeant's—law reports, digests and the like—and some are clearly Halley's—mathematics, astronomy, works of his colleagues, works related to his own studies. However, only a small proportion of the whole can be attributed with any certainty, and most of the books might have been acquired by any person of wide interests and a sound classical education. The total number is much greater than that in the catalogue of Hooke's library in the same volume, 3296. Hooke was an avid collector, often noting purchases in his *Diary*. John Ray's collection, also in the same volume of catalogues, was much smaller, 1240, while Newton's books numbered 1763 (Harrison 1978). If Halley's books were half the total number in the sale, they would be of the same order as Hooke's, if they were only a third they would still be about the same number as Newton's and more than Ray's. On any plausible assumption, Halley was a serious collector. A rather large proportion of the books bear dates after 1700, suggesting that Halley acquired many of them after he became Savilian Professor.

Of the books that are almost certainly Halley's, the following are of particular interest.

Many mathematical and astronomical works, including Copernicus *De revolutionibus*, Aristotle, Archimedes, Briggs *Trigonometria*, Flud *Opera omnia*, Fermat *Opera mathematica*, Flamsteed *Historia coelestis Britannica*, Gilbert *De magnete*, Wallis *Opera mathematica*, de Moivre, Smart *Tables of interest*, Sharp *Geometry*, Streete *Astronomia Carolina*, 1710, Tycho Brahe, Barrow *Lectiones*, Craig on fluxions, Cheyne on fluxions, des Cartes, Huygens, Keill, Kepler, *Astronomia* and *Mysteria cosmographia*, Longomontanus, Horrocks, *Opera posth*, Napier *Mirifici logarith*. All the works of Hevelius were in the sale, and also Richer's *Observations astronomiques et physiques faites en l'isle de Caienne*.

Most of the works of Newton are included, but not copies of the first edition of *Principia*, although it is known that Halley had three at one time or another. There is the 1723 Amsterdam pirated reprint of the second edition, but not the original 1713 Cambridge printing which, according to de Moivre, he had had to buy because Bentley did not set aside copies for him, de Moivre, and Bernoulli (Chapter 6). A folio copy of the third edition of 1726 was in the sale. None of Halley's own works was in the sale and it may be that Halley's daughters kept them and his copies of the first edition of *Principia* in the family. A notable omission was his version of Flamsteed's observations and catalogue, the *Historia coelestis* of 1712.

There are many books relevant to Halley's other interests than astronomy and mathematics: books on voyages, navigation and surveying, Vitruvius on Architecture, Frontinus on Acqueducts, on the Antonine itineraries, on antiquities in England and abroad, especially in Rome. They could have been Halley's. There were also works of Swammerdam, Leeuwenhoek, and Swedenborg, the last perhaps presented by Swedenborg about ten years after his visit to Halley in Oxford. Halley was interested in gardening and the works of Evelyn that were in the sale may have been his, as most probably were the devotional and other books by his friend of many years, Robert Nelson, and the works of Locke.

Runs of scientific journals were in the sale—*Philosophical Transactions*, of course, but also *Journal des Sçavants*, *Acta Eruditorum* and the *Histoire* of the Académie des Sciences.

Some books can be identified as presents to Halley or otherwise acquired by him. Thus on 29 May 1681, he wrote to Hooke that Cassini had presented him with his book of the comet. Again, J. Bernoulli sent de Moivre a copy of his *Théorie de manoeuvre des vaisseaux* for Halley.

A substantial collection of works on surgery, hospital practice, and physic probably came from Halley, having belonged originally to his son, the naval surgeon, who predeceased him in 1736. The books put on sale apparently comprised not two collections but three, those of Halley, of the Sergeant, and of Halley's son.

The greatest part of the books sent for sale cannot be assigned. There were many classical books of the sort that both Halley and the Sergeant might have had at school and at university, many works of literature, classical and modern, many bibles in English, Latin, Greek, and Hebrew, and grammars and dictionaries. There were numerous books on theology and philosophy, including Stillingfleet and many of the Boyle lectures which, on the whole, indicate a serious interest in issues of the day, but are not unorthodox. Ecclesiastical and liturgical works such as missals and a pontifical suggest an interest in or connection with the Roman Catholic Church and there is one that suggests that it belonged to someone with Jacobite leanings.

The unassignable books also include numerous lives and memoirs, many sets of plays, mostly of the period, and a collection of operas by Handel, Bononcini, and others, presumably the libretti sold at performances.

A pair of globes was offered for sale, probably Halley's, but the only item that certainly belonged to Halley is a remarkable model of the Church of the Holy Sepulchre in Jerusalem:

> Wherein is to be Sold, as it was sent from the Holy Land, as a present to the late said Doctor, a most curious and complete MODEL of the whole Temple of CHRIST'S *Sepulchre* at *Jerusalem*: from the showing whereof the Profits Arising have been farmed by the *Turks* at above 3000l. *Sterling per Annum.*

A description which evidently went with the model read:

> A very curious and complete MODEL adorn'd with Mosaic Work, in Mother of Pearl, &c. of the whole TEMPLE of CHRIST'S SEPULCHRE at JERUSALEM, as it was built by HELENA, Mother of the Emperor CONSTANTINE; With all the Chapels, Altars, and other observable Buildings about it. From the Shewing whereof, the Profits have been Farmed at above 3000l. Sterling per Annum. This venerable Curiosity is so artificially and commodiously Contrived, that the Inside and Ground Plan are rendered as visible, as in the Elevation. All the particular Parts or Places in the said Model are distinctly number'd, which Numbers are exhibited together with their Explanations in the following Table.
>
> 1. The Great Door
> 2. The Belfry
> 3. The Chapel of the Egyptians, now the Greeks

4. The Chapel where our Lady stood at the Crucifixion
5. The place where our Lord's body was anointed
6. The tombs of Godfrey of Boulogne and King Baldwin
7. Tombs of other kings
8. Where our Lady stood when Jesus was anointed
9. Stairs to the chapel of the Armenians
10. The cloister of the Armenians
11. The church of the Armenians
12. The High Altar of the Armenians
13. The door to the house of the Abbesses
14. Stairs
15. Tombs of Joseph of Arithmathea and Nicodemus
16. Door to the church of the Syrians
17. The Stone where the angel appeared to Mary Magdalene
18. The Holy Sepulchre
19. The choir of the Franciscans
20. The Greek church
21. The hole that the Greeks call the Centre of the World
22. The choir of the Greeks
23. The High Altar of the Greeks
24. The Crucifix of the Greeks
25. The door to the common cistern
26. The place where our Lord said 'Noli Me tangere'
27. The Chapel of St Mary Magdalene
28. The door to the chapel of the Franciscans
29. Where a dead body was raised to life by a fragment of the True Cross, to convince St Helena that she had found the Cross
30. The altar of the column of Flagellation
31. The high Altar of the Franciscans
32. Another altar
33. The door to the house of the Franciscans
34. Door to the necessary houses
35. The place where Jesus was imprisoned before the crucifixion
36. The door to the cloister
37. Chapel of Longinus
38. The place of casting of lots
39. Stairs to the chapel of St Helena and the Invention of the Cross
40. Chapel
41. Door to the chapel of the Greeks
42. Stairs to Mt Calvary
43. Door to the habitation of the Greeks
44. The cleft in the rock that opened when our Lord gave up the ghost
45. The hole where the Cross was fixed
46. Where Jesus was nailed to the Cross
47. The door to the chapel where Mary stood at the crucifixion.

The model must undoubtedly have belonged to Halley, for he is designated *Doctor Edmund Halley* in the title of the Sale Catalogue and the other owner is not otherwise distinguished than as an anonymous Sergeant-at-Law.

The description and the list of 47 sites identify the model as one of a number made by the Franciscan guardians of the Holy Places from 1600 onwards. A splendid one is in the Franciscan museum in Jerusalem and there are five in London, four in the British Museum and one in the rooms of the Palestine Exploration Society. Wren knew of such a model; a few others are known elsewhere in Europe. They are approximately 0.5 m square by 0.5 m high. As the description indicates, parts of the roof can be lifted off to reveal the internal arrangements. The workmanship is good and the decoration of ivory and mother of pearl is rich. Much of the decoration seems to be highly symbolic and related to the concept of buildings and cities as representations of the City of God. The alignment of the church relative to the cardinal points is especially emphasised. The models show the Holy Sepulchre in about 1600, when it differed from the present state. The basilican part of the church was built at the direction of Constantine, of which the association with St Helena preserves a memory. The title of *Temple* also goes back about one and a half millenia. Those facts were all familiar in eighteenth-century England.

The only hint as to how Halley came by the model is the remark that it was a present to him sent from the Holy Land; might it have come from a relative or acquaintance in the Levant Company?

Appendix 9

The manuscript ULC RGO 1/74 of Flamsteed's catalogue and its implications

The manuscript once in the Radcliffe Observatory, Oxford, and since 1935 in the archives of the Royal Observatory, is in a condition that suggests that it has been much used (Chapter 13). It is effectively two documents, an original catalogue of stars observed by Flamsteed, denoted by M_1, and the version in its present form, denoted by M_2, as modified by rearrangement, additions, and changes, all in a different hand from that of M_1. There are also two printed versions of the catalogue, P_1 in Halley (1712) and P_2 in Flamsteed (1725).

Most of the MS folios are numbered in three ways, two of which correspond to the present arrangement of the folios in the manuscript while the third corresponds to a different arrangement. The numbers of that third set are in a script very similar to that of dates on two of the folios, namely 17 September 1701 on the first folio of the constellation Cygni, and 5 February 1707 on the first folio of Arietis. The two forms of pagination show that the order of constellations was changed at some time after M_1 was assembled.

The stars are grouped in constellations and the constellations in turn are grouped in three sets, the twelve zodiacal constellations, the constellations to the north of the zodiac, and those to the south. In the present arrangement in M_2 the zodiacal stars are placed first, then the northern group, and the southern stars last, but in the original version of M_1 the northern stars came last. In the same way the orders of northern and southern stars are reversed betwwen P_1 and P_2, P_1 agreeing with M_2 and P_2 with M_1. The ordering of constellations within the three groups is the same in all four versions for the zodiacal and southern stars, but in the northern group M_2 and P_1 agree and M_1 and P_2 diverge from them.

The catalogue, P_2, in Flamsteed (1725) includes some constellations that are taken from the catalogue of Hevelius (included in that volume). They are, in the northern group, Camelopardi, Lyncis, Leonis Minoris, Canici Venaticam, Vulpeculae et Anseris, and Lacertae, and in the southern group, Monocerotis and Sextantis. The three other versions omit those constellations.

The ways in which the material is laid out in each version show similar affiliations. The titles of the northern and southern groups have the forms:

M_2 and P_1: Stella Extra Zodiacum Boream Versus
P_2: Fixarum in Constellationibus extra Zodiacum ad Boream Sitis[*]

while the titles of the tables for the constellations are as follows:

M_1: FIXARUM IN CONSTELLATIONE ARIETIS
M_2, P_1 and P_2: IN CONSTELLATIONE DRACONIS.

The columns of data in M_1 are the numbers of the stars in the catalogues of Ptolemy and Tycho, where such numbers exist, a description of the star, the symbols of Bayer, where they exist, the right ascension in time and the angular distance from the vertex, the right ascension and the distance from the north pole in angular measure, the celestial longitude and latitude also in angular measure, the variations of right ascension and polar distance, and the magnitude of the star. The celestial longitude is given as the sign of the zodiac and the angular measure within it. M_2 and both printed catalogues omit the columns of right ascension in time and the distance from the vertex. M_2 adds as the first column a serial number within the constellation. P_1 has serial numbers, P_2 does not. Within each constellation M_1 and P_2 generally group the entries by three's separated by horizontal rules, while M_2 and P_1 invariably group them in fives.

The foregoing general comparisons strongly suggest that P_1 was derived from M_2 and P_2 from M_1, a conclusion that is reinforced by examination of detailed changes beween M_1 and M_2. There are many differences between the descriptions of stars, which are a large part of the changes from M_1 to M_2. Thus, in Perseus, the M_1 version, with alterations, appears on f. 49r. and a fair copy of the M_2 version appears opposite on f. 48v. The descriptions of star *Bayer e* are:

M_1 : *In Planta Pedis ejusdem*
M_2 : *In planta Pedis ejusdem*
P_1 : *In planta Pedis ejusdem*
P_2 : *In Talo dextra Pedis.*

Again for *Bayer η*:

M_1 : *In flexura Brachii superioris*
M_2 : *In superiori Brachio*
P_1 : *In Superiori brachio*
P_2 : none.

[*]It is not clear what the titles were in M_1.

There are not many changes in numerical values; of them a typical example is those made in Perseus *Bayer* η:

	AR ° ′ ″	polar dist. ° ′ ″	Long. ° ′ ″	Lat ° ′ ″
M₁ original	37 08 00	41 32 30	21 32 00	31 40 10
M₁ corrected	37 6 30	35 25 30	24 23 27	37 26 50
M₂	37 6 30	35 25 30	24 23 27	37 26 50
P₁	37 06 30	35 25 30	24 23 27	37 26 50
P₂	37 8 00	41 32 30	21 32 0	31 40 10

'M₁ original' is the original entry in M₁ on f. 49r., 'M₁ corrected' is the correction written over the original, and 'M₂' is the fair copy on f. 48v.

Here, as generally elsewhere, P₁ agrees with M₂ and P₂ with M₁.

M₁ was never completed. The constellations Ursa Major and Cassiopeia show that most clearly, for in both the latitudes and longitudes considerable numbers of stars are missing. The first 70 stars of Ursa Major in P₁ do not appear in P₂, nor do some at the end of the list. M₁ lists all the stars in P₁ but without the longitudes and latitudes for the first 70 and others. M₂, represented by loose and pasted-in sheets, gives the latitudes and longitudes for all the stars printed in P₁.

The conclusion from the analysis of RGO 1/74 and its relation to the printed catalogues must be that P₁ was derived from M₂ and P₂ from M₁. M₂ cannot, however, be the copy from which the printer of P₁ worked for it is in far too messy a condition, indeed illegible or ambiguous in places. A fair copy, M₂′, must have been prepared from it for the printer. The relation of M₁ to P₂ is also not direct, for P₂ follows an arrangement of constellations now preserved in an original pagination of M₁ and retains the Hevelian constellations which correspond to gaps in that pagination. There must therefore have been an earlier form of M₁, M₁*, from which M₁ and P₂ were derived.

Is it now possible to identify the compiler and redactor of M₁? The dates on two folios of RGO 1/74 give some indication; they are 17 September 1701 on f. 70r. (Cygnus) and 5 February 1702 on f. 2r. (Aries). They antedate the appointment of Referees to examine Flamsteed's material and so show that the original version of M₁ was prepared by Flamsteed or one of his assistants. It is possible that it or some of it was the catalogue deposited with Newton at the end of 1704; if so Flamsteed must have kept a copy, M₁* perhaps, when he delivered his sealed and incomplete catalogue to Newton, for otherwise he would not have had the material from which P₂ was printed. The incomplete forms of the catalogues of certain constellations, Ursa Major and Cassiopeia in particular, may give some clue. Flamsteed observed many of the stars in Ursa Major with his mural circle in April 1704 (*Historia coelestis*, Vol. 2), and some at least are among the first 70 in P_1. It may be that Flamsteed gave Newton his sealed copy before he had time to compute the longitudes and latitudes.

When Gregory died in 1708 the sextant observations had been printed. They were the material from which M_1^* would have been produced. Meanwhile Flamsteed was continuing with his observations with the mural arc. It is likely that none of that material contributed to P_2, but the stars observed up to the spring or summer of 1704 were added to M_1. After Gregory's death Halley took over the publication and would have had in his hands M_1 and also evidently the mural arc observations up to 1705, as is plain from the title page of *Historia coelestis*. Many of the alterations to M_1 that were incorporated in M_2 look like Halley's hand, but some could have been by another, perhaps Machin, who had a large part in the preparation of material for the press. In any event M_2 was clearly the basis of P_1 and as such the responsibility of Halley.

Where, finally, do Crosthwait and Sharp come in and why was P_2 such an early form of the catalogue which did not at all represent the sum of Flamsteed's labours? Flamsteed was observing until close to his death, and deliberately or on account of the time he spent observing, did not bring his catalogue up to date. The only complete catalogue approved by Flamsteed and available to Crosthwait and Sharp when he died would then have been M_1^*, twenty years out-of-date though it was.

Appendix 10

Halley's Memoriall to the Emperor (English translation of Halley's Italian original)

Sacred Imperial Royal Majesty

Your Imperial Majesty being pleased not only to grant me the freedom to visit the sea coast and its ports on the Adriatic to ascertain what prospects of support there might be in them, according to the orders given me by the most Serene Queen of England, my Mistress; but also being graciously pleased to have appropriate orders given that I should be provided with every assistance to enable me to carry out the work in the first place for the navigation and security of warships that might come to those seas.

Having then begun to examine the maps that had been printed by various authors and having found in them many departures from the true facts, I wished to produce a similar one of all the coasts of your majesty and of the neighbouring islands that I have been able to survey and to present it to your Majesty.

After that I was obliged to consider what protection there was against the risk of winds that are sometimes felt in those seas, particularly in winter, and I have surveyed all the canal of Morlacchia running along the coast of the mainland subject to your Majesty opposite the islands of Veglia, Arbe and Pago. Vessels might ride well at anchor there, the sea being clean and free from sharp rocks, but the great depth of the water makes it a little difficult to cast anchor there.

Coming then to the ports in the dominion of your Majesty that might be secure against a powerful enemy fleet, I have not found any other than Buccari, which is very good and able to accomodate many warships, extending in length about three Italian miles, with a mouth that does not exceed two hundred paces, so that it may easily be secured with batteries of cannon, making the risk of attempting an entrance unacceptable if every approach is fortified in a similar manner. I wished to lay down the batteries in their proper places on the chart of the same port that, together with the geographical map, I ventured to present to your Majesty; judging that sixty cannon were needed to furnish those batteries, namely forty of eighteen and twenty of twelve [pounds], distributing them according to the sizes of the ball, for small ones do not do sufficient damage to large ships; and the number of cannon is needed because if the entrance is attempted by enemy ships with a fresh wind it would be almost impossible to fire more than one shot. As to the provision of the artillery would your Majesty deign to give the most urgent orders to your ministers so that by taking proper action, nothing may be wanting for the provision of sufficient iron, whether from your own states or elsewhere.

I consider that the construction of the batteries will not entail expense for all that is necessary is to be found on the site and in the neighbourhood, only sand has to be brought from Fiume which takes two hours on the only road. I further consider it necessary that you should be graciously pleased to order a survey of the quality of trees to be made in your woods so that it may be known where everything necessary is to be found.

In this way all the provisions that will be needed for the ships that may pass into the Adriatic will be made ready in good time.

With the hope that I may have the consolation that with this my voyage, beside the obedience rendered to the commands of my most Serene Queen, I have had the fortune to serve to the advantage of your august House and Person, humbly prostrated I bow before your Imperial Throne.

Note:
The London copy is in PRO SP 105/67, ff. 418, 419.
The copy in the Archivio di Stato in Venice was obtained by Loredan in Vienna and sent to the Senate with his despatch of 17 March 1703: ASV Senato Dispacci Germania, f. 185, cc. 317, 318. The Senate at once sent a copy to Mocenigo in London—ASV Senato Segreta, Corti f. 98, 29 March 1703.

Notes to chapters

Abbreviations in the notes

Archives

ASV	Archivio di Stato, Venice
BAB	Bernoulli Archiv, Universitätsbibliothek, Basel
	[Note: Not all letters in the Bernoulli collections are foliated]
BLL	British Library, London
BLO	Bodleian Library, Oxford
BMV	Biblioteca Marciana, Venice
BNP	Bibliothèque Nationale, Paris
FAW	Finanzarchiv, Wien
HHSW	Haus- Hof- und Staatsarchiv, Wien
KAW	Kriegsarchiv, Wien
NAN	Nottingham Archives, Delapre Abbey, Northampton
OdP	Observatoire de Paris
PRO	Public Record Office, London and Kew:
	ADM Admiralty records, especially from 1689
	SP State papers (Secretaries of State)
	T 70 Records of the Royal African Company
RGO	Royal Observatory, Greenwich (in ULC)
RSL	Royal Society, London:
	CMC Council Minutes copy
	JBC Journal Book copy
	LBC Letter Book copy
	LBO Letter Book original
	MM Miscellaneous manuscripts
SLG	Steiermärkisches Landesarchiv, Graz
ULC	University Library, Cambridge

Printed sources

Biogr. brit: 'Edmond Halley'. In *Biographia britannica* (1757), Vol. **4**, pp. 2494–2520. London.

Excerpta J. H.: Olhoff, J. E. (ed.). (1683). *Excerpta ex litteris J. Hevelius*. Danzig.

Flamsteed: Baily, F. (1966). *An account of the Revd. John Flamsteed*. Dawsons, London. Reprint of Baily (1835*b*).

Flamsteed Corresp.: Forbes, E. G., Murdin, L., and Willmoth, F. (ed.) (1995) *The correspondence of John Flamsteed*, Vol. 1. The Institute of Physics Publishing, Bristol and Philadelphia.*

Halley Corresp.: MacPike, E. F. (ed.) (1932). *Correspondence and papers of Edmond Halley*. Clarendon Press, Oxford.

History R. S.: Birch, T. (1756–7). *The history of the Royal Society of London*, 4 Vols. London.

Hooke D1: Robinson, H. W. and Adams, W. (ed.) (1935). *Diary of Robert Hooke, 1672–80*. London.

Hooke D2: Gunther, R. W. T. (1935). Hooke's Diary, 1688–93. In *Early science in Oxford*, Vol. **10**. Clarendon Press, Oxford.

*Vol. 2 appeared too recently to be referred to.

Newton Corresp.: Turnbull, H. W. (ed.) (1959). *The correspondence of Isaac Newton*, 7 Vols. The Royal Society, London.
Oldenburg Corresp.: Hall, A. R. and Hall, M. B. (1977). *Correspondence of Henry Oldenburg*, Vol. 11. Mansell.
Oldenburg Corresp.: Hall, A. R. and Hall, M. B. (1986a). *Correspondence of Henry Oldenburg*, Vol. 12. Taylor and Francis, London.
Oldenburg Corresp.: Hall, A. R. and Hall, M. B. (1986b). *Correspondence of Henry Oldenburg*, Vol. 13.
Paramore: Thrower, N. J. W. (1981). *The three voyages of Edmond Halley in the Paramore 1698–1701*. The Hakluyt Society, London.
Rigaud: Rigaud, S. P. and Rigaud, S. J. (ed.) (1841). *Correspondence of scientific men of the seventeenth century*. 2 Vols. Oxford University Press. (Transcriptions, with notes, of correspondence in the possession of the Earl of Macclesfield at Shirburn Park.)

CSP, D	Calendar of State Papers, Domestic
CTB	Calendar of Treasury Books
CTP	Calendar of Treasury Papers
DSB	*Dictionary of scientific biography*. (1970–1980). 16 Vols. Charles Scribner's Sons, New York.

Notes to Chapter 1

1. Eamon (1994) gives an account of shifts in views of the world.
2. For a general political account of the period and why it remains so significant in British history, see Hill (1980). See also Appendix 4.
3. Savery (1699) showed his steam engine for raising water to the Royal Society in 1699. He had patented it in 1698 and it was the basis for the widely used Newcomen engine (A. Smith 1995).
4. It lay to the west of the present Liverpool Street station and adjoining the Artillery Fields.
5. For a history of the Levant Company, see Wood (1935); see also Woodhead (1965).
6. See Woodhead (1965) and de Krey (1985).
7. For example, John Pearson, author of *An exposition of the creed* (1649), who became Bishop of Chester (see Chapter 2).
8. See Gwynn (1985).
9. See, for example, Scott (1991).
10. For the Tunbridge Wells church and its subscribers, see Cook (1993), Greenwood (1992).
11. For Norwich and Bristol, see Hill (1980); for numbers of ships of various countries trading to London, see Houghton (1727–8).
12. For the great storm of 1703, see Burchett (1720); Book V, Ch. XV, and Lamb (1991); for the Lagos disaster, see Burchett (1720) Book IV, Ch. XI, p. 484 *et seq.*
13. For James II and Cambridge, see Cooper (1845), Vol. 4, pp. 175–9 and Goldie (1996).
14. See Gibson and Johnson (1943) and Carter (1975) for Oxford University Press, and Black (1984) and McKitterick (1992) for Cambridge University Press.
15. See King (1994).
16. Howe (1992) discusses women on the stage.
17. See Robinson (1980).
18. For painting in Stuart times, see articles in Howarth (1993).
19. See Bennett (1982).

20. Selkirk was found by Capt. Edward Cooke in the course of his voyage round the world—see Cooke (1712).
21. Lough (1985, 1987) recounts tales of travellers in France.
22. See Gwynn (1985).
23. Stoye (1994) describes the lands south-east of Austria in relation to the career of Marsigli.
24. See Geiter (1993) for the founding of Philadelphia.
25. For example, two nephews of Robert Nelson—see Secretan (1860).
26. See Cook (1988).
27. For the mural arcs at Greenwich and Paris, see Krisciunas (1988), Ch. 4.
28. Flamsteed to Newton, 24 February 1692, *Newton Corresp.*, Vol. 3, 199–203.
29. For English mathematics in this period, see Bennett (1982).
30. The principal English developments in magnetism were by Gilbert (1600) and Gellibrand (1635). For Wren see Bennett (1982), 46–7. Kircher (1643) made important contributions.
31. See Bernoulli (1708) on the mechanical basis of muscle action; for Stephen Hales, see Clark-Kennedy (1929).
32. Dobbs (1991) provides the leading account of Newton's alchemy.
33. See Hall (1991).
34. The present Accademia dei Lincei is the successor of Prince Cesi's foundation but not without interruptions.
35. On Leibniz and his schemes for academies, see Aiton (1985), Cook (1996).
36. On Tycho Brahe, see Thoren (1985, 1990).
37. For Hevelius and his observatory, see Chapter 4. For Elizabeth Hevelius see Cook (1997).
38. For humanists and their lasting influences, see Kraye (1996). For Gassendi, see Joy (1987); for Halley's estimates, Halley (1691*e*).
39. For the works of Descartes, see Cottingham, Stoothoff, and Murdoch (1985).
40. See Bayle (1697, 1734–41); for his influence on Enlightenment thought, see the selections edited by Popkin (1965).
41. See Ray (1691, 1692). The books were based on sermons that Ray had preached in Cambridge more than thirty years earlier.
42. See Hevelius (1647).
43. From the Revised English Bible.
44. Boyle, quoted in Westfall (1971), p. 115.
45. See Chapman (1994).

Notes to Chapter 2

1. *Memoir* of Halley, reproduced in *Halley Corresp.* See also Aubrey (1958), Vol. 1, p. 282; *Biogr. brit.*, Vol. 4, p. 2494; and Mairan, *Mém. Acad. Roy. des Sciences (Histoire) Ann. 1742*, 172–88; Flamsteed to Towneley, 8 June 1675, *Flamsteed Corresp.*, Vol. 1, p. 352; Halley to Hevelius, 11 November 1678, *Halley Corresp.*, pp. 40–1; for Shoreditch and the City, see maps in Hyde (1981, 1982); for the parents' marriage: register of St Margaret's Westminster, 9 September 1656, see Appendix 2, Note (iv).
2. See Aubrey (1958), p. 282; Beevor, R. A., in *Halley Corresp.*; *Flamsteed* for Luke Leigh.
3. Katherine's marriage is recorded in Metcalfe (1887) and in the marriage register of St Margaret's Barking, Essex, for 24 November 1617 (Appendix 2, Note (ii)). See also Metcalfe (1878–9). James Cook also was married in St. Margaret's.

Katherine's brother Francis held land in Holdenby, Northamptonshire (Metcalfe 1887). The Mewce family derived from a John of Calais. Mewces are not in the Essex *Visitation* (Metcalfe 1878–9) but a family of Meautis, also derived from a John of Calais, is listed there. John of Calais, the common ancestor, was a Norman who was in the service of Henry VII, then Earl of Richmond, in France, and after Bosworth Field he settled in England as the French Secretary to Henry.

4. See Pearson (1649). Pearson (1613–1686), of King's College, Cambridge, was Master of Trinity 1662, FRS 1667, Bishop of Chester, 1673—see *History R. S.*, Vol. 4, p. 506.
5. For Humphrey I see entry in Woodhead (1965) and Chamberlain's Loan Accounts, City of London, 40/1, 40/9, 40/10, 40/16, and 40/17.
6. See CSP,D, 1636.
7. For details see Appendix 2.
8. Quoted in *Halley Corresp.*, pp. 176–7.
9. For Amigoni, see Vertue (1930–50); for the house in St Giles' Cripplegate, PRO Lay Subsidies, 1666, 252–32.
10. Humphrey I paid for 'staking out' two properties in Cannon Street and one in Lombard Street, and Edmond (sen.) paid for one in Thames Street (Jones and Reddaway (1967), Vol. 1, pp. 20, 35, 46, 75).
11. See the schedule to Chancery claims listed in Note 17 for Halley's allowance.
12. For the mother's marriage and death, see Appendix 2.
13. Register, St James, Duke Place for the Chomat marriage.
14. Halley, Edmund vs. Cleeter, Robert and Joane his wife. Money. Answer only. PRO Chancery, C7, Hamilton 181, No. 90, 1694.
15. Will of Humphrey Halley II—PCC Reg. Bench, f. 66, pr. 3, June 1676, PRO PROB 11/351. A Phillip Cawthorne was Master of the Salter's Company in 1670 and a Richard Cauthorne was a Common Councillor in 1687. Another Cawthorne was steward of Denzil Holles, the Earl of Clare, suspected in 1683 of involvement in the Rye House Plot (Chapter 5).
16. CSP, D, 5 April 1682.
17. Cleeter and Cleeter vs. Buckworth, Young, and Halley. PRO Chancery, C10, Whittington, Bundle 222, No. 27, April 1686; Dividend Book of the Tower, 14 September 1664. The father is identified as a yeoman warder by the following:
    (a) The account of his death in the Chancery actions is identical with that of pamphlets (Chapter 5) referring to the warder's death, one of which says that he lived in Winchester Street.
    (b) The item of two quarters' salary from the Tower in his estate.
    (c) Halley's concern when the death of Essex was reopened in 1689 (Chapter 5).
18. See Braddon (1690) and other pamphlets quoted in Chapter 5.
19. See Glass (1966).
20. For evidence see Appendix 3.
21. See Note 9.
22. PRO Close Roll, 4190, 17 April 1665.
23. Halley's will in *Halley Corresp.*, 253–7.
24. PRO Close Roll, Vol. 53, 21 April 1694.
25. Halley vs. Sandwith. PRO Chancery, C7, Hamilton Bundle 171, No. 95, 1670; will of Humphrey II.
26. See the respective wills for these holdings.
27. For visits to Greenwich, see *Hooke D1*, 30 June 1675; 10 July 1676.
28. See Willmoth (1993*a,b*).
29. *Hooke D2*, 7 June 1693; Halley (1686*c*).
30. 'That he was a studious and successful peacemaker ... that I am apt to think that he cemented as many broken Friendships, reconciled as many Quarrels, and adjusted

as many Differences (which other wise might have flamed out into destructive breaches) as most of those Blessed Peacemakers that are gone before him.' Guildhall Library, London, S-Pam. 45; Woodhead (1965).
31. See Glass (1966).
32. *Hooke D2*, 24 July 1693.
33. See Ehrman (1953); for Queen Mary's orders to Rooke (Whitehall 19 May 1693), see PRO ADM 1/4080.
34. PRO, E. I. Coy records; there is a list of Levant Company members from 1652 to 1701 in the pamphlet BLL θ 665, pp. 97,98.
35. See Hamilton (1965), pp. 146–8.
36. BLL Add 31431, Parts 1 and 2.
37. See McDonnell (1977), p. 269 for Halley and p. 258 for Nelson.
38. For hearth tax, St Giles', Cripplegate, See Note 9.
39. For Halley's views on catastrophes, see Halley (1724a,b).
40. For Halley's Ode, See Chapter 6 and Appendix 6.
41. The magnetic observation of 1672 can be found in Halley (1692b). Flamsteed (RG01/50K; f. 233v.) lists the positions of 22 stars, omitted by Tycho, observed by Halley in 1672.
42. See McDonnell (1909); Secretan (1860).
43. Admission Register of the Queen's College.
44. See Fairer (1986) for Anglo-Saxon studies; for Williamson, see DNB, and for his intelligence, Marshall (1994).
45. See Sutherland and Mitchell (1986).
46. See Magrath (1922), Vol. 2.
47. Boucher to Flamsteed, *Flamsteed Corresp.*, Vol. 1, p. 318; Halley to Flamsteed, *Flamsteed Corresp.*, Vol 1, p. 326.
48. Foster (1891–2) lists only one Charles Boucher (or Bouchar); for Boucher and his family in Jamaica, see entries in *Cal State Pap. Colonial*, 2 and 3 September 1678; 19 August 1679; 2 June 1686; 21 September, 6 October 1687; 19 July, 8 August 1688; and many others 1689 to 1692.
49. PRO SP Dom 44, p. 15; Halley to Flamsteed, 10 March 1675, *Halley Corresp.*, 37–9, and *Flamsteed Corresp.*, Vol. 1, pp. 326–9; for the first letters between Boucher and Flamsteed, see *Flamsteed Corresp.*, Vol. 1, p. 318; Flamsteed to Halley, 22 April 1675, the letter is known only from an extract in Flamsteed's papers, *Flamsteed Corresp.*, Vol. 1, pp. 332–3.
    The letter from Boucher to Halley of 10 March 1676, in which he described his fraught voyage to Jamaica, was read to the Royal Society by Oldenburg (*History R. S.*, Vol. 3, p. 318, 8 June 1676) and copied by Flamsteed (ULC RGO 1/43, ff. 19v.–22r.).
50. BLL Add MS. Birch 4393 f. 104r.,v.; CSP, D, Car. II, 1675–6, p. 173.
51. Flamsteed to Towneley, 8 and 22 June 1675, *Flamsteed Corresp.*, Vol. 1, pp. 351–5.
52. 'Interat hisce Observationibus multumque eas sua adjuvabat diligentia ingeniosus Iuvenis Edmundus Hally Oxoniensis': *Philosphical Transactions*, 10, No. 116, pp. 368–70; Flamsteed to Towneley, 3 July 1675, *Flamsteed Corresp.*, Vol. 1, pp. 356–8; Flamsteed to Oldenburg, 24 July 1675, *Flamsteed Corresp.* Vol. 1, pp. 358–64; Oldenburg to Hevelius, *Oldenburg Corresp.*, Vol. 11, pp. 419–21.
53. See Flamsteed (1676); *Hooke D1*, 1675, June 22, 23, 30, July 2, 28.
54. *Hooke D1*, 7 January 1676.
55. Observations of Saturn, 30 October, 1, 5 November 1675; Halley observed lunar eclipse under clear skies, 5 December 1676; observations of sunspots by Halley are all noted in Flamsteed (1725).
56. Flamsteed summarised reports he had received from Halley 'E. literis Ed Halley' of occultations on 12 and 13 March 1676 in ULC RGO 1/43 f. 11r. and of a solar

eclipse of 29 April 1676 in ULC RGO 1/43 f. 15r., v.; see also *Flamsteed Corresp.*, Vol. 1, pp. 438–41; Flamsteed (1725).
57. For the eclipse of 1/11 June 1676, see Flamsteed to Towneley, 7 June 1676, *Flamsteed Corresp.*, Vol. 1, pp. 456–7 and Flamsteed to Oldenburg, 10 July 1676, *Flamsteed Corresp.*, Vol. 1, pp. 478–85.
58. Halley to Flamsteed, 22 June 1676, *Flamsteed Corresp.*, Vol. 1, pp. 473–5; Halley to Flamsteed, 13 July 1676, *Flamsteed Corresp.*, Vol. 1, pp. 486–7; Halley to Flamsteed, 25 July 1676, *Flamsteed Corresp.*, Vol. 1, pp. 496–8.
59. *Hooke D1*, 15 September 1676.
60. Halley (1676a).
61. Halley to Oldenburg, July 0/8 1676, *Oldenburg Corresp.*, Vol. 11, pp. 367–79; also Halley to Oldenburg, July 11 1676, *Oldenburg Corresp.*, Vol. 11, p. 386.
62. Ward (1656).
63. De la Hire (1677).
64. J. J. Ortous de Mairan, Éloge de M. Halley, *Mémoires de l'Académie Royale des Sciences*, Année 1742, Paris 1744–5, 172–88; and *Halley Corresp.*, pp. 15–27.
Halley's method bears some resemblance to Kepler's in his study of Mars: see Kepler (1989).
65. Halley (1676b).
66. Flamsteed to Oldenburg, 30 November 1676, *Flamsteed Corresp.*, Vol. 1, pp. 501–3 and *Oldenburg Corresp.*, Vol. 13, pp. 147–8. Flamsteed wrote 'the Indies' for St Helena.
67. Halley (1676c).
68. Halley also observed the Moon at Oxford on 19 and 20 June 1676 and Saturn and the Moon at Greenwich on 27 February: see ULC RGO 2/9.
69. Boucher to Halley, July ?1675, copy in ULC RGO 1/43, ff. 19v.–22r.
70. Flamsteed to Boucher, 24 January 1678, *Flamsteed Corresp.*, Vol. 1, pp. 595–9 and ULC RGO 1/43 ff. 36r.–38r.
71. Flamsteed had described his equatorial mount in a letter to Towneley, 8 June 1675, *Flamsteed Corresp.*, Vol. 1, p. 351.
72. Halley (1679a); Flamsteed to Boucher, 24 January 1678, *Flamsteed Corresp.*, Vol. 1, pp. 595–7; E. I. C, Court Book, Vol. 30, p. 58; Halley (1683b).
73. Halley (1679a).

Notes to Chapter 3

1. *Catalogus stellarum australium, supplementum catalogi Tychonici*, (London, Thomas James for R. Harford, 1679), dedicated to Charles II.
August Royer published a French translation *Catalogue des estoilles ou supplement du catalogue de Tycho* (Paris, Jean Baptiste Coignard, Imprimeur du Roy, 1679); it accompanies his *Carte du ciel*. Halley's Latin text is printed with the French translation in parallel columns. An anonymous preface explains why it was printed, the dedication to Charles II is omitted but a planisphere of the southern heavens and four figures relative to the transit of Mercury and Halley's method of determining the positions of stars are included. The translator was Jean-Baptiste van Luer—see Hevelius to Halley of 1681, nd. OdP C.1, T. 15, No. 27 (2154), also in BNP LAT 10349, t. xv, ff. 41–.
2. Halley to Oldenburg, 0/8 July 1676, *Oldenburg Corresp.*, Vol. 12, pp. 367–74; Halley to Oldenburg, Oxford, 11 July 1676, *Oldenburg Corresp.*, Vol. 12, p. 386; Halley to Oldenburg, Queen's Coll. Oxon., 8 August 1676, *Oldenburg Corresp.*, Vol. 13, p. 27.
3. CSP, D. Car II, 1676–7, 385, No. 94, September 1676.

4. E.I. Co., Court Book, Vol. **30**, p. 58; *The Observatory*, **51** (1928), 286, 297.
5. E.I. Co, Court Book and correspondence.
6. See Foster (1891–2); *Oldenburg Corresp.*, Vol. **13**, p. 27. The editors of Oldenburg's correspondence suggest that the most likely friend was Thomas Clerk of Cumberland, who was born in 1654 and entered Queen's in 1674; however he took his BA in 1678 and that does not seem to allow time for him to go to St Helena with Halley.
7. See Note 1.
8. 'Observatorem illum in haec palestra minime exercitatum fuisse'.
9. Halley to William Molyneux, 27 March, 1686, *Halley Corresp.*, pp. 64–5.
10. Halley to Oldenburg, 0/8 July 1676, *Oldenburg Corresp.*, Vol. **12**, pp. 367–74; Oldenburg to Hevelius, 27 January 1677, *Oldenburg Corresp.*, Vol. **13**, pp. 194–8; Flamsteed to Hevelius, 23 May 1678, *Flamsteed Corresp.*, Vol. **1**, pp. 622–7; Flamsteed to Moore, 7 March 1678, *Flamsteed Corresp.*, Vol. **1**, pp. 610–14.

    Cassini to Oldenburg, 30 August 1676, *Oldenburg Corresp.*, Vol. **13**, pp. 63–4 and 2 September 1676, *ibid.*, 67–9: he explained that Richer and Meurisse had other duties than stellar observations in Cayenne and that the southern pole could not be seen from it. The St Helena observer would see all the southern hemisphere and would make better observations of stars close to the horizon at Cayenne. He wrote of deriving the lunar parallax from meridian observations at St Helena and France and England. He sent particulars of eclipses of the satellites of Jupiter.

    Cassini afterwards sent Oldenburg a much more detailed account of what Richer and Meurisse had done: Cassini to Oldenburg, 14 October 1676, *Oldenburg Corresp.*, Vol. **13**, pp. 104–7. Halley had probably sailed when the letter reached Oldenburg and his references to the Cayenne expedition in the *Catalogus stellarum Australium* suggest that he never saw it—Oldenburg was dead when Halley returned to London. See Richer (1679) and Cassini (1684).
11. For Newton's theory of the precession, see Westfall (1980).
12. Halley to Hevelius, Oxford, 11 November 1678, *Excerpta J. H.*, pp. 182–4, *Halley Corresp.*, pp. 41–2.
13. Flamsteed to Towneley, 8 June and 3 July 1675, *Flamsteed Corresp.*, Vol. **1**, pp. 351–3, 356–8.
14. Halley (1679a).
15. De L'Isle's sketch of Halley's sextant is in OdP MS E.1.3 MS.
16. See Hooke (1674).
17. Halley to Hevelius, Danzig, 8/18 July 1679, *Excerpta J. H.*, pp. 187–8; *Halley Corresp.*; pp. 44–5.
18. See Thoren (1990).
19. See Gosse (1938), Ch. III.
20. Gill (1878).
21. Copy in Hooke (1678), pp. 75–7.
22. Flamsteed to Boucher, Greenwich, 24 January 1678, *Flamsteed Corresp.*, Vol. **1**, pp. 595–9.
23. *History R. S.*, Vol. **3**, 13 December 1677, 14, 21, 28 February, 7, 14 March 1678; *Journal des Sçavans*, 20 December 1677.
24. Halley (1692b).
25. *Hooke D1*, 30 May 1678.
26. *History R. S.*, Vol. **3**, p. 409, 30 May 1678, and 'Edmund Halley' in Aubrey (1958); it has even been suggested that the husband and wife were Clerk and his wife. There is no evidence that Clerk was married and he was almost certainly not in his 50s.
27. Flamsteed to Hevelius, Greenwich, 23 May 1678, *Flamsteed Corresp.*, Vol. **1**, pp. 622–7; Flamsteed had previously sent Halley's particulars of the transit of Mercury to Bernard: *Flamsteed Corresp.*, Vol. **1**, pp. 606–8.

28. *Hooke D1*, 17 February 1678; ULC, RGO. 2/9.
29. E.I. Co., Court Minutes, 20 February 1678; Gosse (1938), pp. 77, 95.
30. E.I. Co., Court Minutes, 21 August 1678.
31. *History R. S.*, Vol. 3, pp. 433–4, 7 November 1678
32. *History R. S.*, Vol. 3, pp. 441–2, 30 November 1678.
33. ULC, RGO 1/36, f. 61.
34. *Newton Corresp.*, Vol. 4, p. 426.
35. See Halley (1679*a*).
36. See Halley (1679*b*). Kirch (1681*b*) refers to Halley in his preface: 'Portó huic anno 1681 in gratiam Lectoris Astrophili adjeci diu desideratum & summa cura correctum Catalogum Stellarum Australium Clarissimum EDMUNDI HALLEJI Angli'. [For this year of 1681 to gratify my readers who love astronomy, I add the long desired and carefully corrected catalogue of the southern stars of the eminent Englishman Edmund Halley.']

The catalogue is headed:

V.CL.EDMUNDI HALLEI CATALOGUS FIXARUM AUSTRALIUM
Quas in Insulam St Helenae summa cura observit & ad A.C. 1677 completum verificavit.
[Catalogue of the southern fixed stars of the eminent Edmund Halley
As carefully observed in the island of St. Helena and fully checked in the year 1677.]

37. Halley to Hevelius, Oxford, 11 November 1678, *Halley Corresp.*, pp. 41–2; also *Excerpta J. H.*, pp. 182–4.
38. BLL Egerton MS, 2334, f. 32, also BNP, MSS Fonds latin 10349; t. xiii, pp. 84–6.
39. See Cellarius (1660); de la Feuille (1710); Ottens (1729).
40. Flamsteed to Moore, Greenwich, 16 July 1678, *Flamsteed Corresp.*, Vol. 1, pp. 643–6.
41. Flamsteed to Newton, 11 October 1694; *Newton Corresp.*, Vol. 4, pp. 26–33.
42. Gill (1878).
43. Transits in Taton and Wilson (1989); Kepler (1627); J. Gregory (1663).
44. For the Hecker predictions, see Flamsteed, 'Hecker, his large ephemeris for the year 1674', ULC RGO 1/76, 5.
45. See Gallet (1677).
46. Horrocks (1673); see Taton and Wilson (1989), part III, pp. 194–201 for an account of lunar theory before Newton.
47. See Note 32.
48. CSP, D, 1678, p. 517.
49. CSP, D, 1678, p. 528.

Oxford University Archives, Register of Convocation 1671–83, p. 223. The entry in the Register is:

'Die Jovis viz: 28 die Novembr 1678: Cause Convocationis erat ut Literae a Serenissima Regis Matate necnon ab Illustrissimi Cancelarÿ Delegatis ut Senatum Academicum datae publicae legerentur' and, following the copy of the King's letter, 'Unanimini Consensu Venelis Coetus Literae Regiae, quae suprascribantur ratae sunt habitae. Insuper rogantae Dno Vice Cancellario, placuit Senatui Academico ut p'fatus Edmundus Hally in proxima Congregatione ad gradum magistri in Artibus juxta Literas Regias admitteretur.'

Thursday 28 November 1678: The reason for the convocation was that the letters from the King's most serene majesty and from the delegates of the most illustrious chancellory might be read publicly to the academic senate ... By unanimous consent of the meeting the royal letters that are written above were considered. Moreover, by the request of the Lord Vice-Chancellor the academic senate decided that the aforementioned Edmond Halley should be admitted to the degree of master of arts at the next congregation according to the royal letter.

Register of Admissions: Admissiones Magistrorum in Artibus, Term S. Mich. 1678 Edmund Halley è Coll: Reg. creat. Dec 3.

50. Royal mandate for Flamsteed's MA for his useful observations in astronomy: 15 May 1674, *Flamsteed Corresp.*, Vol. 1, pp. 903–4; Flamsteed had been admitted to Jesus College in 1670 and had not taken a BA.
51. Preface to Flamsteed (1680).
52. Flamsteed to Hevelius, 23 May 1678, *Flamsteed Corresp.*, Vol. 1, pp. 622–7; Hevelius to Kirch, Danzig, 20 August 1681, BNP, Lat 10349, T. xv, p. 9.
53. Halley to Hevelius, Oxford, 11 November 1678, *Halley Corresp.*, pp. 42–3; Hevelius (1685).
54. *History R. S.*, Vol. 3, 20 February 1678; see Brown (1934).
55. See de la Caille (1763) and Baily (1847).

Notes to Chapter 4

1. For Hevelius's life, see Béziat (1875); for selected letters, *Excerpta J. H*; for lunar observations, Winkler and van Helden (1993). The Hevelius brewery still operates.
2. *Excerpta J. H.*: see list of abbreviations.
   For transits, see Béziat (1875); for the variation at Danzig, see Hevelius (1670).
   Hevelius himself made a copy of his correspondence. The originals are in the Observatoire de Paris and the copies in the Bibliothèque Nationale. Of the 2700 original letters, 570 were stolen in the nineteenth century, among them four letters from Halley and one hundred of Hevelius; the Observatoire now has no letters from Halley.
3. Hevelius's publications are in the Bibliography.
4. Halley to Flamsteed, 7 June 1679, *Halley Corresp.*, pp. 42–3; *Flamsteed Corresp.*, Vol. 1, pp. 694–5.
5. See Hooke (1674).
6. Hooke (1674); Béziat (1875), 617, 618.
7. For the correspondence between Hevelius and the Society leading to the visit see Hevelius (1685) and *Excerpta J. H.*, pp. 182–7.
8. Halley to Hevelius, Oxford, 11 November 1678, *Excerpta J. H.*, pp. 182–4, *Halley Corresp.*, pp. 41–2; the letters accompanying copies of *Machina coelestis* are in OdP, MSS C1, T. 13, nos. 1940, 1941, 1945 (Greeve), 1947, 1948 (Flamsteed). See also Hevelius to Flamsteed, Danzig, 14/24 April 1679, *Flamsteed Corresp.*, Vol. 1, pp. 684–9.
9. Cluver to Hevelius, 4 March 1679, in *Excerpta J. H.*, pp. 184–5; Croone to Hevelius, 3 April 1679, BNP, Fonds latin 10349, t. xiv, f. 64; *Excerpta J. H.*, p. 185; *History R. S.*, Vol. 4.
10. Hevelius reported Halley's arrival in Danzig in letters to Croone, 10 June 1679, *Halley Corresp.*, pp. 188–9; to Greeve, 21 June 1679, *Halley Corresp.*, pp. 189–90; and to Cluver, 22 June 1679, Hevelius (1685), pp. 100–1. See also Wallis to Hevelius, 8 July 1679, *Excerpta J. H.*, p. 189.
11. *History R. S.*, Vol. 4, p. 475, 8 April 1686; for the letter of 27 March 1686, in which Halley revealed his view to Molyneux, *Halley Corresp.*, pp. 57–60. Halley most likely took with him his small quadrant that he had on St Helena. He may have had a second instrument for it seems that he left one with Hevelius on his departure, whereas he is supposed to have sold his St Helena instrument to Peter the Great (see Note 86 below). For William Molyneux FRS (1656–98) see the DNB.

12. Halley to Hevelius, 8/18 July 1679. This letter is transcribed in BNP, Fonds latins 10349 t. xiv, pp. 25–28 and has been printed a number of times: Hevelius (1685), pp. 101–2; *Biogr. brit.*, Vol. 4, p. 2498, Note [K]; *Halley Corresp.*, pp. 44–5; *Excerpta J. H.*, pp. 187–9.
13. Halley to Flamsteed, 7 June 1679 (os), *Flamsteed Corresp.*, Vol. 1, pp. 694–6. The great merchandise crane of Danzig was the largest in Europe. It may still be seen in its restored form.
14. Hevelius to Flamsteed, 14/24 June 1676, *Flamsteed Corresp.*, Vol. 1, pp. 458–73; also in Hevelius (1685), pp. 67–74.
15. See Note 13.
16. See Note 12.
17. The Latin reads:

    Quippe qui hisce oculis vidi, non unam vel alteram, sed etiam plures observationes Stellarum Fixarum Sextante magno Orichalcico, A diversis etiam Observatoribus, quandoquam etiam a me ipso, licum paret exercitato, peractas ac amota regula repitatas accuratissime, atque fere incredibiliter inter se convenire, ac nunquam nisi temnenda intere se discrepare; quod an majori gaudio an admiratione exceperim nescio.

18. *Hooke D1*, 14 August 1679.
19. For death of Moore, see *Hooke D1*, 2 September 1679; Halley to Olhoff, ? October 1679, *Halley Corresp.*, pp. 45–7.
20. Halley to Aubrey, Oxford, 16 November 1679, *Halley Corresp.*, pp. 47–8; *Hooke D1*, 20 November 1679.
21. Flamsteed to Hevelius, 7/17 October 1681, *Flamsteed Corresp.*, Vol. 1, pp. 822–5.
22. Halley to Hevelius, 5/15 November 1681, and 7/17 April 1682; Hevelius to Halley, 9 January 1682, *Halley Corresp.*, pp. 53–4, 54–5, 195–6.
23. Hearne (1884–), Vol. 4, 13 November 1713. The book Hearne mentions is Hevelius (1685), which was in Bodley's Library.
24. As expressed in the letter to Hevelius—see Notes 12 and 16.
25. Wallis to Flamsteed, 12 February 1686, Molyneux to Flamsteed, 20 February 1686, both *Biogr. brit*, Note p. 2499.
26. Halley to Molyneux, 27 March 1686, *Halley Corresp.*, pp. 57–60; Molyneux to Halley, 8 April 1686, *History R. S.*, Vol. 4, pp. 475–9; Halley to Molyneux, 27 May 1686, *Halley Corresp.*, pp. 64–5.
27. Halley to Hevelius, autumn 1686, *Halley Corresp.*, pp. 72–3. The opening shows that Halley had received two letters from Hevelius on 9 July and 17 September.
28. Halley to Flamsteed, Danzig, 7 June 1679, *Flamsteed Corresp.*, Vol. 1, pp. 694–6.
29. On the life of Cassini, see Cassini (1994). See Halley (1694c) for his criticism of Cassini's view.
30. See Picard (1671).
31. *History R. S.*, Vol. 3, p. 482, 15 May 1679.
32. During 1680 he had continued to make occasional observations of the Moon, among them an occultation of Aldabaran: see ULC, RGO 2/9, ff. 25, 26.
33. See Brokesby (1715).
34. See Secretan (1860).
35. BLO Pamphlet Θ 665, ff. 97, 98.
36. Nelson (1704).
37. For Nelson eliciting subscriptions, see his letter to Pepys in Secretan (1860), pp. 67–9. The church, which probably came out of Wren's office, was torn down at the beginning of the twentieth century. The panelling behind the altar is now in the hall of Selwyn College, Cambridge, and the pulpit is in Lincoln Cathedral: see Royal Commision on Historical Monuments, England, 1959, *An inventory of the historical monuments in the city of Cambridge*, Part II. (H.M.S.O., London.); *History of Lincoln Cathedral* (Cambridge University Press).

38. Nelson to Pepys, Secretan (1860), pp. 67–9.
39. See Goldie (1993); Rose (1993).
40. For letters of Tillotson to Nelson prior to the end of 1680, see BLL Add 4236, ff. 217, 219, 221, 223; see also Birch (1752).
41. BLL Add 4236, f. 216.
42. Halley to Hooke, 5/15 January 1681, *Halley Corresp.*, pp. 48–9 and R.S. Letter Book.
43. BLL Add 4236, f. 225, r.,v.
44. '*Memoir*' in *Halley Corresp.*, pp. 1–13.
45. Flamsteed to Crompton for Newton, 7 March 1680/1, *Flamsteed Corresp.*, Vol. 1, pp. 771-9.
46. ULC Add 4004, ff. 97r., 99r., 101v:

> Decemb 8 stylo veteri Hallius noster tempore matutino Parisias versus iter faciam prope Bolonia ...
> and later,

> Halleius mihi narravit se iter Parisias istituantem Dec 8 stylo veteri caudauer cometae vidissi perpendiculariter ex horizante surgentem ad instar igneae ad longitudinem decem vel majore qui decimi graduum paulo ante ortium solis. Quodque cauda haec non prius disperatet quam sol oriens inciparat conspici: ad solem autem fulgarem mox evanesceret. Et quod cauda a corpore solis exire videretur, ita ut caput comete esset soli proximum. Denique quod ipse quid esset hoc Phaenomenon.

47. Halley to Hooke, Saumur, 19/29 May 1681, *Halley Corresp.*, pp. 49–52; see also *History R. S.*, Vol. **4**, p. 89, 1 June 1681.
48. Tillotson to Nelson, 5 July 1682, à l'hostel Imperial, rue du Four, Faubourg St Germain, À Paris BLL Add 4236, f. 235.
49. Evelyn, *Diary*, December 1643.
50. See Hillairet (1962).
51. See Le Gras (1685).
52. The books that Newton had were Mabillon (1704), Montfaucon (1702), and Muratori (1717); see Harrison (1978).
53. The letter, strictly one to Justel taken by Halley, was read to the Royal Society on 2 December 1680: see *History R. S.*, Vol. **4**, p. 60.
54. See Brown (1934) for the circles of Montmor and Justel.
55. For the cardinal and his brother, see *Nouvelle biographie générale* (Paris, 1859), Vol. **29**, p. 822. For Aubrey's account, see Aubrey (1958). The cardinal had two 14 ft (4.3 m) globes made by the Venetian geographer Coronelli—see letter from Justel, 24 November 1686, *History R. S.*, Vol. **4**, p. 503.
56. Tillotson's letter is in BLL Add 4236, f. 227, r. and v. The comet of 1680–1681 had also been seen by Evelyn who, like Tillotson, anticipated great disaster. For the observations of Hill, see *History R. S.*, Vol. **4**, p. 66.
57. *Halley Corresp.*, p. 18; Mme de Sévigné, *Lettres*, Vol. **II**, p. 893 (Paris, ed. Gérard-Gailly, 1960).
58. Cassini (1681) recorded, p. 9: 'M. Hallei excellent Astronome de la Societé Royale d'Angleterre, qui a esté present à la plupart de mes observations ...' ('Mr Halley, an excellent astronomer of the Royal Society of England, who has been present at most of my observations ...') see also Flamsteed to Caswell, 4 February 1681, *Flamsteed Corresp.*, Vol. **1**, pp. 752-5; Flamsteed to Towneley, 7 February 1681, *Flamsteed Corresp.*, Vol. **1**, pp. 756–9. Halley arrived in Paris on Christmas Eve (ns) and first observed with Cassini shortly afterwards; he was probably present at all the observations from then on until the comet faded. In ULC RGO 2.9 ff. 25–6 there are some observations of Halley's in Paris (not Islington). Cassini implies that Halley computed many of the positions of the comet.
59. Halley to Hooke, 5/15 January 1681, *Halley Corresp.*, pp. 48–9.

60. Halley to Hevelius, Rome, 5/15 November 1681, *Halley Corresp.*, pp. 53–4.
    Halley's collection probably incorporated the material assembled by Cassini (1681). Halley to Flamsteed, 22 January 1681, *Flamsteed Corresp.*, Vol. 1, p. 751. That letter is lost but the observations are given in Flamsteed's letters to ?Caswell of 4 February, *Flamsteed Corresp.*, Vol. 1, pp. 752–5 and to Towneley of 7 February, *Flamsteed Corresp.*, Vol. 1, pp. 756–9. Flamsteed to Newton, 26 September 1685, *Newton Corresp.*, Vol. 2 pp. 421–5.
    Flamsteed stated that Gallet observed in Rome, and so Newton took it in the first edition of the *Principia*. Gallet in fact was in Avignon (Cassini 1681). Newton later changed Rome to Avignon. Surprisingly Halley did not correct the MS of the first edition (Chapter 6).
61. Flamsteed to Halley, 17 February 1681, *Flamsteed Corresp.*, Vol. 1, pp. 760–4.
62. Halley to Hooke, Saumur, 19/29 May 1681, *Halley Corresp.*, pp. 49–52; Cassini (1681).
63. ULC Add 4004 101v.; the gaps are in Newton's note. Richer (1679).
64. See Appendix 8; Harrison (1978).
65. Halley to Hooke, Saumur, 19/29 May 1681, *Halley Corresp.*, pp. 49–52.
66. *Procès-Verbaux de l'Académie Royale des Sciences*, Vol. 9, f. 99v. (Paris).
67. Tillotson to Nelson, 2 June and 7 November 1681, BLL Add 4236, ff. 231,233.
68. For Tillotson to Nelson and Savile's regard for Halley, see Birch (1752).
69. Halley to Hooke, Saumur, 19/29 May 1681, *Halley Corresp.*, pp. 49–52.
70. OdP, MS D.1.8.
71. See Zuber and Theis (1986).
72. For de Moivre, see Schneider (1968). De Moivre was born on 26 May 1667 at Vitry-en-Champagne (le Francois) where his father was a dentist. At twelve he went to the Hugenot Academy at Sedan and stayed until it was suppressed. He moved to Saumur in 1681 and to Paris to the Collège d'Harcourt in 1684.
73. Some of that correspondence is referred to in later chapters.
74. Secretan (1860), pp. 17–18.
75. See Flamsteed (1725), Vol. 1, pp. 281–2. Presumably Flamsteed had had a letter from Halley; see *Flamsteed Corresp.*, Vol. 1, p. 808.
76. Photograph from Dr James Keeler.
77. Halley to Hevelius, Rome, 5/15 November 1681, *Halley Corresp.*, pp. 53–4.
78. For the library of Santa Maria in Vallicella, see Giorgetti Vichi and Mottironi (1961); for the observatory of Pontio, see Rotta (1990).
79. For Christina's library, see Odier (1962).
80. For Christina in Rome, see Akerman (1991), di Palma and Bovi (1990).
81. For Christina's observations of the comet of 1664, see Cassini (1994). For Roman observations of the comet of 1680, see Cassini (1681) and Rotta (1990).
82. See Akerman (1991).
83. Royal Society, unpublished paper in *Halley Corresp.*, pp. 166–7.
84. See Johns (1993), Ciampini (1693).
85. RSL JB, Vol. 7, 4 July 1688 and 13 November 1689.
86. Pullein to Hooke, Rome, May 1680, RSL LBC, Vol. 8, pp. 165–9, also 28 May 1680, pp. 170–5 and 6 November, pp. 202–3.
87. Halley to Hevelius, Rome, 5/15 November 1681, *Halley Corresp.*, pp. 53–4: 'Breve redditum in Patriam meam quam assequi spero circa medium Januarium cumque non sine insigni meo infortunis litterae vestrae quas Londino.'
88. See *Biogr. brit.*
89. See Halley (1731).
90. *Hooke D1*, 24 January 1682.
91. Islington observations in Streete (1716); see Chapter 5; ULC RGO 219, observations in Islington.

92. Swedenborg to Bezelius, London, April 1711, Tafel (1875), Vol. 1, pp. 207–12.
93. For Hevelius's opinion see his letters to Grieve and to Flamsteed, Note 4 and Flamsteed (1680) after observations of 24 May.

    Gregory, in a letter to Newton, (27 August 1691, *Newton Corresp.*, Vol. 3, pp. 165–8) says Flamsteed preferred the work of French Jesuits in *Observations physiques et mathématiques par l'Académie Royale des Sciences* (Paris, 1688); Flamsteed to Newton, (24 February 1691/2, *Newton Corresp.*, Vol. 3, pp. 199–205) refers to incompleteness and haste and defects of sextant observations; Flamsteed to Pound (15 November 1704, *Newton Corresp.*, Vol. 4, pp. 424–8) urges observations of meridional distances of southern stars because Halley left many unobserved, had no good assistant, there were faults in his records, and he did not repeat the calculations; but adds 'without derogating from Mr Halley's previous labours'.
94. OdP, MSS E.1.3, E.1.9.
95. See Chapman (1990) for random errors of 10″. Systematic errors arise from any non-uniform graduation of the scale: if they were not to exceed 1′, the scale would have to be correct to 0.6 mm round the whole quadrant. A precision of 1 mm was beyond the capability of those days. When Graham divided Halley's new mural quadrant just after 1720, he laid it out horizontally on the floor and afterwards raised it vertically for mounting. It could well have distorted by up to 1 mm at some point of the arc.

    There would be another systematic error if the pivot around which the collimating sights rotated were not at the centre of the graduated arc. For a precision of 1′, the axis must be correctly centered to within 30 μm.

    A mural quadrant was supposed to be mounted on a wall aligned in the meridian within millimetres. A correction could be determined from observations of stars, but the wall might shift in time, as Flamsteed found; he had to determine corrections for misalignment every year. When Halley installed his own quadrant at Greenwich, he put the wall on very firm foundations and built it of just nine large stone blocks as may still be seen.

Notes to Chapter 5

1. *History R. S.*, Vol. 4; *Hooke D1*, 24 January 1682; Aubrey (1958), Vol. 1, p. 283.
2. Halley to Hevelius, Rome, 5/15 November 1681, *Halley Corresp.*, pp. 53–4.
3. See also his comment in Halley (1731).
4. *James* Tooke was Auditor of the Court of Wards and Liveries. *John*, the eldest son, matriculated from Pembroke College, Cambridge, was admitted to the Inner Temple 1635–1636, barrister 1644. He was granted the reversion of office of auditor in 1637 (PRO), but may have died before he could take it and it may then have gone to Mary's father. *Charles*, the second son, was admitted to the Inner Temple 1638, barrister 1648. *Edward*, the third son, matriculated from Christ's College, Cambridge, in 1639 (aged 15), admitted to the Inner Temple 1639, barrister 1648, was Librarian to the King. Another Tooke, *Ralph*, second son of *William*, admitted to the Inner Temple 1629, barrister 1637, bencher 1652, may have been a relative. See Venn, (1922–7) and *Members admitted to the Inner Temple 1547–1660*.

    Will of Christopher Tooke, pr. 3 March 1662, PRO PCC reg Juxon, f. 42.

    Complaint of Edmond Halley and Mary, Anthony English and Dorothy and Robert Peirson and Elizabeth *vs*. Charles Tooke and Francis Bostock Fuller: PRO Chancery C7, Bridges 367/19, 27 June 1683.

5. Richard Young vs. Robert and Joane Cleator and Edmund Halley, answer of Halley: PRO Chancery C9, Reynardson 142/57, 28 December 1693.
6. An Edward Tooke, possibly a relative, became a Salter by patrimony in 1698. Benjamin Tooke, a notable City printer, was probably not a relation of Mary. His son Andrew became Gresham's professor of astronomy and the Royal Society met in his rooms at Gresham College. Benjamin printed the account of the trial of Laurence Braddon and Hugh Speke that would figure in Halley family history: see Tooke (1684).
7. Pepys helped to broker such a marriage between Philip, the son of his colleague Sir George Carteret, and Lady Jemima, the daughter of the Earl of Sandwich: *Diary*, July 1665.
8. See Earle (1989). Cookes, a clan probably acquainted with the Halleys, appear very often in the registers of St James's (Phillimore and Cockayne 1900) around 1680. There were Cookes on the Courts of Assistants of the Levant Company, and of the Royal African Company when Halley was diving for them (Chapter 9). By such associations were Halley and his wife linked to influential inhabitants of the City (Chapter 2).
9. Pepys described the rather sober wedding of Philip Carteret and Lady Jemima Montague, *Diary*, 31 July 1665.
10. See Halley (1716a); the appendix to Streete (1716).
11. Schedule to Chancery action, Cleeter, Robert and Joane his wife vs. Buckworth, Sir John, Kt, Young, Richard and Halley, Edmund, All Hallows, London Wall, Personal Estate of Edmund Halley, Bill, answer and schedule: PRO Chancery C10, Whittington 222/27, April 1686.
12. For Golden Lion Court, see Flamsteed to Molyneux, 2 May 1684, in Eggen (1958): Halley may move back into London.
13. Fuller was the brother-in-law of Edward, according to Edward's will. An Edward Bostock Fuller was a colleague of the surveyor Gregory King; in 1687 they produced a manuscript map of 'the Hospital and Precint of St Catherine's near the Tower of London', now in the Guildhall Library.
14. Complaint of Edmond Halley and Mary Anthony English and Dorothy and Robert Peirson and Elizabeth vs. Charles Tooke and Francis Bostock Fuller: see Note 4; Charles Tooke and George Tooke vs. Francis Fuller, answer of Francis Fuller: PRO Chancery C9, Reynardson 94/124, 26 November 1684.

   Will of Christopher Tooke, pr. 3 March 1662: PRO, PCC reg Juxon, f. 42.

   Edward's estate included houses in Moorgate Street, in Drury Lane, Lutenor's Lane, and Crosse Street, and a debt of £200 owed by Christopher. The executors refused to sell the houses and asserted that other legacies, including one owing to Mary's aunt Ursula, had priority.
15. ULC, RGO 2/9, ff. 189–164 (the pages have been reversed in binding), ff. 25, 26. Halley printed the observations in the third edition of Streete (Halley 1716a).
16. See Halley (1716a); the appendix to Streete (1716).
17. See Halley (1731).
18. Election to Council: 30 November 1683, *History R S*, **4**, 231.
19. See Halley (1716a); the appendix to Streete (1716); and Halley (1731).
20. Flamsteed to Molyneux, 12 May 1684, Eggen (1958).
21. Parish Register, St George, Southwark, 2 June 1685.
22. PRO Chancery C10, Whittington, 222/27, as in Note 11.
23. The broadsheet is reproduced in *Halley Corresp.*, pp. 176–7.
24. Originally a manor of the Knights Templar and later in the possession of Denny Abbey to the north of Cambridge; it belonged in the seventeenth century to a Rochester family: Hasted (1778–99), Vol. 1, p. 438. It is now in the care of English Heritage. The poor boy passing by, no doubt beachcombing, was a John Blyew. He

received £20 of the reward and the balance was paid to the overseers of the parish in trust for him—Cleeter and Cleeter *vs.* Halley and Young, 1694, PRO C.7 (Hamilton) 60/53.
25. Ferguson (1684). There were at least two versions in English, and French and Dutch translations. References to the death of Halley's father show that it was written after it, but probably not long after.
26. See, for instance, Ashcraft (1986), McDonald (1991), Greaves (1992). For the 1685 tract and its relation to the death of Charles II, see Braddon (1725). The earl's throat was slit from ear to ear, apparently the typical wound of a suicide.
27. See Tooke (1684).
28. Dr William's Library, London, Vol. Q, pp. 440–1; Morrice was an ejected Puritan minister of the parish of St Giles's Cripplegate, who kept a diary of his times. The *Book* is being edited by Dr Mark Goldie to whom I am indebted for an extract.
29. *Hooke D2*, 22 January 1689; House of Lords Journal for 23 January 1689; House of Lords Committee Book, 23 January 1689 *et passim*. See also Chapter 9, Note 1.
30. CSP, D, 12 April 1681 (os).
31. CSP, D, 1689–1690, 1 March 1690 (os).
32. Braddon (1690); Somers (1801–15), Vol. **10**, pp. 72–174; Braddon (1725).
33. Braddon (1725) gives his name as Broom.
34. For the action before Jeffreys about the boy's reward, see *Halley Corresp.*, p. 178. Joane or Cleeter paid the £100—Halley *vs.* Cleeter, Cleeter, and Young, PRO C.6 Collins 301/45.
35. Hearne (1884–), Vol. 3, p. 473; Vol. **4**, p. 231.
36. Ferguson, a fellow Scotsman with Gregory and Arbuthnott (Chapter 12), later became a Jacobite.
37. June 1684, Edmund Halley [father], Administration by Sir John Buckworth and Richard Young, Prerogative Court of Canterbury, Admon (PRO).
38. See Halley (1731).
39. For the double dealing of those days, especially over the relations with Louis XIV, see Scott (1991).

Notes to Chapter 6

1. Halley is said to have claimed in later years that he was the Ulysses that brought that Achilles (Newton) out of his tent. According to a memoir of de Moivre, now in the Joseph Halle Schaffner Collection of Scientific Manuscripts, University of Chicago Library, reproduced as Plate 1 of Chandrasekhar 1995, 'Dr Halley has often valued himself to me for having been the Ulysses who produced that Achilles'.
   Halley's actions correspond to none of the classical accounts of Ulysses and Achilles, except possibly in the *Achilleid* of Statius (*c.* AD 50–95): see Dilke (1954).
2. For Halley's contributions to the Royal Society, see *History R. S.*, Vol. 4, pp. 240, 244, 260–1, and Halley (1682, 1683*a,b*, 1684).
3. Halley to Newton, 29 June 1686, *Newton Corresp.*, Vol. 2, pp. 441–4.
4. See Westfall (1980), p. 382.
5. Joseph Halle Schaffner Collection, University of Chicago Library, MS 1075–7.
6. Hooke–Newton exchanges, December 1679–January 1680, *Newton Corresp.*, Vol. 2, pp. 304–13; Newton to Halley, 14 July 1686, *Newton Corresp.*, Vol. 2, pp. 444–5.
7. Westfall (1980), p. 403.

8. Edward Paget FRS, 1656–1703, fellow of Trinity College, Cambridge, Mathematical Master at Christ's Hospital 1682.
9. For curvature and its relation to Newton's concept of force in the 1690s, see Brackenridge (1992), pp. 231–60; and Brackenridge (1995).
10. Westfall (1980) on *De motu*, pp. 408–16.
11. *History R. S.*, Vol. 4, p. 347, 10 December 1684.
12. *History R. S.*, Vol. 4, April 1686; Halley (1686*a*).
13. Minutes of the Royal Society: *History R. S.*, Vol. 4, pp. 479–80; Halley to Newton, 22 May 1686, *Newton Corresp.*, Vol. 2, pp. 431–3.
    Nathaniel Vincent (?1639–1697), a friend of James II, was senior fellow of Clare College; he withdrew from the Royal Society at the beginning of 1687.
14. See Cohen (1971), Westfall (1980), Whiteside (1967–81, 1994).
15. Halley to Molyneux, 27 March 1686, RSL LBC, *Halley Corresp.*, pp. 57–60.
16. *History R. S.*, Vol. 4, p. 434, 27 January 1686.
17. *History R. S.*, Vol. 4, p. 435, 27 January and 3 February 1686; see subsequent entries in *History R. S.*, Vol. 4, for Halley's confirmation and his salary. Early in 1687 he asked for a review of his salary (*History R. S.*, Vol. 4, p. 524) and on 6 July it was decided that it should be £50 with a gratuity for the past year of £20, all paid in copies of *The history of fishes* (Willughby 1672): see *History R. S.*, Vol. 4, p. 545. For Sloane and his collections, see MacGregor (1994).
18. In later years Halley had his salary paid in money. See Minutes of the Council of the Royal Society, 22 October 1690, 13 May 1691, 16 November 1692, 21 November 1694, 29 April 1696, 6 July 1698; all those payments were a year or more in arrears.
19. *History R. S.*, Vol. 4, p. 484, 29 November 1686.
20. Letter to Cassini in OdP, B. 4–10. The letter is dated Londres, 8 de Juin, with no year, but probably of 1686—see Débarbat (1990).
21. *History R. S.*, Vol. 4, p. 554, 19 May 1686.
22. Halley to Newton, 22 May 1686, *Newton Corresp.*, Vol. 2, pp. 431–3.
23. *History R. S.*, Vol. 4, p. 486, 2 June 1686 and p. 491, 30 June 1686.
24. For the estimate of 300–400 copies, see Munby (1952); for the second edition, Black (1984), p. 98. Whiteside (1994), who thought that 500 copies might have been printed, estimated that Halley would have made a profit of about £10. Professor Owen Gingerich (personal discussion) thinks as many as 600 may have been printed.
25. Halley to Newton, 5 July 1687, *Newton Corresp.*, Vol. 2, pp. 481–2.
26. Halley to Newton, 7 June 1686, *Newton Corresp.*, Vol. 2, pp. 434–5.
27. Newton to Halley, 20 June 1686, *Newton Corresp.*, Vol. 2, pp. 435–41.
28. Cohen (1971), Chapters 4, 8, and *Supplement VII*; Koyré and Cohen (1972). Whiteside (1994) has maintained that Newton did not see all of Halley's amendments to M until after the printing was finished.
29. Halley to Newton, 29 June 1686, *Newton Corresp.*, Vol. 2, pp. 441–3.
30. Newton to Halley, 14 and 27 July 1686, *Newton Corresp.*, Vol. 2, pp. 444–5, 446–7.
31. Halley to Newton, 14 October 1686, Newton to Halley, 18 October 1686, *Newton Corresp.*, Vol. 2, pp. 452–5.
32. Newton to Halley, 13 February 1687, Halley to Wallis, 11 December 1686, Wallis to Halley, 14 December 1686, *Newton Corresp.*, Vol. 2, pp. 456–66; Wallis (1687); Halley (1691c).
33. Halley to Newton, 24 February 1687, Newton to Halley, 1 March 1687, *Newton Corresp.*, Vol. 2, pp. 469–71.
34. Halley to Newton, 14 March and 5 April 1687, *Newton Corresp.*, Vol. 2, pp. 473–5.
35. Halley to Wallis, 9 April and 25 June 1687, *Halley Corresp.*, pp. 80-2, 85.
36. Halley to Newton, 5 July 1687, *Newton Corresp.*, Vol. 2, pp. 481–2.

37. Débarbat (1990); Molyneux to ?Halley, Dublin, 7 July 1687, RSL LBC 11(1), 101—this letter has no superscription, but is undoubtedly to Halley 'my esteemed friend' for it refers to his chart of the trade winds and comments 'We are beholding to you already for ye Survey of a great part of the Heavens'; Reiselius to Halley, 23 April 1691, RSL LBC 11(1), 175; W. Rooke to Halley, 2 February 1688, RSL LBC 11(1), 114.

38. Newton, *Philosophiae naturalis principia mathematica*, Book I, Prefatio:

> In his edendis Vir acutissimus & in omni literarum genere eruditissimus *Edmundus Halleius* operam navavit, nec solum typothetarum sphalmata correxit & schemata incidi curavit, sed etiam auctor fuit, ut horum editionem aggrederer. Quippe cum demonstratam a me figuram orbium caelestium impetraverat, rogare non destitit, ut eandem cum Societate Regali communicarem, que deinde hortatibus & benignis suis auspiciis effecit ut de eadem in lucem emittenda cogitare inciperem.

The English translation is from Motte (1729).

39. See Appendix 6 for the Latin original; a less exact translation was published by 'Eugenius' in the *General Magazine of Arts and Sciences* (London, 1755, 1, 4: *Halley Corresp.*, pp. 207–8).

40. Leibniz to Justel, Hannover, 3 June 1692, Leibniz (1970). See Conduitt in his Memoirs of Newton, King's College, Cambridge, MS 129, for the remarks on Tycho and others, and for the report of Newton's comment on Bentley.

41. See R. Smith (1995), Ch. 4, for the sublime in Handel's oratorios.

42. Halley to Wallis, 9 April 1687, *Halley Corresp.*, pp. 80–2; Halley to Sturm, after 27 March 1686, *Halley Corresp.*, pp. 62–4; Halley to Reiselius, 19 July 1686, *Halley Corresp.*, pp. 68–9; Flamsteed to Towneley, 14 November 1686, RSL MS 243.68.

43. A copy of the letter to the King was published in London in 1687 and reprinted 'by demand' in *Philosophical Transactions*, 19, (1695–7), 445–57, as *The true theory of the tides, as extracted from that admired treatise of Mr Isaac Newton, intituled* Philosophiae naturalis principia mathematica, *being a discourse presented with that book to the late King James, by Mr Edmund Halley*.

44. Cooper (1845), Vol. 3, p. 614; Goldie (1996).

45. Halley (1687a).

46. See Cohen (1971), Ch. 6, pp. 145–58; Cohen (1990), pp. 91–108.

47. Cohen (1971), as in Note 46.

48. Cotes to Newton, 23 October 1712, *Newton Corresp.*, Vol. 5, p. 351.

49. The copy of the first edition in King's was bought by Maynard Keynes from David, the Cambridge bookseller, when he was an undergraduate. David had found it on a stall on the Old Kent Road, and Keynes paid very little for it. The handwriting of the annotations had often been suspected to be Halley's and that was eventually confirmed (personal communication to the author) by the handwriting expert, Peter Croft, when he was Librarian of King's in the 1980s.

A marginal note implies that the annotations were made from a manuscript of Newton's. They agree closely with the corresponding annotations in Newton's own copy in Trinity College, Cambridge, (E$_{1a}$ in Koyré and Cohen (1972)), but the later sections of Book III are not annotated and the annotations seem to have been made before Newton had completed his revision for the second edition.

50. Bentley to Newton, 30 June 1713, *Newton Corresp.*, Vol. 5, p. 413; de Moivre to Bernoulli, 28 June 1714, *Newton Corresp.*, Vol. 6; N. Bernoulli to Newton, 22 May 1717, *Newton Corresp.*, Vol. 6, pp. 388–9.

51. Halley to Caswell, 9 July 1686, *Halley Corresp.*, p. 67; Halley to Reiselius, 19 July 1686, *Halley Corresp.*, p. 68–9. The reference to the shirt is:

> Nuperrime etiam vidimus Indusium ex India advectum, nullis suturis, solummodo textoris artis concinnatum, cujus generis fortassis erat Vestimentum istud Servatoris nostri Jesu Christi, cujus mentio fit in Sacris Literis.

52. The following letters are printed in *Halley Corresp.*, pp. 55–87: Halley to St George Ashe, 27 March 1686; Halley to Hayley, 1686; Halley to Hevelius, 1686; Halley to Leeuwenhoek, 2 March and another, 1686; Halley to Molyneux, 27 March 1686; Halley to Sturm, 1686; Halley to Valvasar, 1688; Halley to Wallis, 9 July and 11 December 1686, 1 January, 15 February, 9 April, and 25 June 1687.
53. Halley's papers in this period are of 1684, 1686a,b,c,d, 1687a,b,c,d.
54. *Halley Corresp.*, pp. 70, 71; *History R. S.*, Vol. **4**, p. 491. For a review of early geodesy in England, see Cook (1976); Norwood (1676).
55. RSL JBC; *History R. S.*, Vol. **4**, 11 July 1686.
56. Halley to Caswell, 9 July 1686, *Halley Corresp.*, p. 67; Halley to Wallis, 13 November 1686, RSL LBC **11**(1), 73.
57. *History R. S.*, Vol. **4**, 7 June 1681; for Wren's ideas about standards of length, see Bennett (1982), p. 50.
58. *History R. S.*, Vol. **4**, pp. 434–5, 28 October, 25 November 1685; RSL JBC 7, 1688, 4 July.
59. *History R. S.*, Vol. **4**, 28 May 1688, Longitude; 15 July, Equation of time.
60. Flamsteed to Towneley, 4 November 1686, in Eggen (1958).
61. *Flamsteed*, pp. 193–4.
62. Flamsteed to Towneley, Willmoth (1994), 4 November 1686, RSL MS 243.68.
63. Perkins spoke of his work before the Royal Society on three occasions: 29 January, 23 February, 4 March 1680, *History R. S.*, Vol. **4**, pp. 7, 18–19, 21–22. He was asked by the President to collect journals of voyages, something that Flamsteed and Halley both did in later years: 11 March 1680, *History R. S.*, Vol. **4**, p. 24.
64. Halley (1683b).
65. See Eggen (1958).
66. Hawking and Israel (1987), Chen and Cook (1993); for the reaction of Leibniz, see Bertoloni Meli (1993).
67. See Cook (1994).
68. De Morgan (1847).

Notes to Chapter 7

1. For Halley's re-election: RSL JBC 9, 30 November 1700.
2. For the early history of the Royal Society, see for example, Hall (1991), Hunter (1982, 1989). The Académie Royale des Sciences became a very different body from the Royal Society, its members had pensions and were engaged on corporate activities, as fellows of the Royal Society never were.
3. Brouncker, Pepys, and Williamson were administrators, Pepys and Williamson musicians, and Brouncker a courtier who was an even more avid womaniser than Pepys: for the story of Brouncker and two maids of honour, see Chapter 2, n. 35.
4. First Minute Book; Oldenburg to Richard Norwood, 6 March 1664, *Oldenburg Corresp.*, Vol. 2, p. 146.
5. There are full records in the Journal Books, Council Minutes, and Letter Books of the Royal Society.
6. For the letter to Cassini, see Débarbat (1990); for the correspondence with Leibniz, Justel to Leibniz, London, 25 March 1692, Leibniz (1964); Leibniz to Justel, Hannover, 24 May 1692, Justel to Leibniz, London, 21 June 1692, Leibniz (1970); Leibniz sent a rather flowery letter to Halley with the letter of 24 May 1692.
7. W. Hayley to Halley, n.d, RSL LBC **11**(1), 88; W. Hayley to Halley, 24 February 1688, RSL LBC **11**(1), 112; Halley, RSL LBC **11**(1), 107.

8. Bryden and Simms (1993).
9. RSL JBC **9**, 6 November 1700, 26 February, 11 March 1701; 19 March 1701, *Halley Corresp.*, pp. 167, 168.
10. The last paragraph is somewhat puzzling: the bound copies of *Philosophical Transactions* in the University Library at Cambridge contain Numbers 177 and 178 dated December 1685, and Volume 15 runs from Number 167 of January 1685 to Number 178. There is no evidence of interruption.
11. *Halley Corresp.*, pp. 135–70, 210–40.
12. Halley (1697*b*); RSL JBC **9**, 4 October, 4 November 1696, 10 and 31 March 1697 (all dates os as in JBC).
13. Halley (1697*c*,*d*); RSL JBC **9**, 9 May and 27 October 1696 (dates os).
14. Halley (1698); RSL JBC **9**, 8 June 1698.
15. Halley (1696*b*); Wallis to Waller, 2 July 1695, RSL LBC **11**(1), 58.
16. Halley (1693*b*).
17. *Lettres de M. Auzout, Voyages de M. Cassini observation envoyées des Indes et de la Chine.* (P. Mortier, Amsterdam, 1736.)
18. Halley (1687*b*,*d*, 1694*a*); some of the letters in the Macclesfield collection are printed by *Rigaud*, Vol. **1** and 2.
19. Waller to Lloyd, 6 February 1694, RSL LBC **11**(1), 312–3. See also, Waller to Lloyd, 31 July 1694, RSL LBC **11**(2), 42.
20. Halley (1694*a*).
21. Logan to Jones, 31 March 1738, *Rigaud*, Vol. **1**, pp. 323–7, and 4 May 1738, *Rigaud*, Vol. **1**, pp. 327–32; Anderson to Jones, 10 May 1739, *Rigaud*, Vol. **1**, pp. 342–6. Logan was a Welshman who had settled in Pennsylvania and had some mathematical competence. Anderson seems to have been a craftsman.
22. Halley (1695*b*); Wallis to ?Halley, 24 October 1692, RSL LBC **11**(1), 135; Wallis to Halley, 11 November 1695, RSL LBC **11**(1), 146; Wallis to Halley, n.d., RSL LBC **11**(1), 148; the two latter letters show that Halley wrote to Wallis on 7 November and again on 21 November. Halley suspected he had an enemy somewhere and thought it might be Caswell, but Wallis assured him that was not so. Wallis urged Halley to persuade Newton to publish his book on light and colours—it only came out in 1703 after Hooke had died.
23. Anderson to Jones, 21 July 1736, *Rigaud*, Vol. **1**, pp. 293–7.
24. Halley (1696*a*).
25. Halley (1692*a*).
26. Halley (1695*d*).
27. Luttrell (1857) mentioned Halley as 'the famous mathematician' long before he edited *Apollonius*: July 1699, Vol. **4**, pp. 352, 358; 14 September 1700, Vol. **4**, p. 687; 24 October 1702, Vol. **5**, p. 228 and 11 January 1704, Vol. **5**, p. 379.
28. For experiments on the dependence on distance from a lodestone, see *History R. S.*, Vol. **4**, p. 526, 23 February 1687 and *Halley Corresp.*, pp. 135–7; see also Kircher (1643). For a proposal to observe the variation at a winter solstice, *History R. S.*, Vol. **4**, p. 448, 2 December 1685.
29. Halley (1683*b*); Halley to Cassini, n.d., RSL LBC **11**(1), 11.
30. RSL JBC **8**, 4 and 25 November, 2 December 1691; *Halley Corresp.*, pp. 226, 227; Halley (1692*b*).
31. See Bullard (1956*a*,*b*).
32. Celsius to Swedenborg, 23 June 1741, Tafel (1875–7), pp. 578–80.
33. Halley (1686*e*); for the significance of Halley's chart, see Thrower and Glacken (1969) and chapter 10; see also, Wallis to Halley, 14 December 1686, *History R. S.*, Vol. **4**, pp. 514–15.
34. Jeffreys (1926).
35. That was first realised by Hadley (1735).

36. Halley (1694b); *Halley Corresp.*, p. 140.
37. For the circulation of water, see Halley (1691a); for the hail storm, Halley (1697b).
38. Halley (1686d).
39. *History R. S.*, Vol. 4, 472, 14 April 1686.
40. Halley to Sloane, 5 April 1697, RSL Letter Book; Caswell to Halley, 3 August 1686; account of his report to the Royal Society in 1682, RSL LBC 11(1), 70.
41. Halley (1697c); Derham to Sloane, 6 December 1697, RSL LBC 11(2), 215.
42. Halley (1693b).
43. Halley (1693e,f).
44. Halley (1700a); RSL JBC 9, 8 June 1698 (haloes), 5 January 1701 (rainbow).
45. For de Dominis, see Redondo (1989).
46. Newton (1704), Book I, Part II, Prop. 9, Prob. 4.
47. See also Hermann (1704) and Whiteside (1967–81), Vol. 3, pp. 499–503. Hermann's letter to J. J. Scheuchzer of 5 June 1703 is in Zürich, Zentralbibliothek, MS H 318.
48. Halley (1691e); Kaye and Laby (1995), p. 275.
49. Halley (1693a); Heywood (1994); Halley to Neumann, n.d., 1693, RSL LBC 11(2), 1; Neumann to Halley, 1 March 1694, RSL LBC 11(2), 2.
50. Graetzer (1883).
51. Halley (1706a).
52. Heywood (1994). Halley's study of the Breslau tables was probably known to Franklin in Philadelphia: see Cohen (1995).
53. For Sloane and the foundation of the British Museum, see MacGregor (1994).
54. RSL JBC 7, 15, 22 January 1690; Halley (1691b).
55. See Airy (1863).
56. *Halley Corresp.*, pp. 166–7.
57. RSL JBC 8.
58. Vitruvius, *De architectura* (Loeb edition), Book X, c. ix.

Notes to Chapter 8

1. For an account of lunar and planetary astronomy before Newton, see Taton and Wilson (1989), Section III.
2. See, for example, Humphreys (1993); for a comprehensive account of observations of comets and ideas about them, Yeomans (1991); for the identification of Giotto's comet, Olson (1994).
3. See Halley (1705b,c,d); Halley (1726) in Gregory (1726a).
4. See Chapter 4; see also articles on Halley and comets in Part IV of Thrower (1990).
5. *Flamsteed*.
6. Halley to Hooke, 19/29 May 1681, RSL H.3.40. Wren had discussed Kepler's theory in 1663.
7. Flamsteed to Crompton for Newton, 15 December 1680, *Newton Corresp.*, Vol. 2, pp. 315–17.
8. Halley to Hooke, 19/29 May 1681, RSL H.3.40; Cassini (1681).
9. Flamsteed to Halley, 17 February 1681, *Newton Corresp.*, Vol. 2, pp. 336–40. Newton to Crompton for Flamsteed, 28 February 1681, *Newton Corresp.*, Vol 2, pp. 340–7.
10. Newton to Flamsteed, 16 April 1681, *Newton Corresp.*, Vol. 2, pp. 363–7.
11. Halley to Hooke, 19/29 May 1681, *Halley Corresp.*, pp. 49–52.
12. Dörffel (1681); Armitage (1951). The only copy of Dörffel's pamphlet in England appears to be that in the British Library. It is not in Bodley's Library, nor the

University Library, Cambridge, nor the Royal Society, and it was not noticed by Newton in the *Principia* nor by Halley in his accounts of comets. It is likely therefore that it was not known in England.

13. Halley (1726) in Gregory (1726a).
14. Halley to Newton, 7 September 1695, *Newton Corresp.*, Vol. **4**, p. 165. Newton to Flamsteed, 14 September 1695, *Newton Corresp.*, Vol. **4**, 169.
15. For the distance of comets, see Newton (1687), Book III, Lemma 4 and Prop. **40**; for the method of interpolation, Lemma 5; for the initial determination of the orbit, Prop. **41**, Prob. **21**; and for the improvement of an initial orbit, Prop. **43**, Prob. **23**. For a discussion and appreciation of Newton's theorems, see Chandrasekhar (1995).
16. Halley (1705b).
17. Halley to Newton, 21 October 1695, *Newton Corresp.*, Vol. **4**, p. 182.
18. ULC RGO 2.9.
19. Halley to Newton, 28 September 1695, *Newton Corresp.*, Vol. **4**, pp. 171–2.
20. Halley to Newton, 7 September 1695, *Newton Corresp.*, Vol. **4**, p. 165.
21. Halley to Newton, 7 October 1695, *Newton Corresp.*, Vol. **4**, pp. 173–4. Newton to Halley, 17 October 1695, *Newton Corresp.*, Vol. **4**, pp. 180–1.
22. Newton to Halley, 17 October 1695, *Newton Corresp.*, Vol. **4**, pp. 180–1.
23. Newton to Halley, 17 October 1695, *Newton Corresp.*, Vol. **4**, pp. 180–1. Halley to Newton, 21 October 1695, *Newton Corresp.*, Vol. **4**, p. 182.
24. Halley to Newton, 1695/6, *Newton Corresp.*, Vol. **4**, p. 190; the editors of the Newton *Correspondence* had no date for this letter but it clearly refers to Halley's activities on the *Guiney* frigate of the Royal African Company (see Chapter 9).
25. RSL JBC 8, 3 June and 1 July 1696.
26. RSL JBC 9, 18 March 1702. The article published initially in *Philosophical Transactions* in Latin (Halley 1705b) was issued as a separate Latin tract (Halley 1705c) and in an English translation, *A synopsis of the astronomy of comets* (Halley 1705d). David Gregory appended a Latin version to *Astronomia physicae et geometricae elementa*. Gregory's book, first published in 1702, is an account of astronomy on the basis of Newton's celestial mechanics. The University Library, Cambridge, holds a copy (Adv.a.58.1) of the first edition, into which is bound at the back a manuscript entitled *Astronomiae cometicae synopsis/ autor Edmundo Halleio/ apud Oxoniensis Geometriae Professore Saviliano* with the subscription 'Edm Halleus Geomtr. Prof. Savil. Oxon./Dat. Bibliothec. Mathemat. Savil. Jun.8 1705'.

The MS differs from the printed versions in that it does not include the more speculative sections, but the account of the history of comets and of ideas about them, the explanation of the calculation of orbits, general table, examples of its use, and the table of the elements of 24 comets are all identical, as is the section on the identity of the comets of 1532, 1607, and 1682. The manuscript is not a draft of the *Philosophical Transactions* paper, for it is dated later. The ULC copy of Gregory has a book plate and class mark of the Skene Library in Aberdeen. Later editions of Gregory's book in Latin and English include Halley's tract, in Latin or English as appropriate. (Gregory 1726a,b). (See also Section 9.3.)

See also Carter (1975), Appendix 1705; BLO Savile B 13; and Halley to Charlett, London, 23 June 1705, *Halley Corresp.*, p. 125:

> I return you thanks for your repeated favours as well in what relates to my house, wherein I must esteem you my greatest benefactor, as for your kind endeavours to give reputation and value to my small performance about Comets, which no wais deserves a place in your Catalogue, or to bear the badg of the Theater ...

27. For other particulars of the published versions of Halley's tract, see the notes to Chapter 13. John Senex issued an English version, *A synopsis of the astronomy of comets*, in London in 1726.

28. Leibniz to Halley, 14 July 1703 and 8 December 1705, *Halley Corresp.*, pp. 200–1. For Halley's meeting with Leibniz, see Chapter 11 Note 36. Halley first knew of Kirch's observations through a letter from Leibniz that Halley read to the Royal Society in 1701: RSL JBC 9, 21 May 1701. Kirch derived the positions of the comet from its relation to fixed stars (Kirch 1681a). Halley amended Kirch's positions after Pound had better positions for the stars (Halley 1715a).
29. On 22 August 1721, it was noted in Hearne (1884–), Vol. 7, p. 305 that Halley's astronomical tables were to be published.
30. Halley (1749, 1752). The tables for the planets and for comets were reprinted in France by de LaLande (1759) along with Halley's theory of comets and the history of the return of his comet in 1759.
31. Halley to Sloane, Greenwich, 7 November 1722, BLL Sloane 4046, f. 307; Halley to Newton, Greenwich, 16 February 1725, *Newton Corresp.*, *Halley Corresp.*, p. 131.
32. Halley (1715a); see also Kirch (1681a) and Newton, *Principia*, Book III (3rd edn). See also Note 28.
33. See, for example, Schaffer (1990).
34. See Cook (1988).
35. De LaLande wrote (1759):

> Je continuai de faire pour cette révolution, le même calcul des commutations & des distances que j'avois fait les deux autres, mais il faut convenir que cette suite immense de détails m'eût semblé effrayante si Madame Lepaute, appliquée depuis longtemps & avec succès aux calculs Astronomiques, n'est eût partagé le travail.

For Mme Lepaute see *Nouvelle biographie générale* (Paris, 1859), Vol. 29, p. 822, and Cook (1997).
36. For recent studies of chaotic behaviour, see Yeomans (1991).
37. For a recent survey of the problem of the longitude, see Andrewes (1997). Halley's review is in RSL JBC 7, 9 and 16 May 1688.
38. RSL JBC 8, 29 October 1690, *Halley Corresp.*, p. 220.
39. Halley (1694c, 1749, 1752).
40. RSL JBC 8, 16 and 23 November 1692, *Halley Corresp.*, pp. 162–4 and 230.
41. RSL JBC 8, 23 March 1692, *Halley Corresp.*, pp. 161–2 and 228.
42. RSL JBC 8, 23 September 1691; published version in Halley (1716d).
43. See Beaglehole (1974).
44. See Beaglehole (1974).
45. Halley (1693d).
46. *Philosophical Transactions*, **19** (1695), No. 217, pp. 83–110; *Philosophical Transactions*, **19** (1695), No. 218, 124–60; Goodyear's name is given as Aron in the lists of members of the Levant Company. For al-Battānī and the observatory at Tadmor, see Krisciunas (1988), Chapter 2.
47. Halley (1695e).
48. Halley, RSL JBC 8, 19 October 1692; following his final remarks on comets in Prop. 42 of Book III of *Principia*, in which he speculates on an interplanetary medium, Newton adds in the second edition (but deletes from the third):

> Decrescente autem corpore Solis motus medii Planetarum circum Solem paulatim tardescent, & crescente Terra motus medius Lunae circum Terram paulatim augebitur. Et collatis quidem observationibus Aclipsium Babylonicis cum iis Albategnii & cum hodiernis, Halleius noster motum medium Lunae cum motu diurno Terrae collatum, paulatim accelerari, primus omnium quod sciam deprehendit. [Koyré and Cohen (1972), pp. 758, 759].
>
> [As the body of Sun also decreases, the mean motions of the planets around the Sun are slightly delayed, and as the Earth increases, the mean motion of the Moon around the Earth is slightly advanced. And by bringing together the eclipse observations of the Babylonians and of Albategensis with those of today, our Halley showed that the mean motion of the Moon accelerates slightly by comparison with the diurnal motion of the Earth, the first of all, so far as I know, to have discovered it.]

49. Lambeck (1980); for historical records of eclipses, see Stephenson and Morrison (1995).
50. Halley (1695a).
51. Laplace 1798–1823, Vol. 3 Book VIII.
52. Adams (1853).
53. Taylor (1919); Jeffreys (1920).

Notes to Chapter 9

1. *Hooke D2*, 21 January 1689. For other contemporary reports, see Luttrell (1857), Vol. 1, pp. 299 (trial of Braddon and Speke in 1684), 497 (detention of Hawley, 22 January 1689) and 505 (arrest of Holland, 20 February 1689).
2. *Biogr. brit.*, Vol. 4, p. 2507. It was also reported that William interviewed Halley.
3. For *St Albans*: Commissioners of the Admiralty to Navy Board, PRO ADM 1/3557, especially p. 719. A court martial was held on the pilot. Trinity House removed his pilot's qualification, but protested that he had been properly examined and found competent, and that harsh treatment from the ship's officers had contributed to his failure: Trinity House to Navy Board, 28 November 1688, Guildhall Library, MS 30048/2. For a light on the Goodwin Sands: RSL JBC 7, 10 July 1689.
4. RSL JBC 7, 9 and 23 May, 13 June, 1 August 1688, 3 July 1689, *Halley Corresp.*, pp. 212, 213, 215
5. RSL JBC 7, 13 June, 1 August 1688, 6 March 1689; *Hooke D2*, 22 March and 3 April 1689; RSL JBC 7, 3 and 24 July 1689; for the Sussex survey, see RSL JBC 8, 15 November 1693 and *Halley Corrresp.*, p. 233.
6. Sellers (1671).
7. Tanner (1926b), pp. 135, 345.
8. Phineas Bowles, Secretary to the Commissioners of the Admiralty, to Sellers, asking him to bring to the Board maps, charts and plans of coasts, roads, and ports: PRO ADM. 2/377, f. 199v.
9. Tanner (1926b) pp. 188, 189, 324, 388; Collins (1693).
10. Minutes of Court of Trinity House, 21 February 1689, Guildhall Library, London, MS 300004/7.
11. For embargoes and exemptions, see the many orders by the Privy Council in orders to the Admiralty in PRO ADM.1, 5139 and 5140 for 1687–1690.
12. Halley to Southwell, 27 January 1702, RSL; *Halley Corresp.*, pp. 120–2; *Paramore*, pp. 338–40.
13. Snel (1617), pp. 204–6; de la Hire, Sedileau and Pothenot (1693).
14. Minutes of the Court of Assistants of the Royal African Company, PRO T70/83, ff. 5v., 6r., 8r.
15. Minutes of the Court of Assistants of the Royal African Company, 28 May, 25 June, 2 July 1689, PRO T 70/82, ff. 60v., 62r., v.
16. Minutes of the Court of Assistants of the Royal African Company, 15 and 23 October 1689, PRO T 70/82, f. 68r., v.
17. Minutes of the Court of Assistants of the Royal African Company, 27 and 28 February 1691, PRO T 70/83, ff. 5v., 6r.
18. *Black book* of the Royal African Company of England, p. 105, PRO T 70/1433. The *Book* is mostly a record of unsatisfactory behaviour by servants of the Company, but other special circumstances are recorded.

19. Halley to Newton, no date, *Newton Corresp.*, Vol. 4, pp. 190–5. In a footnote the editors say that Halley may be writing of *Paramore*, in which he made his later cruises in the Atlantic and Channel (Chapter 10). *Paramore*, however, was no frigate, but a pink. Halley's reference was clearly to his activities on the wreck of *Guynie*: elsewhere he speaks of his 'Guinea frigatt'. He must have written his letter about the end of March 1696, just after Newton had been offered the post of Warden of the Mint (*Newton Corresp.*, Vol. 4, pp. 195–6, 200).
20. Minutes of the Court of Assistants of the Royal African Company, 6 and 8 April 1691, PRO T 70/83, f. 8r.
21. Halley to Hill, 22 June 1691, Astle (1767), p. 136 & *Halley Corresp.*, p. 88.
22. RSL JBC 7, 6 March 1689 and *Halley Corresp.*, pp. 144–5; RSL JBC 8, 6 and 13 May 1691.
23. Minutes of the Court of Assistants of the Royal African Company, 30 June 1691, PRO T 70/83, f. 16r.
24. CSP,D, 1690–1691.
25. Minutes of the Court of Assistants of the Royal African Company, 9 February and 8 March 1692, PRO T 70/83, ff. 41v. and 43r., v.
26. Minutes of the Court of Assistants of the Royal African Company, 25 April 1693, PRO T 70/83, f. 81r. and 6 February 1694, PRO T 70/84, f. 22v.
27. Minutes of the Court of Assistants of the Royal African Company, 9 April 1695, PRO T 70/84, f. 44r.
28. Houghton (1727–8), Vol. 2, for 28 February 1696.
29. Houghton (1727–8), Vol. 2, for 20 July 1694.
30. Houghton (1727–8), Vol. 1, 21 and 22 May, 7 August 1692; Vol. 1, 20 January, 31 March, 20 April, 5 May 1693; Vol. 2, 17 July 1696.
31. Halley to Newton, n.d., *Newton Corresp*, 4, pp. 190–1.
32. Tanner (1926b), p. 311.
33. RSL JBC 8, 15 November 1693.
34. Halley (1716f); RSL JBC 7, 6 March 1689, *Halley Corresp.*, pp. 144–5, 214; RSL JBC 8, 6 and 13 May 1691, *Halley Corresp.*, pp. 223, 224; RSL JBC 8, 12 August 1691, *Halley Corresp.*, p. 224.
35. RSL JBC 8, 26 August 1691; *Halley Corresp.*, pp. 150–2, 152–3.
36. Newton (1704) Book I, Part II, Prop. 10, Prob. 5.
37. RSL JBC 8, 23 September 1691; *Halley Corresp.*, p. 224.
38. RSL JBC 8, 7 October 1691; *Halley Corresp.*, pp. 224–5.
39. RSL JBC 8, 21 October 1691; *Halley Corresp.*, pp. 154–5, 225.
40. RSL JBC 8, 21, 28 October, 4 November 1691; *Halley Corresp.*, pp. 225–6
41. For Michelangelo as engineer, see Wallace (1994).
42. RSL JBC 8, 28 October, 11 November 1691, 26 October 1692; *Halley Corresp.*, pp. 155–6, 225, 226, 229, 230; *Halley Corresp.*, p. 232; *Sussex chart*.
43. RSL JBC 8, 18, 25 March, 22 April 1691; Unprinted papers, 25 March and 26 April, 1691; *Halley Corresp.*, pp. 147–50, 222–3; the value of the acceleration due to gravity was still somewhat uncertain at that time.
44. RSL JBC 8, 11, 18 November, 23 December 1691; *Halley Corresp.*, pp. 156–9, 226–7.
45. Goldie (1993); Spurr (1991).
46. *Flamsteed*, pp. 667, 668.
47. Halley to Hill, 22 June 1691, Astle (1767), p. 136; *Halley Corresp.*, p. 88.
48. Halley (1724a).
49. RSL JBC 8, 19 October 1692; *Halley Corresp.*, p. 229.
50. RSL JBC 8, 11 November 1691; *Halley Corresp.*, p. 226
51. Fatio to Huygens, 29 April/9 May 1692, *Huygens Oeuvres*. Vol. **XXII**, p. 295.

52. Fatio to Huygens, 8/18 September 1691, *Huygens Oeuvres.*, Vol. X, p. 146; Newton to Charlett, 27 July 1691, *Newton Corresp.*, Vol. 3, pp. 154–5.
53. BLO MS Rawlinson J No. 4.2.
54. Wallis to Flamsteed, 28 December 1698: *Newton Corresp.*, Vol. 4, p. 290; the reference to Halley and Gregory being competitors must relate to the Savilian Chair; Montague was the Lord Treasurer who arranged Newton's appointment to the Mint.
55. For the record of the election on 23 December 1691, see Oxford University Archives, *Records of Congregation*, 1680–1692, Be 20 1680–1692, f. 214r.; Bernard's resignation, f. 213v., is dated 9 November 1691. The Registry's certificate that Gregory had taken the oaths according to the statutes of the realm and of the Savilian Chairs is also on f. 214r.
56. Westfall and Funk (1990).
57. For a general account of the Mint in this period, see Challis (1992).
58. Halley to Newton, ? March 1696, *Newton Corresp.*, Vol. 4, pp. 190–1; for the dating of this letter, see Note 16.
59. Mint to Treasury, 4 June 1696, *Newton Corresp.*, Vol. 4, No. 550.
60. Newton to Halley, 11 February 1697 and Halley to Newton, 13 February 1697, *Newton Corresp.*, Vol. 4, Nos. 562 and 563.
61. Newton to Godolphin, 24 June 1707 and subsequently, *Newton Corresp.*, Vol. 4, No. 724 *et seq.*
62. Halley to Sloane, 12 October 1696, RSL *Guard Book*, H.3, 48.
63. 12 October 1696, CTB, XI.
64. Halley to —, 26 October 1696, RSL *Guard Book*, H.3, 49.
65. Halley to Sloane, 2 and 25 November 1696, RSL *Guard Book*, H.3, 50, 51.
66. Halley to Newton, 28 Novmber 1696, *Newton Corresp.*, Vol. 4, No. 555; this seems on the face of it not a matter for Halley as deputy Comptroller but for the deputy Master, Clarke—but he was mostly absent. See also, Halley to Molyneux, 28 November 1696, RSL MM. 5.40. Molyneux is Thomas Molyneux of the Tower Mint, not the Royal Society's correspondent in Dublin.
67. Lords of the Treasury, 15 March 1697, CTB, XI.
68. Newton to Halley, 21 June 1697, *Newton Corresp.*, Vol. 7, pp. 339–49.
69. Halley to Sloane, 25 October 1697, RSL archives; an account of the lunar eclipse of 19 October is in Halley (1697*d*).
70. Halley to Newton, 13 February 1697, *Newton Corresp.*, Vol. 4, pp. 230–1.
71. Tower Mint to Chester Mint, August 1697, *Newton Corresp.*, Vol. 7, pp. 400–1; Halley to Newton, 30 December 1697, *Newton Corresp.*, Vol. 4, p. 254.
72. Halley to Molyneux, 25 August 1697, RSL MM. 5.46.
73. Halley to Molyneux, 2 August 1697, RSL MM. 5.42.
74. Halley to Newton, 2 August 1697, *Newton Corresp.*, Vol. 4, p. 246.
75. Lords of the Treasury, 9 November 1697, CTB, XIII.
76. Halley to Newton, 30 December 1697, *Newton Corresp.*, Vol. 4, p. 254.
77. Lords of the Treasury, 2 February 1698, CTB, XIII.

    Other letters on the dispute at the Chester are Halley to Molyneux, 21 July 1697, RSL MM. 5.44. and 31 July 1697, RSL MM. 5.45.

Notes to Chapter 10

1. See the article on Kircher in the *DSB*.
2. For the loss of Shovell, see Burchett (1720), Book V, Ch. 24, p. 733.

3. The first mention of the project is by Hooke: Hooke *D2*, 11 January 1693; *Paramore*, p. 250; Middleton's application to the Royal Society was later, in April 1693: RSL JBC 8, 12 April 1693.
4. *Newton Corresp.*, Vol. 4, p. 311.
5. Referred to as *Paramore*.
6. See Venn (1922–7); Greenwood (1992).
7. Admiralty to Navy Board, 12 July 1693, PRO ADM. 2.173, p. 374; *Paramore*, p. 252.
8. Leibniz to Bernoulli, Hanover, 15 January 1704, Gerhardt (1962).
9. *Paramore*, pp. 31, 253–5.
10. *Paramore*, pp. 256–7.
11. Halley to Navy Board, 19 June 1696, NMM ADM/A/1831, *Paramore*, p. 260.
12. Admiralty to Navy Board, 19 August 1696, NMM ADM/A/1833, *Paramore*, p. 261.
13. *Hooke D2*, 19 July 1693; see Chapter 2.
14. *Paramore*, p. 262.
15. Evelyn, *Diary*, 2 January, 6 February, 21 April, 9 June 1698; Wren and the gardener London reported on the damage to the house and gardens, and recommended a payment of £150 to Evelyn for making good.
16. *Biogr. brit.*, Vol. 4, article on Halley; Tafel (1875–7).
17. Harrison (1696); Harrison's career in the Royal Navy is listed by Syrett and DiNardo (1994) as lieutenant 16 February 1691, merchant service 1700. Flamsteed has notes of Harrison's magnetic observations as a merchant officer in ULC RGO 1/41 ff. 125v., 126r.
18. ULC RGO 1/41 ff. 125v., 126r.
19. PRO ADM 2 179, p. 30, ADM 6/5 f. 27v.; *Paramore*, pp. 257, 264, 265, 266; Syrett and DiNardo (1994).
20. PRO ADM 2 25, pp. 155–6; *Paramore*, pp. 268–70.
21. PRO Navy Board Minutes, 17 July 1693, Sergison/37; Deptford Yard Letter Book NMM ADM 106/3291; *Paramore*, pp. 253, 254, 256.
22. RSL JBC 7, 9 May 1688; *Halley Corresp.*, p. 212.
23. RSL JBC 8, 29 October 1690; *Halley Corresp.*, p. 220.
24. For Newton's remarks to the Royal Society, see RSL JBC 9, 19 July and 16 August 1699.
25. RSL JBC 8, 9 and 16 November 1692; *Halley Corresp.*, p. 230.
26. See Waters (1990).
27. ULC RGO 1/41, 125v.
28. Halley (1701*a*).
29. *Paramore*, pp. 88, 89.
30. Halley to Burchett, 1 November 1698, PRO ADM 1.1871; *Paramore*, pp. 89–91, 271.
31. *Paramore*, p. 92.
32. RSL CMC 2, 11 January, 8, 15 February, 8 March 1699.
33. *Paramore*, p. 95.
34. *Paramore*, p. 98.
35. Halley to Burchett, 23 June 1699, PRO ADM 1.1871; *Paramore*, pp. 281, 282.
36. *Paramore*, pp. 98, 99.
37. *Paramore*, p. 101.
38. *Paramore*, p. 104–6.
39. *Paramore*, p. 107–110.
40. Halley to Burchett, 23 June 1699, PRO ADM 1/1871; *Paramore*, pp. 281, 282.
41. Burchett to Halley, 29 June 1699, PRO ADM 2/397, p. 67; orders to Shovell to hold court martial, ADM 2/26, p. 34; Reports of Courts Martial, ADM 1/5261; *Paramore*, pp. 283–6; Luttrell (1857), Vol. 4, pp. 352, 358.

42. Dampier (1939).
43. Admiralty to Navy Board, 21 July 1699, NMM ADM/A/1867; PRO Navy Bd ADM 106/2292 f. 100v.; Lords Letter Book. 3 August 1699, ADM 2.180, p. 7; *Paramore*, pp. 291– 4; Luttrell (1857), Vol. **4**, pp. 533, 538.
44. RSL JBC **9**, 19 July and 9 and 16 August 1699.
45. Halley's Orders and Instructions, PRO ADM 2/2, pp. 128–9; *Paramore*, pp. 301–2.
46. Halley to Burchett, St Jago, 28 October 1699, PRO ADM 1/1871; *Paramore*, p. 305.
47. *Paramore*, pp. 145–9.
48. See Halley to Burchett from St Helena, 30 March 1700, PRO ADM 1/1871; *Paramore*, pp. 306–7.
49. *Paramore*, pp. 153–62.
50. Halley to Burchett, 30 March 1700, PRO ADM 1/1871; *Paramore*, pp. 162, 306–7; Beaglehole (1974), pp. 430, 431.
51. Halley to Burchett, 30 March 1700, PRO ADM 1/1871; *Paramore*, pp. 163–5; Beaglehole (1974), p. 432.
52. *Paramore*, pp. 167–70, 174, 178–80.
53. *Paramore*, p. 186.
54. *Paramore*, pp. 188–9.
55. Halley to Burchett, 8 July 1700, PRO ADM 1/1871; *Paramore*, pp. 194–200, 202, 307–8.
56. *Paramore*, pp. 205–6.
57. *Paramore*, pp. 210–3. Luttrell noted Halley's return on September 14: Luttrell (1857), Vol. **4**, p. 687.
58. RSL JBC **9**, 30 October 1700; see also 9 August 1699 and 5 February 1701.
59. RSL JBC **9**, 4, 11, and 18 June 1701. The chart was engraved by John Harris and published by Mount and Page in London in 1701. Thrower (*Paramore*, Appendix E) gives a list of charts derived from it and from Halley's World Map of 1702. See Halley 1701*c*. 1702*a*.
60. The worst discrepancies were on the first voyage from Africa to Paraibo. Halley's longitudes were more than 10° too low at Paraibo, which he attributed to currents, especially as they were often becalmed. Harrison's values agreed more closely with the astronomical observations at Paraibo. Presumably Harrison made much the same estimates for course as Halley did. Did he know more about the currents, or did he adjust his longitudes after he knew the astronomical value at Paraibo?
61. See Bullard (1956*a*,*b*); Malin (1981), who lists Halley's observations at London of 1672, 1683, 1685, 1692, 1698, 1701, 1702, and 1716.
62. Thrower and Glacken (1969), Thrower (1996); *Paramore*, p. 57. For Borri, See Kircher (1643). Athanasius Kircher (1602–1680) was a notable Roman Jesuit Father who taught at the Collegio Romano. He was closely associated with Eschinardi and others whom Halley probably met in Rome (Chapter 4). He suggested the use of a balance to measure magnetic force—see the *DSB*. Halley knew something of his ideas—Halley (1683*b*), p. 215.
63. *Paramore*, p. 58, Fig. 7, pp. 365–7.
64. RSL JBC **9**, 5 February 1701; *Paramore*, p. 60, Fig. 8; for a list of published versions, see Appendix E of *Paramore*.
65. Chapman (1941); Bemmelen (1899).
66. See Fig. 10.2 and Fig. 5 in *Paramore*.
67. RSL JBC **9**, 30 November, 17 December 1700, 19 November 1701.
68. Admiralty instructions to Halley, *Paramore*, pp. 327–9.
69. Halley (1700*b*, 1701*b*).
70. *Paramore*, pp. 328–9.

71. Halley to Burchett, 11 June 1701, PRO ADM 1/1872; ADM to Halley, 12 June 1701, PRO ADM 2/27, pp. 131–2; *Hooke D2*, 24 March 1693.
72. *Paramore*, pp. 220, 327–9; 221–2.
73. *Paramore*, pp. 221, 235; for the storm of 1703 and the destruction of the Eddystone Lighthouse, see Lamb (1991).
74. *Paramore*, pp. 222–4, 226, 228, 232, 235–6.
75. RSL JBC 9, 18 June, 30 July, 12 November 1701. *Paramore*, pp. 330, 332, 337.
76. *Paramore*, pp. 236, 238–47.
77. *Paramore*, pp. 63, 64, 221, 222; Proudman (1941).
78. See Proudman (1941).
79. Beechey (1848, 1851).
80. Tafel (1875–7).
81. Adm, 6 May 1701, Lords Letter Book, PRO ADM 2/181, p. 125; Adm, 20 April 1702, Lords Letter Book, PRO ADM 2/182, p. 164; *Paramore*, pp. 321, 345.

Notes to Chapter 11

1. Queen's orders, NAN FH 275, pp. 204–6.
2. See Veenendaal (1970); Hattendorf (1983).
3. See Veenendaal (1970).
4. Dollot (1961), pp. 16–18; Tamaro (1924), p. 156
5. Halley to Stepney, 5 August 1703, PRO SP 105/69.
6. Stepney to Hedges, September 1702, PRO SP 80/19, f. 177; Nottingham to Hedges, October 1702, NAN FH 275, pp. 133, 145, 150, 161; see also Hattendorf (1980, 1983).
7. Luttrell (1857), Vol. 5, p. 228; see also pp. 265, 278, 285, 306; Stepney to Hedges, PRO SP 80/20, f. 11.
8. For example, *Biogr. brit.*, Vol. 4, p. 2512; Tamaro (1924).
9. Queen's instructions: NAN FH 275, p. 204; Pass: CSP, D. Entry Book 104, p. 412; £200 for expenses: CSP, D. 1702/3, p. 278.
10. ASV Sen. Disp. Inghil., f. 76, c39,40; Sen. Disp. Germ., f. 185, c175.
11. Nottingham to Shovell, 1703, PRO SP 44/209, 54/5; Stepney to Hedges, 10 March 1703, PRO SP 80/20, f. 196; Stepney to Hedges, 28 March 1706, SP 80/28, f. 120. CSP, D, Entry Book, 208, 82–6, 209, 53–5. Also R. C. Hist. MSS, MSS of the House of Lords, Nos. 518, 524.
12. Lago and Rossi (1981).
13. See Cook (1985). Maps and plans by Weiss are in the Österreichisches Staatsarchiv, Finanz und Hofkammerarchiv, in Vienna. The Venetian manuscript maps are in the archives of the Provedditori della Camera dei Confine and the Provedditori dei Fortezze.
14. CSP,D, 1702/3, pp. 278 and 412; NAN FH 275, pp. 204–6; PRO SP 105/67, ff. 51–4.
15. PRO SP 80/19, f. 434; SP 80/20, ff. 30, 31; SP 105/67, ff. 67, 75.
16. PRO SP 80/20, ff. 104, 107, 109, 114; SP 105/67, ff. 208, 219; SLG HK 1703–I–53, ff. 10–12. For Leibniz's journey, see Leibniz (1923), Ser. I, Vol. 5, p. 682.
17. PRO SP 80/20, ff. 128, 136, 138, 145, 157; SP 105/67, ff. 251, 268, 279, 294, 300; SLG HK 1703–I–53, f. 5.
18. PRO SP 80/20, ff. 147, 149, 193, 216; SP 105/67, ff. 322, 376, 418–9, 420; ASV Sen. Disp. Germ., f. 185, cc317–18.

19. PRO SP 44/209, f. 54; SP 80/28, f. 120.
20. For the loss of Shovell, see Burchett (1720), Book. V, Ch. 24, p. 733; PRO SP 105/67, ff. 418–19.
21. ASV Sen. Disp. Germ., f. 185, cc309–11, cc317–18; Sen. Disp. Inghil., f. 76, cc267, 268r.
22. PRO SP 80/20, f. 109.
23. SLG HK 1703–I–53, ff. 10r.–12r.
24. PRO SP 105/67, ff. 418–19: a translation is in Appendix 10.
25. ASV Sen. Disp. Germ., f. 185, cc309–16, cc317–18.
26. ASV Sen. Disp. Germ., f. 185, cc309–11.
27. See Cook (1985).
28. PRO SP 80/20, ff. 157–8.
29. PRO ADM 7/636–8.
30. PRO SP 105/67, f. 417.
31. BLL Add MS, 7058, 16 March 1703 (o.s.).
32. PRO SP 105/68, f. 83:

> ... cose bene che di mettere il Porto di Buccari in Stato di ricerverli con Sicurezza facendo a tal fine fortificarlo, secondo li dissegni che il Capitano Halley ha havuto l'onore di presentare a Vostra Maestà Ceasarea.

33. CSP,D, 1703, 711, 716.
34. Nottingham to Marlborough, NAN FH 277, 22–3, 27; Nottingham to Stanhope, FH 277, 9; Nottingham to Stepney, FH 277, 12; Nottingham to Grand Pensioner, FH 277, 37. See also Burchett (1720), Book V, Ch. 14.
35. Stepney to Hedges, 19 May 1703, PRO SP 80/20, f. 367.
36. Stepney to Hedges, 16 June 1703, PRO SP 80/2, f. 28.
37. For Leibniz's report of Halley's visit, Leibniz to J. Bernoulli, Hannover, 15 January 1704, Gerhardt (1962), pp. 739–40:

> Gratiam mihi est, quod Newtoniana colorum Theoria prodit. Vellem etiam ut daret Newtonus novam suam Theoriam Lunae, quam apud me laudavit Halleius, in Angliam nunc, ut mihi scribitur, ab Adriatico mare reversus, et ut ipsi mihi narravit, mox in mare pacificum navigaturus, ut perficiat theoriam declinationis magneticae. Laudandum est indefessum Viri studium juvandae scientiae.

For Leibniz and the observations of Kirch, see Chapter 8, Note 28. For the meeting with George I and his sister and a visit to Osnabruck, see *Biogr. brit.*, Vol. 4, p. 2512.

38. ASV Senato Dispacci Germania, f. 186, c186, 28 July 1703:

> È già arrivata e partito verso Bucari il Capitan Hall percorso da due Ingenieri che vanno a rivider e stabilire le operazione già disposte per la siccurezza di quel Porto. Stò procurato che parta ben impresso verso le publiche convenienze nè l'espressioni potevano desiderarsi più favorevoli. Resterà assicurato in valida forma quel ricovero perchè si rilette che la flotta non nè ha nel Mediterraneo altro nel quale abbia pronta e sicura la retirata.

39. SLG HK 1703–IV–62. In 1702 Rauschendorf, with Field Marshal Graf von Heister, the military commander in Inner Austria, made a survey of Istria, especially of the border with Venetian territory. They found the frontier defences in a deplorable condition and the Hofkammer did something to repair them. The records are in the Hofkammer archive in Graz.
40. PRO SP 105/69 (unfoliated).
41. Herberstein seems not to have stayed to the end, for later the commissioners are named as Endtres, Pacher, and Cilli, who brought the money for paying the masons and other skilled workmen.
42. Halley to Nottingham, 11 August 1703, PRO SP 105/69 (unfoliated).
43. Halley to Stepney, 11 August 1703, PRO SP 105/69 (unfoliated).

44. PRO SP 105/70, f. 29; ASV Senato Dispacci Germania, f. 185, c310, 17 March 1703.
45. Halley to Stepney, 5 August 1703, PRO SP 106/69, unfoliated.
46. Weiss: Hofkammerarchiv, Kartensammlung, C 16; Kriegsarchiv, Kartensammlung, G I h 62 and B IX b 89.
47. Anon [Archduke Louis Salvator] (1871).
48. The two passages from Salvator's book are

> [Site D] ... bemerkt man eine verfallene Strandbatterie (Batteria), welche ehemals dazu diente, den Hafeneingang und Porto Ré, welches sie volkommen bestricht, zu verheidigen. Sie liegt auf einem kleinen Vorsprung von zahnartigen Felsen und hat drei Seiten, wovon die zwei längeren, welche nach Südwesten unter einem stumpfen Winkel zusammenstossen, allein mit Schiesscharten versehen sind. Die Seite, zu der man zuerst gelangt, enthält deren fünf, die zweite, mehr der Hafeneinfahrt zugekehrte, neun.
> [Site C] Etwas weiterhin liegt eine Strandbatterie (Batteria), welche der Form des Vorsprunges folgt und auf dessen Felsen aufgebaut wurde, welche, wie auf der gegenüberliegenden Seite, die Mesembryanthemum überwuchern. Sie weist noch vierzehn wohl erhaltene Schiesscharten auf, unter welchem ein Cordonsims verläuft; in der Mitte ist zie eingestürzt, und die in das Wasser hinabgerollten Mauertrümer wereden nun von den schäumenden Wellen umspült.

49. PRO SP 80/28, f. 120.
50. PRO SP 105/68, f. 294; letter from Buccellini at f. 300–1.
51. See Caputo (1982)
52. PRO MPF 352, also SP99/60; see Cook (1985), Note 58.
53. Halley to Southwell, London, 27 January 1701/2, *Halley Corresp.*, pp. 120–2.
54. *Paramore*, pp. 233, 235.
55. For surveys of French coast, see *Paramore*, pp. 223, 228.
56. See Thrower and Glacken (1969) and *Paramore*, pp. 57–8.
57. Halley (1686*b*, 1695*c*).

Notes to Chapter 12

1. CSP,D, Anne, 1703–4, 191. Viscount Hatton, a descendant of Christopher Hatton of Elizabeth's court, was the head of the Hatton-Finch family to which Daniel Finch, the Earl of Nottingham, belonged. Hatton was governor of Guernsey but lived mostly at Kirby Hall in Northampton where he and his brother constructed notable gardens, now being restored.
2. See Cook (1984*b*).
3. ULC RGO, 1/Flamsteed correspondence, 18 December 1703.
4. BLL Add MS 29595, f. 256.
5. Acta Congregationis Universitatis Oxoniensis, 1703–23, f. 251, Certificatoria Professorum in Universitate; Luttrell knew of the election on 11 January: Luttrell (1857), Vol. 5, p. 379. The six electors who signed the certificate were the Archbishop of Canterbury, Thomas Tenison; the Lord Chancellor, Sir Nathan Wright; the Chief Justice of the Queen's Bench, Sir John Holt; the Chief Baron of the Exchequer, Sir Edward Ward; the Dean of the Arches, Sir John Cooke; and Nottingham as principal Secretary of State. For holders of offices, see Haydn (1894).
6. In Act Books of the Archbishops of Canterbury: Edmund Halley 1703–4 Apptd. Savilian Professor of geometrie Oxford, v. 109. 1741–2 Noted as dead viii. 191, 192 (see Dunkin, Jenkins, and Fry, 1929). See also Hiscock (1937), p. 20.
7. Hearne to Rev Dr Smith, 8 June 1704, Bodleian, Rawlinson MSS, Letters 37, No. xi.

8. Hiscock (1937), p. 20.
9. Gregory (1702).
10. Turner (1986).
11. *An account of the visitation of the mathematick professors lands, 1704*, Oxford Univesity Archives, UD 31/10/1, ff. 13, 14. See also a book compiled apparently by S. Rigaud: UD 31/1/2. The *Account* is not by Halley.

    Charters, manorial deeds, and other documents relating to the four estates were originally in the library of the Savilian Professors and are now in the Savilian collection in the Bodleian Library, as follows:

    | Little Hayes | BLO, Savile a 11 |
    | Norlands | BLO, Savile a 8–10 |
    | Purston | BLO, Savile a 1–7 |
    | Moreton Hindmarsh | BLO, Savile a 14. |

    There are plans of Little Hayes and Norlands at Savile a 12(R).
12. See Dunbain (1986).
13. Little Hayes farm in the parish of Stow Mareys in the south-east of Essex lies on the River Crouch between Burnham-on-Crouch and Wickford. (Chapman and André 1774; Ordnance Survey 6-inch map, Essex Sheet 62 SW.) Part of the present house is at least four hundred years old and the farm is still worked. The river to the south was navigable to Gt. Hays, 'so that in one tide a Hoy might be in the Thames', and carriage by water to London was as cheap as by land to Romford (about 25 miles).

    The rent was £100 until 1670. Reduced to £90 in 1671 and to £60 from 1671 to 1700, it was raised to £65 in 1701; there was no lease.

    Norlands was in the parish of Ebeny (now Ebony) in a small enclave on the north of the Isle of Oxney, just north of Rye. It does not appear on any contemporary map, but is mentioned by Hasted (1778–99) as conveyed to Oxford by Sir Henry Savile. It still supported the Savilian Professors at the time of the Tithe Commutation award of 1843. (Canterbury archives: Tithe Commutation Map, Kent, 1843.) There was no farm house. The lands were by water, the tenant owned a sloop and cut osiers.

    The rent was formerly £78 but then £70.

    The rent of Purston was £140 in Savile's day, but then was reduced to £110, later became £118 and was £100 in Halley's time. The house was in poor condition. There was no lease.

    Morton Hindmarsh (or Hendmarsh, now Moreton-in-the-Marsh) is between Stow-on-the-Wold and Chipping Campden. It had a lease running to 1728, after which the rent would be £80 to the University, of which £68 would be for the stipends. Halley's plan is in BLO UD 31/8, f. 118. Luttrell thought Halley's chair was worth £300 but he must have confused it with the stipend of the two professors together—see Luttrell (1857), Vol. 5, p. 379.
14. See Turner (1986); Evelyn, Diary, 10 July 1669.
15. See *Flamsteed*.
16. Halley to Charlett, London, 23 June 1705, Halley to Hudson, London, 16 March 1706, *Halley Corresp.*, p. 125.
17. See Goldie (1993).
18. For Hearne's stories see Hearne, October 1712, Hearne (1884–), Vol. 4, p. 257; 22 January, 18 February, 25 September 1718,, Hearne (1884–), Vol. 6, p. 139.
19. See Goldie (1993).
20. Hearne (1884–), Vol. 3, p. 133.
21. Evelyn, *Diary*, 10 July 1669.
22. Evelyn, *Diary*, 11 July 1669.
23. Hearne (1884–), Vol. 4, p. 429.

24. Gibson and Johnson (1943); for the history of the Press, see Carter (1975).
25. Hearne (1884–), 26 February 1712, Vol. 3, p. 473.
26. Oxford University Archives, Register of Convocation, 1703–10, f. 68v.
27. Hogwood (1984).
28. Hearne (1884–), 24 January 1707, Vol. 1, p. 321.
29. Westfall (1980), p. 624.
30. Hearne (1884–), 24 February 1712, Vol. 3, p. 473. Letter of July 1996 from the Society of Antiquaries.
31. F. Bianchini, *Iter in Britanniam*, Biblioteca Vallicelliana, Rome, T. 46 p2a, f. 21–2; also Costa (1968).
32. Oxford to Newton and Halley, 28 August 1712, *Newton Corresp.*, Vol. 5, pp. 332–3; Newton and Halley to Oxford, 18 September 1712, *Newton Corresp.*, Vol. 5, pp. 340–1.
33. Testimonial from Newton and Halley, 7 January 1709, *Newton Corresp.*, Vol. 7, p. 469.
34. In the Advertisement to the *Opticks*, dated 1 April 1704, Newton wrote 'To avoid being engaged in Disputes about these Matters, I have hitherto delayed the printing'.
35. RSL JBC, 1704.
36. Westfall (1980); for the Leibniz controversy, see also A. R. Hall (1980), Cook (1996).
37. See Keill to Newton, 25 May 1715, *Newton Corresp.*, Vol. 6, p. 142: 'I leave my whole paper to You and Dr Halley to change or take away what you please'; Halley to Keill, 3 October 1715, *Newton Corresp.*, Vol. 6, pp. 242–3: Newton does not want to appear in publishing the *Recensio*, his own comment on the *Commercium epistolicum* and reply to the *Charta volans*.
38. A. R. Hall (1980); J. Bernoulli to A. de Moivre, 23 November 1712, Wollenschläger (1933); J. Bernoulli to Burnet, 17 December 1712, BAB Bernoulli Correspondence, No. 27.
39. Carter (1975), Vol. 1.
40. Halley to Charlett, London, 23 June 1705, Halley to Hudson, London, 16 March 1706, *Halley Corresp.*, pp. 125–6.
41. Review in *Philosophical Transactions*, 23, 1558–60.
42. Carter (1975), Vol. 1.
43. Thomas (1980); Toomer (1990).
44. Bernard's copy is Bodl. 250; there is another Arabic version, written in AD 1235 in Seld. 3140,7; also MS Arch. Seld. A 72(3) No. 32; Hearne (1884–), Vol. 1, p. 18, 23 January 1706. Keill was deputy to the Sedleian Professor of Natural Philosophy and would succeed to the Savilian Chair of Astronomy in 1712. For later work on *Cutting off a section*, see Toomer (1990), p. xii.
45. Gregory, February and March 1705, Hiscock (1937), p. 24; Halley to Charlett, London, 23 June 1705, *Halley Corresp.*, Carter (1975), Vol. 1, p. 214, Appendix for 1706.
46. Heath (1896). Heath's account has been superseded in some respects by later studies—see Hogendijk (1984) and Toomer (1990).
47. For the history of the transmission of the works of Apollonius, see Heiberg (1974), Prolegomena to Vol. 2, Hogendijk (1984), and Toomer (1990). Steinschneider (1960) noted that Ibn Nadīm said that four figures of the eighth book were preserved, but thought that their existence was uncertain.

    It is usual to refer to the Islamic mathematicians as 'Arabs' because they wrote in Arabic, but almost none of them was Arabian; many were Syrian or Persian or Spanish.
48. Heiberg (1974), Prolegomena to Vol. 2.
49. There are three Greek manuscript copies of *Conics* I–IV in the Bibliothèque Nationale, 2355, made in 1558, and 2356 and 2357, both made in the sixteenth

century. The manuscript 2357 also has a version of Serenus, *De sectione*. MS Savile 7 in the Bodleian Library is the copy of BNP 2356 used by Halley's Oxford printer, but there is a second copy, Savile 59, of the first four books of the *Conics* that is also said to have been the printer's copy of 1710. A third MS, Savile 10(16), is incomplete.
50. BLO Cod. Barocciani, 169, f. 1.
51. Heiberg (1974); Heiberg's MSS were of the twelfth to fourteenth centuries, Vatican Gr 206 and 203, Constantinople, palatii veteris No. 40, Paris Gr 2342; see also Heath (1896).
52. See Note 47; the Arabic epitome is by Abū l'Husayn 'Abd al-Malik al-Shīrāzī. For Thābit ibn Qurra, see DSB.
53. The MS is BLO Marsh 667 (now Arch. o.c.3); for its history, see Toomer (1990). Written in AD 1070 at Maragha in Azerbaijan, it has books I–IV of *Conics* in the translation by Hilāl b. abī Hilāl al-Himsī and books V–VII in the translation by Thābit ibn Qurra. It was deposited in the University library at Leiden and eventually sold by Golius's heirs to Narcissus Marsh, the archbishop of Armagh. Marsh lent it to Halley and later presented it to the Bodleian Library. Bernard made a copy while the MS was still in Leiden—now BLO Thurst. 1.
54. The Ravius MS is BLO Thurst 3. The text was reworked from the Banū Mūsā version by Abū l'Husayn 'Abd al-Malik al-Shīrāzī. Other Bodleian manuscripts with Arabic versions are Thurst 3968.1 and Marsh 207, 208, 452, 453, and 720.

Collins wrote to James Gregory in 1671:

> Mr Bernard is a good mathematician and understands the Arabic tongue well; he hath found in the libraries there [Oxford] two entire copies of the seven books of Apollonius his conics and some other tracts of that author, the one of Ben Musa and the other of Abdelmelech, and one of them hath Eutocius his notes.

(Collins to J. Gregory, 25 March 1671, *Rigaud*, Vol. 2, p. 217.)

For the Istanbul MS of 1024 AD (Süleymaniye Aya Sofya 2762) and that in Teheran of 1290 AD (Kitabkhana-i Milli-i Malik 3597), see Toomer (1990).
55. Hiscock (1937), pp. 34,35.
56. *De sectione* is in BNP MS 2357, originally 2152 of the Royal Library, Paris; the copy made for Aldrich in 1704, now in the Bodleian Library, is Savile 58—some corrections on it are apparently by Halley.
57. Gibson and Johnson (1943). Carter (1975) lists no work by Grave from the Oxford Press. Halley seems to have done nothing in the matter, or if he did he did not bring it to a conclusion. The works of Greaves were published more than twenty years later by Thomas Birch (Greaves 1737).
58. In his earlier years Greaves travelled extensively in Italy, Turkey, and Egypt. He made astronomical observations and acquired manuscripts and coins and wrote on navigation. He corresponded with Golius, and produced a lexicon of Persian. His most notable works were his studies of Roman weights and measures and, above all, his investigations of the great pyramids near Cairo. He continues to be held in high regard as an Egyptologist.
59. BLO Seld. 3138,5 has two books of the *Sphaericorum*, and Seld. 3139 A46 and Hunt 308 have all three books. For Nasir al-Dīn al-Tūsī, see DSB.
60. N. Bernoulli to Montmort, No. 10, 30 December 1712, BAB. N. Bernoulli to de Moivre, 30 December 1713, BAB.
61. J. Bernoulli to P. Varignon 26 February 1706, BAB; J. Bernoulli to Halley, 6 April 1707, BAB, Mscr L Ia 674,66. Bernoulli to Halley is not in *Halley Corresp.*: see Appendix 7.
62. Burnet to Bernoulli, 26 March 1712, Bernoulli to Burnet, 24 August 1712, BAB, Nos. 24, 25.
63. Wollenschläger (1933).

64. For example, 'Halley's edition is a monument of scholarship, worthy of the author to which it is devoted' (Toomer 1990, p. xxvi). The recent editions of Books V–VII are those of Nix (1889) and Toomer (1990).
65. For critical assessments of Halley's reconstruction and Ibn al-Haytham's *Completion*, see Hogendijk (1984) and Toomer (1990).
66. *Principia*, Book I, Section I; for Newton's mathematics in *Principia*, see Chandrasekhar (1995).
67. *Principia*, Book I, Prop. 10, Prob. 5 and Prop. 11, Prob. 6; Book I, Section V, Lemma 17.
68. Barrow (1675); see Harrison (1978).
69. Letter to Roger Gale; Hearne (1884–), Vol. 2, p. 312; Vol. 6, p. 366.
70. Halley (1725b).
71. Halley (1714a).
72. Halley (1714b); also (1721b): Halley relates Rogers's results to his own discussion of a likely tract of westerly variation in the South Pacific. Cape St Lucar is the most southerly point of Baja California.
73. Halley (1715c).
74. Halley (1715e).
75. Halley (1715f).
76. Halley (1716b,c; 1719c).
77. Halley (1716d).
78. Halley (1717a,c,d, 1719a).
79. Halley (1717e).
80. Halley (1719b).
81. Halley (1720c): Halley argues that the diffraction image of Cassini's telescope would be too large; in (1720f; 1721d) Halley remarks that the positions of more stars in the Zodiac are needed.
82. Halley (1716f; 1721c).
83. Halley (1720d,e).
84. The geometry on which Halley based his argument was well known. Thus Gregory (23 November 1704: Hiscock (1937), p. 21) refers to an erroneous estimate of Keill. Gregory had discussed the topic with Newton in 1694—*Newton Corresp.*, Vol. 3, pp. 312, 317, 321.
85. Halley (1715b,d); reference to map in (1715h).
86. Newton gave Rémond de Monmort presents for himself and his wife, for which Montmort thanked him, writing of the 'ornaments given by Mr Newton and chosen by Mrs Barton whose wit and taste are equal to her beauty'. Letter of 25 February 1717, King's College Cambridge MS 101. See also Cook (1997).
87. For Cotes's observations, see his own account, Cotes to Newton, 29 April 1715, *Newton Corresp.*, Vol. 6, pp. 218–21; *Flamsteed*, p. 313.
88. Morrison, Stephenson, and Parkinson (1988).
89. J. Bernoulli to W. Burnet, 9 January 1709, No. 7; J. Bernoulli to P. R. de Montmort, 26 January 1719, No. 36, BAB. See Wollenschläger (1933): on 9 April 1710, 'vos plus grands mathématiciens, MM Newton et Halley', and on 23 November 1712, 'vos plus grands géometres, M Newton et M Halley'.

Notes to Chapter 13

1. Halley to Newton, London, 5 July 1687, *Newton Corresp.*, Vol. 2, p. 482.
2. Leibniz to J. Bernoulli, Hanover, 15 January 1704, Gerhardt (1962), pp. 739–40.

3. The posthumous publications, edited by John Bevis, are Halley (1749) (Latin) and (1752) (English and Latin). The lunar and solar tables were reprinted in France by the Abbé Chappe and the planetary and cometary tables by de LaLande (1759). John Bevis FRS (1693–1771), a friend of Halley's, was a medical man and amateur astronomer. He assisted Halley in the observation of the transit of Mercury in October 1736 (*Philosophical Transactions*, **42**, 622) and on other occcasions (see DSB).
4. Halley (1731).
5. The anonymous paper in *Philosophical Transactions*, Halley (1717b), is his, written as editor; see also *Flamsteed*.
6. Halley's original Greenwich observations, or perhaps a copy of them, are in four small quarto manuscript books bound in board and vellum in the archives of the Royal Observatory. Now in the University Library, Cambridge, and designated RGO 2/1,2,3 and 4, they were originally labelled OBS 1,2,3 and 4.

   Explanatory notes by John Bevis are pasted in. The series runs from 11 April 1719 to 31 December 1739. The copy in the Royal Astronomical Society is in one folio volume, MS Add. 38; it was made by order of the Lords Commisioners of the Admiralty for Francis Baily (Baily 1835a).
7. Halley (1731).
8. The books from the archives of the Royal Observatory are now in the University Library at Cambridge with the following identifications:

COMP	1:	RGO 2/	5
	2:		6
	3:		11
	4:		8
	5:		10
	6:		7

   RGO 2/9 is a book of miscellaneous observations and data, including Halley's observations at Islington in 1682–1684, and RGO 2/12 contains a summary of the results in COMP 1–6. The contents of COMP 1–6 are almost entirely rough sheets of calculations of the Moon's position at the times of observations of its right ascension, according to Halley's printed tables.

   John Bevis assembled and bound the sheets, and may have used them in his posthumous edition of Halley's *Tables*. There are additional pages elsewhere, in particular in RGO 2/9, already mentioned in Chapter 5. Most of the lunar entries in RGO 2/9 are observations made by Halley at Islington between 1682 and 1684, as indicated by his own note.
9. Thoren (1985).
10. Halley (1679a).
11. Whiteside (1976) has argued that Newton began the revision in 1695; the correspondence with Cotes shows that he was still engaged upon it until the last days of the preparation of the second edition of *Principia*.
12. Halley (1731); RSL JBC 7, 16 October 1689.
13. Cook (1988).
14. For curvature in Newton's celestial mechanics, see Brackenridge (1992), and generally for Newton's dynamics and celestial mechanics, Brackenridge (1995) and Chandrasekhar (1995).
15. For Newton's theories generally, see Chandrasekhar (1995).
16. See Cook (1988); Whittaker (1927), p. 83; see also Chandrasekhar (1995), p. 184.

    Take polar coordinates (r,θ) about the centre of attraction and put u equal to $1/r$. Let $F(u)$ be the central force. The equation for u is

    $$\frac{d^2 u}{d\theta^2} + u = \frac{F(u)}{h^2 u^3},$$

where h is the constant areal velocity.

Thus

$$F(u) = h^2 u^3 + h^2 u^2 \frac{d^2 u}{d\theta^2}.$$

If an orbit has the same form but rotates at an uniform rate in its plane, the angular variable will be $\kappa\theta$, while the areal velocity will be κh. The central force is then given by

$$F'(u) = \kappa^2 h^2 u^3 + \kappa^2 h^2 u^2 \frac{d^2 u}{\kappa^2 d\theta^2}$$

or

$$F'(u) = \kappa^2 h^2 u^3 + h^2 u^2 \frac{d^2 u}{d\theta^2}.$$

Hence

$$F'(u) - F(u) = \frac{h^2(\kappa^2 - 1)}{r^3}.$$

The difference of forces in the stationary and rotating orbits is proportional to $1/r^3$, and the rate of rotation, κ, is equal to the square root of.

$$1 + [F'(u) - F(u)]\frac{r^3}{h^2}.$$

The essential defect in this and Newton's treatment is the assumption that the rotating orbit has the same form as the fixed one. That is inconsistent with the dynamics.

17. Whiteside (1967–81), Vol. 6, Annual advance of lunar apogee, pp. 508–37.
18. Flamsteed (1680).
19. Conduit reported an occasion when Halley dined at the Mint and told Conduit of his progress with the Moon, with Newton's comment that, unlike other theories, his theory started from the cause (King's College, Cambridge, MS 129).
20. *A new and most accurate theory of the Moon's motion* had appeared first in English in an anonymous pamphlet, now very rare, of 1702 (Newton 1702) and almost simultaneously in Latin, in David Gregory's *Astronomiae physicae et geometricae elementa*. It was included in *Miscellanea curiosa* (Halley 1705a) as *The famous Mr Isaac Newton's theory of the Moon*. I. B. Cohen (1975) discussed the origin of the tract, its relation to Newton's papers, and the identity of the translator, whether into or out of English. Much is obscure. Cohen discounted any involvement of Halley in the publication of 1702. That seems right. There is no evidence that Halley was concerned, and being much occupied with naval matters up to the beginning of 1702, he is unlikely to have had the time to produce either the original English pamphlet or Gregory's Latin version; he was well aware of it, for he told Leibniz of it in Hanover in 1703 (Chapter 11).
21. Newton's 7th equation is a term in the longitude proportional to sin ξ. (Here ξ is the angular distance between the Sun and the Moon.) It is present in modern theories and the coefficient in *The theory of the motion of the Moon*, 2′ 20″, agrees well with the modern value of 2′ 5″. Newton estimated that it varied considerably. He did include it in the second edition of *Principia* but deleted it from the third (Koyré and Cohen 1972), and Halley omitted it: see Kollerstrom (1995).

22. There is a further difference, apart from the use of different angles. The angles used today are mean anomalies or longitudes, and the mean motions are constants of the theory. Newton wrote his theory in terms of true longitudes which change as successive corrections are applied according to Newton's specification. The difference matters for large effects but not for the small terms he gave in the scholium.
23. The entries in this table are taken from Kollerstrom (1995), Table 1.
24. See Cook (1988).
25. For references to Halley's discussions with Gregory and Newton, see Hiscock (1937), 25 March 1707 [1708].
26. For the last minute revision of the lunar theory, see Cohen (1971); for knowledge of Halley's Tables, de Maizeaux to Abbé Conti, 11 September 1720, *Newton Corresp.*, Vol. 7, 99–100.
27. Halley (1749); RG0 1.50I; le Monnier (1746); also Leadbetter (1735).
28. Gregory visited Newton in Cambridge in May 1694 and talked with him on many topics including the theory of the Moon (*Newton Corresp.*, Vol. 3, pp. 311–22, 327–44). A memorandum of July 1694, now in the University Library, Edinburgh, has more details of the developments of the lunar theory that Newton had in mind (*Newton Corresp.*, Vol. 3, pp. 384–9). Gregory's memoranda and Newton's correspondence show that, as early as September 1694, Halley and Gregory as well as Newton were studying the Moon's motion on the basis of Newton's dynamical theory and wanted to see Flamsteed's observations. Flamsteed complained that Newton had communicated the results of his theory to Gregory and Halley in breach of his promise to give Flamsteed a privileged account (for example, *Newton Corresp.*, Vol. 4, pp. 7, 8, 13, 26, 34, 37). By the late summer of 1695 Halley had begun his cometary calculations and seems not to have resumed work on the Moon until 1704, and perhaps not until March 1708 when Newton advised him to 'make tables'. Newton had himself constructed tables of the motion of apogee and the variation of the eccentricity which he had given to Flamsteed (Newton to Flamsteed, 23 April 1695, *Newton Corresp.*, Vol. 4, pp. 105–9).
29. The numerical parameters implied by the tabular values are:

    The sidereal mean period (that is, relative to the fixed stars) is 27.245d.

    The period of rotation of apogee is 8y. 333d. and that of the rotation of the nodes is 18.676y.

    The value for the secular motion of the node was close to observation and modern theory, but that for apogee was less satisfactory, reflecting both the failure of Newton's theory and the few observations that were available away from syzygies and quadrature.

    The annual equations for the corrections to the mean anomaly of the Moon, M_L, and the longitudes of apogee, ω, and node, Ω, are proportional to the sine of the Sun's mean anomaly, M_S. The following are the greatest values of the coefficients in Newton's Scholium and the values on which Halley's tables are based:

	Newton	Halley
δM_L	+11′ 50″	+11′ 49″
$\delta\omega$	−19′ 43″	−20′
$\delta\Omega$	+9′ 24″	+9′ 30″

    The modern value for δM_L is 11′ 8″.

    Halley's tabular values for the correction to the longitude of the Moon (*Aequatio semestris prima*) that depends on the annual argument, the distance of the Sun from the lunar apogee, $(v_S - \omega)$, are:

    $$\delta M_L = -3' 45' \sin 2(v_s - \omega);$$

    Newton's coefficient is also 3′ 45″.

The correction (*Aequatio semestris altera*) depends on the distance of the Sun from the node:

$$\delta M_L = -47'' \sin 2(v_s - \Omega);$$

again the coefficient is Newton's.
The last of this group (*Aequatio quarta lunae*) is Newton's 6th equation:

$$\delta M_L = -2' \ 25' \sin(\xi - \omega - \omega_S);$$

with ω_S the longitude of the Sun's apogee. The coefficient is again Newton's.

Halley showed how to use his tables in the worked example for 5 December 1725 when, at Greenwich, he observed the west limb of the Moon to cross the meridian at 9h 8min 5s, its right ascension being 42° 26' 15" and the distance from the lower limb to the zenith being 34° 09' 15".

The time is entered into a table that gives the true longitude of the Sun. The mean longitudes of the Moon, her apogee and node, all proportional to the mean motions multiplied by the time measured from an appropriate origin, are next found from the respective tables. They are adjusted, 'equated' in contemporary terms, for the effects taken from Newton's Scholium. Further corrections to the apogee and eccentricity of the Moon depend on the annual argument and correspond to the evection.

The extreme values of apogee do not occur symmetrically: the maxima are at 51° and 231° and the minima at 129° and 309°. The eccentricity has maxima of 0.066777 at 0° and 180° and minima of 0.043323 at 90° and 270°.

The true anomaly is found from the mean anomaly by the equation of the centre (see Appendix 1), using the corrected values of the longitudes of apogee and of the eccentricity.

The variation is obtained from a table that is entered with the argument (Moon's true anomaly–Sun's true longitude). The tables are constructed with an amplitude of 35' 10", less than the true values.

30. Halley (1731).
31. Halley (1731).
32. See the Conduitt papers in King's College, Cambridge, MS 129.
33. Baily (1835a).
34. Cohen (1975); Kollerstrom (1995).

Notes to Chapter 14

1. CSP,D, Car. II, 1673–5, 15 December 1674, p. 467; PRO SP Dom. 44/334, pp. 27–8; Forbes (1975).
2. Evelyn, *Diary*, 26 May 1671, 30 November 1680; Scott (1991).
3. CSP,D, Car. II, 1675–6, p. 7, 4 March 1675; PRO SP Dom. Entry Book 44, p. 10.
4. CSP,D, Car. II, 1675–6, p. 173, 22 June 1675; BLL Add. MS Birch 4393, f. 104r.,v.; PRO SP Dom. Entry Book 44 , p. 15.
5. *Hooke D1*, 23 June, 2 July 1675.
6. *Hooke D1*, 7 January 1676.
7. Flamsteed (1725), Vol. 1, p. 187.
8. The Moon was close to a star on 27 July, Halley observed Jupiter on 27 August, and he and Thomas Perkins did so again on 14 September. Halley observed the Moon on 28 September and 15 October: Flamsteed (1725), Vol. 1, pp. 31–3, 132, 231–3.

9. Flamsteed (1725), Vol. 1, pp. 234–8; Flamsteed also told Hevelius and Cassini of the results, and that the times differed by $9\frac{1}{4}$ min from those corresponding to an eclipse of 21 December 1673: Flamsteed to Hevelius, 9 January 1679, ULC RGO 1/43, ff. 56v.–57v.; Flamsteed to Cassini, 9 January 1679, ULC RGO 1/43, f. 58v.
10. Flamsteed (1725), Vol. 1, p. 149:

> invitatus aderat D. Halleius, qui mecum frequenter & Fabro meo easdem repetens certissimas semper & accuratissimas pronuntavit; testem adhibere in tante subtilitatis omino duxi accessarium, nec magis idoneum quenquam putavi.
> D. Halley, being invited, was there, who with me and my Smith often repeated the same (observations) very reliably and reported them most accurately; having that evidence of his acuteness, I took him as an assistant nor did I think anyone more worthy.

Jonas Moore reported the observations to the Royal Society: *History R. S.*, Vol. 3, p. 458, 23 January 1679.
11. Halley (1720c).
12. For Römer's visit to the Observatory, see *History R. S.*, Vol. 3, p. 482, 15 May 1678; Halley to Flamsteed, 7 June 1679, *Flamsteed Corresp.*, Vol. 1, pp. 694–5.
13. *Halley Corresp.*, p. 51; for the copy of Cassini's map of the Moon that Halley, with some difficulty, obtained for Flamsteed, see Flamsteed to Halley, 17 February 1681, *Flamsteed Corresp.*, Vol. 1, pp. 760–4, and Halley to Hooke, *Halley Corresp.*, p. 51.
14. *Flamsteed*, p. 123.
15. Flamsteed (1725), Vol. 1, pp. 288–90.
16. Eggen (1958).
17. Flamsteed to Halley, 21 November 1682, ULC RGO 1/36, f. 63 r. and v. The body of the letter is a report of some observations; it ends: 'Yr affectionate ffreind & servant, My most humble Service to yr Mrs'.
18. Flamsteed (1725), Vol. 1, pp. 110, 354, 357, 308, 112.
19. *Hooke D1*.
20. Flamsteed *vs.* Halley and Hooke—see Chapter 6.
21. *Flamsteed*, p. 160; also *Newton Corresp.*, 1695.
22. Wallis to Newton, *Newton Corresp.*, Vol. 4, pp. 300–1, Flamsteed to Newton, pp. 302–3.
23. Stukeley (1936). On another occasion, Conduitt noted.

> Dr Halley told me he pressed Sir I to compleat his theory of the Moon, saying nobody else would do it. Sir I said it has broke my rest so often I will think of it no more but afterwards told me that when Halley had made six years observations he would have t'other stroke at the Moon.
> (Conduitt papers, King's College, Cambridge, MS 129.)

24. *Newton Corresp.*, Vol. 4, p. 311.
25. *History R. S.*, 16 and 23 January 1679.
26. *Flamsteed*, p. 746.
27. For Arbuthnott see Beattie (1935), Beeks (1987), Gibson (1987), and R. Smith (1995).
28. Arbuthnott to Newton, 30 July 1706, *Newton Corresp.*, Vol. 4, p. 475.
29. *Flamsteed*, pp. 259 and 319.
30. Recommendations by Keill, Newton, Halley, and Pound, 12 May 1713, *Newton Corresp.*, Vol. 5, p. 408; Machin to ?Halley, 2 July 1706, *Rigaud*, Vol. 1, p. 255.
31. RSL JBC 10, 14 December 1710.
32. *Flamsteed*, pp. 322, 323; the copy was recently (1994) bought by Professor Owen Gingerich. The volume consists of sheets of the 1712 catalogue which have been bound in blue paper, similar to that found as binding of other Flamsteed manuscripts.

Towards the end of her life, Mrs Flamsteed wrote to the Vice-Chancellor of Oxford asking for the 1712 version to be removed from the Bodleian Library

because it was not Flamsteed's work (Mrs Flamsteed to Vice-Chancellor, 22 March 1726, *Flamsteed*, p. 363).
33. See *Flamsteed* for the printing of the first two volumes of Flamsteed (1725).
34. Flamsteed gave two different accounts: *Flamsteed*, pp. 97, 228-9, and 294-5.
35. *Flamsteed*, p. 98, Flamsteed's autobiographical notes.
36. For Newton's accounts and payments to Halley, see Newton to Oxford, 14 February 1712, *Newton Corresp.*, Vol. 5, pp. 224-5: 'the hard work of recalculating, cataloguing and reducing 500 additional places took him a year's hard labour'; Newton recommended a payment of £150; see Newton's account for the 1712 edition, 13 January 1716, *Newton Corresp*, Vol. 6.
37. Westfall (1980).
38. *Historia coelestis* (1712): see Bibliography, Halley (1712).
39. Flamsteed (1725), see Bibliography.
40. He wrote bitterly about it to Sharp on 29 August 1730: *Flamsteed*, p. 363.
41. *Flamsteed*, pp. 286–90.
42. Flamsteed to Pound, 15 November 1704, *Newton Corresp.*, Vol. 4, p. 424.
43. ULC RGO 1/74.
44. *Newton Corresp.*, Vol. 4, pp. 420–2.
45. *Flamsteed*, pp. 344, 345, 348, 352, 353. The progress of Sharp's calculations can be followed in a number of letters of Crosthwait from 24 July 1721 onwards: For Halley's anger, see Crosthwait to Sharp, 13 October 1722, *Flamsteed*, p. 352.
Crosthwait and Sharp clearly intensely resented Halley's appointment to succeed Flamsteed; thus, Crosthwait to Sharp, 18 March 1720, *Flamsteed*, p. 335: I believe you conjecture right about Dr. Halley; for some of his greatest admirers (that I have met with) cannot help saying that he will make a sine cure of it. I think what I heard him give for a reason some years ago, viz, that no person ought to enjoy that place after he was 60 years of age, may now very justly be returned upon him, who is almost 64 complete.
46. Chapman and Johnson (1982) have published an English translation, along with an English translation of the *Prefatio* to Halley's 1712 *Historia coelestis*.
47. Hearne (1884–), Vol. 7, 11 December 1719, p. 79, 10 January 1720, p. 88.
48. *Halley Corresp.*, pp. 11–12; Macclesfield to Jones, 14 January 1742, *Rigaud*, pp. 366–7; Macclesfield to Lord Hardwicke, n.d., *Rigaud*, pp. 368–70:

> Dr Halley could not hold out longer than a day or two ... it was upon this foot that my father, when in the post which you now enjoy, took it upon him to recommend Dr Halley to the royal professorship at Greenwich, and Mr Bradley to the Savilian [of astronomy] at Oxford and succeeded in both his recommendations; and he always thought it for his honour to have recommended two so able men.

49. A number of letters from Crosthwait to Sharp set out the dispute over Flamsteed's instruments: *Flamsteed*, pp. 333 ff.
50. ULC, Reports, Royal Observatory Visitors, Vol. 1, pp. 21, 23.
51. Howse (1975), 32–4.
52. ULC, Reports, Royal Observatory Visitors, Vol. 1, pp. 23–4.
53. Howse (1975), 126–8.
54. Whiston and Ditton (1713).
55. Journal of the House of Commons, June 1714.
56. Quill (1966).
57. Beaglehole (1974).
58. Halley (1732). The passage lasted from 2 February to 5 May and variations of more than 20° E were found in the Indian Ocean.
59. Halley (1721*a*); the table is Newton's, not hitherto published.
60. Halley (1725*a*).
61. Halley (1722*a*, 1737); the second paper is just one page of times.

62. Whoever is greedy for life does not wish to die on attaining his own world. Halley (1724*a,b*).
63. See Force (1985).
64. Halley (1727*a,b*); for Newton's chronological studies, see Westfall (1980), especially pp. 805–15.
65. See, for example, Humphreys (1993).
66. Halley (1731).
67. R. S. Club: the Club may have met in the Queen's Arms—'We heard yesterday you were well and drank to your good health at the Queen's Arms': Philip Miller to Halley, 21 December 1739, ULC RGO 2/17.
68. Halley to Newton, 16 February 1725, *Newton Corresp.*, *Halley Corresp.*, pp. 131–2; also perhaps, Halley to Newton, *Halley Corresp.*, p. 132; Bevis to Halley, 12 October 1739, ULC RGO 2/17; Nicholas to Halley, Ordnance Office, 26 October 1739, ULC RGO 2/17; Bouguer and la Condamine to Halley, May 1738, *Rigaud*, Vol. 1, pp. 333–5.
69. Stukeley (1936).
70. For the visit of Queen Caroline, see Halley to Sloane, 1 September 1729, *Halley Corresp.*, pp. 132–3; *Biogr. brit.*, Vol. **4**, pp. 2515–6; Cook (1997).
71. For Mary's death, see Boyer (1736), p. 215; for her burial, 14 February 1736, see Register of St Margaret's, Lee, and *Notes and Queries*, **153**, (1927), 212–13. On 25 December 1739 Gael Morris wrote to Halley about some calculations he had done for Halley and added 'Pray my service to Mrs Halley. I wish she had happened of a luckyer Partner for our ticket prove a Blank.' (ULC RGO 2/14) 'Mrs Halley' must be Halley's daughter Margaret.
72. Flamsteed to Halley—letter as in Note 17.
73. *Halley Corresp.*, pp. 258–60.
74. I am indebted to Dr James Keeler for a photograph of the rue Halley at Avignon.

Notes to Chapter 15

1. For an appreciation of Halley 250 years after his death, see Hide, Wolfendale *et al.* (1993).
2. RSL JBC 24 July 1689; *Halley Corresp.*, p. 216. Halley repeated the story on 3 January 1714 (RSL JBC 10) but then put it in 1683.
3. Halley to Sloane, 7 November 1722, BLL Sloane MSS, 4046, f. 307, *Halley Corresp.*, p. 131; Halley to Newton, 16 February 1725, *Newton Corresp.*
4. Rigaud (1844).
5. Fatio de Duilléer to Huygens, 29 April/9 May 1692, *Huygens Oeuvres*, Vol. **XXII**, p. 295. Note on letter of Wallis to Flamsteed, 28 December 1698, *Newton Corresp.*, Vol. 1, p. 290.
6. See Conduitt's Memoirs, King's College, Cambridge, MS 129.
7. See Luce (1952), pp. 53–102.
8. Luce (1952), p. 112.
9. Stock (1776).
10. Spence (1964), pp. 56, 99.
11. *Biogr. brit.*, Vol. 3, pp. 2130–7; *Dict. nat. biogr.*: Garth, Samuel.
12. Pope (1966), 6, 99; Sherburn (1956), Vol. 2, p. 25.
13. Spence (1964): Garth a papist, p. 35; for Towneley, p. 88; Ramsay, p. 61. The Chevalier Ramsay resided at the Court of France and was the author of the *Travels of Cyrus*. See also Fosenberg, A., 1959, The last days of Sir Samuel Garth, *Notes and Queries*, **204**, 272–4.

14. See Hunter (1995), p. 232.
15. For the Newtonian Revolution, see Cohen (1980).
16. Schama (1989).
17. Bullard (1956a,b).
18. For Laplace's tidal theory, see Laplace (1798–1823). For G. Darwin's work on tidal theory and analysis, see Darwin (1907). Very recent developments will be found in Parker (1991).
19. Bullard, Freedman, Gellman, and Nixon (1950).
20. Celsius to Swedenborg, 23 June 1741, Tafel (1875–7), pp. 578–80.
21. BAB: D. Bernoulli to Jallabert, 17 June 1750, Jallabert to D. Bernoulli, 26 July 1750.
22. Tanner (1926b), p. 420.
23. See Halley (1700b).
24. De Morgan (1847).

Bibliography

Sources

Much of Halley's scientific work is readily available in his books, his numerous papers in the *Philosophical Transactions* of the Royal Society, and in unpublished notes for talks in the archives of the Royal Society, many of them printed by MacPike (1932). There is other material in the Halley papers of the Royal Observatory, now in the University Library, Cambridge (ULC RGO.2) and in the Flamsteed papers (ULC RGO.1).

Halley's surviving correspondence is not extensive; MacPike (1932) listed that available to him and printed selections. A supplementary list is given in Appendix 7. His official activities are recorded in State Papers, Domestic and Foreign, in Admiralty papers, in the archives of the Senate of Venice and of the Innerösterreichische Hofkammer in Graz, and in the records of the East Indian and Royal African Companies.

Halley appears in some of the diaries of the period and in the correspondence of others, in particular in the diaries of Hooke, Pepys, Evelyn, and the memoirs of Hearne, and in the correspondence of Oldenburg, Flamsteed, Newton, Charlett, and Hevelius.

A: Primary sources (unprinted)

Basel, Universitätsbibliotek, Bernoulli correspondence

Cambridge,
  King's College library;
  University Library:
    Flamsteed papers, RGO.1
    Halley papers, RGO.2
    Newton MSS

Chelmsford, Essex Record Office, registers of St Margaret's, Barking

Graz,
  Steiermärkisches Landesarchiv:
    Records of the Innerösterreichische Hofkammer, HK 1703, I, IV, VII, IX, X

London,
  British Library:
    Stepney Papers, Add. MSS. 7058
    Finch Papers, Add. MSS. 29591/2
  Greater London Public Record Office:
    Parish Registers
  Guildhall Library, City of London:
    Parish Registers
  Public Record Office:
    Actions in Chancery: Bridges, Collins, Hamilton, Reynardson, Whittington
    Close Rolls, 53
    State Papers, Foreign, SP 42,44,80,84,105; MPF 352(4)
    Wills, Prerogative Court of Canterbury, Regs, Eure, Dycer, Admon Book 1684
  Royal Society:
    Collectanea Newtoniana
    Halley MSS

Journal Books
Council Minutes
Miscellaneous Manuscripts
Tower of London, Warders' Dividend Book

Northampton,
Northamptonshire County Record Office, Finch-Hatton papers FH 275

Oxford, Bodleian Library:
Acta Congregat. Univ. Oxon. 1703–1723, f. 251
Savile papers, now University Archives
Rigaud papers

Paris,
Bibliothèque Nationale:
Fonds latin, 10347, 10348, 10349
Observatoire:
MSS C, E

Venice,
Archivio di Stato:
Senato, Dispacci, Germania, filze 185, 186
Senato, Dispacci, Inghilterra, filza 76
Senato, Segreta, Corti, for 1703
Disegni, Provedditori della Camera dei Confini, Provedditori dei Fortezze
Biblioteca Marciana

Vienna,
Haus-, Hof-, und Staatsarchiv
Hofkammerarchiv:
Kartensammlung
Kriegsarchiv:
Kartensammlung

Note

Correspondence of Johann Hevelius: 95 of his letters are printed (see Béziat (1875); Gassendi, 1658, *Epistolae*, Vol. VI (Lyon); *Philosophical Transactions of the Royal Society*, 1668, 1684). Hevelius's own collection of the originals is in the Observatoire de Paris (Group C1). There were once 2700 items of which at least 570 were stolen by a notorious plunderer, Libri, including 4 letters of Halley. Hevelius himself made a copy of his letters, *Epistolarum clarissimorum virorum ad Dn Johanneum Helvetium*, now in the Bibl. Nat. Paris, Fonds latin, 10347, 10348, 10349. Four letters from Halley are there. For the misadventures of the correspondence, see Lalanne et Bordier (1851–3), *Dictionnaire des pièces volées aux bibliothèques*.

B: Printed sources (see also below, for abbreviations)

Biographia britannica. (1757). Vol. 4, pp. 2494–520. London.
Calendar of State Papers, Domestic, Charles II, Anne 1702–1703 and 1703–1704. London: HMSO.
Calendar of Treasury Books, XVII (1702), XVIII (1703), London: HMSO.
Correspondence of Isaac Newton ed. H. W. Tumbull. (1959). Cambridge University Press.
Éloge by Mairan, printed by MacPike (1932).
Memoir (anonymous), printed by MacPike (1932).

Preface to de LaLande (1759).
Halley in Cunningham (1836).
De Morgan (1847).
Dictionary of national biography, article on Edmond Halley, by Agnes Clark.
Reports of the Royal Commission of Historical Manuscripts: MSS of the House of Lords, (n.s., V), 1702–1704.

Notes
The *Éloge* by Mairan was written shortly after Halley's death and derives in part, though not completely, from the *Memoir* printed by MacPike (1932). The *Memoir*, which is anonymous, was found by S. P. Rigaud in the Bodleian Library. He observed that the author was a Cambridge man who had known both Halley and the professor of Hebrew at Cambridge, Sykes. MacPike thought the author was probably Martin Folkes, the President of the Royal Society at that time. The *Éloge* and the *Memoir* both show a good acquaintance and sound judgement of Halley's contributions to science, although later studies have shown that they are incomplete and in places somewhat imaginative.

The article in the *Biographia britannica* was based in part on information from Martin Folkes, a President of the Royal Society, who knew Halley well in his later years. It also drew on a manuscript memoir by Henry Price, Halley's son-in-law. Price is not always reliable and the accounts of some of Halley's early life show animus against the stepmother Joane and others. Price would not have had first-hand knowledge of Halley's younger days and his account appears at times to be a family tradition; it should not be accepted uncritically without supporting evidence. The style of articles in the *Biographia britannica* follows that of Bayle's *General dictionary* (Bayle 1697).

Despite gaps and deficiencies, the three earliest memoirs do give a good conspectus of Halley's achievements and include a few items not otherwise known. Along with de LaLande's *Preface* they show the esteem in which Halley was held abroad as well as at home. The notes to the *Biographia britannica* article show that the author was well informed of the state of astronomy in Halley's day and include comments on Halley's contributions.

Publications of Edmond Halley

(*P.T.: Philosophical Transactions of the Royal Society of London*).

1676a. Methodus directa et geometrica, cujus ope investigantur aphelia, eccentricitates, proportionesque orbium planetarum primariorum, absque supposita aequalitate anguli motūs, ad alterum ellipses focum, ab astronomis hactenus usupartā. Auth Edmundo Halley jun. è Collegio Reginae Oxon. *P.T.*, 11, No. 128, 683–6.

1676b. An extract of an account given by Mr Flamsteed of his own and Mr Edmund Halley's observations concerning the spots in the sun appearing in July and August, 1676. *P.T.*, 11, No. 128, 687–8.

1676c. Mr Edmund Halley's observations concerning the same occultation of Mars by the Moon, made at Oxford, Anno 1676, Aug. 21, P.M. *P.T.*, 11, No. 129, 724.

1677. Halley to Hooke, St Helena, 22 November 1677, with an observation on the Transit of Mercury. In Hooke: *Lectures and collections*; *Cometa* (1678), pp. 75–7.

1679a. *Catalogus stellarum Australium, supplementum catalogi Tychonici*. London: Thomas James for R. Harford.
[There is a brief factual summary in *P.T.*, 12, No. 141 (1678), 1032–4, printed 1679, which may be by Halley.]

1679*b*. *Catalogue des estoilles ou supplement du catalogue de Tycho.* Paris: Jean Baptiste Coignard, Imprimeur du Roy. It accompanies the *Carte du ciel* published by August Royer.

1679*c*. *A plain declaration of the vulgar new heavens flatform, Serving not onely fore this age but also fore the future age of 100 year.* (Written to accompany a dialling instrument.)? 1679.*

1681. Southern catalogue in Kirch (1681*b*).

1682. Observations made at Ballasore, in India, serving to find the longitude of that place, and rectifying very great errors in some famous modern geographers. *Philosophical Collections*, No. 5, 124–6.

1683*a*. A correction of the theory of the motion of the satellite of Saturn by that ingenious Astronomer Mr Edmund Halley. *P.T.*, **13**, No. 145, 82–8.

1683*b*. A theory of the variation of the magnetical compass *P.T.*, **13**, No. 148, 208–21.

1684. An account of the course of the tides at Tonqueen, in a letter from Mr. Francis Davenport July 15, 1678, with a theory of them, at the barr of Tonqueen by the learned Edmund Halley Fellow of the Royal Society. *P.T.*, **13**, No. 162, 677–88.

1685*a*. Johannis Hevelii Consulis Dantiscani, *Annus Climactericus*, Gedani 1685, Wherein (amongst other things) he vindicates the justeness of his celestial observations, against exceptions made by some against the accuracy of them. *P.T.*, **14**, No. 175, 1162–83. [Anonymous review but probably by Halley as editor.]

1685*b*. Edwardus Bernardus, *De mensuris e ponderibus*. *P.T.*, **14**, 1242–3. [Anonymous review but probably by Halley as editor.]

1686*a*. Advertisement to *P.T.*, **16**, 1.

1686*b*. A discourse concerning gravity, and its properties, whereby the descent of heavy bodies, and the motion of projects is briefly but fully handled; together with the solution of a problem of great use in gunnery. *P.T.*, **16**, No. 179, 3–21.

1686*c*. Two astronomical observations of the eclipses of the planet Jupiter, by the Moon in March and April, 1686, at London. *P.T.*, **16**, No. 181, 85–7. This paper, which is anonymous, reports observations made jointly by Halley and Hooke. In a later paper, *P.T.*, **16**, No. 183, 175–84, Halley summarised a number of foreign observations.

1686*d*. A discourse of the rule of the decrease of the height of the mercury in the barometer, according as places are elevated above the surface of the Earth; with an attempt to discover the true reason of the rising and falling of the mercury, upon the change of weather. *P.T.*, **16**, No. 181, 104–16.

1686*e*. An historical account of the trade winds and monsoons observable in the seas between and near the tropicks, with an attempt to assign the cause of the said winds. *P.T.*, **16**, No. 183, 153–68.

1686*f*. *Voiage de Siam des Peres Jesuites envoyez par le Roy aux Indes et la Chine. A Paris 1686 4°.* With *a remark concerning the Longitude of the Cape of Good Hope*. *P.T.*, **16**, No. 185, 249–54.

[Anonymous review but by Halley as editor: see Halley 1719*b*.]

1686*g*. *Ephemeris ad annum 1686.* London: Typis J. Heppenstall, impensis S. Cooper.

1687*a*. *Philosophiae naturalis principia mathematica*, Autore Is Newton Trin. Coll. Cantab. Soc. Matheseos Professore Lucasiano & Societatis Regalis Sodali. 4to Londini. *P.T.*, **16**, No. 186, 291–7. [Review by Halley.]

1687*b*. De constructionem Problematum solidorum, sive aequationum tertiae vel quartiae potetatis unica data parabola ac circulo efficienda. *P.T.*, **16**, No. 188, 335–43.

1687*c*. An estimate of the quantity of vapour raised out of the sea by the warmth of the sun: derived from an experiment shown before the Royal Society, at one of their late meetings. *P.T.*, **16**, No. 189, 366–70.

---

*Another version is *A declaration of the earthly flatform*.? 1679. They may not be by Halley.

1687d. De numero radicum in aequationibus solidis ac biquadraticis sive tertiae ac quartae potestatis, carumque limitibus. *P.T.*, **16**, No. 190, 387–402.

1687e. An account of some observations at Nuremburg by Mr P. Wurtzelbauer, showing that the latitude of that place has continued without alteration for 200 years last past; as likewise the obliquity of the eclipticks; by comparing them with what was observed by Bernard Walther in the year 1487, being a discourse read before the Royal Society in one of the late meetings. *P.T.*, **16**, No. 190, 403–6.

1687f. Letter to James II: 'May it please the King's Most Excellent Majesty ...' London, 12pp.

1687g. *Ephemeris ad annum 1687*. London: Typis J. Heppenstall, impensis S. Cooper.

1688. *Ephemeris ad annum 1688*. London: Typis J. Heppenstall, impensis S. Cooper.

1690. *Stellarum fixarum hemisphericum australe*. ?London.

1691a. An account of the circulation of the watery vapours of the sea, and of the cause of springs. *P.T.*, **17**, No. 192, 468–73.

1691b. A DISCOURSE tending to prove at what time and place, Julius Cesar made his first descent upon Britain. *P.T.*, **17**, No. 193, 495–9.

1691c. De visibili conjunctione inferiorum planetarum cum Sole, dissertatio astronomica. *P.T.*, **17**, 511–22.

1691d. Emendationes ac notiae in tria locata vitiose edita in textu vulgato Naturalis Historiae C. Plinii. *P.T.*, **17**, No. 194, 535–40.

1691e. An account of the measure of the thickness of gold upon gilt-wire, together with a demonstration of the exceeding minuteness of the atoms or constituent particles of gold. *P.T.*, **17**, No. 194, 540–2.

1692a. An account of the several species of infinite quantity and of the proportions they bear one to the other, as it was read before the Royal Society. *P.T.*, **17**, No. 195, 556–8.

1692b. An account of the cause of the change of the variation of the magnetick needle, with an hypothesis of the structure of the internal parts of the Earth. *P.T.*, **17**, No. 195, 563–78; {MC}.

1693a. An estimate of the degrees of mortality of mankind; drawn from curious tables of the births and funerals at the city of Breslaw; with an attempt to ascertain the prices of annuities upon lives. *P.T.*, **17**, No. 196, 596–610, 654–6.

1693b. An account of several experiments made to examine the nature of the expansion and contraction of fluids by heat and cold, in order to ascertain the divisions of the thermometer, and to make that instrument, in all places, without adjusting by a standard. *P.T.*, **17**, No. 197, 650–6.

1693c. A discourse concerning the proportional heat of the Sun in all latitudes, with the method of collecting the same. *P.T.*, **17**, No. 203, 878–85.

1693d. Emendationes ac notae in vetustas Albatenii observationes astronomicas cum restitutione tabularum lunisolarium ejusdem authoris. *P.T.*, **17**, No. 204, 913–21.

1693e. An instance of the excellence of the modern algebra, in the resolution of the problem of finding the foci of optick glasses universally. *P.T.*, **17**, No. 205, 960–9.

1693f. Some queries concerning the nature of light, and diaphanous bodies. *P.T.*, **17**, 998–9.

1693g. Some further considerations on the Breslaw Bills of Mortality. By the same hand. *P.T.*, **17**, No. 197, 654–6.

1694a. Methodus nova accurata et facilio inveniendi radices aequationum quarumcumque generaliter sine praeria reductione. *P.T.*, **18**, No. 210, 136–48.

1694b. An account of the evaporation of water as it was experimented in Gresham College in the year 1693. With some observations thereon. *P.T.*, **18**, No. 212, 183–90.

1694c. Monsieur Cassini his new and exact Tables for the eclipses of the first satellite of Jupiter, reduced to the Julian stile, and meridian of London. *P.T.*, **18**, No. 214, 237–56.

1694*d*. *Key of the mathematicks*, W. Oughtred, translated by Edmond Halley.
1695*a*. A discourse concerning a method of discovering the true moment of the Sun's ingress into the Tropical Signs. *P.T.*, **19**, No. 215, 12–18.
1695*b*. A most compendious and facile method for constructing the logarithms, exemplified and demonstrated from the nature of numbers, without any regard to the hyperbola, with a speedy method for finding the number from the logarithm given. *P.T.*, **19**, No. 216, 58–67.
1695*c*. A proposition of general use in the art of gunnery, showing the rule of laying a mortar to pass, in order to strike any object above or below the horizon. *P.T.*, **19**, No. 216, 68–72.
1695*d*. Propositio generalis arearum dimensionem exhibens in universo illo curvum genere quae revolutione aequabili circuli super basii quamvis vel rectilineam vel circularum describi possint; nempe omnium cycloidium vel epicycloidium quovis modo genitearum. Cum demonstratione quadraturae portionis epicycloidis Domino Caswell inventae. (*P.T.*, **19**, No. 217, 114); *P.T.*, **19**, No. 218, 125–8.
1695*e*. Some account of the ancient state of the City of Palmyra with short remarks upon the inscriptions found there. *P.T.*, **19**, No. 218, 160–75.
1696*a*. An easie demonstration of the analogy of the logarithmic tangents to the meridian line or sum of the secants: with various methods for computing the utmost exactness. *P.T.*, **19**, No. 219, 202–14.
1696*b*. Part of a letter from Mr Halley at Chester Oct. 26th 1696 giving an account of an animal resembling a whelp voided per anum by a male greyhound, and of a Roman altar found there. *P.T.*, **19**, No. 222, 316–18.
1697*a*. The true theory of the tides, extracted from that admired treatise of Mr Isaac Newton, intituled *Philosophiae naturalis principia mathematica*, being a discourse presented with that book to the late King James, by Mr Edmond Halley. *P.T.*, **19**, No. 226, 445–57.
1697*b*. A letter from Mr Halley at Chester, giving an account of an extraordinary hail in those parts on the 29th of April last. *P.T.*, **19**, No. 229, 570–2.
1697*c*. A letter from Mr Halley of June the 7th 97, concerning the Torricellian experiment tryed on the top of Snowdon-hill and the success of it. *P.T.*, **19**, 582–4.
1697*d*. Part of a letter from Mr Halley, dated Chester October 25 1697, giving an account of his observation there of the eclipse of the Moon on the 14th of the last month. *P.T.*, **19**, No. 235, 784.
1698. An account of the appearance of an extraordinary iris seen at Chester in August last. *P.T.*, **20**, 193–6.
1700*a*. De iride, sive de arca coelesti, dissertatio geometrica qua methodo directa iridis utriusque: diameter data ratione refractionis, obtinetur; cum problematis, solutioni inversi sive inventione rationis istius ex data arcus diametro. *P.T.*, **22**, No. 267, 714–24.
1700*b*. An advertisement necessary for all navigators bound up the Channel of England. *P.T.*, **22**, No. 267, 725–6; BLO Rigaud MS 88, f. 13. [Note: The single sheet, pp. 725, 726, is incorrectly bound after p. 666 in the copy of *Philosophical Transactions* in the University Library, Cambridge.]
1701*a*. An account of Dr Robert Hook's invention of the marine barometer, with its description and uses. *P.T.*, **22**, No. 269, 791–4.
1701*b*. *An advertisement necessary to be observed in the navigation up and down the Channel of England*, Communicated by a Fellow of the Royal Society. London: Printed for Sam. Smith and Benj. Welford, Printers to the Royal Society, at the Prince's Arms in St Paul's Church-Yard. Price Two Pence. BLL, 816.m.7.(95).
1701*c*. *A new and correct chart showing the variations of the compass in the Western and Southern Oceans as observed in ye year 1700 by his Maties Command by Ed. Halley.* London: Mount and Page.

1702*a*. *A new and correct sea chart of the whole world shewing the variations of the compass as they were found in the year m.d.c.c.* London: Mount and Page.

1702*b*. An account of the appearance of several unusual parahelia, or mock-suns, together with several unusual arches lately seen in the air. *P.T.*, 23, No. 278, 1127–8.

1704. The catalogue and character of most books of travel. In Churchill, A. and J. (1704).

1705*a*. *Miscellanea Curiosa*, 3 vols. London: Wale and Senex.*

1705*b*. Astronomiae cometicae synopsis. *P.T.*, 24, 1882–99.

1705*c*. *Astronomiae cometicae synopsis*. Oxon., fol.

1705*d*. *A synopsis of the astronomy of comets.* London: John Senex; 8vo.

1706*a*. Of compound interest and annuities, Chapter IV in Sherwin (1706).

1706*b*. A most compendious and facile method for constructing the logarithms, exemplified and demonstrated from the nature of numbers, without any regard to the hyperbola, with a speedy method for finding the number from the logarithm given. Reprint of Halley (1695*b*) in Sherwin (1706).

1706*c*. *Apollonii Pergaei de Sectione Rationis Libri Duo ex Arabico MS Latine versi. Accedunt ejusdem de Sectione Spatii Libri Duo restituti ... Opera et Studio Edmundi Halley.* Oxonii; 8vo.

1707. Methodus nova accurata et facilio inveniendi radices aequationum quarumcumque generaliter sine praeria reductione. Reprint of Halley (1694*a*) in Newton (1707).

1710. *Apollonii Pergaei Conicorum Libri Octo et Sereni Antissensis de Sectione Cylindri et Coni Libri Duo ... edidit Edmundus Halleius. Gr. & Lat.* Oxoniae; fol.

1712. *Historia coelestis libri duo. Quorum Prior Exhibet/ Catalogorum Stellaram Fixam Britannicum/ Novum et Completissimum/ una cum/ earundum/ Planetarum Omnium/ Observationibus/ Sextante, Micrometro &c. habitis/ Posterior/ Transitus Per Planum Arcus Meridionalis/ et Distantius eorum a Vertice/ Compleatiter/ Observante Johanne Flamstedis A.R./ In Observatorio Regio/ Grenovicensi/ Continua serie/ Ab Anno 1676 ad Annum 1705 Completum/ Londini/ Typis J Matthews MDCCXII.* The frontispiece is a portrait of Prince George of Denmark and there is a *Memorial* of him.

1714*a*. An account of several extraordinary meteors or lights in the sky. By Dr Edmund Halley Savilian Professor of Geometry at Oxon and Secretary to the Royal Society. *P.T.*, 29, No. 341, 159–69.

1714*b*. Some remarks on the variation of the magnetical compass published in the Memoirs of the Royal Academy of Sciences, with regard to the General Chart of the variations made by E. Halley; as also concerning the true longitude of the Magellan Streights. *P.T.*, 29, No. 341, 165–8.

1715*a*. Observationes quaedam accurate insignis cometae sub finem anni 1680 visi Coburgi Saxoniae a Domino Gottfried Kirch habitae, decimo tertio ante quam à quoque alio observatus sit. *P.T.*, 29, No. 342, 170–2. [Anonymous, but evidently by Halley.]

1715*b*. Observations of the late total eclipse of the Sun on the 22nd of April last past, made before the Royal Society at their house in Crane Court in Fleet Street London, by Dr Edmund Halley, Reg.Soc.Secr. With an account of what has been communicated from abroad concerning the same. *P.T.*, 29, No. 343, 245–62.

1715*c*. A short account of the cause of the saltness of the oceans, and of the several lakes that emit no rivers; with a proposal by help thereof, to discover the age of the

---

*None of the three editions of *Miscellanea Curiosa*, 1705, 1708, gives any indication of the identity of the compiler. Cohen (1975), who had the advice of Sir Edward Bullard in the matter, considered that there was no reason to contradict the assertion in the *Biographia britannica* (1757) that Halley was the compiler. Even if Halley did not assemble it, it is reasonable to include it among his publications since it contains a large proportion of his papers up to 1705.

world. Produced before the Royal Society by Edmund Halley, R.S.,Secr. *P.T.*, **29**, No. 344, 296–300.

1715*d*. Some accounts of the late great solar eclipse on April 22 1715. mane. Communicated to the Royal Society from abroad. *P.T.*, **29**, No. 345, 314–16.

1715*e*. A short history of the several new stars that have appeared within these 150 years; with an account of that in Colli Cygni, and of its continuance observed this year 1715. *P.T.*, **29**, No. 346, 354–6. [Anonymous, but evidently by Halley.]

1715*f*. An account of several nebulae or lucid spots like clouds, lately discovered among the fix'd stars by help of the telescope *P.T.*, **29**, No. 347, 390–2.

1715*g*. An account of Mr Dodwell's Book *De cyclis* in a letter to Robert Nelson, Esq. In Brokesby (1715), pp. 611–38.

1715*h*. *The Black Day or a prospect of Doomsday exemplified in the great and terrible eclipse which will happen on the 22nd of April 1715 ... and explaining the schemes thereof according to . . calculations by Mr Halley ... Mr Whiston etc.* London. [Anonymous, but evidently by Halley.]

Another version, undated: *A description of the passage of the shadow of the Moon over England in the total eclipse of the Sun, on the 22nd day of April 1715 in the morning.* London: J. Senex.

1716*a*. A series of observations on the planets, chiefly of the Moon, made near London ... being a proposal to find the longitude, &c. In Streete (1716).

1716*b*. An account of the late surprizing appearance of lights seen in the air, on the sixth of March last; with an attempt to explain the principal phaenomena thereof; As it was laid before the Royal Society by Edmund Halley J.V.D. Savilian Professor of Geometry, Oxon, and Reg. Soc. Secr. *P.T.*, **29**, No. 347, 406–28.

1716*c*. A description of the phenomenon of March 6 last as it was seen on the ocean, near the coast of Spain; With an account of the return of the same sort of appearance on March 31 and April 1 and 2 following. *P.T.*, **29**, No. 348, 430–2.

1716*d*. Methodus singularis quâ Solis parallaxis sive distantia à Terra, ope Veneris intra Solia conspiciendae, tuto determinare potuit; proposita coram Regia Societate ab Edm. Halleio J.U.D. ejusdem Societatis Secretario. *P.T.*, **29**, No. 348, 454–64. [Anonymous, but by Halley as editor.]

1716*e*. An account of the cause of the late remarkable appearance of the planet Venus, seen this summer, for many days together, in the day time. *P.T.*, **29**, No. 348, 466–8.

1716*f*. The art of living under water; or, a discourse concerning the means of furnishing air at the bottom of the sea, in any ordinary depths. *P.T.*, **29**, No. 349, 492–9. [see also 1721*c*, *P.T.*, **31**, 177–80.]

1717*a*. Postcript in vindication of his sea-chart made to show the variations of the compass, &c. In Frezier (1717).

1717*b*. Observationes stellae fixae in Geminis à corpore Jovis occultatae, Januarii 11mo. St.vet. 1717 & transitus arctissimi Martis infra Borealem in fronte Scorpii Febr.5 mane. *P.T.*, **30**, No. 351, 546.

1717*c*. An advertisement to astronomers of the advantages that may accrue from the observation of the frequent appulses to the Hyades during the three next ensuing years. *P.T.*, **30**, No. 354, 692.

1717*d*. An account of a small telescopical comet seen at London on the 10th June 1717 by Edm Halley, LL.D. R.Soc.Secr. *P.T.*, **30**, No. 354, 721–3. [Anonymous, but by Halley as editor of *Philosophical Transactions.*]

1717*e*. Considerations on the change of the latitude of some of the principal fixt stars. By Edmund Halley R.S.Sec. *P.T.*, **30**, No. 355, 736–8.

1719*a*. An account of an extraordinary METEOR seen all over England, on the 19th of March 1718/9. With a demonstration of the uncommon height thereof. By Edm Halley, LL.D. and Secretary to the Royal Society. *P.T.*, **30**, No. 360, 978–90.

1719b. An observation of the end of the total lunar eclipse on the 5th March 1718, observed near the Cape of Good Hope, serving to determine the longitude thereof with remarks thereon. By E. Halley R.S.Secr. *P.T.*, **30**, No. 361, 992–4; see Halley (1686f).

1719c. An account of the phenomena of a very extraordinary Aurora Borealis seen at London on November 10, 1719, both morning and evening. By Dr Edmund Halley R.S.Secr. *P.T.*, **30**, No. 363, 1099–100.

1719d. Zodiacus stellatus fixus omnes hectenus cognitas ad quas lunae appulsus ubi terrarum telescopii observari poterunt complexus.*

1720a. A new, exact, and easy method, of finding the roots of any aequations generally, &c. In Newton (1720).

1720b. *A new and correct chart, showing the variations of the compass in the Western Atlantic and Southern Oceans, as observed in ye year 1700* ... by E. H. London: Mount and Page.

1720c. Some remarks on a late essay of M.Cassini wherein he proposes to find by observation, the parallax and magnitude of Sirius. By Edmund Halley, LL.D. R.S.S. *P.T.*, **31**, No. 364, 1–4.

1720d. On the infinity of the sphere of the fix'd stars. By Edmund Halley, L.L.D. R.S.S. *P.T.*, **31**, No. 364, 22–4.

1720e. Of the number, order and light of the fix'd stars. By the same. *P.T.*, **31**, No. 364, 24–6.

1720f. Some remarks upon the method of observing the differences of right ascension and declination by cross hairs in a telescope. By Dr. Edm. Halley, Astr. Reg. R.S.S. *P.T.*, **31**, No. 366, 113–16.

1720g. A proposal for measuring the height of places by help of the barometer of Mr Patrick, in which the scale is greatly enlarged. *P.T.*, **31**, No. 366, 116–19.

1721a. Some remarks on the allowances to be made in astronomical observations for the refraction of the air. By Dr. Edm Halley R.S.S. Astronomer Royal. With an accurate table of refractions. *P.T.*, **31**, No. 368, 169–72. [The table (Newton's) not previously published.]

1721b. The variation of the magnetic compass observed by Capt. Rogers, Commander of the ship *Duke*, in his passage from Cape St Lucar in California to the Isle of Guam or Guana, one of the Ladrones, with some remarks thereon. Communicated by the same [Halley, that is]. *P.T.*, **31**, No. 368, 173–6.

1721c. An addition to the description of the art of living under water, publish'd in Phil. Transact. No. 349. By the same. *P.T.*, **31**, No. 368, 177–80.

1721d. On the method of determining the places of the planets by observing their near appulses to the fixed stars. By Edmund Halley, LL.D. Astron. Reg. & Reg.Soc.Soc. *P.T.*, **31**, No. 369, 209–11.

1721e. Observation of a parahelion, Oct 26th. 721. *P.T.*, **31**, No. 369, 211–12.

1722a. The longitude of Buenos Aires, determined from an observation made there by Pere Feuillèe. By Edm Halley LL.D. Astronomer Royal and FRS. *P.T.*, **32**, No. 370, 2–4.

1722b. Observatio eclipsis Solaris ab Edmundo Halleio, LL.D., RSS, Astronom. Reg. & Geom. Prof. Savil. Oxon. Novem. 27° 1722.p.m. Grenovici. *P.T.*, **32**, No. 374, 197.

1723a. Observations on the eclipse of the Moon, June 18, 1722 and the longitude of Port Royal in Jamaica determined thereby. By Dr Halley, Astronomer Royal FRS. *P.T.*, **32**, No. 375, 235–7.

---

*This chart of stars close to the orbit of the Moon was apparently engraved by Senex who concealed Halley's authorship (Crosthwait to Sharp, 8 October 1720, *Flamsteed*, pp. 339–40). Monmort received a copy from Halley in 1719—'un Planisphere curieux de M. Halley' (Monmort to N. Bernoulli, 8 June 1719, Bernoulli correspondence, BAB).

1723*b*. The longitude of Carthagena in America by the same. *P.T.*, **32**, No. 375, 237–8.

1724*a*. Some considerations about the cause of the universal deluge, laid before the Royal Society, on 12th December 1694 by Dr Edmund Halley FRS. *P.T.*, **33**, No. 383, 118–23.

1724*b*. Some further thoughts upon the same subject, delivered upon the 19th of the same month. By the same. *P.T.*, **33**, No. 383, 123–5.

1725*a*. An account of the appearance of Mercury, passing over the Sun's disk, on the 29th of October 1723, determining the mean motion, and fixing the nodes of the planet's orb. *P.T.*, **33**, 228–38.

1725*b*. Lectures used in the School of Geometry in Oxford, concerning the geometrical construction of algebraic equations and the numerical resolution of the same by the compendium of logarithms. 4 lectures. Appendix to Kersey (1725).

1726. Cometographia. In Gregory (1726*a* and *b*).

1727*a*. Remarks upon some dissertations lately publish'd at Paris, by the Rev. P. Souciet, against Sir Isaac Newton's chronology by Dr Edmund Halley, Astronomer Royal FRS. *P.T.*, **34**, No. 397, 205–10.

1727*b*. Some further remarks on P. Souciet's dissertations against Sir Isaac Newton's chronology, by Edmund Halley, L.L.D. Astron. Reg. In a Letter to Dr Jurin Coll.Med. and S.R.Soc. *P.T.*, **35**, No. 399, 296–300.

1727*c*. Astronomical tables. In Whiston (1727).

1728*a*. Astronomical observations at Vera Cruz by Mr Joseph Harris. Revised and communicated by Edm Halley L.L.D. Astron. Reg. & R.S.S. *P.T.*, **35**, No. 401, 388–9.

1728*b*. Atlas maritimus & commercialis or A general view of the world so far as it relates to trade and navigation. London: for James and John Knapford.

1731. A proposal of a method for finding the longitude at sea within a degree or twenty leagues By Dr Edmund Halley Astr. Reg. Vice-President of the Royal Society. With an account of the progress he hath made therein by a continued series of accurate observations of the Moon, taken by himself at the Royal Observatory at Greenwich. *P.T.*, **37**, No. 421, 185–95.

1732. Observations of latitude and variation, taken on board the Hartford in her passage from Java Head to St. Helena, Anno Dom. 1731/2. Communicated by Edmund Halley, LL.D. Regius Astronomer at Greenwich. *P.T.*, **37**, No. 424, 331–6.

1737. Observations made on the eclipse of the Moon on March 15, 1736. *P.T.*, **40**, 14.

Posthumous works

1749. *Edmundi Halleii Astronomi dum viveret Regii tabulae astronomicae accedunt de usu tabularum praecepta*, ed. J. Bevis. London.

1752. *Astronomical tables with precepts both in English and in Latin for computing the places of the Sun, Moon, &c.* London. [Translation of Halley (1749); see also, de LaLande (1759).]

1758. *Menelaus Sphaericorum libri III, quos olim, collatis MSS Hebraeis et Arabicis, Typis exprimendos curavit Vir Cl. E. HALLEIUS L.L.D.,R.S.S.& Geometriae Professor Savile.* With *Praefatio* added by G. Costard. Oxon.

Printed works other than those of Halley

Adams, J. (1780). *Essay concerning self-murther*. London: for T. Bennet.

Adams, J. C. (1853). On the secular variation of the Moon's mean motion. *Philosophical Transactions of the Royal Society*, **143**, 397–406.

Airy, G. B. (1863). *Essays on the invasion of Britain by Julis Caesar* ... London: privately printed.

Aiton, E. J. (1985). *Leibniz, a biography*. Bristol and Boston: Adam Hilger.
Akerman, J. Y. (1851). *Moneys received and paid for secret services of Charles II and James II*. London: Camden Society.
Akerman, S. (1991). *Queen Christina of Sweden and her circle*. Leiden, New York: Brill.
Andrewes, W. J. H. (ed.) (1997). *The quest for longitude*. Cambridge, Mass.
Anon. (1685). *An inquiry into and detection of the barbarous murder of the late earl of Essex*. Holland.
Anon. [Archduke Louis Salvator] (1871). *Der Golf von Buccari–Porto Re, Bilder und Skizzen*. Prague.
Armitage, A. (1951). Master Georg Dörffel and the rise of cometary astronomy. *Annals of Science*, 7, 303–15.
Armitage, A. (1966). *Edmond Halley*. London: Nelson.
Ashcraft, R. (1986). *Revolutionary politics and Locke's two treatises of government*. Princeton University Press.
Astle, T. (1767). *Familiar letters which passed between Abraham Hill Esq. and several eminent and ingenious persons of the last century*. London.
Aubrey, J. (1958). *Brief lives*, 3rd edn, ed. O. L. Dick. Vol. 1, pp. 282, 283. London.
Baily, F. (1835*a*). Some account of the astronomical observations made by Dr Edmund Halley, at the Royal Observatory at Greenwich. *Memoirs of the Royal Astronomical Society*, 8, 169–90.
Baily, F. (1835*b*). *An account of the Revd. John Flamsteed*. London: Lords Commissioners of the Admiralty. Reprinted 1966 by Krips Reprint Co., Holland.
Baily, F. (ed.) (1847). *A catalogue of 9766 stars in the southern hemisphere for the beginning of the year 1750, from the observations of the Abbé de Lacaille at the Cape of Good Hope in the years 1751 and 1752*. London.
Barrow, I. (1675). *Archimedis opera: Apollonii Pergaei Conicorum libri IIIi. Theodosii sphaerica: methodo novo illustrata, & succintè demonstrata*. 3 parts. London.
Bayle, P. (1697). *Dictionnaire historique et critique*. Rotterdam.
Bayle, P. (1734–41). *General dictionary*. London.
Beaglehole, J. C. (1974). *The life of Captain James Cook*. London: A. and C. Black.
Beattie, L. M. (1935). *John Arbuthnot, mathematician and satirist*. Cambridge, Mass.: Harvard University Press.
Beechey, F. W. (1848). Report of observations made upon Tides in the Irish Sea and upon the great similarity of the Tidal Phenomena of the Irish and English Channels ... *Philosophical Transactions of the Royal Society* for 1848, 105–16. (no volume nor part number)
Beechey, F. W. (1851). Report of the further observations upon the tidal streams of the North Sea and English Channel with remarks upon the laws by which those streams appear to be governed. *Philosophical Transactions of the Royal Society* for 1851, 703–718. (no volume nor part number)
Beeks, G. (1987). 'A Club of Composers': Handel, Pepusch and Arbuthnot at Cannons. In *Handel tercentenary collection*, (ed. S. Sadie and A. Hicks) pp. 209–21. Basingstoke: MacMillan.
Bemmelen, W. van. (1899). Die Abweichung der Magnetnadel; Beobachtungen, Säcular-Variation, Wert und Isogonensysteme bis zur Mitte des XVIII ten Jahrhunderts. Supplement to Royal Magnetical and Meteorological Observatory at Batavia, Observations, Vol. 21.
Bennett, J. A. (1982). *The mathematical science of Christopher Wren*. Cambridge University Press.
Bernoulli, J. (1708). *De motu musculorum*. London, apud R. Smith. (With a prefatory notice to the reader by Richard Mead)
Bertoloni Meli, D. (1993). *Equivalence and priority: Newton versus Leibniz*. Oxford: Clarendon Press.

Béziat, L. C. (1875). La vie et les travaux de Jean Hévélius. *Bulletino Bibliografia e Storia Scienzie Mathmetica e Physica*, **8**, 497–558, 589–669.

Biographia britannica. (1757). Article on Edmond Halley, Vol. 4, pp. 2494–520. London. Reprinted by Olm Verlag, Hildesheim, 1970.

Birch, T. (1752). *The life of John Tillotson, Lord Archbishop of Canterbury*. London.

Birch, T. (1756–7). *History of the Royal Society of London for Improving Natural Knowledge*. 4 vols.

Black, M. H. (1984). *Cambridge University Press, 1584–1984*. Cambridge University Press.

Boyer, A. (1736). *The political state of Great Britain*. Vol. 51, p. 215. London.

Brackenridge, J. B. (1992). The critical role of curvature in Newton's developing dynamics. In: Harman and Shapiro (1992).

Brackenridge, J. B. (1995). *The key to Newton's dynamics. Kepler's problem and the Principia*. Berkeley, Los Angeles, London: University of California Press.

Braddon, L. (1690). *Essex's innocency and honour vindicated. As proved before the Right Honourable (late) Committee of Lords, or ready to be deposed. In a letter to a friend, Lawrence Braddon of the Middle Temple, Gent. 1690*. In Somers (1809–15), Vol. 10, pp. 72–174.

Braddon, L. (1725). *Vindication of the Earl of Essex from Bp. Burnet's false charge of self-murder*. London.

Bradley, J. (1726). The longitude of Lisbon and the Fort of New York from Wansted and London determined by eclipses of the first satellite of Jupiter. *Philosophical Transactions of the Royal Society*, **34**, No. 394, 85–90.

Brahe, Tycho. (1666). *Historia coelestis*. Aug, Vind.

Brokesby, F. (1715). *Life of Mr Henry Dodwell*. London.

Bromley, J. S. (1970). *The new Cambridge modern history*, Vol. 6. Cambridge University Press.

Brown, H. (1934). *Scientific organizations in seventeenth century France (1620–1680)*. Baltimore: Williams and Wilkins.

Bryden, D. J. and Simms, D. L. (1993). *Annals of Science*, **50**, 1–32.

Bullard, E. C. (1956a). Edmond Halley, (1656–1741) *Endeavour*, **15**, 189–99.

Bullard, E. C. (1956b). Edmond Halley, the first geophysicist. *Nature*, **178**, 891–2.

Bullard, E. C., Freedman, Cynthia, Gellman, H., and Nixon, Jo. (1950). The westward drift of the earth's magnetic field. *Philosophical Transactions of the Royal Society*, A, **243**, 563–78.

[The estimate of the rate of the westerly drift in this paper was the first since Halley's (Halley 1692b)].

Burchett, J. (1720). *A complete history of the most remarkable transactions at sea*. London: Walthoe and Walthoe.

Caputo, F. (1982). *Le carte del impero*. Venice.

Carter, H. (1975). *A history of the Oxford University Press*, Vol. 1. Oxford: Clarendon Press.

Cassini, Anna. (1994). *Gio. Domenico Cassini Uno scienziato del seicento*. Torino and Commune di Perinaldo.

Cassini, J. D. (1681). *Observations sur la comète qui a paru au mois de Décembre 1680. Et en Janvier 1681. Présentées au Roy*. Paris.

Cassini, J. D. (1684). *Les elemens d'astronomie verifiez par Monsieur Cassini*. Paris: de l'Imprimerie royale. Reprinted in: Académie Royale des Sciences. (1693). *Recueil d'observations faites en plusieur voyages par ordre de Sa Majesté*. Paris: de l'Imprimerie royale.

Cellarius, Andreas. (1660). *Atlas coelestis seu harmonia macrocosmica*. 2nd edn, 1661. Amsterdam.

Challis, C. E. (1992). Lord Hastings to the great silver recoinage, 1464–1699. In *A new history of the Royal Mint* (ed. C. E. Challis). Cambridge University Press.

Chandrasekhar, S. (1995). *Newton's* Principia *for the common reader.* Oxford: Clarendon Press.
Chapman, A. (1990). *Dividing the circle.* London, New York: Ellis Horwood.
Chapman, A. (1994). Edmond Halley's use of historical evidence in the advancement of science. *Notes and Records of the Royal Society of London*, **48** (2), 167–91.
Chapman, A. and Johnson, A. D. (1982). *The Preface to John Flamsteed's Historia coelestis Britannica.* Maritime Monographs and Reports, No. 52. Greenwich: National Maritime Museum.
Chapman, J. and André, P. (1774). *A map of the county of Essex.* Essex Record Office reproduction.
Chapman, S. (1941). Edmond Halley as a physical geographer and the story of his charts. *Occasional Notes of the Royal Astronomical Society*, No. 9, 122–34.
Chen, Y. T. and Cook, A. H. (1993). *Gravitational experiments in the laboratory.* Cambridge University Press.
Chester, J. L. (1876). *The marriage, baptismal and burial registers of the collegiate church or abbey of St Peter's Westminster.* London: Harleian Society, Vol. 11.
Chester, J. L. (1886–). *Westminster marriage licences.* London: Harleian Society, Vols 23,30,31,33,34.
Churchill, A. and J. (1704). *A collection of voyages and travels etc.* With a general Preface by John Locke and 'The catalogue and character of most books of travel' by Edmund Halley. London.
Ciampini, G. (1693). *De sacris aedificiis a Constantino Magno constructis.* Rome.
Clarendon, Hyde, E., first earl of. (1702–4) *The history of the rebellion and civil wars in England.* 3 vols. Oxford University Press.
Clark-Kennedy, A. E. (1929). *Stephen Hales, D.D.,F.R.S.* Cambridge University Press.
Cohen, I. B. (1971). *Introduction to Newton's* Principia. Cambridge University Press.
Cohen, I. B. (1975). *Isaac Newton's* Theory of the Moon's motion, *(1702).* London: Wm. Dawson.
Cohen, I. B. (1980). *The Newtonian Revolution.* Cambridge University Press.
Cohen, I. B. (1990). Halley's two essays on Newton's *Principia.* In Thrower (1990).
Cohen, I. B. (1995). *Science and the founding fathers.* New York and London.
Collins, Greenville. (1693, 1723). *Great Britain's coastal pilot.* First Part, the Thames westward; Second Part, the Thames northward. London: Mount.
Cook, A. H. (1976). The history of geodesy and geophysics in Britain. *Fisica e Geologia Planetaria, Atti. Convegni Lincei*, **25**, 259–80.
Cook, A. H. (1984a). Halley in Istria, 1703; navigator and military engineer. *The Journal of Navigation*, **37**, 1–23.
Cook, A. H. (1984b). The election of Edmond Halley to the Savilian Professorship of Geometry. *Journal of the History of Astronomy*, **15**, 34–6.
Cook, A. H. (1985). An English astronomer on the Adriatic. *Mitteilungen des österreichischen Staatsarchivs*, **38**, 123–62.
Cook, Alan. (1988). *The motion of the Moon.* Bristol and Philadelphia: Adam Hilger.
Cook, A. H. (1990). Halley, surveyor and military engineer: Istria, 1703. In Thrower (1990).
Cook, Alan. (1991). Edmond Halley and Newton's *Principia. Notes and Records of the Royal Society of London*, **45**, 129–38.
Cook, Alan. (1993). Halley the Londoner. *Notes and Records of the Royal Society of London*, **47**, 163–77.
Cook, Alan. (1994). *Observational foundations of physics.* Cambridge University Press.
Cook, Alan. (1996). Leibniz and the Royal Society. *Notes and Records of the Royal Society of London*, **50**, 151–63.
Cook, Alan. (1997). Ladies in the scientific revolution. *Notes and Records of the Royal Society of London*, **51**, 1–12.

Cooke, E. (1712). *A voyage to the South Seas and round the World*. 2 vols. London.
Cooper, C. H. (1845). *Annals of Cambridge*. Cambridge.
Costa, G. (1968). Documenti per una storia dei rapporti anglo-romani. *Saggi sul Settecento*, 437. Naples.
Cottingham, J., Stoothoff, R., and Murdoch, D. (1985). *The philosophical writings of Descartes*, Vol. 1. Cambridge University Press.
Cudworth, W. (1889). *Life and correspondence of Abraham Sharp*. London.
Cunningham, G. G. (1836). *Lives of eminent and industrious Englishmen*, Vol. 4. London.
Dampier, W. (1939). *A voyage to New Holland* (ed. J. A. Williamson). London: Argonaut Press.
Darwin, Sir G. (1907). *Scientific papers*. Vol. 1. *Oceanic tides*. Cambridge.
Davis, R. (1967). *Aleppo and Devonshire Square*. London.
Débarbat, S. (1990). Newton, Halley and the Paris Observatory. In Thrower (1990), pp. 27–52.
de Krey, G. S. (1985). *A fractured society: the politics of London in the first age of party, 1688–1715*. Oxford: Clarendon Press.
de la Caille, N. L. (1763). *Coelum Australe stelliferum*. Paris.
de la Feuille, Jacob. (1710). *Atlas*. Paris.
de la Hire, P., (1677). Extrait d'une Lettre de M de la Hire à l'Auteur du Journal touchant la Problème contenu dans la Methode Geometrique de M Hally dont il a esté parlé dans le Journal precedent. *Journal des Sçavans*, 93–5.
de la Hire, P., Sedileau and Pothenot, L. (eds.) (1693). *Divers ouvrages de mathématique et physique. Par messieurs de l'Académie royale des Sciences*. Paris: Académie des Sciences.
de LaLande, J. J. (1759). *Tables astronomiques de M Halley pour les planetes et les cometes ... et l'histoire de la comete de 1757*. Paris: Durand.
Delambre, J. B. J. and Méchain, P. F. A. (1806–10). *Base du système métrique décimal ...* 3 vols. Paris.
De Morgan, A. (1847). 'Halley'. In *Cabinet portraits of British worthies* (ed. Charles Knight), Vol. 11, p. 12.
Dilke, O. A. W. (1954). *Statius Achilleid*. Cambridge University Press.
di Palma, W. and Bovi, T. (1990). *Cristina di Svezia, scienza ed alchimia nella Roma barocca*. Roma: Dedalo.
Dobbs, B. J. T. (1991). *The Janus faces of genius*. Cambridge University Press.
Dollot, R. (1961). *Trieste et la France, (1702–1958), Histoire d'un consulat*, pp. 16–18. Paris: a Pedone.
Dörffel, G. M. (M.G.S.D.) (1681). *Astronomische Betrachtung des grossen Cometen ...* Plauen.
Dunbain, J. P. D. (1986). College estates and wealth. In Sutherland and Mitchell (1986).
Dunkin, E. H. W., Jenkins, C., and Fry, E. A. (1929). *Index to the Act Books of the Archbishops of Canterbury, 1663–1859*. London: British Record Society.
Eamon, W. (1994). *Science and secrets of nature*. Princeton.
Earle, P. (1989). *The making of the English middle class: business, society and family life in London, 1660–1730*. London: Methuen.
Earle, P. (1994). *A city full of people: men and women of London, 1650–1750*. London: Methuen.
Eggen, O. J. (1958). Flamsteed and Halley. *Occasional Notes of the Royal Astronomical Society*, 3, 211–21. Better versions of the letter are in *Flamsteed Corresp.*, Vol. 2.
Ehrman, J. P. W. (1953). *The Navy in the war of William III, 1689–1697: its state and direction*. Cambridge University Press.
Evelyn, J. (1950–2). *Diary and correspondence of John Evelyn FRS* (ed. W. Bray), 4 vols. London: Henry Colburn.

Fairer, D. (1986). Anglo-Saxon studies. In Sutherland and Mitchell (1986).
Ferguson, R. (1684) *An inquiry into and detection of the barbarous murther of the late Earl of Essex*. Two issues in English, a French and a Dutch translation.
Field, J. V. and James, F. A. J. L. (eds.) (1993). *Renaissance and revolution*. Cambridge University Press.
Flamsteed, J. (1676). Mr Flamsteed's letter concerning his observations and those of Mr Towneley, and Mr Halton, of the late eclipse of the Sun. *Philosophical Transactions of the Royal Society*, 11, No. 127, 662–4.
Flamsteed, J. (1680). *The doctrine of the sphere, grounded on the motion of the Earth and the antient pythagorean and Copernican system of the World. In two parts*. London: A. Godbid and J. Playford.
Flamsteed, J. (1725). *Historia coelestis Britannica*. London.
Forbes, E. G. (1975). *Greenwich Observatory*, Vol. 1, p. 18. London: Taylor and Francis.
Force, J. E. (1985). *William Whiston, honest Newtonian*. Cambridge University Press.
Foster, J. (1889). *The Register of Admissions to Grays Inn, 1521–1889, with marriages in Grays Inn Chapel*. London: Hansard.
Foster, J. (1891-2). *Alumni Oxonienses, 1500–1714*. Vols 1–4, early series. Oxford: Parker.
Frezier, A. F. (1717). A *voyage to the South-Sea ... in ... , 1712. '13 & '14*. London; translated from the French.
Gallet, J.-C. (1677). *Mercurius sub Sole visus Anenione die 7 Novembriis 1677*. Avignon.
Geiter, M. K. (1993). The incorporation of Pennsylvania and late Stuart politics. Cambridge Ph.D. thesis, unpublished.
Gellibrand, H. (1635). *A discourse mathemticall on the variation of the magneticall needle*. London.
Gerhardt, C. I. (1962). *G. W. Leibniz Mathematische Schriften*, Vol. III/2. Hildesheim: Georg Olms.
Gibson, E. (1987). The Royal Academy of Music and its directors, 1719–28. In *Handel tercentenary collection* (ed. S. Sadie and A. Hicks), Basingstoke: MacMillan. pp. 209–21.
Gibson, S. and Johnson, J. (1943). *The First Minute Book of the delegates of the Oxford University Press*, Oxford University Press.
Gilbert, W. (1600). *De magnete magneticisque corporibus et de magno magnete tellure ...* London: Petrus Short.
Gill, I. S. (1878). *Six months in Ascension*. London.
Giorgetti Vichi, A. M. and Mottironi, S. (1961). *Catalogo dei manoscritti della biblioteca Vallicelliana*, Vol. 1. Roma: Istituto Bibliografico dello Stato.
Glass, D. V. (1966). *London inhabitants within the walls, 1695*. London Record Society, publ. 2. Leicester.
Goldie, M. (1993). John Locke, Jonas Proast and religious toleration 1688–1692. In Walsh, Haydon, and Taylor (1993).
Goldie, M. (1996). Joshua Basset, popery and revolution. In *Sidney Sussex College, Cambridge historical essays* (ed. D. E. D. Beales and H. B. Nisbet). Woodbridge: Boydell Press.
Gosse, P. (1938). *St Helena, 1502–1938*. London: Cassell.
Graetzer, J. (1883). *Edmund Halley und C. Neumann (Ein Beitrag zur Geschichte der Bevölkerung-Statistik Beilagen)* Breslau.
Greaves, J. (1737). *Miscellaneous works of Mr John Greaves, Professor of Astronomy in the University of Oxford*. 2 vols. Published by Thomas Birch MA FRS. London: J. Hughes for Brindley and Corbett.
Greaves, R. (1992). *Secrets of the kingdom: British radicals from the Popish plot to the Revolution of 1688–89*. Stanford University Press.
Greenwood, Fiona. (1992). Undergraduate thesis, Faculty of History, University of Cambridge, unpublished.

Gregory, D. (1702). *Astronomiae physicae et geometricae elementa*. Oxford
Gregory, D. (1726a). *The elements of physical and geometrical astronomy. To which is annexed Dr Halley's synopsis of the astronomy of comets*. Revised by Edmund Stone FRS, 2 vols. London.
Gregory, D. (1726b). *D Gregorii astronomiae elementa physica & geometricae. Cometographia Halleiana in modum appendicis*. Geneva.
Gregory, J. (1663). *Optica promota*. London: J. Hayes for S. Thompson*
Gunther, R. W. T. (1935). *Early science in Oxford*, Vol. 10, Oxford.
Gwynn, R. D. (1985). *Huguenot heritage*. London: Routledge, Keegan, Paul.
Hadley, G. (1735). *Philosophical Transactions*, **39**, 39, 58ff.
Hall, A. R. (1980). *Philosophers at war*. Cambridge University Press.
Hall, A. R. and Hall, M. B. (1977). *Correspondence of Henry Oldenburg*, Vol. 11. Mansell
Hall, A. R. and Hall, M. B. (1986a). *Correspondence of Henry Oldenburg*, Vol. 12. London: Taylor and Francis.
Hall, A. R. and Hall, M. B. (1986b). *Correspondence of Henry Oldenburg*, Vol. 13. London: Taylor and Francis.
Hall, M. B. (1991). *Promoting experimental learning*. Cambridge University Press.
Hamilton, A. (1965). *Count Gramont at the court of Charles II* (ed. and trans. N. Deakin). London.
Harman, P. M. and Shapiro, A. E. (eds.) (1992). *The investigation of difficult things*. Cambridge University Press.
Harrison, Edward, Lt. (1696). *Idea longitudinis, being a brief definition of thr best known axioms for finding the longitude*. London: Sellers, Mount and Leigh.†
Harrison, J. (1978). *The library of Isaac Newton*. Cambridge University Press.
Hasted, E. (1778–99). *The history and topographical survey of the County of Kent*, 4 vols. Canterbury.
Hattendorf, J. D. (1980). The machinery for the planning and execution of English Grand Strategy in the War of Spannish Succession, 1702–1713. In *Changing interpretations and new sources in naval history (Papers from the Third United States Naval Academy History Symposium)* (ed. R. W. Love Jnr), pp. 80–95. New York and London.
Hattendorf, J. D. (1983). English Grand Strategy and the Blenheim Campaign of 1704. *History Review*, **5**, 3–19.
Hawking, S. and Israel, W. (1987). *300 years of gravitation*. Cambridge University Press.
Haydn, J. (1894). *Dignities*. London.
Hearne, T. (1884–). *Remains and collections*. Oxford Hist. Soc.
Heath, T. L. (1896). *Apollonius of Perga treatise on the conic sections*. Cambridge University Press.
Heiberg, J. L. (1974). *Apollonii Pergaei quae Graece exstant cum commentariis antiquis*, 2 vols, Prolegomena to Vol. 2. Leipzig, 1891–3; reprinted Stuttgart: B. G. Teubner.
Hermann, J. (1704). Méthode géometrique et générale de determiner le diamètre de l'Arc-en-Ciel. *Nouvelles de la Republique des Lettres*, June 1704, **4**, 658–71.
Hevelius, J. (1647). *Selenographia*. Danzig.
Hevelius, J. (1662). *Venus in Sole visu*. Danzig.
Hevelius, J. (1665). *Prodromus cometicus*. Danzig.
Hevelius, J. (1668). *Cometographia*. Danzig.
Hevelius, J. (1670). *Philosophical Transactions of the Royal Society*, **5**, 2060.
Hevelius, J. (1673). *Machina coelestis*, Part I. Danzig.
Hevelius, J. (1679). *Machina coelestis*, Part II. Danzig.

---

*The copy in the University Library at Cambridge bears Flamsteed's signature on the title page.
†The copy in the Royal Society library has a MS dedication to the Society by the author.

Hevelius, J. (1685). *Annus climactericus.* Danzig.
Hevelius, J. (1690). *Prodromus astronomiae.* Danzig. (posthumous)
Heywood, G. (1994). Edmond Halley—Actuary. *Quarterly Journal of the Royal Astronomical Society,* **35**, 151–4.
Hide, R., Wolfendale, A., Ronan, C., Chapman, A., Cook, A., Hughes D. W., and Malin, S. R. C. (1993). Edmond Halley—A commemoration. *Quarterly Journal of the Royal Astronomical Society,* **34**, 135–149.
Hill, C., (1980). *The century of revolution 1603–1714,* 2nd edn. London: Nelson.
Hillairet, J. (1962). *Dictionnire historique des rues de Paris,* 2 vols. Paris: Les Éditions de Minuit.
Hiscock, W. G. (ed.) (1937). *David Gregory, Isaac Newton and their circle. Extracts from David Gregory's Memoranda, 1677–1708.* Oxford, for the Editor.
Hogendijk, J. P. (1984). *Ibn Al-Haytham's completion of the Conics.* New York: Springer-Verlag.
Hogendijk, J. P. (1986). Arabic traces of the lost works of Apollonius. *Archive for the History of the Exact Sciences,* **35**, 187–253.
Hogwood, C. (1984). *Handel.* London: Thames and Hudson.
Hooke, R. (1674). *Animadversions on the first part of the Machina Coelestis of the honourable, learned and deservedly famous astronomer, Johannes Hevelius, Consul of Danzig; together with an explication of some instruments made by Robert Hooke, Professor of Geometry in Gresham College and fellow of the Royal Society.* London: John Martyn, Printer to the Royal Society.
Hooke, R. (1678). *Lectures and collections, cometa, microscopium.* London: J. Martyn.
Hooke, R. (1935). *Diary, 1672–80,* transcribed and ed. H. W. Robinson and W. Adams. London.
Hooke, R. *Diary, 1688–93.* In Gunther (1935).
Hooper, W. (1774). *Rational recreations,* 4 vols. London
Horrocks, J. (1673). *Jeremiae Horroccii opera posthuma.* Collected by John Wallis. London: G. Godbit for J. Martyn.
Houghton, J. (1727–8). *Improvement of husbandry and trade* (ed. R. Bradley), 4 vols. London.
Howarth, D. (1993). *Art and patronage in the Caroline Courts.* Cambridge University Press.
Howarth, R. G. (1933). *Letters and second diary of Samuel Pepys.* London.
Howe, Elizabeth. (1992). *The first English actresses.* Cambridge University Press.
Howse, D. (1975). *Greenwich Observatory,* Vol. 3, *The buildings and instruments.* London: Taylor and Francis.
Humphreys, C. (1993). The Star of Bethlehem. *Science and Christian Belief,* **5**, 83–101.
Hunter, M. (1982). *The Royal Society and its fellows, 1660–1700.* Monographs, British Soc. Hist. Sci.
Hunter, M. (1989). *Establishing the new science, The experience of the Royal Society.* Woodbridge: Boydell.
Hunter, M. (1995). *Science and the shape of orthodoxy.* Woodbridge: Boydell.
Hyde, R. (1981). *The A to Z of Georgian London.* Lympne Castle: Harry Margary.
Hyde, R. (1992). *The A to Z of Restoration London.* Lympne Castle: Harry Margary.
Inner Temple. (1877). *Students admitted to the Inner Temple 1547–1660.* London: The Inner Temple.
Jeffreys, H. (1920). Tidal friction in shallow seas. *Philosophical Transactions* A, **221**, 239.
Jeffreys, H. (1926). On the dynamics of geostrophic winds. *Quarterly Journal of the Royal Meteorological Society,* **52**, 85–104.
Johns, C. M. S. (1993). *Papal art and cultural politics.* Cambridge University Press.

Jones, P. E. and Reddaway, T. F. (1967). *The survey of building sites in the City of London after the Great Fire of 1666 by Peter Mills and John Oliver*, Vol. 1, Index to Vols 1–5. London Topographical Society, No. 103.
Joy, Lynn S. (1987). *Gassendi the atomist*. Cambridge University Press.
Kaye, G. W. C. and Laby, T. H. (1995). *Tables of physical and chemical constants*, 16th edn. Harlow: Longman.
Kepler, J. (1618). *Epitome astronomiae Copernicanae*. Lentijs.
Kepler, J. (1627). *Tabulae Rudolphinae*. Ulm.
Kepler, J. (1989). *The new astronomy* (translation of *Astronomia nova* by W. Donahue). Cambridge.
Kersey, J. (1725). *The elements of that mathematical art called algebra*. London: Page and Mount.
King, R. (1994). *Henry Purcell*. London: Thames and Hudson.
Kirch, G. (1681a). *Neue himmels Zeitung, Novus nuncius coelestis*. Nuremberg.
Kirch, G. (1681b). *G Kirchii annus 1 ephemeridum motuum coelestium ad annum ... 1681 Cui acessit catalogus novus stellarum australium ... E. Hallei.* *Leipsius, 1680–9.
Kircher, A. (1643). *Magnes, sive de arte magnetica opus tripartium*. Cologne.
Kollerstrom, N. (1995). A reintroduction of epicycles: Newton's 1702 lunar theory and Halley's Saros correction. *Quarterly Journal of the Royal Astronomical Society*, 36, 357–68.
Koyre, A. and Cohen, I. B. (1972). *Isaac Newton's Philosophiae naturalis principia mathematica. The third edition (1726) with variant readings*, 2 vols. Cambridge University Press.
Kraye, J. (ed.) (1996). *The Cambridge companion to Renaissance humanism*. Cambridge University Press.
Krisciunas, K. (1988). *Astronomical centers of the world*. Cambridge University Press.
Lago, C. and Rossi, L. (1981). *Descriptio Histriae*. Trieste: Lint.
Lalanne, M. L. C. et Bordier, H. L. (1851–3). *Dictionnaire de pièces autographes volées aux bibliothèques publiques de France*. Paris.
Lamb, H. H. (1991). *Historic storms of the North Sea, British Isles and North-west Europe*. Cambridge University Press.
Lambeck, K. (1980). *The Earth's variable rotation*. Cambridge University Press.
Laplace, P. S. (1798–1823). *Traité de mécanique céleste*. Books I and IV. Paris.
Leadbetter, C. (1735). *Uranoscopia*. London.
Le Gras, N. (1685). B, *Description nouvelle de ce qu'il y a de plus remarquable dans la Ville de Paris*. Paris. Translated as *A new description of Paris, Translated out of French*. London for Henry Bowicke; 1687.
Leibniz, G. W. (1923). *Sämtliche Schriften und Briefe*. Berlin: Akademie Verlag.
Leibniz, G. W. (1964). *Allgemeiner politischer und historischer Briefwechsel*, Vol. 7. Berlin: Akademie Verlag.
Leibniz, G. W. (1970). *Allgemeiner politischer und historischer Briefwechsel*, Vol. 8. Berlin: Akademie Verlag.
le Monnier, P.-C. (1746). *Institutions astronomiques*. Paris.
Levine, J. H. (1991). *The battle of the books*. Cornell.
Lough, J. (1985) *France observed in the seventeenth century by British travellers*. Stocksfield: Oriel Press.
Lough, J. (1987). *France on the eve of revolution*. London and Sydney: Croom Helm.
Luce, A. A. (1952). *The works of George Berkeley Bishop of Cloyne*, Vol. 4. London: Nelson.

---

*The title page of the copy in the British Library begins '*Annus 1* ...', has no author, omits the reference to Halley and gives no printer. The copy is bound in with ephemerides with later years that bear Kirch's name and that of the printers in Leipzig, but which omit Halley's catalogue.

Luttrell, N. (1857). *Brief historical relation of State affairs*, Oxford.
Mabillon, J. (1704). *Librorum de re diplomatica supplementum* ... Luteciae-Parisiorum.
McDonald, M. (1991). The strange death of the Earl of Essex, 1683. *History Today*, **41**, Nov., 7–18.
McDonnell, M. F. J. (1909). *A history of St Paul's School*. London.
McDonnell, M. F. J. (1977). *The register of St Paul's School 1509–1748*. London: for the Governors.
MacGregor, A. (ed.) (1994). *Sir Hans Sloane*. London: Trustees of the British Museum.
McKitterick, D. (1992). *A history of Cambridge University Press*. Cambridge University Press.
MacPike, E. F. (1932). *Correspondence and papers of Edmond Halley*. Oxford.
MacPike, E. F. (1937). *Hevelius, Flamsteed and Halley*. London.
MacPike, E. F. (1939). *Dr Edmond Halley, (1656–1742), A bibliographic guide to his life and work, arranged chronologically*. London: Taylor and Francis.
Magrath, J. R. (1922). *The Queen's College*, Vol. 2. Oxford: Clarendon Press.
Mairan, J. J. Ortous de. (1742). Éloge de M. Halley. *Mémoires de l'Académie Royale des Sciences*, 1744–5, 172–88. Paris.
Malin, S. R. C. (1981). The direction of the Earth's magnetic field at London, 1570–1975. *Philosophical Transactions of the Royal Society*, A **299**, 357–423.
Marshall, A. (1994). *Intelligence and espionage in the reign of Charles II*. Cambridge University Press.
Metcalfe, W. C. (ed.) (1887). *The visitations of Northamptonshire made in 1564 and 1618–9*, p. 114. London.
Metcalfe, W. C. (1878–9). *The visitations of Essex by Hawley 1552, Hervey 1558, Raven 1612, and Owen and Lilley 1634*, 2 vols. London: Harleian Society Publications, 13, 140.
Molyneux, W. (1716). *Dioptra nova*. London: Ben Tooke.
Montfaucon, B. de. (1702). *Diarium italicum. sive monumentorum veterum, bibliothecarum, musaeorum, &c. Notitiae singulares in itinerario italico collectae*. Paris.
Moore, J. (1660). *Arithmetick*. London.
Morrison, L. V., Stephenson, F. R., and Parkinson, J. (1988). Diameter of the Sun in AD 1715. *Nature*, **331**, 421–3.
Motte, A. (1729). *The mathematical principles of natural philosophy*, 2 vols. London: Motte. English translation of Newton (1687).
Munby, A. N. L. (1952). Distribution of the first edition of Newton's *Principia*. *Notes and Records of the Royal Society of London*, **10**, 28–39.
Munby, A. N. L. (1975). *Sale catalogues of eminent persons, Vol. 11, Scientists*. London: Mansell and Sotheby Parke Bernet.
Muratori, L. A. (1717). *Delle antichità estensi ed italiane, Pt 1*. Modena.
Nelson, R. (1704). *A companion for the festivals and fasts of the Church of England*. London. (More than 35 subsequent editions.)
Newton, I. (1687). *Philosophiae naturalis principia mathematica*. London: Royal Society.
Newton, I. (1702). *A new and most accurate theory of the Moon's motion*. London: A. Baldwin; (facsimile in Cohen (1975)).
Newton, I. (1704). *Opticks*. London.
Newton, I. (1707). *Arithmetica universalis*. Cambridge: Typis Academicis; London: Benj. Tooke.
Newton, I. (1720). *Universal arithmetick*. London.
Nix, L. (1889). *Das fünfte Buch der Conica des Apollonios von Perga in der arabischen Übersetzung des Thäbit ibn Corrah (Teilweise) Herausgegeben* ... Leipzig.
Norwood, R. (1676). *The sea-man's practice containing a fundamental problem in navigation experimentally verified namely, touching the compass of the Earth and Sea and the quantity of a degree in our English measures*, 12th edn. London: William and Eliza Hurlock.

Odier, J. B. (1962) Le fonds de la reine à la bibliothèque Vaticane. *Studi e testi*, **219**, 159–89. Città di Vaticano: Biblioteca Apostolica Vaticana.
Olhoff, J. E. (1683). *Excerpta ex litteris ... ad J Hevelium*. Danzig.
Ollard, R. (1984). *Pepys, a biography*. Oxford University Press.
Olson, R. J. M. (1994). Much ado about Giotto's comet. *Quarterly Journal of the Royal Astronomical Society*, **35**, 145–8.
Osler, M. J. and Farber, P. L. (eds.) (1985). *Religion, science and worldview*, pp. 275–95. Cambridge University Press.
Ottens, R. (1729). *Atlas maior*. Amsterdam.
Oughtred, W. (1694). *Key of the mathematicks*, trans. by Halley from *Clavis mathematica*.
Owen, J. H. (1938). *War at sea under Queen Anne, 1702–8*. Cambridge.
Parker, B. B. (ed.) (1991). *Tidal hydrodynamics*. New York etc: Wiley.
Pearson, J. (1649). *An exposition of the creed*. London: R. Daniel for J. Williams.
Pepys, S. (1970–83). *The diary of Samuel Pepys*. (ed. R. Latham and W. Matthews), 11 vols. London: Bell and Hyman.
Phillimore, W. P. W. and Cockayne, G. E. (eds.) (1900). *London parish registers*, Vol. 1, *Marriages at St James Duke Place from 1668 to 1683*. London.
Picard, J. (1671). *Mesure de la Terre*. Paris: Imprimerie royale.
Pope, A. (1966). *Poetical works*, ed. H. Davis. London.
Popkin, H. (1965). *Pierre Bayle, historical and critical dictionary, selections*. Indianapolis: Bobbs Merrill.
Proudman, J. (1941). Halley's tidal chart. *Geographical Journal*, **100**, 174–6.
Quill, H. (1966). *John Harrison, the man who found longitude*. London: John Baker.
Ray, J. (1691). *The wisdom of God manifested in the works of the creation*. London: for J. Smith.
Ray, J. (1692). *Miscellaneous discourses concerning the dissolution and changes of the world*. London.
Redondo, P. (1989). *Galileo: heretic*. Princeton University Press.
Rigaud, S. J. (1844). *A defence of Halley against a charge of infidelity*. Oxford: Ashmolean Society.
Rigaud, S. P. and S. J. (eds.) (1841). *Correspondence of scientific men of the seventeenth century*, 2 vols. Oxford University Press.
Richer, J. (1679). *Observations astronomiques et physiques faites en l'isle de Caienne*. Reprinted in: Académie Royale des Sciences. (1693). *Recueil d'observations faites en plusieur voyages par ordre de Sa Majesté*. Paris: de l'Imprimerie royale.
Robinson, H. W. and Adams, W. (eds.) (1935). *Diary of Robert Hooke 1672–80*. London.
Robinson, N. H. (1980), *The Royal Society catalogue of portraits*. London: The Royal Society.
Ronan, C. (1970). *Edmond Halley, genius in eclipse*. London: Macdonald.
Rose, Craig. (1993). The origins and ideals of the SPCK, 1699–1716. In Walsh, J. Haydon, C., and Taylor, S. (eds.) *The Church of England c.1689–c.1833. From toleration to tractarianism*. Cambridge University Press.
Ross, A. M. (1956). The correspondence of Dr John Arbuthnot. Ph.D. thesis, University of Cambridge. Unpublished.
Rotta, S. (1990). L'accademia fisico-matematica Ciampiniana: un'iniziativa di Cristina? In di Palma and Bovi (1990).
Royal Commision on Historical Monuments, England. (1959). *An inventory of the historical monuments in the city of Cambridge, Part II*. London: H.M.S.O.
Royer, A. (1679). *Cartes du Ciel reduites en quatre tables ...* Paris.
Sandford, F. (1687). *The coronation of James II*. London.
Savery, T. (1699). *Philosophical Transactions of the Royal Society*, **21**, 228.
Schaffer, S. (1990). The making of the Comet. In Thrower (1990).

Schama, S. (1989). *Citizens. A chronicle of the French revolution.* London: Viking.
Schneider, I. (1968). Der Mathematiker Abraham de Moivre. *Archive for the History of the Exact Sciences,* **5**, 177–317.
Scott, J. (1991). *Algernon Sidney and the Restoration Crisis, 1677–1683.* Cambridge University Press.
Secretan, C. F. (1860). *Life and times of the pious Robert Nelson.* London: John Murray.
Sellers, J. (1671). *The English pilot,* Books 1 and 2. London.
Sherburn, Q. (ed.) (1956). *The correspondence of Alexander Pope.* Oxford.
Sherwin, H. (1706). *Mathematical tables.* London: Mount and Page.
Smith, A. (1995). Engines moved by fire and water. *Transactions of the Newcomen Society,* **66**, 1–25.
Smith, Ruth. (1995). *Handel's oratorios and eighteenth century thought.* Cambridge University Press.
Snel (Snellius), W. (1617). *Eratosthenes batavus, De Terrae ambitus vera quantitae.* Lugduni Batavorum: apud Ioducum a Colster.
Somers, J. (1809–15). *A collection of scarce and valuable tracts,* 2nd edn (ed. W. Scott), 13 vols. Vol. 10, pp. 72–174. London.
Spence, J. (1964). *Anecdotes, observations and characters of books and men* (ed. S. J. Springer and B. Dobrée). London: Centaur Press.
Sprat, T. 1685, *True account and declaration of the horrid conspiracy against the late King, his present Majesty and the government.* (London: in the Savoy.)
Spurr, J. (1991). *The restoration Church of England 1646–1689.* New Haven and London: Yale University Press.
Steinscheider, M. (1960). *Die arabishen Uberzetzungen aus dem Griechischen* (reprint of 7 articles of 1889). Graz.
Stephenson, F. R. and Morrison, L. V. (1995) Long-term fluctuations in the Earth's rotation: 700 BC to AD 1990. *Philosophical Transactions Royal Society London,* **351**, 165–202.
Stock, J. (1776). *An account of the life of G. Berkeley late bishop of Cloyne.* London.
Stoye, J. (1994). *Marsigli's Europe.* New Haven and London: Yale University Press.
Streete, T. (1716). *Astronomia Carolina, 3rd. ed. corr.*[*] *To which is added a series of observations on the planets, chiefly of the Moon, made near London with a sextant of near six foot radius, in order to find out the lunar theory a posteriori. Being a proposal how to find the longitude &c. By Dr Edmund Halley.* London: for S. Briscoe and R. Savile.
Stukeley, W. (1936). *Memoirs of Sir Isaac Newton's life* (ed. H. White). London.
Sutherland, L. S. and Mitchell, L. G. (1986). *The history of the University of Oxford.* Vol. *V, The eighteenth century.* Oxford: Clarendon Press.
Syrett, D. and DiNardo, R. L. (1994). *The commisssioned sea officers of the Royal Navy.* Aldershot: Scolar Press for the Navy Records Society.
Tafel, R. L. (1875–7). *Documents concerning the life and character of Emanuel Swedenborg,* 2 vols. London: Swedenborg Society.
Tamaro, Attilio. (1924). *Storia di Trieste,* Vol. II, p. 137. Rome.
Tanner, J. R. (ed.) (1926*a*). *Private correspondence of Samuel Pepys,* 2 vols. New York.
Tanner, J. R. (ed.) (1926*b*). *Samuel Pepys's Naval Minutes.* London: Navy Records Society.
Taton, R. and Wilson, C. (eds.) (1989). *Planetary astronomy from the Renaissance to the rise of astrophysics. Part A: Tycho Brahe to Newton. The general history of astronomy,* 2A. Cambridge University Press.
Taylor, G. I. (1919). Tidal friction in the Irish Sea. *Philosophical Transactions* A **220**, 1.
Thomas, I. (trans.) (1980). *Selections illustrating the history of Greek mathematics,* Vol. 2. London: Heinemann.

[*]The corrections to Streete's text were by Halley.

Thompson, E. M. (1878). *Correspondence of the family of Hatton*. London: Camden Soc.
Thoren, V. E. (1985). Tycho Brahe as the dean of a Renaissance research institute. In Osler and Farber (1985).
Thoren, V. E. (1990). *The Lord of Uraniborg*. Cambridge University Press.
Thrower, N. J. W. (ed.) (1981). *The three voyages of Edmond Halley in the 'Paramore', 1698–1701*. Hakluyt Society Publications, 2nd series, Vols 156–7. London.
Thrower, N. J. W. (ed.) (1990). *Standing on the shoulders of giants*. Berkeley, Los Angeles, Oxford: California University Press.
Thrower, N. J. W. (1996). *Maps and civilization*, 2nd edn. Chicago and London: University of Chicago Press.
Thrower, N. J. W. and Glacken, C. J. (1969). *The terraqueous globe*. Los Angeles: William Andrews Clark Memorial Library.
Tooke, B. (1684). *The tryal of Lawrence Braddon and Hugh Speke, Gent. Upon an information of high misdemeanor, subornation and spreading false reports before Sir George Jeffreys holden at Westminster Hall on Friday the 7th of February 1683[4]*, by order of Sir George Jeffreys. London: Printed for Benjamin Tooke, at the Shop in St Paul's Church-Yard.
Toomer, G. J. (1990) *Apollonius Conics books V to VII*. New York: Springer-Verlag.
Turner, G. L'e. (1986). The physical sciences. In Sutherland and Mitchell (1986).
Veenendaal, A. J. (1970). The War of the Spanish Succession in Europe. In Bromley (1970), pp. 410–45.
Venn, J. (1922–7). *Alumni Cantabrigensii; Pt 1 from the earliest times to 1751*, 4 vols. Cambridge.
Vertue, G. (1930–50). *Note Books*. Oxford: The Walpole Society.
Vitruvius, M. V. P. (1935). *De architectura*. Books 6–10. (ed. and trs. F. Granger) Cambridge Mass. and London: Loeb Classical Library.
Waliszewski, K. (1897). *Peter the Great*, trans. Lady Mary Loyd. London.
Wallace, W. E. (1994). *Michelangelo at San Lorenzo*. Cambridge University Press.
Wallis, J. (1658a). Review of Hevelius's *Annus climactericus*, with extract in English of Halley's letter to Hevelius at 9/18 July 1679. *Philosophical Transactions* of the Royal Society, 15, No. 175, 1162–83.
Wallis, J. (1685b). Concerning the collection of secants. *Philosophical Transactions of the Royal Society*, 15, No. 176, 1193–201.
Wallis, J. (1687). A discourse concerning the measure of the airs resistance to bodies moved in it. *Philosophical Transactions of the Royal Society*, 16, 269–80.
Walsh, J. Haydon, C., and Taylor, S. (eds.) (1993). *The Church of England c.1689–c.1833. From toleration to tractarianism*. Cambridge University Press.
Ward, S. (1654). *Idea trigonometria demonstrate. Item praelectio de cometis et inquisitio in Bullialdi astronomiae philolaicae fundamentae*. Oxford.
Ward, S. (1656). *Astronomia geometrica*. London.
Waters, D. W. (1990). Captain Edmond Halley, F.R.S., Royal Navy, and the practice of navigation. In Thrower (1990).
Westfall, R. S. (1971). *The construction of modern science*. New York. Wiley and Sons; republished, Cambridge University Press, 1977.
Westfall, R. S. (1980). *Never at rest*. Cambridge University Press.
Westfall, R. S. (1986). Newton and the acceleration of gravity. *Arch Hist. Exact Sci.*, 35, 255–72.
Westfall, R. S. and Funk, G. (1990). Newton, Halley and the system of patronage. In Thrower (1990), pp. 3–13.
Whiston, W. (1727). *Astronomical lectures read in the publick schools in Cambridge* (with astronomical tables of Flamsteed, Halley, Cassini and Streete), 2nd edn. Cambridge.
Whiston, W. and Ditton, H. (1713). *The Guardian*, No. 107, 11, July 1713.

Whiteside, D. T. (1967–81). *The mathematical papers of Isaac Newton*, 8 vols. Cambridge University Press.

Whiteside, D. T. (1976). Newton's lunar theory: from high hope to disenchantment. *Vistas in Astronomy*, **19**, 317–28.

Whiteside, D. T. (1994). The prehistory of the 'Principia'. *Notes and Records of the Royal Society of London*, **45**, 11–61.

Whittaker, E. T. (1927). *A treatise on the analytical dynamics of particles and rigid bodies*. 3rd edn. Cambridge.

Willmoth, Frances. (1993a). *Sir Jonas Moore, practical mathematics and restoration science*. Woodbridge: Boydell.

Willmoth, Frances. (1993b). Mathematical sciences and military technology: the Ordnance Office in the reign of Charles II. In Field and James (1993).

Willmoth, Frances. (1994). Transcripts of Flamsteed's correspondence. Unpublished*.

Willughby, F. (1672). *De historia piscium libri quarter*. 2 vols. Oxonii.

Winkler, M. G. and van Helden, A. (1993). Johannes Hevelius and the visual language of astronomy. In Field and Frank (1993).

Wollenschläger, K. (1933). Der mathematische Briefwechsel zwischen Johann I. Bernoulli und Abraham de Moivre. *Verhandl. Naturforsch. Gesells. Basel*, **43**, 151–317.

Wood, Anthony à. (1691–2). *Athenae Oxoniensis*, 2 vols. London.

Wood, A. C. (1935). *A history of the Levant Company*. Oxford

Woodhead, J. R. (1965). *The rulers of London*. London: London and Middlesex Archaeological Society.

Yeomans, D. H. (1991). *Comets. A chronological history of observation, science, myth and folklore*. New York: Wiley.

Zuber, R. and Theis, L. (1986). *La Révocation de l'Édit de Nantes (Colloque de Paris, 15–18 Oct. 1985)*. Paris: Soc. Hist. Protest. Français.

*The second volume of *Flamsteed Corresp.* appeared too recently to be referred to.

Wills

Halley, Humphrey (I), PCC Reg Eure, f. 122, pr. 23 October 1672.

Halley, Humphrey (II), PCC Reg Bench, f. 66, PRO PROB 11/351, pr. 3 June 1676.

Halley, Humphrey (III), PCC Admon Bk. 1684, PRO PROB 6/59.

Halley, William, PCC Reg Dyer, f. 145, pr. 6 March 1675.

Kinder, Gilbert, 1645.

Tooke, Christopher, PCC, PROB 11/158, f. 71 (l+r), 72 (1), pr. 6 August 1630.

Tooke, Christopher, PCC Reg Juxon, f. 42, PROB 11/310, p. 328, pr. March 1662.

Tooke, Edward, PCC Reg Hone, f. 80, PRO PROB 11/327, pr. January 1668.

Tooke, Margaret, *Notes and Queries*, 10S, 8, 221–2.

Tooke, Ursula, PCC Reg Young, f. 221, pr. 11 October 1711.

Parish registers

Essex:
St. Margaret's, Barking. Essex Record Office.

Kent:
: St. Margaret's, Lee. Duncan, L. L. and Barron, A. O. Lewisham Antiquarian Society (1888)

London:
: St. Benet's, Paul's Wharf. Copy in Society of Genealogists.
: St. Giles's, Cripplegate. Guildhall, London, MS 6418.
: St. Helen's, Bishopsgate. Harleian Society (Register Series) vol. 31 (1904), and Guildhall, London, MS 6830/1.
: St. James's, Duke's Place. Phillimore and Cockayne (1990).

Surrey:
: St. George, Southwork. Boyd's Marriage Register (in the Guildhall, London).

Westminster:
: The Abbey and St. Margaret's—Chester (1876, 1886–).

Index

Note: Some proper names of people and places have alternative spellings, as shown in the index entries. Bold roman capital figures are numbers of plates.

aberration of light 349
Abington, Earl of 351
academies 25, 182
   Académie royale des Sciences 25, 66, 116
   Accademia fisico-matematico (of G. G. Ciampini) 121
   Accademia dei Lincei 18, 25, 120
   Accademia reale (of Queen Christina) 121
Achilles 147, 472, n. 1
Acta eruditorum 169, 448
actresses 16
Adams, John Couch 227
Addison, Joseph 16, 28, 396, 409
Adriatic 236, 293, 313
age of the world 346
air, density 196
Airy, Sir George Biddle 201
al-Battānī (Albategnius) 225, 226
Alberghetti, Signor 185
alchemy 24, 25, 122, 412, 460, n. 30
Alconbury 14, 33, 34, 37, 45, 148, 431
Aldermen, Court of 6
Aldrich, Henry 331, 334, 336
Aleppo 226
Alfrey, George, surgeon 264, 281
Al-Hasan ibn al-Hasan ibn al-Haytham (Alhazen) 341
al-Ma'mūn, caliph 226, 337
altar, Roman 187
Alexandria 202
American colonies 19, 207
Analyst 408, 410
Ancelin, Pierre 282
Ancona 293
Anderson, Robert 188, 189
Andreassi, Signor 300, 305
Anglois, Cape 257, 280
Anguilla 257, 280
Anne (Stuart), Queen 50, 291, 292, 295, 297, 298, 305, 307, 318, 322, 386, 397, 424
annuities 199
Antarctic 277, 329
Antigua 257, 274
Antioch 225, 226, 227
antiquities 14, 123, 171
Antonine itineraries 123, 201, 330, 416
Apian, Peter 214
Apocalypses 29
Apollonius 188, 190, 334, 336
   Conics 334, 336, 338, 340, 341, 342
   De sectione rationis 335
   De sectione spatii 335
Appian (classical historian) 226
appulses 105, 130, 398

apse line, *see* Moon, apogee
Aquilea 293, 294, 406
Arabic *Epitome* 338
Arbe 293, 303
Arbuthnott (Arbuthnot), John 271, 384, 385, 386, 389, 399, 406, 407; **XI**
areal velocity theorem 150, 342, 343
Arianism 25, 412
Aries, first point of 67
Argonauts 29, 400, 401
Aristotle 26, 198
Arundel 232
Arundel, Earl 328
Arundel marbles 243, 328
Ascension Island 81
Ashe, St George 19
Asia Minor 184
Aston, Francis 151, 384
astrology 122
astronomy 20–3, 65–9
atheism 171, 307, 412
Atkins, Sir Robert 248
Atlantic
   chart **IX**
   ports 257
atoms
   dimensions 198, 229
   hypotheses 26, 127, 198, 229
Aubrey, John 32, 75, 108
Augustine, St 27, 30
aurorae 346–8, 418
Auzout, Adrien 23, 114, 208
Avignon 74, 83, 115, 118, 119, 207, 381, 404
azimuth 66, 67
Azores 270

Babylon 227
Bacon, Francis 26, 198
Bagdad 226
Baily, Francis 373
ball lightning 171
Baltic Company 9
Bank of England 10
Banks, Sir Joseph 224
Bantry Bay, battle 234
Banū Mūsā 337
Barbados 257, 268, 273, 279
Barker's equation 430
Barking, Essex, 11, 12, 33, 46, 134, 431, 432
Barocci collection 333, 336
barometer 24, 269
Barton, Catherine ('Kitty', Mrs Conduitt) 351, 385, 406, 407, 491 n. 86

Basel 19
Basset, Joshua 168
Batavia 283
Bathurst, Sir Benjamin 236
battles, heard in London 46, 47
Bavaria 299
Bayer, Johann 389, 453
Bayes, Thomas 200
Bayeux tapestry and Halley's comet 206
Bayle, Pierre 27, 127
Baynard's Castle 4, 5
Beachy Head 290
  battle 236
Beechy, F. W. 290
Bemmelen, W. van 283
Benbow, John, Admiral 263, 271, 291
Bentley, Richard 165, 170, 247, 408
Berkeley, George, Bishop 408
  criticism of Newton 408
Bermuda 257, 268, 270, 280
Bernard, Edward, Savilian professor 52, 246, 322, 334, 336
Bernoulli family 19
  Daniel 418
  Jacques 119, 339
  Johann I (Jean) 24, 150, 170, 333, 340, 353
  Nicholas 170, 331, 333, 339, 340
Bethlehem Hospital (Bedlam) 36, 39
Bevis, John 213, 402, 492 n. 8
Bianchini, Francesco 122, 321
Bible, interpretation 29
Bibliothèque Universelle 169
Billingsgate 37, 47
Biographia britannica 27, 32, 49
biology 24, 30
bird flight 244, 245
Blaeu, Wilhem J. 65
Blenheim, battle and campaign 292, 315
Bodleian Library, 15, 330
Books, Battle of 165, 422
booksellers 44
Borri, Giuseppe Francesco 282
Boucher (Bouchar), Charles 53, 59, 60, 81
Bouguer, Pierre 402
Boulogne 109, 287
Bowles (Chester Mint clerk) 252, 253
Boyle, Robert 24, 30, 39, 44, 198, 418, 424
Braddon, Lawrence 135–9, 141, 144
Bradley, James 349, 393, 397
Brahe, Tycho, *see* Tycho Brahe
Brazil 272, 273, 277, 279
Breslau (Wroclaw) 198–200
Brest 43, 314
Bridgewater Street 5, 37
Briggs, Henry 24
Britain, coastal surveys 233, 234
Broad Street 9
Broad Street ward, 6
Brouncker, William Viscount 42, 45, 54, 55, 60, 61, 66, 102, 182, 377
Browne, Capt. Thomas 233, 234,
Bruinss, Peter 282

Bruyinx, Count de 306
Bryant, Humphrey 280
Buccari (Bakar) 236, 293, 297, 299, 300, 302–6, 313, 314, 316
  fortification 305, 306, 309–13
  plans and views 310, 311, 313
Buccarizza (Bakarac) 302, 303, 304, 314
Buckworth, Sir John 36, 37, 41, 142, 143, 261, 435
Bullard, Sir Edward 415, 417
Bullialdus (Boulliau), Ismael 349
bullion composition 252
Burchett, Josiah 260, 274, 276, 277, 279, 280, 291
Burnet, Gilbert, bishop 140
Burnet, Thomas 399
Burnet, William 340, 353
Bushey, Herts. 37
Buthner, Nathaniel 93, 94, 95, 98, 102
Butterfield (astronomer, Ingenieur du Roy) 77

Caesar, Julius, landing in Kent 200, 201, 401
calculus dispute 332
calendars xvi
Cambridge 12, 14, 45
  university 10, 15
Cambridge platonists 15, 29
Campbell, Revd. Colin 260, 383
Canal du Midi 117
Canary Islands 65, 76, 220, 270, 276
Cannon (Canning) St. property 35, 36, 37, 47
Canterbury 108, 207
Canterbury, Prerogative court 10, 142
Canvey Island 232
Cap d'Antifer 287
Cape Verde Islands 65, 76, 105, 270
Capell, Sir Henry, Earl of Essex 136, 140
Carbery, John, Earl 185
Carlisle, Lady 409
Carlopago 293, 300, 302, 305
Carlstadt (Karlovac) 310
Carniola (Slovenia) 171
Caroline of Anspach, Queen 400, 402
Carribean islands 268
cartography 297, 314, 315
Caskett (Casquett) rocks 285, 290
Cassini, Gian-Domenico (Jean-Dominique) 65, 66, 72, 80, 84, 86, 92, 112, 119, 290, 381; V
  career 105
  correspondent of Halley 153, 184
  comets 114, 115, 122, 126, 207, 208
  geodesy 21
  Halley's visit 115, 116, 118, 120, 126
  instruments 59, 60
  Jupiter's satellites 23, 105, 218
  lunar studies 105, 354
  Queen Christina 121, 122, 207
  refraction 53
  solar parallax 81, 381
  sunspots 58
Cassini, Jacques, *le fils* 187

Cassius, Dion 201, 402
Caswell, John 170, 172, 190, 196, 330, 407
Cawthorne, Anne (see also, Chomat, Joe) 34, 35, 129, 431, 432
Cawthorne, Elizabeth 34, 431
Cawthorne, John 34, 431
Cayenne 65, 66, 105
celestial mechanics 203–5, 344
Cellarius, Andreas 79
Cellio, Marco Antonio 115, 122, 207
Celsius, Anders 192, 418
Cesi, Prince Federico 25, 120
Chancery actions 129, 130, 132, 141, 142, 143, 144, 152
Channel Islands 288
Chantrell, Capt. William 236, 237
Chapels Royal 7
Chapman, Sidney 283, 284
Charles I, execution 6
Charles II 61, 77, 79, 182, 230
   recommendations for degrees 85
   restoration, 8, 16
   portrait 16
   Royal Observatory 377, 378
Charles the Martyr, church, Tunbridge Wells 8, 41, 261
Charlett, Arthur 330, 334, 384
charts, Atlantic 268
charts, magnetic 281, **IX**
   errors 282
Chatham 231, 261, 314
Cheiron (Chiron) centaur 28, 29, 400, 401
Chelsea college 10
chemistry 24
Cherbourg 287
Cherso (Kres), island 303
Chester 195, 196
   mint 11, 44, 211, 212, 239, 249–54
Chichester 237
Chomat, Joe 34, 35, 37, 129, 134, 431, 432
Christina, Queen of Sweden 120, 121, 123, 127, 208, 399
Christology 411
chronology, ancient 400
chronometer
   Harrison 397
   Kendal 397, 398
Church of England 7
churches and parishes
   All Hallows, Lombard Street 34
   All Hallows, London Wall 36
   St Clement's, Eastcheap 33
   St Giles's, Cripplegate 34, 35, 47, 142
   St Helen's Bishopsgate 5, 6, 128, 134
   St James's, Duke Place 5, 128, 129, 432
   St Margaret's, Barking 33, 34, 134, 431, 432, 460 n. 3
   St Margaret's, Lee 403, 404, 432
   St Margaret's, Westminster 35, 432
   St Peter-le-Poer 6
Ciampini, Giovanni Giustino 121, 122, 123
Clairaut, Alexis-Claude 204, 211, 215, 365, 415

Clarendon Building 328
Clarendon, Earl 114
Clarendon Press 328
Clarke, Samuel 252, 253, 396
Cleeter (Cleater), Robert 36, 132, 143, 152, 431, 432, 435
Clerk (Halley's friend) 64, 73, 76
clipping, coins 249
clocks
   astronomical 22, 131
   Flamsteed's 379
   Halley's 69, 380
   Hevelius's 91, 126
   pendulum 59, 126, 173
   Royal Observatory 380
clothes 44
Cluver, Detlerus 93
Coal (Cole) Office 11, 37, 143, 435, 436
Cocker, Edward 49
coffee house
   Garraway's 41
   Jonathan's 39
colatitude 20, 66, 67
Collegio Romano 120
collimation errors 104, 126
Collins, John 62
Collins, Capt. Greenville 233, 234
colours, submarine 241
Colson, John 381
comets 205, 206
   associations 205
   attraction of Jupiter and Saturn 211, 213–15
   chaotic behaviour 217
   effects 216
   'Halley's' comet returns 215, 216, 217
   influence 315, 416
   observations 109, 113, 207, 209
   observed in New England 19, 207
   orbits 115, 160, 207, 208, 210–14
   periodic 211–14
   predictions 213
   1301 ('Halley's') 216
   1305 214
   1456 ('Halley's') 214
   1531 (1532) ('Halley's') 212
   1607 ('Halley's') 211
   1664 121, 211
   1680 102, 109, 113, 114, 115, 119, 122, 148, 207, 209, 392
   1682 ('Halley's') 116, 206, 209, 211, 382, 415
   1683 209, 382
   1910 ('Halley's') 216
   1986 ('Halley's') 216
command at sea 259
Commandinus, Fredericus 336
Common Council 6
commoners (undergraduates) 51
Compton, Henry, bishop of London 50, 246, 248
Conduitt, John 165, 406, 407, 408
conic sections 150, 341–3
Constable, Captain 231

Constantinople 202
constellations 79, 389
　Aries 400
　Cassiopeia 455
　Chameleon 66
　Robur Carolina (Charles's Oak) 77, 79
　Taurus 389
　Ursa major 389, 455
　see also Historia coelestis, Historia coelestis britannica
Cook, James 31, 220, 259, 265, 278, 280, 317, 397, 398
coordinate systems 66
Coriolis force 194, 258
Coronelli, Marco Vincenzo, cartographer 297
Costard, G 339
Cotes, Roger 352, 396, 492 n. 11
Covell, John 114
Cowes 287
Crane, Capt. Gilbert 233, 234
Cripplegate 4
Cromer 289
Cromleholme, Samuel 48
Croone (Croune), William 93, 113
Crosse, Thomas 128
Crosthwait, Joseph 88, 388, 390, 391, 392, 455
Crutched Friars 36
Cudworth, Ralph 15
currents, oceanic 256, 258, 417
curvature 150, 343
Cystat, Johann Baptist 81
Czar Street, Deptford 263

d'Alembert, Jean le Ronde 204, 215, 217, 365
Dalmatia 293, 296, 297, 313
Dampier, William 275, 276
Dante 27, 30, 65
Danzig (Gdansk) 25, 89, 94
　observations 94–100
'Dead's part' 142
Deane, Sir Anthony 108, 248
declination 66
de Dominis, Antonius 197
Defence of free-thinking in mathematics 409
Defoe, Daniel 11, 14, 16, 46
de la Brière Lepaute, Mme Nicole-Reine Étable 215, 216, 479 n. 16
de la Condamine, Charles Marie 402
de la Hire, Phillippe 57
de LaLande, Joseph Jérome le Francais 215, 216
de la Roche, Abbé 77
de L'Isle, Joseph-Nicolas 70, 71, 125, 215, 224
Democritus 127
demography 198–200
de Moivre, Abraham, 8, 113, 119, 149, 170, 333, 353
de Montesquiou, Joseph comte d'Artagan 111
de Montmort, Rémond 351, 406, 491 n. 86
de Morgan, Augustus 178, 423
Deptford 256, 263, 276, 280, 314
Derham, William 196, 352

Descartes, René 26, 30, 127, 198, 342
Dieppe 287
disasters, naval 258
Ditton, Humphrey 331, 396
diversions 43, 44, 45
diving 211, 212, 240
　bell 38, 238–42
　suit 241, 242
Dodwell, Henry 108
Dog Tavern 37, 143, 435, 436
Dolfin, Daniele III 296
Dörffel, Samuel 208
Dover 12, 13, 200, 201
Dover, Straits 286, 289
Downs 11, 13
Drury Lane 6
Dryden, John 16, 409
duels 44, 253
Dungeness 280, 288, 290
Dutch wars 46, 231
dynamics
　chaotic, 177, 217
　Newtonian 177

Earth
　dimensions 21, 105, 201
　earthquakes 416
　figure 87, 159, 160, 402
　interior 191, 192, 347, 399
　magnetic field 24, 75, 190–2; see also magnetic variation
East Anglia 12
East India Company 9, 43, 60, 61, 62–4, 76, 226
Eaton, John (rector of Sawtry) 37, 148
eccentric angle 426, 427
eclipse observations
　lunar 53, 74, 75, 118, 119, 226, 349, 381
　solar 54, 74, 75, 171, 226, 351
　1715 351–3; **XVI**
ecliptic 21
Eddystone lighthouse 287
Edict of Nantes, revocation 8, 17, 113, 119, 145
Ely 14
elliptical orbit, elements 204, 426, 428, 429
embargoes on shiping 234
Emperor, Holy Roman, *see* Leopard I, Holy Roman Emperor
England 11
　agriculture and industry 11
English embassy, Paris 111
English Channel 232, 233, 287
　chart 285–90; **X**
　navigation 284, 285, 290
　tides 286, 288, 289
Enlightment 27, 29
Epicurus 26, 198
equation
　of the centre 428
　of time 428

equinox
  precession, 160
  times 226
errors
  astrometric, 100, 104, 126, 370, 371–3, 441
  systematic 98, 100, 126
Eschinardi, Francesco P. 117, 122, 123
Essex, coastlands 14
Essex, countess of 136, 140
Essex, Arthur Capell, Earl 8, 134, 135, 136, 137, 138, 139, 140, 144, 145, 414
  pamphlets about death 135–9
Essex, Lady 409
Estrées, Admiral d' 114
Estrées, Cardinal d' 114
Eudoxus 400
Euler, Leonhard 204, 215, 365, 398
Euphrates 225
Europe 17
Euston 14
Eutocius 336, 337
Evans, Sir Stephen 238
evection, lunar 357
Evelyn, John 10, 11, 14, 16, 17, 39, 46, 48, 263, 377
  Oxford 327, 328
  Paris 111
  Royal Society 182
  works 448
exclusion crisis 8

Falkland Islands 278, 283, 345
Fallopio, Gabriele 30
Fatio de Duillier, Nicholas 214, 247, 407
Fell, John 333, 334
fens 2, 14, 38
Ferguson, Robert (the Plotter) 135, 137, 138, 139, 141
Fernando Loronho (de Noronha) 257, 270, 272, 273, 277
Field, C. Gregory (Governor of St Helena) 76
Fiennes, Celia 11, 14, 45
First Point of Aries 429
Fiume (Rijeka) 293, 296, 297, 302
Flamsteed, John 15, 32, 41, 52, 53, 58, 66, 91, 92, 119, 246, 290, 352, 424; **IV**
  Astronomer Royal 54, 55, 377, 378–80, 381
  Boucher 53, 59
  catalogue 384
  clocks 60
  comets 115, 207, 392
  complaints against Halley 175, 259, 327
  correspondence 45, 53, 59, 66
  criticisms of southern observations 77, 78, 79, 87, 126, 381
  death 393
  Doctrine of the Sphere 86
  equatorial mount 70
  geodetic measurements 415
  Halley, relations 165, 247, 284, 380–7, 407
  Harrison, Lt. 264, 269, 275, 283, 483 n. 17

Historia coelestis, *see* separate entry
Historia coelestis Britannica, *see* separate entry
  instruments 38, 54, 60, 69, 379, 380
  lunar observations 369, 370
  lunar tables 83, 367, 376
  magnetic observations 415
  Newton, relations 382–7, 392
  refractions 53, 60, 81
  reluctance to publish 383
  sextant 23, 38, 70, 87, 379
  sunspots 58
  see also referees, Flamsteed's catalogue
Flamsteed, Margaret 379, 380, 388, 393
Fleming, Henry 51, 52
Florence 118
Florjancic (cartographer) 314
flowers, scent off land 277
Folkes, Martin 402
foot, values 122, 173, 201
Forbin, Comte de 294, 315
Fort Nelson (Hudson's Bay) 224
Foulness 232
foundationers 51
Fowey 287
France 17
Francis, Fr. 168
Freius 118
French coast 285
frigates 237
Frio, Cape 257, 270, 277
fundamental measurements, astrometric 69

Gale, Thomas 48, 108, 151, 184, 344
Gale, Roger 344
Galen 28
Gallet (Galet), Jean Charles 74, 77, 83, 115, 118, 119, 149, 207, 381
Galileo 18, 23, 26, 30, 120
Ganges 224
Garth, Sir Samuel 16, 409, 410
Gassendi, Pierre 26, 30, 81, 127, 198, 229
Gellibrand, Henry 24
General lottery 37, 143, 435
Gennes (Genoa) 118
gentlemen commoners (undergraduates) 51
geodesy 416; *see also* Earth, dimensions; Flamsteed, geodetic measurement 5
geomagnetism 174
geophysics 31, 416
George I (Elector of Hanover) 19, 307, 318, 325
George Prince of Denmark 291, 384
Gibraltar 293, 309, 313
Gilbert, William 24
Gill, David 73, 81
Gill, Isobel 73
Giotto spacecraft 216
Gloucester, Duke of 384
God 25, 27, 29, 416
Godolphin, William 292
Gold Coast 236
Golden Lion Court 37, 130, 153

Goldie, Dr Mark 472 n. 28
Golius, Jacobus 337
Good Hope, Cape 278, 279, 284, 349
Goodwin Sands 12, 13, 231
Goodyear, Anson 226
Graham, George 380, 394, 397, 402
Graves, Benjamin 239
Gravesend 256, 270
gravitation 41, 77
Graz 293, 294, 295, 297, 299, 314, 318
Great Fire 4, 5, 34, 35, 43, 46, 47
Greatorex, Ralph 38
Great Plague 4, 43, 46
Greatzer, J. 199
Greaves (Grave), John 338, 490 n. 58
Greek mathematicians 333, 334, 336, 339, 341, 342
Greenwich 53, 54, 119, 378, 379
Gregory, David 204, 247, 251, 382, 384
   Apollonius edition 334, 337
   Historia coelestis 384, 385, 386, 388, 455
   Principles of physical and geometrical astronomy 176, 249, 355
   religion 244, 327
   Savilian professor 323, 330, 407
   see also Historia coelestis
Gregory, James (elder) 66, 82, 190
Gregory, James (younger) 204, 260, 383
Gresham('s) college 5, 10, 15, 25, 39, 40, 46, 151, 184
Gresham('s) professors 24, 385
Grey, Ralph 279
greyhound, anomalous birth 187
Griggs, Mrs 117
Guernsey 288
Gunfleet 12, 13, 231
gounfounding, Austrian 309, 313
gunnery 185, 317

Hackney 6
Hadley, John 398
hail 195
Hales, Stephen 24
Halifax, Charles Montague, Earl, *see* Montague, Charles
Halifax, George Savile, Marquis 117, 145
Hall, Thomas 250
Halley, Anne, mother (née Robinson) 34, 35, 431, 432
Halley Bay 278
Halley, Dr 32
Halley, Edmond, astronomer 41, 431, 432;
   Frontispiece, VIII, XIV, XV, XVII, XVIII
   acquaintances 406
   Adriatic expedition 145, 292–318, 421, 438
      Memoriall 300, 303, 456–7
      records 295, 296
   adventurer 425
   age of the World 246, 346
   almanacks 187
   ancestors 32

antiquarian 171, 200, 330
Apollonius 323; *see also* Apollonius
Arabic 335
Astronomer Royal 3, 16, 124, 355, 356, 393–5, 402, 421, 438
astronomical observer 124–6
astronomical tables 170, 355, 366–8, 373–5, 376
Atlantic cruises 182, 251, 255, 256–81, 421, 438
   officers 272–5; *see also* Harrison, Edward
atomic dimensions 198, 229
aurorae 346–8, 418
barometer 187
birth 32, 437
books 44, 123, 405, 447–51
Captain RN 256, 259, 260, 262, 263, 264, 265, 276, 291, 295, 304, 307, 402, 406
cartographer 31, 315–17
Catalogue des estoilles australes 78, 107, 117
Catalogus stellarum australium 65, 78, 413
character 315, 317, 406
charts
   Adriatic 300, 304, 305
   Atlantic 260, 281
   English Channel 232, 233, 235, 236, 260, 288–90, 421, **X**
   magnetic 281, 283, 284, **IX**
   Sussex coast 243
   Thames approaches 232, 233, 234, 289
Chester 187
Chester mint 182, 239, 250–4, 438
children 130, 152, 387, 403, 404, 431, 432, 438, 439
chronology 437, 438
classical scholar 318
Clerk to Royal Society 151, 152, 153, 170, 182, 184–7, 254, 271, 284, 423, 437
clocks 69, 394, 395
cometary collisions 399
cometary orbits 115, 210–15, 414, 415, 438
commissions 262, 276
correspondence 444–6
correspondents 123, 124, 153, 170, 184, 315
councillor, Royal Society 132, 144, 147, 405
court patronage 61, 230, 248
Danzig 70, 72, 92, 93, 381
LL D (Civil Laws) (Oxon) 329, 438
death 403, 438
demography 198–200
diplomat 92, 259, 318
diving 211, 212, 238–243, 244, 398, 421, 438
Earth, interior 191, 347, 399
eclipse observations, *see* eclipse, lunar, solar and 1715
engineer 242, 243, 315, 318, 421
equipment grant 393, 394
estates visits 323, 324, 488 nn. 11, 13
father's death, 132–4, 140, 141, 437
Flamsteed; for relations with Flamsteed *see* Flamsteed
FRS 77, 84, 284, 437, 438
FSA 112, 438

Halley (cont.)
  geodetic plans 171–3
  geophysicist 31, 414, 415–18
  Hanover 19, 213, 307
  health and sickness 279, 407
  Hearne's stories 326, 327
  Hevelius, opinion of 103
  Historia coelestis 385–7, 438, 452–5
  historian 200, 201, 225–7, 228, 349
  Houghton 239, 240
  houses 77, 130, 325, 404, 405; *see also* Golden Lion Court and Islington
  hydrodynamics 244, 245
  hydrographic surveys 182, 231, 236, 261, 267, 316, 417, 437, 438
  income 130, 143, 153, 324, 435, 436
  instructions 265, 276, 285, 292, 297, 298
  instruments 59, 62, 71, 72, 87, 93, 96, 218, 219, 266, 268, 269, 355, 356, 369, 379, 393–5
  Islington 6, 71, 124, 130–2, 209, 382
  interests 405
  itinerary in France 117, 118, 437
  last papers 401
  lectures 345
  Leibniz 19, 164, 213, 262, 307
  linguist 49, 200, 335
  longitude, *see* separate entry
  lunar computations 369, 370
  lunar observations 355, 356, 368, 369
  lunar studies 31, 84, 227, 228, 354, 369–73, 413, 414, 415
  lunar tables, *see* Halley, astronomical tables
  MA degree 77, 84, 85
  magnetic experiments 87, 147
  magnetic variation 49, 174, 187, 190–2, 345, 398, 417
  management of estate 143, 148
  mathematician 187–90, 318, 353, 414
  Memoir 32, 460 n. 1, 502
  Memoriall to the Emperor 303, 304, 456, 457
  memorials 403, 404
  mercury, density 195
  meteorology 195, 416, 417
  metrology 122, 173, 195, 305
  mural arc 393, 394, 395
  navigator 317, 419–21
  observations, accuracy 88, 96–9, 125, 195, 372
  observations at school and Oxford 49, 53, 54, 58, 187
  ode in *Principia* 163–5, 442, 443
  originality 229
  Oxford activities 344, 345
  pass for Vienna 298, 299
  patent for diving 238
  patriotism 62
  patrons 61, 248, 249, 251, 322
  Paris 105–16, 124, 437
  Peter the Great 263
  planisphere 77
  plans for cruises 262, 307

  politics 326, 327, 412, 413
  portraits xv
  Principia 149, 151, 152–63, 178, 248, 366, 437; *see also* separate entry
  property 144, 435, 436
  publication, speedy 87
  publications, undergraduate 54–8
  publications, various 185, 346–9
  public service 321, 329, 330, 423, 424
  rainbows 197–8
  relatives 32, 33
  religion 171, 326, 407–13
  reputation 144, 295, 306, 318, 353, 393, 405
  review of *Principia* 168, 169
  rewards 291
  ring, Emperor's 300, 306
  Rome 119–24, 437
  St Helena 59, 60, 279, 281, 401, 421, 437
    barometric observations 74, 87
    climate 74, 76, 194
    eclipses 74
    instruments, 70, 71, 73
    length of pendulum 87
    magnetic observations 75, 87
    occultations 76
    transits 81–3
    voyages, use 64, 76
    woman with child 75
  salinity of lakes and seas 346
  salvage 236–42
  Savilian professor, 438
    astronomy 245
    geometry 11, 322
  schooldays 46, 48, 49, 437
  scientific method 228, 229
  sea captain 317
  seagoing 231
  seamanship 64, 261, 420
  Secretary, Royal Society 332, 423, 438
  sextant 71, 125, 129, 209
  Snowdon 187, 196
  southern catalogue 86, 88, 92, 124, 391, 439–41
  'southern Tycho' 86, 87
  'spys' 286
  surveyor 315–17, 421
  tidal survey 272, 284–90, 438
  tides 175, 284, 286
  trade winds 87, 192–5
  transit telescope 393, 394
  'Ulysses' 178, 423, 472 n. 1
  undergraduate 50–2, 59
  Universal Deluge 246, 399
  versifier 49, 165
  Vice-President, Royal Society 401
  viewed site of Observatory 54
  Visitor, Royal Observatory 386
  Visitors' reports 393, 394, 402
  visitors to Oxford 331
  visits to Newton 148
  youth 20
  see also transits, Mercury

Halley, Edmond, father 4, 34, 431, 432
  'afraid of son' 141, 326
  burial 134, 432
  disappearance 124, 132, 133, 137–40, 144, 148
  estate 34, 142, 143, 435, 436
  Great Fire 47
  identity 137, 138
  Salter 6, 34
  soap boiler 6, 34
  support of Halley 107, 130, 144
  yeoman warder 4, 8, 36, 137, 138
Halley, Edmond, son 130, 387, 431, 432, 449
Halley family position 60
Halley, Humphrey, of Bakewell 32
Halley, Humphrey I, grandfather 33, 34, 37, 431, 432
  loss in Fire 47
  Vintner, 6, 33
Halley, Humphrey II, uncle 34, 35, 36, 37, 148, 431, 432
Halley, Humphrey III, brother 34, 142, 143, 431, 432
Halley, Joane, stepmother 35, 129, 132, 133, 140, 141, 142, 143, 431, 432, 435, 436
Halley, Katherine, daughter 130, 387, 403, 431, 432
Halley, Katherine, grandmother 33, 431, 432; *see also* Mewce, Katherine
Halley, Katherine, sister 34, 431, 433
Halley, Margaret, daughter 130, 387, 403, 431, 432
Halley, Mary (née Tooke), wife 37, 128, 129, 130, 131, 382, 387, 405, 431, 432, 433, 434
  death 403
  family 128
Halley, William, uncle 34, 37, 431, 432
'Halley's Mount' (St Helena) 73
Halleyan lines 283
Hampton Court 14
Handel, Georg Frederick 16, 18, 385
Hanover 213, 262, 318
Hansteen, Christopher 282
Hardwick, Mr 279
Harriot, Thomas 188, 208
Harrison, Edward, Lt. 264, 269, 273, 274, 284
  court martial 275
Harrison, James 397
Harrison, John 22, 397
Harvey, William 30
Harwich 12, 231, 232
Hatton, Viscount 321
Hawley (Alternative to Halley) 32, 36
Hawley, Major Thomas 33, 136, 137, 139, 140
Hayley, Capt. 171
Haynes, Edward 381
Hearne, Thomas 50, 51, 247, 325, 326, 344, 386, 405, 412
  stories of Halley 10, 141, 326, 327, 393, 407
Heberstein, Count Johann Ernst von 299, 300
Hecker, Johannes 82

Hedges, Sir Charles 292, 306, 307
Heiberg, J. L. 336
height, barometric 196
Heinsius, Anthonie 298, 299, 318
heliocentric system 27
Hevelius, John 17, 23, 25, 28, 32, 53, 58, 65, 66, 69, 82, 86, 87, 89, 90, 107, 123, 124, 170, 290
  Annus climactericus 86, 91, 92, 100
  Cometographia 91, 101, 206
  correspondence 89, 92
  discoveries 89
  FRS 89
  Halley's visit 92
  open sights 91
  instruments 91
  Machina coelestis 70, 90, 91, 92, 100, 164
  observations 70, 96–9
  observatory 89, 101
  Oxford portrait 102, 326
  Prodromus astronomiae 91
  report of death 101, 102
  Selenographia 28, 101
  status 89
Hevelius, Elizabeth 25, 44, 70, 89
  gown 101
  observer 89, 90, 93, 95, 96
  relations with Halley 101, 102, 327
Hexaemeron 401
Hilal b. abi Hilal al-Himsi 337
Hill, Abraham 9, 236, 237, 238, 242, 246, 386
Hill, of Canterbury 114, 207
Hipparchos 226, 400
Historia coelestis 327, 385, 386, 387, 388–92, 455
  Prefatio 392
Historia coelestis britannica 78, 88, 387, 388–92, 400, 455
  Prolegomena 392
Historia piscium (*The history of fishes*) 152, 155, 183
History of the Great Rebellion 328
Hoare, James 250
Hodgson, James 392
Holland, John 238
Holt, Sir John 248
Holy Roman Empire 293
Holy Sepulchre, Halley's model 449–51
Hooke, Robert 10, 17, 23, 38, 39, 41, 60, 74, 75, 108, 112, 114, 123, 171, 208, 405, 424
  Animadversions 70
  architect 36, 39
  barometer 269
  City surveyor 39
  criticisms of Hevelius 70, 91, 100
  curator, Royal Society 39
  diary 11, 54, 100, 137, 232
  earthquakes 416
  gravitation 147, 148
  mercury, density 195
  Royal Observatory 377, 378, 380
Horrocks (Horrox), Jeremiah 15, 23, 52, 82, 83, 84, 357, 362, 367, 413

Hoskins (Hoskyns), Sir John 151, 262, 288, 378
Hoskins, Lady 378
Houblon, family 7, 42
Hough, bishop 247
Houghton, John 239, 240
Hourman, Frederic 65
households 43, 44
Hudson, John 50, 331, 334
Hudson's Bay 224, 400
Huguenots 7, 8, 17, 19
humming bird, flight 245
Hunt, H. (operator of the Royal Society) 186
Huntingdon 11, 12, 14, 33, 45
hurricanes 194
Huygens, Christian 19, 30, 105, 113, 147, 203
Hyde, Lady 409
hydrological cycle 195

icebergs 278
inclination 429
Indian Ocean 193, 194
'infidel mathematician' 408–10
Innerösterreichischen Hofkammer 295
Inns of Court 10
instruments, astronomical 23, 60
interest, compound 199
International Geophysical Year 278
interpolation 210
Invalides 111, 112
inverse square law 29, 41, 148
Irish Sea 228
Islington 6, 34, 37, 46, 124, 130, 131, 144, 153, 209
isarithms, *see* isolines
isogones 282
isolines 31, 282, 283, 317
Istria 293, 296, 297
Italy 18

Jamaica 53, 270
James I 164
James II (formerly Duke of York) 35, 38, 137, 145, 166–8, 230, 248
Jeffreys, George, judge 135, 144
Jeffreys, Sir Harold 227
Jenkin, Sir Leoline 333
Jennings, Frances 45
Jersey 288
Johnson, Manuel John (astronomer) 73
Jones, Inigo 6
Jones, Israel 271
Journal des Sçavans 169, 448
Jupiter 53, 204, 211, 213, 214, 215
    cometary impact 400
    oblateness and satellite perturbations 218, 229
    tables 373–5
    see also occultations; satellites
Jurin, James 349
Justel, Henri 8, 113, 184, 198

Kant, Emmanuel 228, 351
Keill, John 330, 332, 385
Kepler, Johannes 55, 65, 81, 164, 165
    equation 427
    laws 25, 55, 148, 149, 203
    problem 214, 427
    Rudolphine tables 82
    transits 81
Kinder, Gilbert 128, 433, 434
Kinder, Margaret 128, 433, 434
King's College, Cambridge 170, 207, 208
Kirch, Gottfried 78, 213, 214, 465 n. 36
Kircher, Athanasius 120, 191, 256, 282
Kit-Kat Club 16, 406, 409
Klobucaric, Ivan, cartographer 297
Kneller, Sir Godfrey 16

La Caille, Nicholas Louis 88
Lagos disaster 13, 42, 43, 251, 263
Lagrange, Joseph Louis 365
Lanoy, Timothy 226
Laplace, Pierre-Simon, Marquis de 29, 200, 227
latitude 21, 67, 118, 125, 147, 219, 267, 287
Laud, William 333
law courts, 10
lectures 52
Leeuwenhoek, Antoni van 24, 30, 170, 171
Legorne 118
Leibniz, Gottfried Wilhelm 19, 25, 123, 184, 203, 228, 299; **XIII**
    criticism of Newton 26, 203, 228
    Halley 164, 213, 262, 307
    priority dispute 332, 333, 385
Leiden 19, 336
Leigh, Luke 32, 392
Leipzig 299
Leithullier, family 9, 33, 62
Lely, Sir Peter 16
le Monnier, Pierre Charles 367
Leopold I, Holy Roman Emperor 292, 295, 296, 299, 300, 303, 306, 310, 318, 331
Levant Company 9, 41, 42, 61, 143, 226
Lewis, Ed 252, 253, 254
life tables 199
light, speed 105
Linnaeus, Carl 24, 30
Liouville, Joseph 351
Little Hayes 12, 323, 488 nn. 11, 13
Livery companies, 6
Lizard 287, 289
Ljubljana (Laibach) 299
Lloyd, Owen 188
Locke, John 16, 24, 26, 114, 118, 123, 135, 145, 169, 249
Logan, James 188
logarithms 188, 189
log line 267
London, 4–10
    churches 7; *see also* churches and parishes
    commerce 6

London (*cont.*)
  Great fire 4, 5, 46
  financial centre 10
  plan 5
  Plague 4, 46
  population 116, 198, 199
  port 6
  religion 7
  trade 9
  walls 4
  wards 6
longitude 20, 21, 22, 67, 86, 173, 201, 217, 256, 267, 377, 395–8
  Board 397
  effect of currents 258, 267, 268, 273, 274
  errors in magnetic charts 282
  Jupiter's satellites 202, 218, 268, 274
  lunar position 131, 217, 218, 268, 273, 398
  measurements 58, 119, 131, 202, 218, 257, 258, 277, 349, 381, 396
  prize 396, 397
  Whiston's proposal 396
Longomontanus (Christian Severin) 214
Lord Mayor 7
Lords House committees 136, 140
Loredan, Francesco 296, 303, 304, 310
Louis XIV 113, 115, 292
Louis Salvator, Archduke 312, 313
Louise de la Kéroualle (Duchess of Portsmouth) 377
Low Countries 236
Lowndes, William 249
Lowther, William 208
loxodrome 189
lunar theory (orbital) 23, 131, 159, 160, 356–66, 398
Luttrell, Narcissus 276, 295

Mabillon, Jean 112
Macclesfield, Earl of 188, 393
Machin, John 385, 397, 402
Madeira 65, 76, 257, 270, 271, 276
Maestlin, Michael 213
Magellan, straits 278
Magellanic clouds 76
magnetic variation 131
  charts 283
  observations 271, 283, 287, 288
Mairan, Jean-Jacques Ortous de 58, 114
Malamocco 315
Malpighi, Marcello 24, 30
Margate 12, 270
Marlborough, Duke of 48, 292, 298, 299
marriage customs 129
marriages, private 129
Mars, parallax 75, 81, 381; *see also* occultations
Marseille 118
Martin Vaz 257, 279
Mary II (Stuart), Queen 50, 246, 260, 261
Maskelyne, Nevil 398

Mason, Charles 250, 254
Massachusetts 19
massacre of St Bartholomew 7
mathematics 24
Maupertuis, Pierre-Louis Moureau de 402
May, Island 257, 272
Mayer, Johann Tobias 398
Mead, Richard 386, 393, 410
mean anomaly 427
Mediterranean, length 23, 202
Mediterranean, naval control 310, 313, 315
Menaechmus 341
Menelaus *Sphaericorum* 188, 338, 339
mercury (metal) density 195
Mercury (planet) 60, 61, 66, 74, 75, 80, 81–3, 89, 380, 398; *see also* transits
meridian line 189, 190
Messier, Charles 216
metaphysics, Judaeo-Christian 30
metrology 122, 173, 304, 305
Meurisse 65, 66, 80
Mewce, Katherine 33, 431
Mewce, Nicholas 33, 431
Michaelangelo 242, 243
micrometer, eyepiece 24, 59, 60, 69
microscopes 24
Middleton, Benjamin 260, 261, 262
Middleton, Thomas, col. 261
millenarianism 29
Milton, John 48
Mint, Edinburgh 251
Mint, Royal 249, 250
mints, temporary 249; *see also* Chester, mint
Mint, officers 250, 254
Mira 89
Mocenigo, Alvise III 296
Molineux, Samuel 352
Molyneux, Thomas 250, 254
Molyneux, William 66, 93, 103, 119, 161, 170, 172, 174, 175, 380, 381
Monmouth rebellion 145
Montague, Charles, Earl of Halifax 239, 247, 253, 406, 409
Montanari, Geminiano 207
Montpellier 118, 119, 120
Monument (London) 196
Moon
  acceleration 173, 225, 227–9, 349, 419
  apogee 226, 362
  as clock 22
  diameter 99
  libration 354, 419
  map 105, 354, 381
  mean motion 226
  node 226
  observations 71, 116, 132, 227
  orbit 83, 84, 88, 203
  parallactic inequality 84
  parallax 66, 84
  see also appulses; evection; lunar theory; node; occultations; saronic cycle; variation

Moore, Sir Jonas, 4, 14, 38, 63, 74, 75, 76, 233, 238, 383
  death 101
  Flamsteed's patron 39
  Halley's patron 60, 61, 77
  mathematician 38
  Royal Observatory 377, 378
  surveyor 38, 233, 316
Moorfields 4, 5, 46, 47
Moray, Sir Robert 182
More, Henry 15
Morgan, Sir Henry 53
Morlacco, canal di 303, 456
Morland, Sir Samuel 377
Morrice, Roger 136
Morton Hindmarsh (Moreton-in-the-Marsh) 323, 324, 488 nn. 11, 13
mural arc 69
  Flamsteed's 59, 60, 378, 379, 380
  Halley's 394, 395
  quadrant 59, 60, 100, 105
Muratori, Ludovico Antonio 112
music, 15, 16, 45, 407

Nantucket 280
Napoleon Bounaparte 73
Narbonne 118
Nāsir al-Dīn al-Tūsī 339
natural philosophy 24, 26, 27, 415
natural theology 27, 411
navigation 20, 203, 258
  Atlantic 21
  Channel 290
  errors 273
Neale, Thomas 250, 254
nebulae 346; *see also* Magellanic clouds
Nelson, Delicia (née Roberts) 50, 107
Nelson, Robert 17, 42, 50, 107, 119, 129, 236, 248, 405, 412
  friend of Halley 50, 107, 248, 405, 412
  friend of Tillotson 108
  nonjuror 107, 248
  travels with Halley 107, 117, 120
Nelson, Theophile 107, 120
Netherlands 18
Neumann, Caspar 198, 199
New England 19
  comets observed 19
Newfoundland 280
Newmarket 14, 45, 134
Newton, Isaac, 15, 23, 24, 25, 26, 41, 52, 116, 123, 172, 227, 237, 380, 382, 405, 424; **VI**
  biblical studies 27
  Chester mint 239, 249–54
  chronology 28, 400
  colours under sea 241
  comets 207–10; *see also Principia*
  correspondence with Halley 154–61, 210–12
  De motu corporum 151
  De motu corporum in gyro 149
  experimenter 25

Historia coelestis britannica 382, 383, 384, 385, 386, 387
knighthood 330
longitude prize 396, 397
lunar theory 307, 355, 357–66, 367, 372, 373, 375
Opticks 27, 241, 307
optics 25
patron of Halley 248, 250, 251, 322
precursors 30
President, Royal Society 384
Principia, see separate entry
refraction 80
relations with Halley 163, 212, 407, 408; *see also Principia*
religion 27, 387
Royal Mint 249
Savilian elections 247, 248, 322
The theory of the Moon's motion 170, 364, 375
New Zealand 224
Nicholas, Robert 238
noblemen (undergraduates) 51
node 67, 363, 429
nonconformists 7
nonjurors 107, 108, 248
Norlands 323, 488 nn. 11, 13; *see also* Oxney
North Atlantic currents 258
North Equatorial current 272, 273, 274
North Sea 11, 13
  storms 13
Norwood, Richard 21, 171, 172, 201
Nottingham, Earl (Daniel Finch) 16, 292, 298, 299, 300, 306, 309, 318, 424
  Savilian elector 248, 321, 322
novae 346

Oak Apple Day 79
Observatoire de Paris 21, 22, 26, 105, 106, 112, 113, 126, 127, 381; V
occultation 54, 76, 130
  Jupiter 93, 104, 171
  Mars 58
  Saturn 104
Ogilby's wheel 201
Olbers's paradox 350, 351, 416
Oldenburg, Henry 42, 53, 55, 61, 66, 113, 182; **II**
Olhoff, J. E. 89, 93, 96, 100, 101, 102
optics 197
orbits, determination 55–8
Ordnance office 38
Ormonde, Duke of 329, 424
Ottens, Reiner 79
Ottoman Empire 293, 309
Oughtred, William 38
Oxford 15
  politics 325
  scholarship 330
  university 10, 15, 51, 327, 328
  University Press 51, 328, 330, 333

Oxford, Earl of (Harley) 331
Oxney 12; *see also* Norlands

Padua 17, 18
Paget, Edward 149
Pagham 236, 237, 238, 240
painting 16
Palais du Luxembourg 111, 113
Palazzo Corsini 120
Palazzo Riario 120, 122
Pallavicini (cartographer) 314
Palmyra (Tadmor) 226
Papal States 293
Papin, Denis 152
Pappus 188, 190, 334, 336, 337, 338
parabola 429, 430
Paraiba 257, 273
parallax
  annual 68
  lunar 66, 84
  solar 80, 81
Paramore (*Paramour*, pink) 13, 256, 260, 265, 266, 281, 284
  building 260, 262, 263, 288
  defects 259, 263, 266, 270
  laid up 262
  modified 276
  name 262
Pargiter, Samuel 239
Pargo 293
Paris 18, 110–13, 118
  colleges 19
  extent 116
  population 116, 198
  see also Observatoire de Paris
Pay Office 42
Pearson, John, bishop of Chester 33
Pell, Dr John 188, 377
Pendennis Castle 288
pendulum, length 87, 105, 116
Penn, William, quaker 19
Penn, Sir William, admiral 53
Pepusch, Johann Christoph 385
Pepys, Samuel 11, 16, 36, 37, 42, 47, 102, 114, 240, 261, 306
  charts 233, 234
  Great Fire 46
  Great Plague 46
  music 45, 407
  naval administration 13, 259, 275
  Naval Minutes 240
  nonjuror 107, 248
  opinion of Halley 405, 419
  Principia 150, 162, 168
  President, Royal Society 150, 172, 183
Perkins, Peter 174, 176, 191, 381, 382
Pernambuco (Recife) 257, 270, 279, 280
Peter the Great (Tsar) 124, 263
Peterborough 11, 12, 14, 34, 45
Pevensey 201
physics 197, 198

Picard, Jean 21, 23, 69, 84, 105, 113, 172, 201, 381
pigeon, flight 244, 245
pilots 231
pink (ship) 265, 266
pirates 272, 276
planets
  observations 54
  orbits 54–7; *see also* Jupiter, Saturn, Venus
Plato 26
Plato Tiburtius 225
playwrights 16
Pliny the Younger 226
Plymouth 235, 236, 271, 274, 280, 287
Poincaré, Henri 217
Poisons, affair of 113
polar distance 67
political state 145, 181
Pollexfen, Sir Henry 248
polynomial equations 188
Pontio (Ponzio, Ponthaeus), Giuseppe Dionisio 122, 148, 207
Pope, Alexander 17, 409, 410
populations, London and Paris 116
Portland 270, 271, 287
Port Mahon 293, 309, 313
Porto Re (Kraljevica) 302, 303, 309, 311
Portsmouth 11, 236, 261, 271, 286
Pothenot, Laurent 235
Pound, James 78, 352, 382, 385, 389
Prague 299
precession 68, 80, 160
Principia (*Philosophiae naturalis principia mathematica*) 19, 31, 49, 55, 68, 116
  acknowledgement to Halley 163
  Amsterdam reprint (1723) 170
  annotated copies 169, 170
  announcements by Halley 160, 165
  comets 115, 149, 208–10
  cost and price 155
  early forms 150, 151
  geometry 342, 343
  Halley's copies 170
  Halley's ode 163, 164, 442, 443
  hydrodynamics 243
  influence 176–8, 203, 229, 419
  limitations 419
  lunar theory 359–66
  planetary perturbations 374, 375
  presentation copies 161, 162, 163, 169
  presentation to James II 166–8
  printing 153, 155–9, 421
  publication 145, 157, 161, 421
  reviews 168
  revision 157
  second edition 169, 170
  stimulus 147, 149
  suppression of Book III 157, 159
  writing 145
Proast, Jonas 325, 326, 327, 412
proper motions 348, 349
prophecies 29

psalms 28, 29
Ptolemy 65, 201, 225, 226, 227, 344, 389
Pullein, Octaviari 123
Punta Gavranic 303, 312
Purcell, Henry 15, 62, 407
Purston 323, 324, 488 nn. 11, 13
Puy de Dome 196

quadrature, lunar 361
quantum mechanics 177
Queen's College, Oxford 50
Quietanus, Remus 81

Radcliffe Observatory manuscript 390, 452–5
rain 195
rainbows and haloes 187, 197, 198
Ranelagh, Katherine, Lady 39
Raqqa (Arracta) 225, 226
Ravenna 293
Ravius, Christianus 337
Ray, John 24, 27, 30
recoinage 249
referees, Flamsteed's catalogue 384
reference system, astronomical 68
refraction, atmospheric 53, 60, 63, 72, 80, 81, 91, 126, 398
Reiselius (Reisel), Solomon 161, 165, 170
relativity, special 177
resection 235
Revolution, Glorious 146, 186, 413
revolution, Newtonian 146, 413
rhumb line 189
Richborough 201
Richer, Jean 61, 65, 66, 80, 87, 105, 113, 116
right ascension 66, 67
Rio Janeiro 257, 270, 277, 281
Riprapps 287
risks 243
Roberts, Francis 384, 386
Roberts, Sir Gabriel 9, 42, 45, 50, 107, 236, 407
Robinson, Anne, *see* Halley, Anne
Robinson, Tancred 151, 152
Roman catholics
  at court 7
  in universities 15
Roman science 123
Rome 18, 107, 118, 119, 120–2, 124, 202
  antiquities 123
Römer, Olaus 105, 113, 381, 385
Rooke, Sir George 43, 288, 291, 293
Rotterdam (English church) 107, 476 n. 37
Royal African Company 9, 41, 236, 239, 279
  court 236, 237, 238, 239, 240
  trade 237
  see also ships, *Guynie*
Royal Navy
  Pay Office 9
  scientific expeditions 31, 290

Royal Observatory 22, 26, 38, 39, 41, 379
  archives 390
  equipment 378, 379, 393–5
  foundation stone 378
  plans and construction 53, 54, 378
  order for 378
  Visitors 386, 393, 402
  see also Flamsteed; Halley
Royal Society 11, 15, 18, 19, 25, 75, 77, 92, 108, 181, 261, 332
  charts presented 234, 276, 281, 288
  early years 182, 183
  election of Clerk 151, 152
  Journal Book 186
  meetings 183, 240
  Philosophical Transactions 183, 185, 186
  Secretaries 151, 152; *see also* Jurin, James; Sloane, Sir Hans
  transit of Venus 224
Royer, Auguste 78
rue de Buci 110, 111
  du Four 111
  Halley 404
  St Jacques 110, 113
  de l'Université 110, 111
  de Seine 110, 111
  Vivien 110, 116
Russell, Earl 8, 36, 134, 135, 140, 145
Rye House plot 8, 134, 145
Ryswick, Peace of 187, 263, 292

St Christopher's 257, 274, 279
St Germain-des-Prés, abbey and district 18, 110–12
St Helena 22, 38, 47, 50, 55, 61, 223, 270, 279, 281
  climate 73, 74, 75, 194
  disturbances 73, 76
  gravity 159
  island 73, 257, 270
  see also Halley, entries for St Helena
St Helens (Isle of Wight) 271
St Jago (Iago) 257, 276
St Maur congregation 112
St Paul's cathedral 4, 5, 47
St Paul's School 4, 43, 46, 47, 48
  curriculum 48, 49
St Pierre, le Sieur de 377, 378
St Thomas, Cape 257, 277
salinity of seas and lakes 346
Salisbury (Clerkship candidate) 152
Sandwich, Susanna 37, 148
sanitation 44
Sanson, Nicholas 297
Santa Maria in Vallicella 120, 121, 122
saronic cycle 130, 131, 349, 357, 369, 372, 373
satellites
  artificial 218
  Jupiter 22, 23, 66, 105, 122, 218
  regression of nodes 218
  Saturn 104, 105

Saturn 24, 53, 204, 211, 213, 214, 215
  tables 373–5
  see also occultations; satellites
Saumur 117, 118, 119
Savile, George, *see* Halifax, Marquis
Savile, Sir Henry (Provost of Eton) 322, 323
Savile, Sir Henry (ambassador to France) 112, 117, 145
Savilian professors 52, 245, 323
  astronomy 237, 245
  duties 324
  estates 323, 324
  geometry 321, 322
Sawtry 37
Sawtry, rector of 37, 148
Sceptics 26
Scilly Isles 21, 252, 258, 274, 280
seamless shirt 171, 411
Seething Lane 9
Seine 287
Selkirk, Alexander 16
Seller(s), John 233
Selsey 232, 287
Seng 293, 302
Sept Isles 288
Serenus (mathematician) 338
Sévigné, Mme de 114
sextant 23, 59, 69, 70, 125
  diagram 71, 125
  equatorial mount 59, 70
  Flamsteed's 378, 379
  Halley's 379
Shaftesbury, earl of 35
Sharp, Abraham 88, 352, 385, 387, 390, 455
  revision of Halley's southern catalogue 391, 392, 440, 441
Sheerness 232
Sheldonian theatre 328
sherrifs 7
Sherwin's Tables 199
shipowners 9
ships
  Falconbird 276
  Golden Fleece 76
  Guynie (*Guinea*) 232, 236–40, 243, 261, 286
  Hannibal 279
  Paramore (*see* separate entry)
  Roeburck 275
  Swiftsure 275
  Triumph 287
  Unity 9, 62, 63, 64
shipwreck 21
shipyards 11, 314
Shovell, Sir Cloudesley 21, 258, 274, 275, 287, 291, 296, 300, 313
Sidney, Algernon 8, 35, 36, 114, 118, 134, 135, 138, 145
Siena 118
sights
  open 70
  telescopic 71, 87, 103, 104
Silchester (*Calleva atrebatum*) 330

Sisson, Jonathan 394
Sloane, Sir Hans 16, 107, 152, 153, 196, 213, 251, 332, 406; **VII**
Smith ('Faber') 381
Smyrna 171
Snel (Snellius, Snell) 21, 235
Snowdon 11, 187, 196
Scoiety for the Promotion of Christian Knowledge 107
Society for the Propagation of the Gospel in Foreign Parts 107
Soissons 171
Solent 286
Souciet, Fr Étienne 440
Southampton 286
South Foreland 287
Southwell, Sir Robert 16, 50, 235
sparrow hawk, flight 245
Spectacle Makers Company 185
Speke, Hugh 135, 138, 139, 144
Spinkes, Nathaniel 107
Spithead 286
star charts 79
Star of Bethlehem 206
stars, descriptions 389
Start Point 271
stationers 112, 269
steam engines 11
Steiermark (Styria) 293
Stepney, George 16, 292, 294, 297, 298, 300, 306, 309; **XII**
stereographic projection 190
Stillingfleet, Edward, bishop 247, 408
Stoics 26
Stonehenge 14
storms 13, 278
Stourbridge fair 14, 15
Strasoldo, Vito di, Count 299, 300
Streete, Thomas 53, 59, 60, 66, 82, 269
Strong, Capt. 345
Stroud, Kent 132
Stukeley, William 383, 402, 406
Sturm, Johann Christoph 165, 170, 184
Sumatra 65
Sun
  dimensional stability 352, 353
  noonday observations 267
  parallax 80, 81, 159, 220–5, 381
  perturbation of Moon 359–61, 363
  rotation 58, 381
Sunderland, Lord 393
sunspots 58, 60, 381
Swedenborg, Emanuel 124, 192, 263, 267, 290, 331, 418
Swift, Jonathan 17, 406
syzygies 361

Tabula Peutingeriana 297
Tahiti 223, 224
Tangier expedition 38, 238
'tarpaulins' 259

Index

Taylor, Brooke 402
Taylor, G. I. 227
technology 24
Temple Farm 133, 134
Tenison, Thomas, archbishop 322
Terrae Filius 327
Terra Incognita 265
Thābit ibn Qurra 337
Thames, River 11
   approaches 12, 13, 231, 232, 233, 236, 316
theatre 16, 44
thermometer 24, 197, 269
Thoynard, Nicholas 114
three-body problem 204, 215
Thurlow, John 50
Thyssen, Francis 238
tidal friction 227, 228
tides 160, 166, 167, 175, 256, 316; *see also*
   Halley, tides
Tillotson, John, archbishop 108, 109, 114, 117,
   119, 207, 237, 246, 412
Titus, Col. Silius 377
Toad's Cove (Bay) 258, 280, 281
Tompion, Thomas 379
Tonquin (Tonkin), tides 131, 132, 167
Tooke, family 42, 128, 130, 433, 434
Tooke, Margaret, mother-in-law 37, 128, 433,
   434
Tooke, Mary, wife, *see* Halley, Mary
Torbay 288
Toulon 118, 313, 314
Toulouse 118, 119
Toupion, Thomas 239
Tower of London 4, 5, 38
   Ordnance Office 10
Towneley, Richard 23, 32, 165, 174
trade winds 192–5
trained bands 7
transits 82, 219–25
   Mercury 60, 81, 82, 219
   Venus 81, 82, 184, 220
travel 13, 14
treason, accusations 253
triangulation of Britain 173
Trieste (Tergestum) 236, 293, 294, 296, 300,
   315, 318
   maps 314, 316
   survey 299–302
Trinidada (Trinidad, Trinidade) 270, 272, 279
Trinity House 231, 233
Tristan da Cunha 258, 270, 278
Trojan War 400, 401
turbid water 272
Tycho Brahe 22, 25, 72, 79, 91, 165, 208, 290,
   349, 385, 386, 392
Typographus (printer) 93, 96, 98, 99

Ulysses 423, 472 n. 1
Universal Deluge (Noachian Flood) 246,
   399
Uraniborg 25, 72, 104

Ushant 287

Valvasar 171
van Luer, Jan Baptiste 78
variation, lunar 354, 357, 361
Veglia 293, 303
Venice 18, 293, 296
   Archivio di Stato 296, 297, 300, 303
   Arsenal 314
Venus, 130
   transits 31, 81, 88, 89, 220, 395, 398
Vermuyden, 14, 38
Vernada, Antonio de 312
Versailles 113
Via Carolina 312
victualling 306
Vienna 184, 213, 299, 300, 318
   Hofkammerarchiv 310
   Kriegsarchiv 310, 314
Ville de Venise (hotel) 110, 111
Vincent, Nathaniel 151
Vitruvius 202, 243
Viviani, Vincenzo 163
Voltaire, François Marie Arouet de 406
vortices 169, 204

Waller, Richard 188
Wallis, John, Savilian professor 42, 55, 83, 93,
   383; **III**
   death 321
   Greek editions 334
   Halley correspondent 159, 160, 170, 172,
   188
   mathematician 190, 321
   Savilian professor 52, 321, 327
Ward, Seth, bishop 55, 377
watch, parish 7
water, pressure 243
wars 17, 19
   Nine Years 233
   Spanish Succession 279, 291, 292, 293
Washingtons of Sulgrave 33
Weald of Kent 11
Weddall, Robert 252, 253, 254
Weiss, Matthias Antonio 297, 310, 314
Weisskarte von Buccari 311, 312, 313
westerly drift of magnetic field 194, 417
West Indies 19, 194, 265, 271, 274, 279, 280
Westminster Abbey 404
Westminster Hall 138, 139
Whichcote, Benjamin 15
Whiston, William 331, 410
White, Manly (boy) 276
Whitehall 6
Wight, Isle 235, 236, 270, 271
Wilkins, John, bishop 108, 172
William of Orange, William III, 13, 113, 230,
   231, 232, 246, 424
Williamson, Sir Joseph 16, 38, 50, 60, 61, 62,
   84, 183, 333, 377, 407; **I**

Willoughby, Francis 152
Winchelsea 232
Winchester Street 4, 5, 6, 9, 35, 39, 46, 435, 436
womanisers 45
Wood, Anthony à 50
Worcester, battle 79
Wratislaw count 298, 299
Wren, Mr Christopher 386
Wren, Sir Christopher 16, 60, 86, 108, 113, 263, 377
  architect 39
  Flamsteed referee 384
  Gresham professor 24

President, Royal Society 147, 183
Visitor, Royal Observatory 386

yard, standard 173
Yate, Thomas 333
yeoman warders 7, 35, 36
York, Duke of, *see* James II
Young, Richard 41, 143, 435

Zadar 294
Zodiac, signs of 67, 68